AI-Driven Research Innovations in Computing and Information Science

Patrick Ngulube
https://orcid.org/0000-0002-7676-3931
University of South Africa, South Africa

IGI Global
Scientific Publishing
Publishing Tomorrow's Research Today

Vice President of Editorial Melissa Wagner
Director of Acquisitions Mikaela Felty
Director of Book Development Jocelynn Hessler
Production Manager Mike Brehm
Cover Design Jose Rosado

Published in the United States of America by
IGI Global Scientific Publishing
701 East Chocolate Avenue
Hershey, PA, 17033, USA
Tel: 717-533-8845 | Fax: 717-533-7115
Website: https://www.igi-global.com E-mail: cust@igi-global.com

Library of Congress Cataloging-in-Publication Data

CIP Data Pending
ISBN13: 9798260004340
Isbn13Softcover: 9798260004357
EISBN13: 9798260004364

British Cataloguing in Publication Data
A Cataloguing in Publication record for this book is available from the British Library.

Editorial Advisory Board

Table of Contents

Detailed Table of Contents

Digital labour platforms increasingly rely on algorithmic decision-making systems to manage and allocate work. However, these algorithms remain largely opaque to researchers, regulators, and workers. This black-box environment complicates the study of how algorithmic management influences labour conditions, access to opportunities, fairness, and worker well-being. This chapter explores new research approaches developed to investigate algorithmic injustice on digital labour platforms, especially when direct access to backend data, source code, or allocation logic is restricted. It highlights an expanding range of empirical methods, such as black-box audits, data donation initiatives, worker-centred sensing, platform reverse engineering, simulated task experiments, and collaborative ethnography, that allow researchers to infer system behaviour from observable outputs and lived experiences. The chapter discusses the advantages and disadvantages of these methods and underscores the importance of interdisciplinary approaches combining computational techniques with qualitative insights.

This study explores everyday GenAI use through an empirical, interpretivist investigation involving eighty-eight qualitative survey responses. Drawing inferences from Davis's (1989) Technology Acceptance Model (TAM) and the socio-technical perspective (Trist & Bamforth, 1951). The research offers two key contributions. First, it identifies three thematic patterns in GenAI use: (1) A Journey of Discovery, (2) Workplace Partner—Friend or Foe, and (3) AI's Ethical Crossroads. The findings also illustrate that, contrary to concerns of knowledge dilution, findings suggest that GenAI serves as a cognitive amplifier, enabling users to reallocate effort toward

This chapter delves into key AI tools and methodologies that propel advancements across various domains. ML techniques enable systems to learn from data autonomously by employing supervised learning and unsupervised learning. Deep learning, leveraging neural networks, excels in medical image analysis to enhance diagnostic accuracy. Reinforcement learning enables AI for tasks like robot navigation with Q-learning optimizing strategies. Generative AI innovates by generating new data, aiding drug discovery through synthetic protein structures. Transfer learning enhances NLP tasks like sentiment analysis by adapting models efficiently. Big data analytics tools manage extensive datasets, facilitating genomic research in identifying disease markers. High-performance computing resources like GPUs and TPUs accelerate AI computations, while explainable AI frameworks ensure transparency in applications such as healthcare diagnostics. These tools drive interdisciplinary AI innovation, emphasizing ethical and impactful applications from healthcare to climate modeling.

Artificial intelligence (AI) tools are rapidly transforming research practices in information science, enabling new approaches to literature review, hypothesis development, academic writing, referencing, and data analysis. This study provides a critical examination of the ethical issues surrounding the use of AI applications such as ChatGPT, Elicit, Scite, Grammarly, and Semantic Scholar in scholarly work. Using a conceptual research design, the paper synthesizes evidence from recent literature, policy documents, and grey sources to explore the opportunities and risks of AI adoption. The analysis identifies five major ethical concerns: transparency and disclosure, authorship and intellectual ownership, data privacy and confidentiality, bias and inclusivity, and institutional responsibility. To address these challenges, the study proposes ethical guidelines that include mandatory AI usage statements,

This chapter explores the integration and advancement of Indigenous knowledge systems within Library and Information Science (LIS) research, emphasizing the importance of Indigenous epistemologies and research methodologies. It critically examines how traditional LIS paradigms can be expanded to incorporate Indigenous ways of knowing, valuing cultural context, oral traditions, and community-based approaches. Special attention is given to the cultural context, oral traditions, and community-centered approaches that underpin these epistemologies. The chapter advocates for the development of inclusive research frameworks that respect Indigenous epistemologies, fostering more equitable and culturally responsive LIS scholarship. By advancing these paradigms, the chapter aims to promote greater recognition, validation, and utilization of Indigenous knowledge in information practices and research through providing guidance on how researches biased towards indigenous knowledge systems can be effectively conducted in the LIS discipline.

This chapter outlines the development of a context-sensitive conceptual framework to enhance the trustworthiness of digital records in Zimbabwe's public sector. Centred on accuracy, authenticity, reliability, and usability, it employs a qualitative approach, combining literature review and Delphi method validation. Findings indicate that global frameworks like ISO 15489, MoReq, and InterPARES often mismatch Zimbabwe's resource-constrained context. To address this, the proposed framework incorporates both technical and human factors, such as metadata standards, version control, capacity building, and secure data recovery. Expert validation refined the framework, ensuring relevance and sustainability. The chapter concludes by highlighting the model's implications for theory development, practical application, and research in computing and informatics.

This chapter emphasizes on different methods analyzing the data and presents the results in effective way. These methods also enable the users to focus on the required variables and visualize the results in effective manner. Data collection (integration) is the primary task and data can be collected from multiple heterogeneous sources. Therefore, once the data is collected, it requires application of the 'data cleaning' step. In the initial phase, researchers must divide the given variables into 'dependent' and 'independent' variables. Once the 'data cleaning' task completes next is to work on 'qualitative', 'quantitative', 'descriptive', 'predictive' and 'prescriptive' analysis. There are many sectors where data analysis plays very important role like healthcare, space, education, manufacturing, finance, supply chain etc. Integrating data from multiple sources is very critical task and has to face few challenges like highly configured machines are required, skillful human resource and most important powerful algorithms.

Preface

Artificial intelligence (AI) is growing rapidly and changing the way we conduct research, generate knowledge, and develop new ideas in the fields of computer science and information science. Researchers face new challenges and theoretical problems as AI systems grow more common in daily life, workplaces, public organisations, and scientific study. This edited volume, *Artificial Intelligence-Driven Research Innovations in Computing and Information Science,* addresses this dynamic landscape by compiling interdisciplinary scholarship that illuminates the evolving intersections among AI technologies, research methodologies, and domain-specific applications.

The chapters in this book show that AI is more than just a set of tools. It is a way of looking at issues, data, and new ideas that changes everything. This book includes voices from computing, information science, data analytics, engineering, and social research. It discusses new methods, ethical issues, and real-world uses that are transforming scholarship now and in the future. The chapters show how powerful AI can be in accelerating discovery, but they also underscore the importance of carefully considering fairness, transparency, and epistemic validity.

CHAPTER OVERVIEW

The first chapter pushes the boundaries of how we research algorithmic injustice in digital work marketplaces. Due to the lack of transparency in algorithmic management, which makes it hard to access backend systems, researchers are increasingly using more indirect or creative empirical methods. These include black-box audits, data donation programs, simulated task tests, collaborative ethnographies, and platform reverse engineering. The chapter emphasises the necessity for multidisciplinary methodologies that harmonise computational precision with qualitative understanding by rigorously assessing the advantages and drawbacks of each technique. Its significance lies in showing how researchers can examine fairness, worker well-being, and systemic inequity even when little information is available.

Generative AI has become a useful tool and a cultural trend. The second chapter examines the integration of generative AI into daily life through a qualitative analysis of survey data. Drawing from the Technology Acceptance Model and socio-technical theory, the study delineates patterns of discovery, workplace integration, and ethical tension. It proposes the AI Assisted Thematic Approach (AITA), a structured framework that combines qualitative analysis with AI-driven support, such as quick engineering and human-in-the-loop evaluation. The chapter emphasises that generative AI acts as a cognitive enhancer, necessitating ethical contemplation over its impact in creative and professional spheres.

Chapter three explores the realm of professional sports, focusing on how NBA teams utilise AI in their strategic planning. It details how AI-driven analytics improve scouting accuracy, optimise game strategies, and assist in financial decision-making through the lens of Resource-Based Theory. The chapter emphasises that AI enhances human expertise rather than replacing it; instead, it strengthens organisational capabilities by transforming data into actionable intelligence. This study contributes to the growing body of literature on sports analytics by connecting AI implementation to competitive advantage and organisational strategy.

Chapter four discusses the impact of AI on the research process itself. It looks at how automated literature mining, adaptive modelling, intelligent hypothesis generation, and simulation-driven validation are transforming the planning and execution of research in computer science. While these approaches accelerate discovery and facilitate the reproduction of results, the chapter also highlights concerns about algorithmic opacity, excessive automation, and ethical issues in scholarly workflows. By using case examples from informatics and software engineering, the chapter presents a balanced perspective on how to safely integrate AI into research pipelines.

Chapter five provides a broad overview of modern AI tools and methods, including supervised and unsupervised machine learning, deep learning, reinforcement learning, transfer learning, and explainable AI frameworks. It shows how these methodologies have led to advancements across many fields, including healthcare, genomics, climate prediction, and autonomous systems. The chapter stresses the need for ethical, open, and human-aligned AI development in modern research ecosystems by linking algorithmic methods to real-world uses, including medical imaging, drug discovery, and high-performance computing.

The sixth chapter discusses the ethical issues that arise when AI is used in academic work. It examines concerns about transparency, authorship, intellectual ownership, privacy, bias, inclusion, and institutional responsibility, focusing on technologies such as ChatGPT, Elicit, Scite, Grammarly, and Semantic Scholar. The chapter employs a conceptual research approach to integrate contemporary policy and literature, proposing ethical standards that encompass required disclosure of AI use, human-in-the-loop authorship frameworks, and equity-oriented AI literacy. It shows how important it is to use AI responsibly to keep research honest.

Chapter seven introduces a hybrid Taguchi–RSM framework enhanced by AI techniques, including neural networks and genetic algorithms. Combining statistical design with computational intelligence makes models more accurate, reduces the work required for experiments, and improves the reliability of multi-objective optimisation. Applications encompass manufacturing, civil engineering, energy systems, and materials processing, demonstrating how intelligent optimisation fosters sustainable innovation in the context of Industry 4.0. The chapter presents the integrated framework as a paradigm for data-driven decision-making in intricate contexts.

Chapter eight, which focuses more on information science and education, calls for child-centred methods in research on how children seek information. The chapter criticises conventional methodologies that often fail to accurately reflect genuine child viewpoints, emphasising participatory approaches such as storytelling, flower maps, and observational techniques that enable children to act as co-researchers. A case study conducted in Ugandan primary schools demonstrates that children see libraries as safe, imaginative, and social environments, while simultaneously highlighting deficiencies in library infrastructure. The chapter illustrates the significance of participatory research in producing more nuanced and precise conclusions.

In chapter nine, the focus turns to epistemic diversity and methodological pluralism. The chapter examines the integration of Indigenous knowledge into Library and Information Science research, arguing for frameworks that respect cultural context, oral traditions, and community-rooted ways of knowing. It critiques conventional LIS paradigms and promotes inclusive methodologies that elevate Indigenous epistemologies to equal scholarly status. This contribution offers essential guidance for conducting culturally responsive, ethically grounded research within knowledge systems shaped by indigenous worldviews.

The tenth chapter discusses problems with managing digital records in Zimbabwe's public sector. It shows that global standards such as ISO 15489 and MoReq don't always align with the resources available in Zimbabwe. The chapter presents a contextually relevant conceptual framework focused on authenticity, accuracy, reliability, and usability, developed through qualitative design and Delphi validation. The chapter gives practical advice on how to construct trustworthy digital record systems in public places with few resources. It focuses on the interaction of technological and human variables, such as metadata standards, capacity development, and safe record recovery.

The last chapter provides a clear and thorough overview of the main data analysis methodologies and ways to deliver results that work. It starts with basic procedures such as combining data, cleaning it, and classifying variables. Then it goes on to several types of analysis, including qualitative, quantitative, descriptive, predictive, and prescriptive. The chapter emphasises the importance of computational infrastructure, trained staff, and algorithmic innovation in managing heterogeneous and complex datasets, and showcases applications across domains such as healthcare, finance, manufacturing, and space science. It is a useful reference for students and researchers who work with large amounts of data.

The chapters in this volume collectively illustrate the significant and far-reaching impact of AI on research methodologies, professional practices, and knowledge systems in computing and information science. This book highlights the richness of AI-driven scholarship across various fields, including algorithmic fairness, sports analytics, participatory methodologies, Indigenous epistemologies, optimisation frameworks, and ethical principles. It encourages readers to critically evaluate both the positive and negative aspects of AI and to envision new research opportunities accessible to everyone and responsive to societal needs.

As the editor, I am confident that this volume will make a significant contribution to academic discourse and serve as a valuable resource for researchers, practitioners, educators, and students seeking to navigate the rapidly evolving frontiers of artificial intelligence.

Patrick Ngulube
University of South Africa, South Africa

Acknowledgment

Many people and organisations have helped make *Artificial Intelligence-Driven Research Innovations in Computing and Information Science* possible by sharing their knowledge, hard work, and dedication to scholarship. I formally thank them for their contributions with deep respect and gratitude.

I would like to express my heartfelt gratitude to all the chapter authors. Their thorough research, methodological rigour, and intellectual engagement are the foundation of this volume. Their readiness to contribute original scholarship, respond thoughtfully to editorial feedback, and maintain the highest academic standards has resulted in a substantive and coherent contribution to the field.

I would also like to formally acknowledge the invaluable contributions of the peer reviewers. Their thorough evaluations, constructive feedback, and expertise in their respective fields greatly improved the clarity, coherence, and scholarly rigour of each chapter. We deeply appreciate their commitment to academic excellence.

Institutional support has been crucial in the development of this publication. I am grateful to the universities, research centres, and academic units that provided the resources, intellectual environments, and administrative assistance necessary for both the authors and editors to carry out their work effectively.

I would like to express my gratitude to the editorial and technical teams for their professionalism in supporting the production of this manuscript. Their careful attention to detail, effective coordination of the editorial workflow, and dedication to quality were essential for the successful completion of this volume.

Finally, I express my deep gratitude to colleagues, mentors, and family members for their personal support. Their encouragement and understanding were vital in sustaining the editorial process. Their patience and unwavering support laid the foundation for this work.

To all who contributed directly or indirectly, I express my sincere gratitude. This volume serves as a testament to your combined expertise, dedication, and commitment to advancing research in an era influenced by artificial intelligence.

Chapter 1
Methodological Innovations in Studying Algorithmic Injustice in Digital Labour Platforms

Tom Kwanya
http://orcid.org/0000-0002-6306-2669
The Technical University of Kenya, Kenya

ABSTRACT

Digital labour platforms increasingly rely on algorithmic decision-making systems to manage and allocate work. However, these algorithms remain largely opaque to researchers, regulators, and workers. This black-box environment complicates the study of how algorithmic management influences labour conditions, access to opportunities, fairness, and worker well-being. This chapter explores new research approaches developed to investigate algorithmic injustice on digital labour platforms, especially when direct access to backend data, source code, or allocation logic is restricted. It highlights an expanding range of empirical methods, such as black-box audits, data donation initiatives, worker-centred sensing, platform reverse engineering, simulated task experiments, and collaborative ethnography, that allow researchers to infer system behaviour from observable outputs and lived experiences. The chapter discusses the advantages and disadvantages of these methods and underscores the importance of interdisciplinary approaches combining computational techniques with qualitative insights.

DOI: 10.4018/979-8-2600-0434-0.ch001

INTRODUCTION

Digital labour platforms have rapidly expanded into areas such as ride-hailing, food delivery, home services, and online freelancing. They are becoming increasingly significant in global labour markets (Eurofound, 2025; Lee, 2025). According to recent estimates, digital labour platforms grew from 193 in 2010 to over 1,070 by 2023, engaging potentially hundreds of millions of workers globally, with informal work, often platform-mediated, accounting for over 60% of employment in the Global South (approximately 2 billion people) (International Labour Organization. 2024a). These platforms utilise algorithmic management systems to assign tasks to workers, monitor their productivity, set rates, and determine who can remain on the platform. Platforms often describe algorithmic processes as efficient, impartial, and data-driven. However, recent studies suggest that these processes may heighten employer control, increase precarity, and exacerbate existing inequalities (Azevedo & Mendonça, 2023; Kadolkar et al., 2024).

One major challenge in understanding how these systems affect people is that platform systems remain very difficult to comprehend (Kwanya & Wakunuma, 2023). Companies often keep proprietary algorithms, operating procedures, and backend data hidden from external scrutiny by claiming they are safeguarding trade secrets, safety, and business confidentiality. This limited access creates clear information gaps, where platforms can monitor everything workers do and how the market is performing, but workers find it difficult to understand changes in pay or deactivation (Jabagi et al., 2024; Mirbabaie et al., 2025). Even the EU Platform Work Directive and other regulatory measures struggle to ensure that people can truly access algorithmic decision-making systems (Lee, 2025).

The black-box environment significantly challenges standard methods in computing and informatics research that rely on revealed datasets, well-defined system rules, and stable APIs. Platforms may restrict researchers' access, provide filtered data that conceal contentious procedures, or alter system parameters without notification, thereby reducing scientific reliability and reproducibility (International Labour Organization. 2021; Azevedo & Mendonça, 2023). These issues are particularly critical in the Global South, where regulatory oversight is developing, research resources are limited, and workers are more vulnerable to algorithmic exploitation (Adekoya & Chen, 2023; Kwanya & Wakunuma, 2023; Ndunda et al., 2024).

In response, researchers are developing new methodological frameworks to investigate algorithmic injustice without direct access to backend systems. These innovations include adversarial audits that experimentally examine platform behaviour (Eticas Foundation, 2023), worker-led data donation and sensing tools that capture real-time job allocation patterns (International Labour Organization. 2021), and mixed-methods approaches that combine computational inference with ethnography to clarify lived labour experiences (Azevedo & Mendonça, 2023). These methodologies aim to reconstruct system logic from observable outputs while emphasising workers' perspectives on fairness, autonomy, and dignity within algorithmic governance (Mirbabaie et al., 2025).

This chapter positions these emerging ideas within the broader discourse on algorithmic injustice and employment rights. It argues that platform opacity is not merely a technological constraint but a deliberate barrier that protects digital business models from accountability. Therefore, research methods must be devised to remain effective in situations of intentional confidentiality, while also respecting ethical duties towards vulnerable labour groups. This chapter demonstrates how innovative methodological approaches can uncover hidden inequalities in work allocation, pay distribution, and deactivation processes, thereby fostering more equitable digital labour futures. Drawing on examples from both global and Global South contexts, it shows how these approaches can uncover hidden inequalities in work allocation, pay distribution, and deactivation processes, thereby fostering more equitable digital labour futures.

EVOLUTION OF ALGORITHMIC GOVERNANCE IN PLATFORM LABOUR

Digital labour platforms have progressed from early tests in task distribution, such as Amazon Mechanical Turk in the mid-2000s, to advanced algorithmic systems that now underpin large parts of informal and gig work worldwide. Foundational research on algorithmic opacity pointed out how opaque systems concentrate power in the hands of platform owners. Pasquale (2015) described these as "black box" decision-making structures in the "Black Box Society", where secrecy in algorithms controlling credit, search, and reputation allows unaccountable power over economic opportunities. Similarly, Zuboff (2019) introduced the idea of behavioural surplus extraction in the age of surveillance capitalism, where platforms unilaterally claim human experience as raw material for predictive products traded in behavioural futures markets, which is directly relevant to how ride-hailing and delivery platforms monetise worker data and behaviour. Foundational critiques of algorithmic opacity and emergent harm, such as those by Tufekci (2015), highlighted how opaque computational systems can concentrate power and produce unintended consequences, laying thc groundwork for later analyses of platform labour governance (Pasquale, 2015; Zuboff, 2019). Kellogg et al. (2020) posit that the increasing integration of algorithmic technologies in workplaces creates a new, contested environment for organisational control, challenging traditional understandings and practices.

These early critiques, mainly rooted in Global North settings, have been expanded by more recent work specifically addressing platform labour. Gillespie (2010) investigated how platforms influence public discourse through algorithmic curation, establishing the foundation for understanding governance via code. In the gig economy, Rosenblat (2018) and others documented how algorithmic management alters labour rules, often worsening precarity. By the 2020s, research increasingly concentrated on the Global South, where platforms became part of informal economies with minimal regulation (Heeks, 2022; Adekoya & Chen, 2023). Adverse digital incorporation (Heeks, 2022) describes how workers engage with digital networks under exploitative conditions, often experiencing greater income volatility due to opaque allocation and pricing practices. These trends emphasise ongoing information asymmetries and the necessity for methodological innovation to examine opaque systems without direct access to them.

THE BLACK BOX PROBLEM IN PLATFORM RESEARCH

Studying algorithmic unfairness on digital labour platforms is extremely challenging because the algorithms that drive these platforms are like "black boxes" that conceal their workings. Researchers, employees, and regulators face limited access to backend data, decision-making protocols, and optimisation models that control the distribution, assessment, and remuneration of work. This lack of transparency creates information gaps, allowing platforms to hide potential forms of discrimination or exploitation. As a result, traditional research methods often prove inadequate. This section explores the origins of algorithmic opacity, its implications for evidence and justice, and why new approaches are necessary in data-limited contexts.

Algorithmic Opacity as a Design Feature

Digital labour platforms use computational methods to assign jobs, set wages, rate workers, and enforce penalties such as suspensions or deactivations. Although these systems are technically complex, their lack of transparency is not merely a consequence of their technical sophistication; it is often a conscious choice made during design and governance processes. Platforms frequently claim that their algorithms and data constitute trade secrets, citing intellectual property rights, security concerns, and the need to maintain a competitive edge as reasons for withholding information (Lee, 2025; Wood et al., 2024). This deliberate opacity creates an informational black box in which workers, regulators, and researchers cannot see the inputs, internal logic, or optimisation criteria behind job distribution algorithms (Azevedo & Mendonça, 2023). Workers are only exposed to partial outputs, such as ride offers, salary estimates, acceptance rates, and ratings. They remain unaware of why specific jobs are assigned, why their visibility changes, or how their performance scores are determined (Eurofound, 2025; Ndunda et al., 2024).

From a research perspective, this indicates that traditional methodologies, which depend on access to system documentation or complete datasets, are becoming less viable. Instead, scholars operate within a system where critical elements such as the matching function, the weighting of performance metrics, and the criteria for setting prices are opaque, mutable, and strategically shielded.

Data Asymmetries and Power Imbalances

The black box operates because of significant differences in how people access and control data. Platforms often gather detailed information about customers and workers, including GPS traces, timestamps, acceptance and cancellation rates, customer comments, device metadata, and behavioural trends. These data streams are essential for generating revenue and enhancing platforms, but those who produce them often lack access to the raw data (International Labour Organization. 2021; Rosenblat, 2018). In contrast, workers access only a limited subset of what platforms collect. They might see restricted personal histories, such as travel logs, earnings summaries, or basic statistics, but lack the broader context needed to understand their position relative to other workers or how the entire system functions (Adekoya & Chen, 2023; Veen et al., 2024). This unequal flow of information exacerbates power imbalances, as platforms can monitor and penalise workers on a large scale, while workers are unable to effectively scrutinise or challenge platform decisions.

For researchers, data asymmetry manifests in at least three ways:

a) **Limited access to raw data**. Requests for APIs or data-sharing agreements are often rejected or accepted only under strict conditions that hinder comprehensive analysis (Eurofound, 2025).
b) **Selected datasets**. When data is provided, it may be heavily filtered, anonymised, or combined in ways that obscure differences, such as gender-based pay disparities, geographic discrimination, or changes in income over time (Azevedo & Mendonça, 2023).
c) **Dynamic and unclear system updates**. Algorithms and business rules are frequently altered without warning, making it difficult to reproduce results or observe how fairness evolves (Kadolkar et al., 2024).

These inequalities impede the scientific study of algorithmic unfairness, as researchers must speculate on how different data sources influence outcomes.

Legal, Commercial, and Technical Barriers to Access

Many legal and business restrictions exacerbate the black-box problem. Researchers struggle to systematically monitor platform operations because their terms of service often prohibit scraping, reverse engineering, or automated data collection (Eticas Foundation, 2023; Lee, 2025). Legal protections for trade secrets and intellectual property can also be used to prevent the disclosure of algorithmic processes, even in regulatory or judicial settings (Hacker, 2021). Even where data protection and labour laws aim to promote transparency, such as the EU Platform Work Directive or requirements for information on automated decision-making, platforms may only provide superficial descriptions or general documents that fail to reveal how things truly operate (Hacker, 2021; Eurofound, 2025). Workers are often told that "multiple factors" influence allocation or pricing, though the details of how these factors are prioritised and how they interact with profile and reputation data remain hidden.

There are also technical challenges. Machine learning models are used to develop algorithms, but they are difficult to understand because their functioning depends heavily on the context (Burrell, 2016; Mirbabaie et al., 2025). Furthermore, the constantly evolving nature of these systems, the 'moving target', means that any snapshot of transparency can quickly become outdated. From a research perspective, this reduces the feasibility of investigations based solely on disclosures and encourages methods that can adapt to dynamic, adversarial environments.

Consequences for Evidence, Policy, and Worker Rights

The lack of transparency in platform systems directly affects how evidence is created, how policies are formulated, and how workers' rights are safeguarded. Black-box conditions hinder researchers' ability to determine causal links between algorithmic decisions and observed labour outcomes, including income fluctuations, workload distribution, and varying deactivation rates (International Labour Organization. 2021; Wood et al., 2024). It is challenging to substantiate many claims about fairness or discrimination to the level required by courts, regulators, or public debate without robust access to data and system logic. This makes it difficult for policymakers and regulators to fully understand what is happening. Regulators may lack the independent data and technical expertise needed to verify compliance with regulations requiring non-discrimination, fair working conditions, or explanations for automated decisions (Hacker, 2021; Lee, 2025). If available, platform-generated transparency reports or audits are often used as self-reporting tools that prioritise reputation management over genuine accountability.

The black box undermines basic workers' rights. First, it breaches the right to information: workers cannot decipher how performance is assessed, how ratings are calculated, or why their access to jobs varies over time (Ndunda et al., 2024). Second, it hampers workers' ability to challenge decisions: appeals against deactivation or pay changes are usually handled through opaque, automated channels, with limited evidence to support workers' claims. Third, a lack of transparency makes it more difficult for workers to collaborate to understand how the system affects them and where they might implement changes (Veen et al., 2024; Wood et al., 2024). In the Global South, these issues are exacerbated by additional systemic weaknesses, such as fragile social safety nets, high unemployment, and limited enforcement capacity. Research on ride-hailing and delivery platforms in Kenya, Nigeria, and South Africa shows that opaque algorithms result in unpredictable incomes, increased dependence on incentives, and difficulties in planning livelihoods (Adekoya & Chen, 2023; Kwanya & Wakunuma, 2023; Ndunda et al., 2024). Here, the black box is not merely a technical challenge but also a structural barrier that keeps workers in precarious, low-visibility employment.

Methodological Challenges and the Need for Innovation

The black box problem significantly transforms research on digital labour platforms. Researchers must deal with incomplete, noisy, or strategically curated data while respecting legal and ethical limits on data collection. This creates several methodological challenges:

a) **Validity and reliability**. When data is collected through scraping, self-reporting by workers, or small-scale studies, questions arise about sample representativeness, selection bias, and whether observed patterns reflect wider platform behaviour (Eticas Foundation, 2023; International Labour Organization. 2021).
b) **Reproducibility**. Frequent system updates and undocumented changes hinder the ability to replicate studies over time or across locations, which impairs the development of knowledge (Kadolkar et al., 2024).
c) **Ethical risks**. Researchers must prevent workers from being exposed to reprisals, such as collecting data that platforms could use to identify "difficult" workers or organisers, and must not breach terms of service in ways that jeopardise participants (Azevedo & Mendonça, 2023; Mirbabaie et al., 2025).

These issues highlight the need for methodological innovation in platform labour research. Instead of abandoning comprehensive research, experts are developing methods to handle unclear situations and hostile environments, recognising these as significant design limitations. Black-box audits, worker-centred data architectures, and mixed-method triangulation are strategies to address the reality that direct backend access is unlikely to happen soon. The rest of this chapter explores these ideas further by proposing new approaches to investigate algorithmic injustice within digital labour platforms, especially where data access is highly restricted.

THEORETICAL UNDERPINNINGS FOR EXAMINING ALGORITHMIC INJUSTICE

Researchers need distinct theoretical frameworks to investigate algorithmic injustice in digital labour markets. "Unfairness" or "bias" alone are too vague to support serious research, especially when platforms obscure their operations. This section outlines the core intellectual foundations that enable methodological innovation (See Table 1). It begins with exploring algorithmic injustice in platform work, then discusses the well-known fairness–accountability–transparency (FAT) triad in algorithm studies. It continues with the concepts of data justice and labour/data justice, which link algorithmic systems to broader issues of power, rights, and redistribution. The focus then shifts to perspectives from the Global South and decolonial viewpoints. Finally, it considers socio-technical and worker-centred approaches that see platforms as systems subject to contestation and negotiation, rather than mere technological objects.

Defining Algorithmic Injustice in Digital Labour Platforms

Algorithmic injustice on digital labour platforms refers to consistent patterns of disadvantage caused or worsened by algorithmic management, rather than isolated "errors" or individual complaints. Schulze et al. (2022) describe "algorithmic unfairness" as management strategies that result in repeated devaluation, restriction, and exclusion of workers. Examples include giving certain workers less priority in job allocation, making pay schemes difficult to comprehend, or using unclear evaluations to justify deactivation.

This unfairness relates to relationships and systems. It originates from how algorithms convert platform priorities, such as customer satisfaction, speed, and profit, into metrics like acceptance rates, cancellation thresholds, or customer ratings, which influence workers' access to income and the platform itself (Schulze et al., 2022). Workers frequently perceive these procedures as capricious or punitive, even though they lack the requisite information to substantiate any systematic harm.

Algorithmic injustice does not only arise from biased data or flawed models. It also demonstrates how design decisions influenced by power shape whose interests are protected and whose risks are shifted onto others. Research on platform work shows how algorithms can normalise unpaid waiting times, increase surveillance, and penalise workers who refuse low-wage or dangerous jobs (Mbare, 2024). For academics, this means that "fairness" must be seen not only as a statistical characteristic of models, but also as a matter of whose livelihoods are consistently affected by algorithmic management.

Fairness, Accountability, and Transparency (FAT) in Algorithmic Management

The FAT trinity functions as a common framework for understanding responsible algorithmic decision-making (Lepri et al., 2018; Shin & Park, 2019). Fairness means there are no unjust differences in outcomes between people or groups. Accountability means you can hold someone responsible and address the harm they have caused. Transparency means affected stakeholders can understand how systems, data, and decisions function (Crespo, 2022; Cheong, 2024).

When it comes to digital labour platforms, fairness involves issues such as: Do workers with similar performance receive the same job opportunities and pay? Are certain locations, genders, or types of contracts consistently assigned lower ratings or less favourable allocations? (Jabagi et al., 2025; Schulze et al., 2022).

When workers are underpaid, unfairly deactivated, or placed in dangerous situations, questions of accountability arise regarding who is responsible: the platform, the customer, or the individual who created the algorithm (Singhal, 2024).

Transparency is really difficult to grasp. As Crespo (2022) points out, algorithmic transparency might mean several things, such as being able to see the data used, the model's structure, the decision-making rules, or simply the results. In platform work, workers rarely achieve more than outcome-level transparency, if at all. This raises the question, "Who should be transparent, about what, and why?" (Cheong, 2024).

FAT principles help define the measurements or inferences researchers seek in their methodological design. For instance, they include differences in results (fairness), the presence or absence of justifications or appeals processes (accountability), or the amount and quality of information provided to workers and regulators (transparency).

Data Justice and Work/Data Justice

FAT examines the features of decision systems, but data justice frameworks consider how these systems are integrated into broader social and political contexts. Taylor (2017) states that data justice concerns ensuring fairness in how digital data makes people visible, represented, and treated. This includes concerns about individuals impacted by datafication, including beneficiaries and those adversely affected. McCann and Cruz-Santiago (2022) explore this idea in the workplace by introducing the concept of labour/data justice. They argue that it is essential to analyse how datafication alters labour relations, especially regarding the extraction, commodification, and use of workers' data within regulatory frameworks (such as worker classification, targeted inspections, or automated enforcement).

Labour/data justice links traditional labour law issues, such as fair work, collective bargaining, and freedom from discrimination, to the management of platform data infrastructures (McCann & Cruz-Santiago, 2022). Recent policy initiatives for data justice in platform labour emphasise the rights to data access, portability, and co-governance. The ability for workers to access their individual logs, ratings, and performance histories, and to utilise them collectively in negotiations or litigation, is considered vital for addressing structural imbalances (Leslie et al., 2022).

Data justice offers researchers a framework for creating tools that do more than merely identify bias. It prompts researchers to consider whether research empowers workers with meaningful information. Does it oppose data extraction practices, or does it inadvertently reinforce them? In this way, worker-centred data collection, data donation systems, and participatory audits can be seen as efforts to ensure that methodological practices align with the principles of data justice.

Global South, Negative Digital Integration, and Decolonial Viewpoints

To truly understand algorithmic unfairness in digital labour markets, we need to examine global inequities. Heeks (2022) introduces the concept of "adverse digital incorporation" to describe instances where individuals are integrated into digital networks under exploitative or disempowering conditions.

For many workers in the Global South, engagement in platform labour does not mean an escape from precarity; rather, it signifies a reconfiguration of precarious conditions through digitally mediated piecework, rigorous surveillance, and reliance on opaque grading systems.

Decolonial and indigenous perspectives on data justice highlight the deep roots of platform and AI systems in colonial exploitation and racial capitalism. Arora (2024) formulates a "creative data justice" framework that emphasises southern and indigenous epistemologies, questioning the valuation or erasure of creativity and labour within AI-driven economies. De Souza (2024) also champions decolonial data law and governance that challenge hegemonic systems and embrace pluriversal ideas of rights and accountability.

For digital labour platforms in cities across Africa, Asia, and Latin America, these perspectives emphasise the intersection of algorithmic injustice with persistent disparities in labour markets, infrastructure, and regulation (Heeks, 2022). Methods that ignore this background may mistake reported discrepancies as simple technical faults rather than indicators of broader political and economic institutions.

In theory, the Global South and decolonial frameworks shift research from merely "bias correction" to addressing issues of structural power, such as who develops platform systems, their integration into global value chains, and how they are held accountable to workers and communities most impacted.

Perspectives on Socio-Technical and Worker-Focused Issues

Finally, we need to examine algorithmic injustice through socio-technical perspectives that see workers as active interpreters and strategists, not just as individuals controlled by algorithms. Schulze et al. (2022) utilise a socio-technical framework to show how employees perceive injustice as layered disadvantages across different processes, including task allocation, assessment, and conflict resolution.

Recent research on freelancers' sense-making and coping techniques shows how workers collectively aim to "reverse engineer" platform behaviour through online forums, social media, and informal data sharing (Upwork, 2025). Workers share advice on maintaining high ratings, avoiding deactivation, and gaining more publicity, which reveals the system's lack of clarity and their attempts to regain control through creative methods.

Worker-centred approaches to algorithmic fairness consider not only objective differences but also subjective perceptions of fairness and legitimacy. Research shows that employees' perceptions of fairness and transparency significantly influence job satisfaction, trust, and their willingness to follow algorithmic directives (Shin & Park, 2019; Jabagi et al., 2025). Mbare (2024) links certain algorithmic management techniques, such as dynamic pricing and ongoing surveillance, to psychosocial stress and decreased well-being.

From a methodological perspective, socio-technical and worker-centred approaches endorse qualitative and participative methods, such as interviews, focus groups, digital ethnography, and co-design of data tools, as essential additions to quantitative audits. They support research approaches that see worker narratives, resistance strategies, and informal knowledge-sharing networks as vital data for understanding the production, perception, and challenge of algorithmic injustice in everyday situations.

Research Gaps and Contribution

Previous methods for studying algorithmic management in digital labour platforms often rely on assumptions that limit their use in opaque, rapidly changing environments. Many studies depend on disclosed datasets, stable APIs, or insider access to backend logic (International Labour Organization. 2021; Azevedo & Mendonça, 2023), which are rarely accessible due to proprietary protections and frequent system updates. This reliance decreases reproducibility, especially when platforms change parameters without warning or restrict researcher access (Jabagi et al., 2024). Approaches focused on the Global North, such as those by the Eticas Foundation (2023), typically emphasise statistical audits but pay less attention to workers' lived experiences and power imbalances in settings with little regulation or data scarcity. Similarly, while FAT frameworks provide statistical benchmarks (Lepri et al., 2018; Shin & Park, 2019; Dwork et al., 2012), they frequently overlook socio-political dimensions of labour control and exploitation, as critiqued by data justice scholars (Taylor, 2017; McCann & Cruz-Santiago, 2022).

These limitations are particularly evident in the Global South, where regulatory oversight is weaker, informal employment prevails, and workers face greater vulnerability to algorithmic decisions (Adekoya & Chen, 2023; Ndunda et al., 2024). Existing methods seldom incorporate worker-centred data collection or participatory approaches that empower participants rather than treat them as passive subjects.

This chapter advances the field by proposing a set of opacity-resilient, interdisciplinary methodological innovations that do not require direct backend access. By combining black-box audits, worker-led data donation, reverse engineering, simulated experiments, and collaborative ethnography, the proposed approaches infer system behaviour from observable outputs and lived experiences. These methods explicitly address gaps in reproducibility and representativeness by triangulating computational inference with qualitative insights and foregrounding workers' perspectives on fairness, autonomy, and dignity. Table 1 below maps key theoretical frameworks to prior studies, their limitations, and how this chapter extends them.

Table 1. Theoretical Frameworks and Advancements

Theoretical Framework	Key Prior Studies	Limitations in Prior Work	How This Chapter Advances It
Fairness–Accountability–Transparency (FAT)	Lepri et al. (2018); Shin & Park (2019)	Focus on statistical fairness; limited attention to power in labour contexts	Integrates FAT with data justice and Global South perspectives for worker empowerment
Data justice / labour-data justice	Taylor (2017); McCann & Cruz-Santiago (2022)	Broad conceptual; less operationalised for platform research	Proposes worker-centred data donation and participatory methods as practical justice tools
Adverse digital incorporation	Heeks (2022)	Descriptive of structural exclusion	Links to methodological innovations that uncover hidden inequalities in opaque systems
Surveillance capitalism	Zuboff (2019)	Broad critique; less focus on gig work specifics	Applies to algorithmic management in platforms, emphasising opacity-resilient methods

These contributions position the chapter within ongoing debates on algorithmic governance and platform labour by offering practical tools for researchers operating under intentional confidentiality, while maintaining ethical responsibilities toward vulnerable workers.

INNOVATIVE METHODOLOGICAL APPROACHES FOR DIGITAL LABOUR PLATFORM RESEARCH

Researchers have regarded opacity as a key design constraint rather than merely an annoyance, given digital labour platforms' refusal to provide meaningful access to backend data and decision logic. This has led to a surge of methodological innovation focused on deducing, reconstructing, or challenging algorithmic management through observable traces and workers' lived experiences. This section examines some of the most important methods, including adversarial and black-box audits, worker-centred data collection and donation, reverse engineering and technological probing, simulation and synthetic task trials, and collaborative or participatory approaches involving workers and unions (See Table 2). These methodologies demonstrate that rigorous research on algorithmic injustice is possible even with limited data.

Black Box and Adversarial Audits

Black box audits examine the platform as an opaque system that can still be investigated by analysing its inputs and outputs (See Table 2). Researchers design controlled trials or quasi-experiments that systematically alter key factors such as time, location, worker profile, and task features. They then observe how the system assigns employment, sets prices, or sanctions individuals.

There is a growing number of hostile algorithmic audits of ride-hailing and delivery platforms. These studies typically involve users or workers submitting controlled requests to the platform, followed by statistical analysis to identify patterns in how individuals are treated differently based on factors such as location, time of day, or worker status. Recent guides from civil society organisations outline detailed steps for conducting these audits. These plans include forming hypotheses, designing scenarios, selecting samples, evaluating risks, and informing affected communities and authorities of the findings (Eticas Foundation, 2023; Eticas Foundation, 2024).

Fairness analyses based on outcomes are also being applied directly to platform data when available. For example, research on ride-sharing markets uses extensive trip datasets to examine variations in allocation or waiting times across neighbourhoods or passenger demographics, and to assess the trade-offs between platform profitability and driver earnings (Binns et al., 2025). These efforts often employ innovative access mechanisms, such as data subject access requests (DSARs), which require organisations to provide personal data in accordance with data protection legislation and can be compiled from many employees.

However, black box audits have their limitations. They rarely disclose the exact model architecture or all the features utilised. They can be costly to implement on a large scale, and platforms might alter their systems in response to criticism. Nevertheless, they remain a very valuable method for documenting distributional harms and testing certain fairness assumptions when uncertainty prevails.

Worker-Centred Data Collection and Data Donation

Another significant area of innovation is worker-centred data infrastructures (See Table 2). Researchers collaborate with workers' groups, unions, or NGOs to collect comprehensive logs of work, pay, ratings, and interactions with the app, rather than relying on platforms.

For example, Calacci and Pentland (2022) discuss how the "Shipt Calculator" was co-designed and implemented. This browser extension and data tool was developed with US grocery delivery workers to monitor their actual income, time, and deductions for each order. The technology enabled both individual workers and organisers to question what the platform reported about pay and to measure unpaid work and fluctuating piece rates. Recent research broadens this methodology to include larger "worker data collectives," incorporating application logs, screenshots, and self-reports to influence policy debates and support legal actions (Hsieh et al., 2024). Worker-centred data collection can use multiple methods:

a) **Digital diaries and mobile ethnography**, where workers record trips, waiting times, cancellations, and experiences in real time (Arubayi, 2022).
b) **Survey panels** that periodically collect structured data on hours, income, algorithmic changes, and perceived fairness.
c) **Topic modelling and text analysis** of large corpora of worker discussions on forums, Reddit, or WhatsApp groups to capture experiences at scale (Ütük Bayılmış, 2025).

These methods centre workers in data and knowledge production, aligning with concepts of labour/data justice. Methodological challenges include selection bias (only some workers have the time or literacy to participate), data quality issues, and the need for robust privacy and security measures to prevent retaliation.

Reverse Engineering and Technical Probing

Reverse engineering techniques enhance black box audits by analysing the actual functionality of platform applications on devices and networks (See Table 2). Researchers and engineers do not merely infer logic from what they observe. Instead, they examine program binaries, network traffic, and interface behaviour to understand how data flows and to identify hidden variables or metrics (Isbah, 2022).

Unions and technical activists have begun systematically documenting techniques for mobile reverse engineering in the gig economy. Some methods include intercepting and decrypting network traffic, analysing API calls, and monitoring the evolution of app versions (Zhou et al., 2018). These activities can uncover, for example, which data points are collected (such as battery level, device model, and GPS accuracy), how often location data is recorded, or whether "shadow ratings" exist that workers cannot see (Lettieri et al., 2022).

Reverse engineering presents numerous complex legal and moral challenges, especially when terms of service prohibit such analysis or when decrypting data might violate laws against intercepting communications. However, in some jurisdictions, it can be lawful if there are exceptions for security research or if workers' rights to access their data, such as the right to understand automated decision-making processes, are upheld (Zhou et al., 2018). Reverse engineering is a meticulous process that demands significant time, effort, and specific skills. Nevertheless, its findings can be compared with worker accounts and audit data to gain a clearer understanding of algorithmic management.

Simulation and Synthetic Task Experiments

Another technique for analysing job allocation and pricing on digital labour markets is simulation and synthetic experiments (See Table 2). In this context, researchers use computational models to simulate platform behaviour or conduct experiments on actual or near-actual platforms to assess the impacts of various allocation policies (Brooke & Rao, 2024).

In economic and operations research, discrete-event simulations and agent-based models are utilised to examine how various job allocation algorithms influence matching efficiency, platform profitability, and worker income, for example, by simulating scenarios contrasting "uniform allocation" with priority rules in online labour markets (Bokányi et al., 2020). Researchers can use these models to evaluate situations that would be impractical or unethical to test on live platforms. For example, they might assign higher importance to workers with lower past wages or manage safety-related refusals differently (Brooke & Rao, 2024).

Experimental designs embedded within matching platforms, such as job-search and digital-job platforms in Sub-Saharan Africa, have utilised randomised access to platform features or incentives to use the platforms to estimate causal effects on employment and income (Jones & Sen, 2022; Jones, 2025). Although these studies often focus on labour market outcomes rather than internal algorithmic logic, similar designs can be adapted to assess fairness-oriented solutions, such as fairness requirements in recommendation algorithms.

The main advantages of simulation and synthetic experiments are their ability to control variables and identify causes. Their limitations relate to external validity. Behaviour in simulations may not match real-world platforms, and worker methods might change in ways that models do not capture (Cooke & Shope, 2017). Therefore, it is essential to combine simulation with real-world audit data and workers' insights.

Participatory, Collaborative, and Co-Design Methods

Finally, more people are recognising that new methods must be participatory, involving workers, unions, and community groups as partners rather than merely sources of data. Co-design and community-based participatory research (CBPR) frameworks provide valuable approaches.

Co-design workshops effectively involve marginalised groups in identifying challenges, developing design ideas, and evaluating tools, as shown by participatory projects across various fields (Bromark et al., 2023; Turuba et al., 2024; Istanboulian et al., 2023). In the platform-work sector, similar approaches have been used to create worker-centred data-sharing systems, transparency dashboards, and advocacy tools.

Hsieh et al. (2024) collaborated with gig workers across sectors to develop prototypes for data collectives that support individual decision-making and policy advocacy.

Participatory approaches can be integrated with technical methods in several ways:

a) **Co-defining audit questions**: Workers help identify which aspects of algorithmic management most affect them, such as deactivation, bonus schemes, and guiding the focus of black box audits.
b) **Interpreting findings**: Quantitative audit or reverse-engineering results are presented back to workers for interpretation, contextualisation, and validation.
c) **Co-governance of data**: Worker organisations co-own the data tools and infrastructures developed, ensuring longevity beyond a single research project.

These approaches align with decolonial and data-justice perspectives because they empower workers and help them develop their skills. They also introduce real-life challenges, such as maintaining participant engagement, ensuring fair compensation, managing expectations, and balancing scientific rigour with immediate organisational needs. Participatory designs are essential to ensure that methodological improvements genuinely serve workers' interests and strengthen collective power, rather than merely collecting new data under the guise of research.

Table 2. Comparative Overview of Methodological Approaches

Method	**Data Access Requirements**	**Ethical/Legal Risk**	**Scalability**	**Evidentiary Strength**	**Recommended Contexts**
Black-Box / Adversarial Audits	Low (outputs only)	Medium	High	Medium (distributional)	Initial pattern detection; Global South audits
Worker-Centred Data Donation	Worker logs/ screenshots	High (retaliation)	Medium	High (lived experience)	Union/collective action; income volatility
Reverse Engineering	App/network traffic	Very High	Low	High (technical detail)	Advanced technical probing; high expertise
Simulation / Synthetic Experiments	Model assumptions	Low	High	Medium (counterfactual)	Policy testing; external validity limited
Collaborative Ethnography	Worker narratives	Medium	Low	High (contextual)	Intersectional vulnerabilities; Global South

STRENGTHS, LIMITATIONS, AND ETHICAL CONSIDERATIONS

The methodological innovations discussed earlier (black-box audits, worker-centred data collection, reverse engineering, simulation, and participatory co-design) show that comprehensive research into algorithmic unfairness is feasible despite platforms withholding backend data. However, these methods are not without bias. Each has distinct strengths and limitations, and all raise moral and legal issues, particularly when workers risk losing their jobs or facing penalties. This section critically assesses the effectiveness of specific approaches as evidence, highlights potential biases or misuse, and outlines essential ethical principles for responsible research. It concludes by stressing that triangulation and reflexive, worker-focused practices should be the primary means of protecting workers.

Evaluating Black Box and Adversarial Audits

Black-box and adversarial audits provide a reliable way to empirically examine platform behaviour in situations characterised by opaqueness. Their main benefit is causal leverage: by altering inputs such as location, profile attributes, and request timing, and by observing outputs like job offers, pay, and wait times, researchers can test specific ideas about discriminatory patterns or unfair treatment (Eticas Foundation, 2023, 2024). When conducted properly, these audits provide regulators, judges, and advocacy groups with strong evidence, especially when they identify systematic differences across protected or vulnerable groups (Binns et al., 2025).

However, audits can only offer limited insights. First, they typically only capture a small part of how the platform functions across different times and locations; protocols often focus solely on specific cities, time periods, or types of work. Consequently, the results can quickly become outdated if the platform dynamically modifies its algorithm. Second, black box audits might not be effective because platforms can take measures to prevent them. For example, after an audit is made public, platforms might "clean up" behaviour or present metrics in a more favourable way, but they might also shift exploitation to other areas (Burrell, 2016).

Third, audits risk oversimplifying fairness by focusing on a limited set of observable outcomes, such as acceptance or waiting times, while ignoring more complex forms of unfairness, including unpaid labour, surveillance, or psychological distress (Schulze et al., 2022; Mbare, 2024). Lastly, audits that employ fake accounts or scripted conversations may breach the platform's terms of service. This raises legal and ethical issues that must be considered during institutional review processes.

Assessing Worker-Centred Data and Data Donation

Worker-centred data collection through diaries, browser extensions, survey panels, and data donation systems redistributes epistemic authority to workers. It enables the documentation of labour process dynamics that platform-provided data often overlooks, such as unpaid waiting, cancellations due to unsafe conditions, or interruptions caused by algorithmic changes (Calacci & Pentland, 2022; Arubayi,

2022). These approaches also embody the principles of labour/data fairness by viewing workers as co-researchers and co-owners of data (McCann & Cruz-Santiago, 2022; Hsieh et al., 2024).

Richness and context are among its key features. Worker logs and narratives can connect algorithmic events, such as a sudden drop in offers, to tangible consequences such as income disruptions, indebtedness, or health impacts. Mixed-methods approaches that combine structured logs with qualitative interviews are highly effective in revealing how workers perceive and respond to algorithmic management (Azevedo & Mendonça, 2023).

However, these methods have many issues. Participation is often selective: employees with better connectivity, higher digital literacy, or stronger organisational ties are more likely to provide data. This can result in samples that lack sufficient representation of the most marginalised workers, including women who care for others or migrants without legal status. This reflects the same kind of inequality that research aims to reveal (Heeks, 2022). Self-reported data may also be influenced by recall bias or selective reporting, especially in highly volatile labour markets.

Another significant issue is data security and privacy. If platforms can identify players and penalise them by deactivating accounts or imposing "shadow" punishments, it becomes risky to maintain detailed travel logs, screenshots, or pay statements (Ndunda et al., 2024). Projects must invest in encryption, secure infrastructure, and robust governance plans that clearly specify who can access worker data, when, and for what reasons. Without these measures, worker-focused infrastructures might unintentionally compromise safety.

Technical Approaches: Reverse Engineering and Simulation

Reverse engineering and technical probing provide rare insights into how data flows and hidden metrics operate within platform apps. Researchers can identify variables and tracking methods that are not visible at the interface level by analysing network traffic and code. For instance, they can detect continuous background location tracking or undisclosed performance scores. These discoveries can support workers' concerns about surveillance or covert ratings and strengthen legal claims under data protection or labour legislation.

The main benefit of reverse engineering is its level of detail. It can reveal specific endpoints, parameters, and event logs that help form more accurate assumptions about how work allocation and pricing are decided. This information, along with audits and worker data, allows us to shift from "black box" to "grey box" explanations of how algorithmic management functions.

However, reverse engineering requires significant effort and specialised expertise, making it difficult for small, well-funded research teams to pursue. Another significant issue is that legislation is often unclear: in some areas, reverse engineering can be challenged under laws on circumvention, computer misuse, or trade secrets, even when it benefits the public interest (De Souza, 2024). As a result, ethical review boards may be reluctant to approve such activities.

Conversely, simulation and synthetic experiments closely resemble real platforms. Their advantage is that researchers can systematically modify allocation rules, pricing systems, or fairness constraints to evaluate the trade-offs between efficiency, profit, and equity (Pricing and Job Allocation in Online Labour Platforms, 2019; Kadolkar et al., 2024). Simulations help develop and test policy ideas, such as guaranteeing a minimum wage or prioritising workers with a history of low income, before implementing them in the real world.

The primary limitation is external validity. Simulation models might rely on unrealistic assumptions due to a limited understanding of the platform's algorithms and worker behaviour. If they are not grounded in real-world data and workers' experiences, they could create "toy" worlds that hide rather than reveal structural unfairness.

Ethical and Legal Considerations in Research Practice

Ethical and legal considerations are essential across all methodologies. The fact that workers often face precarious situations, platforms monitor them, and the environment can be hostile makes research on algorithmic management particularly sensitive. Researchers must, at the very least, follow fundamental principles of respect, doing good, and fairness in ways appropriate to this context (Markham & Buchanan, 2012).

First, informed consent must be thorough and ongoing. Workers should clearly understand the risks and potential benefits of participation, including the possibility that platforms may reveal their research efforts. Consent procedures must be accessible to individuals with varying reading abilities and languages and must permit withdrawal without repercussions.

Second, minimising risk requires careful methodological choices. For instance, black box audits that use fake accounts may reduce risk compared to audits that use genuine worker accounts, but they also alter the type of evidence. Data minimisation, pseudonymisation, and strong encryption should be standard for worker-centred data initiatives. They should also avoid storing unnecessary identifiers (Hsieh et al., 2024).

Third, researchers must consider collective harms rather than only focusing on individual harms. Algorithmic injustice often affects groups defined by gender, region, or migration status, even when individual identities are anonymised. Projects that reveal some populations are systematically worse off might unintentionally stigmatise them or shift risks (for example, by quietly deprioritising high-risk areas).

This requires working closely with worker groups and being cautious about how information is shared (Arora, 2024).

Fourth, there are legal requirements and ambiguities. Scraping, reverse engineering, and automated testing may breach platform terms of service, even if used for public-interest research (Eticas Foundation, 2023). Researchers must navigate conflicts between contractual obligations and fundamental rights, including research freedom, workers' rights to knowledge, and data protection rights, often seeking legal advice. In some cases, unions or NGOs may act as protective mediators, taking on legal and political responsibilities that individual workers cannot properly manage.

Fifth, epistemic and situational ethics are crucial. Researchers affiliated with well-funded institutions, often based in the Global North, must consider how their positionality influences research questions, interpretations, and benefits. Decolonial and data justice frameworks emphasise the importance of sharing authority over agenda-setting, co-authorship, and data governance among worker communities and organisations in the Global South (Heeks, 2022; McCann & Cruz-Santiago, 2022).

Finally, researchers, particularly those in the Global North or with institutional privileges, must reflexively consider how their position influences agenda-setting and data governance in the global digital economy (Heeks, 2022)

Toward Reflexive, Triangulated Research Designs

When weighing the pros and cons of each combined strategy, it is clear that triangulated, reactive designs are crucial. No single approach fully captures how algorithmic injustice operates on digital labour platforms. Black box audits effectively highlight differences in outcomes, but they offer limited insight into how individuals actually experience these issues. Data focusing on workers' experiences can reveal what happens in daily life, but such data may not be suitable for large-scale use or be representative. Reverse engineering and simulation provide technical and counterfactual understanding, but they rely on a solid foundation of real-world facts.

A reflective research programme integrates these methodologies, leveraging each to validate and refine the others while persistently scrutinising its own assumptions. It is also crucial that it incorporates ethical and political commitments to prevent suffering, enhance workers' informational and organisational power, and oppose, rather than inadvertently support, the systemic inequalities that render platforms opaque.

Evidence Pyramid for Methodological Approaches

Building on the earlier discussion of methodological strengths, weaknesses, and the importance of reflexive, triangulated designs to address platform opacity, an evidence pyramid offers a structured way to rank these approaches based on their ability to establish causality, uncover hidden mechanisms, and provide reproducible insights into algorithmic injustice. This hierarchy, tailored to the specific challenges of black-box platform research, ranks methods from the most robust (at the top) to more exploratory (at the bottom), guiding researchers in choosing or combining tools for thorough analysis.

At the pinnacle lies reverse engineering combined with worker data triangulation, offering the strongest causal inference by uncovering hidden algorithmic metrics while integrating lived impacts for comprehensive validation. Next, black-box audits with large samples provide strong distributional evidence, as seen in studies reporting quantifiable disparities, such as 7% pay gaps (Binns et al., 2025). Worker-centred logs and diaries follow, delivering high contextual validity through real-time personal narratives. Simulation and synthetic experiments enable counterfactual testing, though they remain assumption-dependent. At the base, qualitative ethnography functions as an essential exploratory foundation, rich in depth but limited in generalisability when used alone.

The pyramid not only emphasises the relative strengths of each method but also highlights the importance of integration, paving the way for the subsequent examination of policy implications and future directions in promoting equitable digital labour ecosystems.

Figure 1. Evidence pyramid of methodological approaches (Author's Design)

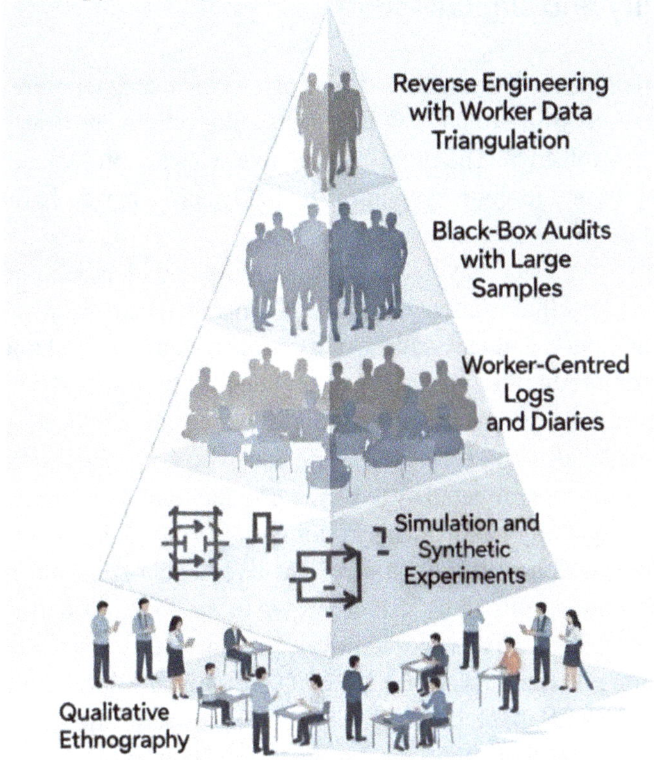

CASE INSIGHTS FROM EMERGING CONTEXTS

This section of the chapter explores how digital labour platforms have rapidly expanded in the Global South, where informality is high, social security is limited, and youth unemployment is significant. Drawing on recent studies from Kenya, Nigeria, South Africa, India, and other regions, this study shows how algorithmic injustice appears through issues such as income insecurity, unclear performance criteria, gender-based exclusion, and the punishment of migrants and refugees. These examples illustrate what black box audits, worker-centred data infrastructures, and participatory methods can and cannot accomplish. They emphasise that methodologies must be adapted to the institutional context, legal risks, and the diverse vulnerabilities of workers across sectors, genders, and migrant statuses.

Kenya: Platform Work at the Intersection of Informality and Digitalisation

Kenya exemplifies how digital labour platforms integrate into an already informal labour market. According to an ILO Prospects study, there are nearly 180 digital labour platforms in Kenya. The number of people working online and in person is increasing rapidly, and foreign companies are highly active in ride-hailing, delivery, and freelancing.

This growth is seen as a way to boost productivity and include more people, but survey statistics show that women who work at home or in beauty jobs still earn very little, work long hours, and lack social safety nets (Kwanya & Wakunuma, 2023).

Recent mixed-methods research on motorcycle ride-hailing and delivery drivers shows how algorithmic management leads to daily economic instability. Brailovskaya et al. (2025) find that drivers are mainly young men whose incomes are highly unpredictable and significantly affected by dynamic pricing, bonus systems, and penalties embedded in opaque algorithms. Many people take out loans to buy motorcycles and smartphones, which makes them very sensitive to small changes in demand or platform regulation. Additionally, Brailovskaya et al. (2025) found that full-time Kenyan motorcycle gig drivers worked over 66 hours per week on average, with net hourly earnings around KES 46.5 (≈ USD 0.36) during July–December 2024, scaling to roughly KES 9,368 monthly, highlighting extreme income volatility and dependence on opaque algorithmic incentives and penalties.

Qualitative research involving ride-hailing drivers in Nairobi reveals how they perceive algorithmic control through the lens of "hustling" culture and informal-economy norms. They develop collective strategies, such as WhatsApp groups, informal unions, and information sharing at various stages, to counter the effects of unpredictable rankings, deactivations, and fare fluctuations.

For scholars, Kenya emphasises several methodological necessities. First, triangulation is crucial. To understand how algorithmic changes cause missed rent payments, loan defaults, and food insecurity, platform-facing black-box audits (such as scripted transport requests across neighbourhoods) must be linked to household-level and asset-finance data. Second, logs and diaries created by workers are essential for capturing issues such as unpaid waiting time, cancelled trips, and refusals due to safety concerns that backend platform data often overlooks. Third, models of allocation and income volatility should incorporate the regulatory and financial infrastructure environment, including the widespread use of mobile money, rather than treating it as a neutral background.

Nigeria and West Africa: Algorithmic Management in Congested Urban Labour Markets

In Nigeria and other West African cities, platform work has grown in densely populated areas with high unemployment and poor transport systems. Adekoya et al. (2023) demonstrate that ride-hailing platforms in Lagos use algorithmic management to carefully organise workers' labour processes, employing acceptance-rate thresholds, star-rating systems, and automatic sanctions, while officially classifying drivers as independent contractors. Drivers find platform interactions very transactional because algorithms control access to clients, set pay, and enforce one-sided rule changes.

These conditions determine which methods are feasible. Black-box audits that employ fake rider accounts can detect patterns of discrimination in space (such as systematically under-servicing specific neighbourhoods). However, they may not capture the intricate discussions drivers need to have in person to deal with traffic, police harassment, and sudden price hikes. Therefore, worker-centred ethnographies and interview-based panel studies are crucial for linking algorithmic signals (such as surge alarms and destination concealment) to time utilisation, "dead kilometres," and informal side-deal activities.

The Nigerian cases highlight three key methodological points. First, the importance of detail in timing: the same allocation algorithm can have varying effects on distribution during rush hours, fuel shortages, or security curfews. Second, scholars should be cautious not to attribute all unfairness solely to the logic of formal algorithms. This is because platform opacity often interacts with brokers, unions, automobile owners, and state authorities in ways that make it challenging to identify who is responsible. Third, there are legal and political risks, particularly concerning data scraping and reverse engineering. To mitigate these risks, companies need to collaborate with worker organisations that can assess the likelihood of retribution and help develop safer audit methods.

Gendered Experiences of Platform Work in the Global South

There is now substantial evidence that digital labour platforms in the Global South are unfair to both men and women. Anwar (2022) studied female platform workers in African cities. He demonstrated that gender norms, caregiving duties, and safety concerns influence not only access to platform employment but also income, mobility, and vulnerability to harassment. Chibanda et al. (2022) argue that women's experiences of cleaning and domestic work in South Africa are best analysed through a "freedom-based" perspective, showing how algorithmic ratings and surveillance undermine dignity and autonomy, even when income seems to increase.

The ILO's (2024) study on women's employment on digital labour platforms in India shows that these platforms often continue occupational segregation. Women mainly work in beauty, tutoring, domestic tasks, and specific online activities, where algorithms may encourage availability and responsiveness, potentially conflicting with unpaid caregiving duties. Women often cover their own social protection and safety measures, while the protections provided by platforms are sometimes incomplete and conditional. Additionally, women on digital labour platforms in the Global South often face occupational segregation and conflicting demands from caregiving, with many earning below viable thresholds and lacking social protections (International Labour Organization. 2024b).

From a methodological standpoint, these findings highlight the limitations of gender-neutral black box audits. If researchers only examine unequal allocation or pricing based on nominal gender indicators, they might miss more subtle mechanisms such as:

a) Algorithms that reward constant availability, which harms carers.
b) Rules for allocating space that favour areas considered high-risk or during the night.
c) Rating systems that reflect clients' gendered expectations regarding "emotional labour."

To understand these dynamics, researchers need sample frameworks that account for different groups and qualitative aspects, focusing on women's experiences of safety, harassment, and limited choices. Time-use diaries, participatory mapping of safe and risky areas, and gender-divided focus groups can generate data not captured in platform logs. Research designs should also anticipate under-representation. For instance, Kwanya and Wakunuma's (2023) and Brailovskaya et al.'s (2025) studies showed that extreme male dominance in the transport and delivery sectors can make quantitative gender comparisons statistically difficult, requiring oversampling or sectoral diversification.

Refugees, Migrants and Other Highly Vulnerable Workers

Digital labour platforms increasingly serve as intermediaries for refugees, migrants, and others forced to leave their homes. A 2024 joint report on refugees and digital labour platforms discusses both the potential of remote work and the heightened hazards, including legal status, documentation issues, and cross-border payment systems (UNHCR Innovation Service, 2024). Refugees and undocumented migrants may be more likely to work in high-risk, low-protection sectors within platform labour, such as delivery, household chores, and microtasks.

The International Trade Union Confederation's (2025) dossier on platform workers and migrant labour shows that algorithmic monitoring and rapid deactivation are frequently used against migrant workers, who have no opportunity to appeal or seek due process. The absence of clear risk scoring, computerised identity checks, and language barriers makes them even more vulnerable.

These situations create significant ethical and methodological challenges. Standard black-box audits might reveal that accounts with foreign-sounding names or certain documentation statuses receive fewer job offers or experience more cancellations. However, they cannot simply demonstrate the risks associated with deportation or the instability of the law. Therefore, worker-centred data collection must be designed with strict confidentiality and data-minimisation safeguards, developed collaboratively with organisations that understand the context of asylum and migration.

Methodologically, refugee and migration cases require mixed-jurisdictional studies. Allocation rules might be established in one country, data stored in another, and personnel based in a third. Researchers may need to undertake various tasks to understand how algorithmic decisions impact national immigration and labour law. For instance, they might have to reverse-engineer app flows, examine data-protection laws, and hold workshops with affected individuals.

Cross-Cutting Lessons for Research Design in Emerging Contexts

From these various situations, a set of overarching insights emerges for methodological innovation in investigating algorithmic unfairness (See Table 3). First, digital labour platforms are not merely tools; they are also institutional actors that transform the way people work, often exploiting institutional gaps in the Global South (Elias & Gomez, 2025). Consequently, research strategies must link micro-level allocation outcomes to macro-level issues, such as investment patterns, regulatory shortcomings, and deeply rooted informal economy practices.

Second, there is evidence that structural inequality is strongly associated with platform growth. A recent study indicates that ridesourcing platforms perform well in cities with high economic disparity, where cheap labour meets demand for rapid fulfilment (de Ruijter et al., 2024). Similarly, research on the expanding gig economy and sustainable development shows that platforms tend to shift risks onto workers and the public, even though they are promoted as tools for inclusion (Ayentimi et al., 2025). Therefore, research methodologies must effectively link algorithmic regulations to the distributive consequences in income, workload, and risk.

Third, context-specific case studies from Kenya and other nations show that research about decent labour is closely linked to algorithmic considerations. The World Employment and Social Outlook report and more recent country studies reveal that low and inconsistent incomes, limited social safety nets, and high levels of unpaid work are common on both online and location-based platforms (International Labour Organisation, 2021, 2024).

This suggests that audits of allocation algorithms should be incorporated into comprehensive research initiatives that also monitor working-time policies, social security contributions, and cases of violence or harassment.

These stories finally highlight the significance of collaboration and involving others. Worker-led data infrastructures, jointly established audit methods, and partnerships with unions and community organisations can enhance data quality, improve interpretative accuracy, and minimise harm (Brailovskaya et al., 2025; International Labour Organisation, 2024).

In emerging environments characterised by limited legal safeguards and widespread retaliation, such collaborations are not only ethical but also methodologically vital. They help ensure that new techniques not only unlock the black box but also empower workers to challenge and alter algorithms' influence.

Table 3. Comparative Summary of Case Insights on Algorithmic Injustice in Digital Labour Platforms

Case / Scenario	Platform / Context	Location / Region	Key Insights on Algorithmic Injustice	Methodological Approach(es) Highlighted	Broader Implications / Lessons
Kenyan Motorcycle Drivers	Shipt-like / ride-hailing and delivery	Kenya (Nairobi and others)	Extreme working hours (>66 h/week), very low net earnings (~KES 46.5/hour ≈ USD 0.36), high income volatility due to opaque incentives/penalties	Worker-centred logs, black-box audits, ethnographic observation (Brailovskaya et al., 2025)	Highlights precarity and opacity in Global South gig work; need for worker-led data collection to expose pay gaps
Gendered Experiences	Various platforms (delivery, freelancing)	India and the Global South	Occupational segregation, below-viable earnings for women, lack of social protections, and conflicting caregiving demands	Collaborative ethnography, qualitative interviews (International Labour Organization. 2024b)	Reveals intersectional vulnerabilities (gender + platform work); calls for gender-sensitive policy interventions
Lagos Platform Workers	Ride-hailing and delivery apps	Nigeria (Lagos)	Opaque deactivation, variable pay, surveillance, and control mechanisms disadvantage workers	Black-box audits, worker diaries, reverse engineering elements (Adekoya et al., 2023)	Demonstrates power asymmetry and need for regulatory oversight in high-informal contexts

continued on following page

Table 3. Continued

Case / Scenario	Platform / Context	Location / Region	Key Insights on Algorithmic Injustice	Methodological Approach(es) Highlighted	Broader Implications / Lessons
Refugee and Host Community Access	Various digital labour platforms	Multiple, Global South	Barriers to entry, poor working conditions, and limited access to opportunities for refugees	Participatory methods, collaborative ethnography (UNHCR Innovation Service, 2024)	Emphasises inclusion and equity; platforms must address the vulnerability of marginalised groups
General Global South Platforms	Multiple (ride-hailing, freelancing, delivery)	Kenya, Nigeria, India	Widespread income volatility, deactivation without explanation, and algorithmic bias amplifying inequalities	Mixed: audits, data donation, ethnography (Ndunda et al., 2024; World Economic Forum, 2025)	Reinforces the need for opacity-resilient methods and cross-regional policy learning

IMPLICATIONS FOR POLICY AND PLATFORM GOVERNANCE

Accountability frameworks for digital labour platforms must require algorithmic transparency, mandating the disclosure of job distribution criteria and performance metrics, similar to the EU Platform Work Directive. Platforms should undertake independent black-box audits and adversarial testing to detect biases in pay disparities or deactivations, with mandatory third-party verification by regulators (Schulze et al., 2022). Worker-focused data donation systems and participatory co-design enable unions to collaboratively oversee data infrastructures, thereby strengthening appeals for automated decisions that comply with labour data justice principles.

Regulators should demand triangulated evidence from black-box audits, worker diaries, and reverse engineering to validate claims of algorithmic injustice, such as income volatility in regions like Kenya and Nigeria (Adekoya & Chen, 2023). Labour advocates should require accessible datasets from data subject access requests and collective logs to challenge platform surveillance asymmetries and support litigation efforts (Calacci & Pentland, 2022). Combining computational inference with ethnography through mixed-methods approaches may provide reproducible evidence for policy changes and decent work standards, emphasising ethical safeguards to protect workers from retaliation (Heeks, 2022).

CONCLUSION AND FUTURE DIRECTIONS

Despite innovations such as black-box audits and worker-centred data donation, methodological gaps persist in studying algorithmic injustice, particularly regarding reproducibility amid platform updates and the ethical risks of worker retaliation in data-scarce contexts of the Global South. Selection bias in participatory approaches and the limited external validity of simulations hinder the development of comprehensive evidence on causal links between algorithms and outcomes such as income volatility. Future research opportunities include triangulating mixed methods with AI-driven inference tools and longitudinal worker collectives to enhance validity, while developing decolonial frameworks to address the adverse effects of digital integration.

Methodological advancements are driving policy advocacy for FAT (fairness, accountability, transparency) principles, necessitating algorithmic disclosures and independent audits to reduce black-box opacity. Labour data justice promotes worker data co-governance, empowering unions with data portability for collective bargaining and litigation. These initiatives help create fairer markets by revealing power imbalances, supporting decent work standards, and shaping regulations such as the EU Platform Work Directive. Ultimately, this ensures that algorithmic management prioritises protecting workers' dignity over maximising profit.

REFERENCES

Adekoya, O. A., & Chen, W. (2023). Implications of algorithmic management on careers and employment relationships in the gig economy: A developing-country perspective. *The Journal of Industrial Relations*, *65*(4), 587–604.

Anwar, M. A. (2022). Platforms of inequality: Gender dynamics of digital labour in Africa. *Gender and Development*, *30*(3), 747–764. DOI: 10.1080/13552074.2022.2121059

Arora, P. (2024). Creative data justice: A decolonial and indigenous framework to assess creativity and artificial intelligence. *Information Communication and Society*. Advance online publication.

Arubayi, D. E. (2022). *Algorithmic management and everyday mobilities: A case study of ride-hailing drivers in Lagos, Nigeria* (Doctoral dissertation, University of Manchester).

Ayentimi, D., Amankwaa, A., & Burgess, J. (2025). The Emerging Gig Economy and Sustainable Development in Sub-Saharan Africa. *Societies (Basel, Switzerland)*, *15*(10), 274. DOI: 10.3390/soc15100274

Azevedo, É. S. F., & Mendonça, J. R. C. (2023). Algorithmic management on digital labour platforms: A systematic literature review. *International Journal of Human-Computer Studies*, *178*, 103048.

Binns, R., Stein, J., Datta, S., Van Kleek, M., & Shadbolt, N. (2025, June). Not Even Nice Work If You Can Get It: A Longitudinal Study of Uber's Algorithmic Pay and Pricing. In *Proceedings of the 2025 ACM Conference on Fairness, Accountability, and Transparency* (pp. 1484-1497). DOI: 10.1145/3715275.3732099

Bokányi, E., Bokányi, E., Hannak, A., Hannak, A., & Hannak, A. (2020). Understanding Inequalities in Ride-Hailing Services Through Simulations. *Scientific Reports*, *10*(1), 6500. DOI: 10.1038/s41598-020-63171-9 PMID: 32300117

Brailovskaya, V., Onkundi, F. K., & Atela, M. (2025). *Economic lives of digital platform gig workers: Case of motorbike passenger and delivery drivers in Kenya*. IDInsight.

Bromark, K., Knutsson, O., & Spånberger Weitz, Y. (2023). Co-designing a dynamic tool to enhance participation for young people: A participatory design project with young service users and social workers. *Children and Youth Services Review*, *149*, 106856. DOI: 10.1016/j.childyouth.2023.106856

Brooke, S., & Rao, A. H. (2024). Designing for justice in freelancing: Testing platform interventions to minimise discrimination in online labour markets. *Big Data & Society, 11*(1), 20539517241232631. Advance online publication. DOI: 10.1177/20539517241232631

Burrell, J. (2016). How the machine "thinks": Understanding opacity in machine learning algorithms. *Big Data & Society, 3*(1), 1–12. DOI: 10.1177/2053951715622512

Calacci, D., & Pentland, A. (2022). Bargaining with the black-box: Designing and deploying worker-centric tools to audit algorithmic management. *Proceedings of the ACM on Human-Computer Interaction, 6*(CSCW2), 1-24. DOI: 10.1145/3570601

Cheong, B. C. (2024). Transparency and accountability in AI systems: Safeguarding wellbeing in the age of algorithmic decision-making. *Frontiers in Human Dynamics, 6*, 1421273. DOI: 10.3389/fhumd.2024.1421273

Chibanda, R., Tsibolane, P., & Nkohla-Ramunenyiwa, T. (2022). Gendered inequality on digital labour platforms in the Global South: Towards a freedom-based inclusion. In Zheng, Y. (Eds.), *ICT4D 2022* (pp. 1–14). Springer. DOI: 10.1007/978-3-031-19429-0_4

Cooke, N. J., & Shope, S. M. (2017). *Designing a Synthetic Task Environment.* Routledge., DOI: 10.4324/9781315243771-21

Crespo, M. L. (2022). Fair, transparent and accountable algorithmic decision-making: The premise, the proposed solutions, and the open challenges. *iSCHANNEL, 17*(1), 28–41.

de Ruijter, A., Cats, O., & van Lint, H. (2024). Ridesourcing platforms thrive on socio-economic inequality. *Scientific Reports, 14*(1), 7371. DOI: 10.1038/s41598-024-57540-x PMID: 38548793

De Souza, S. P. (2024). Decolonial data law and governance. *Technology and Regulation, 6*, 1–20. DOI: 10.71265/rvwfyt51

Dwork, C., Hardt, M., Pitassi, T., Reingold, O., & Zemel, R. (2012). Fairness through awareness. *Proceedings of the 3rd Innovations in Theoretical Computer Science Conference (ITCS '12)*, 214–226. DOI: 10.1145/2090236.2090255

Elias, A., & Gomez, G. (2025). Digital platforms as institutional actors in the global south–Evidence from India's informal sector. *Digital Geography and Society, 9*, 100146. DOI: 10.1016/j.diggeo.2025.100146

Eticas Foundation. (2023). *Adversarial algorithmic auditing guide.*

Eticas Foundation. (2024). *Adversarial audit of ride-hailing platforms.* Taxi Project.

Eurofound. (2025). *Platform work: Algorithmic management and working conditions in Europe*. Publications Office of the European Union.

Gillespie, T. (2010). The politics of 'platforms'. *New Media & Society*, *12*(3), 347–364. DOI: 10.1177/1461444809342738

Hacker, P. (2021). Teaching fairness to artificial intelligence: Existing and novel strategies against algorithmic discrimination under EU law. *Common Market Law Review*, *58*(4), 1143–1186.

Heeks, R. (2022). Digital inequality beyond the digital divide: Conceptualizing adverse digital incorporation in the global South. *Information Technology for Development*, *28*(4), 688–704. DOI: 10.1080/02681102.2022.2068492

Hsieh, J., Zhang, A., Rasetarinera, M., Chou, E., Ngo, D., Lightman, K., Lee, M. K., & Zhu, H. (2024). Supporting gig worker needs and advancing policy through worker-centered data-sharing. *Proceedings of the ACM on Measurement and Analysis of Computing Systems*, *8*(4), 111.

International Labour Organization. (2021). *The role of algorithmic management in the gig economy*. ILO Working Paper Series.

International Labour Organization. (2024a). *Digital labour platforms in Kenya*. ILO Prospects series.

International Labour Organization. (2024b). *Understanding and improving women's work on digital labour platforms – Stakeholder consultation workshop in India.*

International Trade Union Confederation. (2025). *Time to deliver: Platform workers and migrant labour.*

Isbah, M. F. (2022). Algorithmic Exploitation: Understanding Labor Process and Control among RideHailing Platform Workers. *Jurnal Sosioteknologi*, *21*(2). Advance online publication. DOI: 10.5614/sostek.itbj.2022.21.2.5

Istanboulian, L., Rose, L., Yunusova, Y., & Dale, C. (2023). Adapting co-design methodology to a virtual environment: Co-designing a communication intervention for adult patients in critical care. *Research Involvement and Engagement*, *9*(1), 103. DOI: 10.1186/s40900-023-00514-6 PMID: 37957776

Jabagi, N., Croteau, A., Audebrand, L. K., & Marsan, J. (2024). The effects of algorithmic transparency on gig workers. *Information Systems Journal*, *34*(2), 345–367.

Jabagi, N., Croteau, A. M., Audebrand, L. K., & Marsan, J. (2025). Do algorithms play fair? Analysing the perceived fairness of HR-decisions made by algorithms and their impacts on gig-workers. *International Journal of Human Resource Management, 36*(2), 235–274. DOI: 10.1080/09585192.2024.2441448

Jones, S., & Sen, K. (2022). Labour market effects of digital matching platforms: Experimental evidence from Sub-Saharan Africa (IZA Discussion Paper No. 15409). Institute of Labor Economics.

Jones, S., & Sen, K. (2025). The gendered impact of digital jobs platforms: Experimental evidence from Mozambique. *The World Bank Economic Review, 39*(1), 143–172. DOI: 10.1093/wber/lhae019

Kadolkar, I., Ali, S., & Vieira, D. (2024). Algorithmic management in the gig economy: A systematic review. *New Technology, Work and Employment, 39*(1), 102–121.

Kellogg, K. C., Valentine, M. A., & Christin, A. (2020). Algorithms at work: The new contested terrain of control. *The Academy of Management Annals, 14*(1), 366–410. DOI: 10.5465/annals.2018.0174

Kwanya, T., & Wakunuma, K. (2023). Regulation of digital platforms for a socially-just gig economy in Kenya. *HIIG Impact Publication Series.* DOI: 10.5281/zenodo.7588795

Lee, D. S. (2025). *Governing global gig platforms in the age of AI.* Oxford University Press.

Lepri, B., Oliver, N., Letouzé, E., Mehrabi, N., Oliver, N., & Pentland, A. (2018). Fair, accountable, and transparent algorithms in the public sector. *Proceedings of the 1st ACM SIGCAS Conference on Computing and Sustainable Societies (COMPASS '18).* https://doi.org/DOI: 10.1145/3209811.3209812

Leslie, D., Katell, M., Aitken, M., Singh, J., Briggs, M., Powell, R., . . . Mazumder, A. (2022). Advancing data justice research and practice: An integrated literature review. *arXiv preprint arXiv:2204.03090.*

Lettieri, N., Malandrino, D., Guarino, A., & Zaccagnino, R. (2022). The Eye of the Rider. Visualization and data-driven heuristics for the critical analysis of gig economy. *International Conference on Information Visualisation*, 373–378. DOI: 10.1109/IV56949.2022.00068

Markham, A., & Buchanan, E. (2012). Ethical decision-making and internet research (Version 2.0): Recommendations from the AoIR Ethics Working Committee.

Mbare, B. (2024). Algorithmic management, wellbeing and platform work. *Labour & Industry*, *34*(2), 145–166.

McCann, D., & Cruz-Santiago, A. (2022). Labour/data justice: A new framework for labour/regulatory datafication. *Journal of Law and Society*, *49*(4), 555–578. DOI: 10.1111/jols.12392

Mirbabaie, M., Stieglitz, S., & Marx, J. (2025). Transparency fallacy: Perceived fairness in algorithmic management. *Computers in Human Behavior*, *155*, 107347.

Ndunda, E., Ouma, S., & Mwangi, J. (2024). *Digital labour platforms in Kenya: Working conditions and wellbeing*. ILO.

Pasquale, F. (2015). *The Black Box Society: The Secret Algorithms That Control Money and Information*. Harvard University Press. DOI: 10.4159/harvard.9780674736061

Rosenblat, A. (2018). *Uberland: How algorithms are rewriting the rules of work*. University of California Press. DOI: 10.1525/9780520970632

Schulze, L., Cai, Z., Trenz, M., & Tan, C.-W. (2022). Algorithmic unfairness on digital labor platforms: How algorithmic management practices disadvantage workers. In *Proceedings of the International Conference on Information Systems (ICIS 2022)*. AIS.

Shin, D., & Park, Y. J. (2019). Role of fairness, accountability, and transparency in algorithmic affordance. *Computers in Human Behavior*, *98*, 277–284. DOI: 10.1016/j.chb.2019.04.019

Singhal, A. (2024). Toward fairness, accountability, transparency, and ethics in AI systems. *Journal of the American Medical Informatics Association : JAMIA*, *31*(5), 1012–1025.

Taylor, L. (2017). What is data justice? *Big Data & Society*, *4*(2), 1–14.

Tufekci, Z. (2015). Algorithmic harms beyond Facebook and Google: Emergent challenges of computational agency. *Journal of the Association for Information Science and Technology*, *66*(11), 2195–2205. DOI: 10.1002/asi.23499

Turuba, R., Katan, C., Marchand, K., Brasset, C., Ewert, A., Tallon, C., & Barbic, S. (2024). Weaving community-based participatory research and co-design to improve opioid use treatments and services for youth, caregivers, and service providers. *PLoS One*, *19*(4), e0297532. DOI: 10.1371/journal.pone.0297532 PMID: 38635804

UNHCR Innovation Service. (2024). *Emerging practices to improve access to and working conditions on digital labour platforms for refugees and host communities*.

Ütük Bayılmış, O., & Orhan, S. (2025). Decoding digital labor: A topic modeling analysis of platform work experiences. *Systems*, *13*(9), 819. DOI: 10.3390/systems13090819

Veen, A., Barratt, T., & Goods, C. (2024). Platform-mediated work and the politics of algorithmic control. *Work, Employment and Society*, *38*(2), 432–450.

Wood, A. J., Lehdonvirta, V., & Graham, M. (2024). Workers of the gig economy unite? Algorithmic management and collective action. *Socio-economic Review*, *22*(1), 109–132.

World Economic Forum. (2025). *Pathways for decent work in Kenya's digital economy*.

Zhou, Y., Kumar, D., Bakshi, S., Mason, J., Miller, A., & Bailey, M. (2018). Erays: Reverse Engineering Ethereum's Opaque Smart Contracts. *USENIX Security Symposium*, 1371–1385. https://www.usenix.org/system/files/conference/usenixsecurity18/sec18-zhou.pdf

Zuboff, S. (2019). *The Age of Surveillance Capitalism: The Fight for a Human Future at the New Frontier of Power*. PublicAffairs.

Chapter 2
Adopting an AI-Assisted Thematic Approach:
Exploring Everyday Use of Generative AI Through Qualitative Data Analysis

Valerie McTaggart
http://orcid.org/0000-0003-4930-5326
Atlantic Technological University, Ireland

John McCarthy
http://orcid.org/0009-0003-7285-7214
Atlantic Technological University, Ireland

ABSTRACT

This study explores everyday GenAI use through an empirical, interpretivist investigation involving eighty-eight qualitative survey responses. Drawing inferences from Davis's (1989) Technology Acceptance Model (TAM) and the socio-technical perspective (Trist & Bamforth, 1951). The research offers two key contributions. First, it identifies three thematic patterns in GenAI use: (1) A Journey of Discovery, (2) Workplace Partner—Friend or Foe, and (3) AI's Ethical Crossroads. The findings also illustrate that, contrary to concerns of knowledge dilution, findings suggest that GenAI serves as a cognitive amplifier, enabling users to reallocate effort toward more complex and creative tasks. This study advances our understanding of GenAI's socio-technical implications and offers actionable methodological guidance for qualitative researchers. Second, it proposes a novel methodological framework, the AI-Assisted Thematic Approach, which supports qualitative data analysis

DOI: 10.4018/979-8-2600-0434-0.ch002

INTRODUCTION

The artificial intelligence (AI) revolution is here and gathering momentum (Hitch, 2024), with generative AI (Gen AI) predicted to have a significant impact on organisations, societies, and individuals (Cotton et al., 2023). These developments are important because AI provides inestimable possibilities for enhancing people's lives in a variety of areas, including their homes, healthcare, education, employment, entertainment, safety, and transportation (Rahwan et al., 2019.)

While some believe these developments are part of a fourth industrial revolution (Oosthuizen, 2022), as GenAI increasingly permeates individuals' everyday lives (e.g. Crusoe et al., 2024; Neyazi et al., 2024), debates have emerged over the potential benefits, risks and challenges of using GenAI. Therefore it is unsurprising that there have been calls for additional research into the topic (e.g. Gregor, 2024; Jarvenpaa & Klein, 2024) and in doing so for researchers to embrace and capitalise on this new digital development (Sarker et al., 2023) whilst committing to undertaking responsible research practices (Dwivedi et al.,2023) to foster a more inclusive and dynamic research ecosystem (Yoo, 2024).

Hence, our objectives are twofold: first, to investigate the everyday use of GenAI and second, to undertake responsible research practices using GenAI. In doing so, we collected empirical evidence from eighty-eight respondents through the distribution of a qualitative online survey.

To support qualitative researchers capitalising on the opportunities that GenAI offers we took guidance from leading scholars (e.g., Eisenhardt, 1989; Braun & Clarke, 2006; Corbin & Strauss, 2008; Punch,2014; Miles et al., 2020; Saldaña, 2021) and present a novel approach to qualitative data analysis which we have labelled AI-Assisted Thematic Approach.

This chapter is structured as follows: first, we introduce the reader to the evolution of GenAI, and then, we consider the use of GenAI in the Academic Community. An outline of the research design and method follows this. We then present a detailed account of how data analysis was conducted. This is followed by a discussion of the findings and their relationship to the existing literature. The final section presents the study's limitations and opportunities for future research. In the appendix, we provide a clear chain of evidence (Yin, 2018) of how this data analysis was conducted in practice.

EVOLUTION OF GENAI

Although it is acknowledged that AI is an umbrella term (Kemp, 2024), a broad concept (French & Shim, 2024) and always nascent, liminal, and emerging there are different forms of AI (Gregor, 2024) with the term AI often used interchangeably with machine learning, neural networks, and deep learning (Shim et al., 2020).

Typically, discussions about AI centre around several key areas, including machine learning and its deep learning and reinforcement learning abilities (Deng & Yu, 2014); Large Language Models (LLMs) that can generate textual content and extract key information from text, thus offering a powerful tool and various business process automation technologies, including robotic process automation. More recently, commentators have been paying increasing attention to Generative Artificial Intelligence (GenAI), which can drive innovation through new product discovery and development (Mariani & Dwivedi, 2024).

Although early generative models were available in the 1990s and 2000s, two key developments marked significant milestones (Yoo, 2024). Firstly, the introduction in 2014 of generative adversarial networks (GANs) enabled the creation of high-quality, realistic images (Cao et al., 2023), laying the foundation for further advancements in generative models (Goodfellow et al., 2016). The Transformer followed this (Vaswani et al., 2017), which proposed a new simple network architecture which helped revolutionise natural language processing (NLP) from text identification to text understanding (French & Shim, 2024); thus, the generative capabilities of GenAI lie in the Information Systems (IS) architecture (Yoo, 2024).

GenAI is a type of AI based on machine learning that can generate text, images, audio, code, videos, and synthetic data (Kanbach et al., 2023). In short, GenAI creates new, original content by going beyond recognising patterns and making inferences to generate new content based on prompt inputs that mimic the training data of various forms. In contrast, traditional AI primarily makes predictions and classifications based on existing datasets (Wessel et al., 2023). In particular, Chat Generative Pre-Trained Transformer (ChatGPT), a form of OpenAI has sparked waves of excitement and concern worldwide (French & Shim, 2024; Sarker et al., 2023; Sartori & Theodorou, 2022).

This advancement, at its core, is powered by LLMs and adds to their existing functionality. LLMs are programmed to use statistical techniques to predict the network in a sentence. Conversely, ChatGPT utilises a transformer-based model that enables the processing of vast amounts of data in parallel (Alberts et al., 2023). Simply put, when generating new text, GenAI makes predictions based on the context they are given, choosing the next word in a sentence by calculating the probability of each possible next word based on the words that came before it (Yoo, 2024). Adding a chat interface has made using LLMs intuitive, even for those with basic levels of digital literacy (Kietzmann & Park, 2024).

Nevertheless, despite its long history, there is still no universally accepted definition of AI (Collins et al., 2021), although several attempts have been made. For example, Haenlein and Kaplan suggest that "AI is commonly defined as a system's ability to interpret external data correctly, to learn from such data, and to use those learnings to achieve specific goals and tasks through flexible adaptation" (2019, p.5). Similarly, Someh et al. (2022, p.143) posit that "AI is the ability of machines to perform human-like cognitive tasks, including the automation of physical processes such as manipulating and moving objects, sensing, perceiving, problem-solving, decision making and innovation".

Conversely, Berente et al. (2021, p.1435) suggest that AI is "whatever we are doing next in computing", while Russel and Norvig's (2003) book titled 'Artificial Intelligence: A Modern Approach' is considered the most frequently cited source for defining AI (Collins et al., 2021; Smyth et al., 2024). Here, the authors define AI as "the study of (intelligent) agents that receive precepts from the environment and take action. Each such agent is implemented by a function that maps percepts to actions, and we cover different ways to represent these functions, such as production systems, reactive agents, logical planners, neural networks, and decision-theoretic systems" (Russell & Norvig, 2010, p. viii)

AI holds many benefits, including providing powerful approaches for collecting and analysing data at a scale not possible before (Hinings et al., 2018), streamlining business operations (Yu et al.,2023), and business model transformation improved customer service, (Hanelt et al., 2021; Correani et al., 2020) better health care provision (Dwivedi et al., 2023) more efficient supply chains (Faraj et al., 2018) and a new way to provide education (Chan & Hu, 2023).

However, the use of AI depends on workers' trust in AI technology (Glikson & Woolley, 2020; Neyazi et al.,2024), and researchers have identified that AI is rarely a perfect substitute for people (Furr et al., 2022). Consequently, there is an ongoing need for human intervention and human expertise (Faraj et al., 2018).

Moreover, according to Sartori and Theodoru (2022), when examining two fundamental sources of inequalities - such as gender and race - there is evidence that AI systems are far from being neutral, let alone fair. Consequently, the development and implementation of AI will require ongoing monitoring, governance and ethical oversight (e.g. Mergel et al.,2019; Wade & Shan, 2020; Sandberg & Alvesson, 2021) to combat digital inequality as AI systems are more probably blind to social complexity (inequality, diversity, power structures) (Sartori & Theodorou, 2022; Wessel et al., 2023), and provide information assurance including risk management, information quality, intellectual property (Chi et al., 2020) privacy protection that goes beyond an individual's data privacy (Yu et al., 2023) and respects a user's expectations on how their data will be used and a user's expectations for how the system will influence their life and the contexts surrounding them (Redmiles, 2022) and compliance with regulations, and continuity of operations (Mahmood et al., 2010).

Furthermore, the impact of AI on opportunities available to individuals as a result of the digital automation of routine tasks, which may lead to widespread job losses

(Frey & Osborne, 2017), has raised significant concerns (Krauss et al., 2021) as AI tends to excel at activities that people do not, such as spotting patterns in large data sets and following instructions exactly (Moravec, 2008). For example, Kelly (2023) argues that AI has already caused a 50%-70% decrease in wages in the United States, creating income inequality and threatening millions of jobs. Those most at risk are those who undertake low-value, high-volume tasks, otherwise known as routine workers (Alberts et al., 2023).

To avoid futuristic oversimplification (Floridi, 2020), AI needs a multidisciplinary community (Dwivedi et al., 2023) and an interdisciplinary approach when researching the topic (Verhoef et al., 2021) as AI is not a technology or set of technologies, but a continually evolving frontier of emerging computing capabilities (Gerling et al., 2022).

Consequently, AI researchers, economists, and public intellectuals have argued that AI is radically different from prior technologies in its potential to transform the landscape of work and employment (Bailey & Barley, 2020), and managing AI is unlike IT management in the past. Therefore it is unsurprising that there has been a growing concern that research on AI could experience a lack of cumulative building of knowledge (Collins et al., 2021) as public knowledge on what AI consists of and what its implications are is poor (Zhang & Dafoe, 2019) as AI means "different things to different people; ranging from data aggregation and manipulation to magic" (Theodorou & Dignum, 2020, p. 2).

GENERATIVE AI AND THE ACADEMIC COMMUNITY

According to Haenlein and Kaplan (2019), AI will become as much a part of everyday life as the Internet or social media did in the past and will forever change the way we work in all sectors of the economy (Hitch, 2024) with this trend set to continue sparking transformative change to scholarship (Sarker et al., 2023; O'Dea, 2024). In the academic community, GenAI transforms four major educational domains: learning, teaching, assessment and administration (Chiu, 2024), resulting in increased student engagement, collaboration, and accessibility.

Conversely, the challenges are twofold: student academic integrity and research integrity. Firstly, growing concerns over academic dishonesty involving students using AI risks the credibility of assessment approaches within higher education worldwide (O'Dea, 2024). However, higher education providers are also responsible for training students to be future-ready for employment in a society powered by GenAI (Chiu, 2024). Therefore, higher education providers need to transform and revisit their approaches to learning and assessment using GenAI as a learning tool to be embraced rather than something to be prohibited (Dwivedi et al., 2023).

Secondly, research employing GenAI is rapidly expanding across fields and is anticipated to accelerate and transform scientific knowledge. This was evidenced in a survey of Scopus conducted by Bano et al. (2023), which revealed that 587 papers reference ChatGPT in their titles or abstracts, with a distribution spanning diverse domains: 247 from medicine, 147 from the social sciences, 116 from computer science, and 83 from engineering.

Given the natural relationship between AI and statistical analysis, it is unsurprising that big data and machine learning algorithms have enabled AI technologies to provide new opportunities for quantitative research (Zhang & Feng, 2021). However, qualitative and quantitative traditions have unique skills to their approach and offer valuable insights into the phenomena associated with AI (Benbya et al., 2021). Therefore, qualitative researchers are also using AI, particularly GenAI, in qualitative data analysis in several ways, including text vectorisation to transform the text into numbers (Cheligeer et al., 2023) or to identify key themes and underlining meanings in interview data (Chang et al., 2021) as GenAI can identify patterns and predict outcomes in massive unstructured data (Iansiti & Lakhani, 2020).

Although interpretive qualitative researchers are more sceptical towards the use of GenAI (Ngulube, 2023), arguing that it might overlook the subtleties of human communication, which should be nuanced, contextual, and human-centred (Lennon et al., 2021), GenAI is just an extension of previous technological developments (Prem,2024), for example, using NVIVO© or Computer-assisted qualitative data analysis software (CAQDAS) (Saunders et al., 2019) which have become commonplace in qualitative research (Elliott, 2018) to support qualitative data analysis. These platforms, which have benefitted academics for many years, began as a mechanism to better organise and code qualitative data and now include analytic tools such as word frequencies, word clustering, sentiment analysis and thematic analysis (Williams, 2024). More recently, two of the major software packages for computer-assisted qualitative data analysis, ATLAS.ti and MAXQDA, have both developed partnerships with OpenAI, the developer of ChatGPT, while ATLAS.ti's approach to AI relies on automatic coding rather than user-based querying of the data (Morgan, 2023; Paulus et al.,2025).

Researchers have also presented several convincing arguments in support of the use of GenAI in qualitative research. Firstly, as GenAI can analyse large amounts of data quickly and accurately (Anis & French, 2023), it accelerates research outputs, saving time, improving self-efficacy and self-esteem, and reducing stress and obsolete knowledge (Cheligeer et al., 2023).

Furthermore, by bringing AI 'into the loop' as a research assistant, it reduces human bias (Benbya et al., 2021) and researchers “study how these tools expand human ideas, prompts, reflections, and interpretations” (Jarvenpaa & Klein, 2024, p. 110) to generate highly original outputs (Chan & Hu, 2023) which can then contribute to an iterative process of idea formulation and assessment of the quality, novelty, or the state of related research (Susarla et al., 2023). The challenge for the research community is differentiating between human and AI authorship within academic and educational communities (Dwivedi et al., 2023).

To mitigate these challenges and provide reassurance to detractors (Wang & Liang, 2024), the literature offers some helpful guidelines to support the use of AI in qualitative research, which we have acronymised as RATED (Responsibility, Accountability, Transparency, Explainability, Disclosure). While these are not to be considered a panacea, they all contribute to ensuring human control in novel practices that include requirement, design and development methodologies for fairer AI usage (Dwivedi et al., 2023; Sartori & Theodorou, 2022).

To begin with, there must be Responsibility on the part of the researcher to conduct the research responsibly (Andersen et al., 2025) to safeguard against potential biases or discriminatory practices embedded in the AI (Mogaji & Jain, 2024). Accountability is also needed (Benbya et al., 2021; Dennis et al., 2023) for any possible consequences that the AI outputs may have. In addition, there must be Transparency (Berente et al., 2021; Gupta et al., 2024), a frequent criticism of qualitative research, as it is difficult to replicate due to a lack of transparency (Bryman & Bell, 2011). Thus, in the context of AI, the goal is to provide sufficient information to ensure attribution of accountability and, therefore, human control (Bryson & Theodorou, 2019). There must also be Explainability (Reis et al., 2020; Sartori & Theodorou, 2022) that facilitates the explanation of the outcomes generated by AI (Benbya et al., 2021) to mitigate against individuals contesting these outputs. Finally, there must be Disclosure that AI has been used (French & Shim, 2024) to maintain trust (Gupta et al., 2024; Aoki et al., 2024), a necessary criterion by which qualitative research should be judged (Lincoln & Guba, 2000).

The revolution is here, and GenAI has the potential to significantly enhance and transform the work of qualitative researchers from multiple traditions and approaches (Hitch, 2024). Therefore, the question is not whether GenAI should be adopted in qualitative research, but rather how best to adopt it (Williams et al., 2024).

RESEARCH DESIGN AND METHOD

This study examines how individuals use GenAI in their everyday lives. To give context to the study, thereby ensuring its relevance and appropriateness (Gupta et al., 2024), the qualitative survey instrument was developed by integrating items from established literature, with careful customisation to align with Davis's (1989) Technology Acceptance Model (TAM) which emphasises perceived usefulness (the degree to which a potential user believes that technology will help to perform a task) and ease of use (the perceived utility of the effort to use the new technology) as the main determinant of a users' attitude, behaviour and intention to use and accept the information system (IS) (Davis, 1989). More recently, researchers have added the concept of trust as a predictor of technology acceptance (e.g., Ghazizadeh et al., 2012; Hoff & Bashir, 2015), a key consideration for those using GenAI (Dwivedi et al., 2023; Hitch, 2024).

Careful consideration was also given to the socio-technical perspective (Trist & Bamforth, 1951), which is often regarded as one of the foundational viewpoints for the IS discipline, contributing to its uniqueness and distinctiveness (Sarker et al., 2018a). In other words, the mutual relationship between technology and social structure (Trist & Bamforth, 1951; Verhoef et al., 2021) or, as Hansen et al. (2011) suggest, the "silent technology and the human spirit" (2011, p. 473).

The socio-technical perspective is important when considering AI (Yu et al.,2023) as different people have different perceptions about AI therefore, it is important to bring together all the alternative frames of thinking—from the various communities of developers, researchers, business leaders, policymakers, and citizens, acknowledging that new insights are needed on how to facilitate best the fundamental changes brought about by this digital development (Waardenburg & Huysman, 2022; Collins et al., 2021; Sartori Theodorou, 2022; Wessel, 2023).

Data Collection

Although qualitative data collection has been employed in existing studies on the use of AI (e.g., Collins et al., 2021; Someh et al., 2022; Chiu, 2024; Monod et al., 2024), according to Anis and French (2023), further qualitative research is needed to gain a deeper understanding of AI. Consequently, empirical evidence was collected through the distribution of an online qualitative survey. There were several reasons for this decision.

Firstly, traditional qualitative research approaches, such as individual interviews, often face limitations regarding small sample sizes and limited generalisability (Silverman, 2020). Conversely, quantitative research approaches are considered time-consuming, expensive, and unsuitable for providing an in-depth understanding of an issue (Marsland et al., 2014), as they cannot probe further when answers are ambiguous (Dengel et al., 2023).

However, the use of qualitative surveys to collect empirical evidence, although underutilised prior to COVID-19, has gained momentum (Bielska et al., 2024), as they capture what is important to participants (Braun & Clarke, 2013) and access their language and terminology, both of which are frequently claimed advantages of qualitative research (Frith, 2000). When viewed in their entirety, the resulting dataset can provide richness and depth, even if individual responses might be brief (Braun et al., 2021).

The online survey instrument consisted of open-ended questions to gain insights into how participants used GenAI in their everyday lives. Additional questions regarding gender, age, and geographic location were included for descriptive purposes only. Using Microsoft Forms, we worked with Survey Circle. This global survey agency adheres to the industry codes of practice laid out by the European Society for Opinion and Marketing Research (ESOMAR) when distributing the online questionnaire to its members.

Out of 102 (n=102) responses received, the authors received 88 (n=88) usable responses. To ensure the participants' eligibility, we included a screening question: "Are you using GenAI in your everyday life?" Respondents were then provided with a link to a recording in the opening statement explaining what GenAI is to ensure they understood the topic under discussion. Ethical approval for the study was granted through the researcher's University, and the online survey was administered between January 2025 and March 2025.

Data Analysis

Data analysis is considered one of the most challenging aspects of qualitative research, as, unlike quantitative studies, prescriptions of data analysis techniques are not as clearly defined (Cohen, 2007; Myers, 2020; Ngulube, 2023; McTaggart & McLaughlin, 2024) and "there are very few fixed formulas or cookbook recipes as guides" (Yin, 2018, p.165).

The practical process of analysing qualitative data allows the researcher to organise and make sense of the data to produce findings or a theory and develop an overall understanding of the topic under investigation (Simons, 2009). The data analysis techniques in qualitative research follow a similar trajectory, including data display, memo/note writing (e.g., Braun & Clarke, 2006; Corbin & Strauss, 2008; Saldaña, 2021), data reduction, where the researcher constantly compares the data to identify similarities or differences (Eisenhardt, 1989), some form of coding strategy to highlight themes or patterns (Miles et al., 2020), grouping these themes or patterns (Braun & Clarke, 2006) and drawing conclusions or verification (Simons, 2009; Yin, 2018), However, data analysis needs to be "systematic, disciplined and able to be seen (and to be seen through, as in transparent) and described" (Punch, 2014, p. 169)

With the next generation of researchers frequently employing GenAI to support them in their research endeavours (Andersen et al., 2025) and a growing integration of GenAI in research activities (Shaw et al., 2023), we propose and adopt a novel approach, which we have labelled "AI-Assisted Thematic Approach" (AITA) to analyse the qualitative data.

Although others have endeavoured to analyse qualitative data using AI by following the Thematic Analysis guidelines presented by Braun and Clarke (2006) and Braun & Clarke (2021), (e.g. Chang et al., 2021; Perkins & Roe, 2024; Morgan, 2023; Chiu, 2024; Hitch,2024), according to Perkins and Roe (2024), clear guidelines on how to effectively and ethically integrate GenAI tools into research activities must be provided to ensure good practice (Benbya et al., 2023; Gupta et al., 2024) whilst methodologically tailored guidelines for AI-augmented analysis would support its wider uptake, while also promoting rigour and quality (Hitch, 2024). Therefore, the following paragraphs present a detailed account of the data analysis approach adopted in this study, as "clarity on process and practice of method is vital" (Braun & Clarke, 2006, p. 80).

Figure 1. AI- Human Collaboration

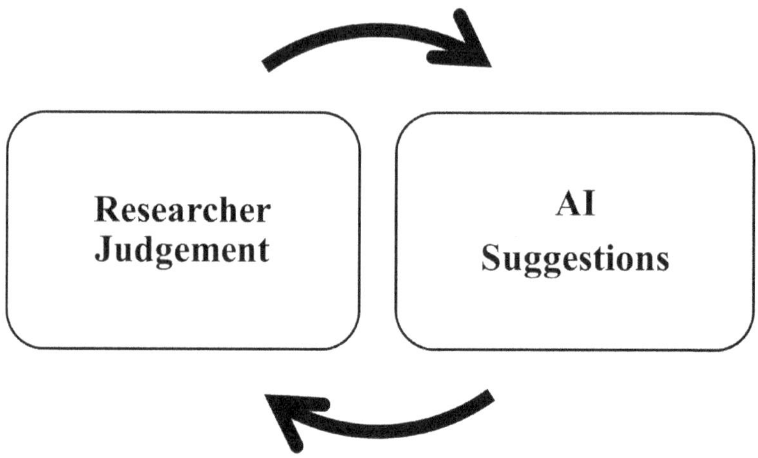

AI-Assisted Thematic Approach

AITA can be considered an interpretive methodological approach that integrates GenAI with qualitative data analysis to enhance thematic identification, coding, and synthesis (e.g., Chi et al., 2020; Collins et al., 2021; Lou & Wu, 2021; Gupta et al., 2024). In this context, AITA supports researchers in engaging with the data, refining emerging themes, and identifying patterns that may not be immediately evident through manual data analysis (Morgan, 2023).

By blending human interpretive judgment with AI-driven augmentation, AITA can support analytical rigour and adaptability, offering a novel yet methodologically sound enhancement to qualitative research data analysis, which is often criticised for its lack of rigour (Gill & Gill, 2020). This human-AI collaboration (Andersen et al., 2025), as illustrated in Figure 1, has the potential to make a significant contribution to the research community, particularly as it combines innovation with a robust methodological foundation.

While some may consider that by presenting accessible guidelines for others to use, in doing so, we have succumbed to proceduralism, similar to Braun and Clarke (2006), we suggest that this is not the case. Instead, the tenet of providing valuable guidelines for interpretive researchers to follow is not unusual in the evolution of the IS research field. For example, Geoff Walsham (1995) wrote what is considered one of the most important and widely cited papers on interpretivism in IS research. This paper represented a cornerstone for interpretive research discourse and is the most cited paper published in the *European Journal of Information Systems* (Stahl, 2014). This paper was particularly useful to IS researchers as it proposed guidelines for conducting empirical work. Similarly, Klein and Myers (1999) produced a set of principles for conducting and evaluating interpretive field studies in IS to support further advancement of interpretivism in the IS field, acknowledging that it does not follow that there should be no standards at all by which interpretive research can be judged, as long as the proposed principles are consistent with the philosophical literature on interpretivism. Therefore, we propose six stages in **Table 1** for the researcher to undertake when adopting an AI-Assisted Thematic Approach to analyse qualitative data.

Table 1. AI-Assisted Thematic Approach

Step 1. Data familiarisation and preparation	Reading and re-reading the qualitative data, immersing yourself in the data and making notes of interest Pre-process and cleanse the data.
Step 2. Create a Prompting Frame	Adopt CLEAR (Concise, Logical, Explicit, Adaptive, and Reflective) principles to develop a prompting frame.
Step 3. AI-Assisted Qualitative Coding	Using the prompts in the prompting frame. Generate Level 1 Codes (Descriptive) Generate Level 2 Codes (Pattern) Employ the principle of coding saturation
Step 4. AI-Assisted Theme Identification	Using the prompts in the prompting frame. Generate Themes Employ the principle of theme saturation Generate a thematic map

continued on following page

Table 1. Continued

Step 5. Humans in the loop quality check	Employ Investigator Triangulation Review the study in conjunction with the RATED principles (Responsibility, Accountability, Transparency, Explainability, Disclosure) Complete quality check framework
Step 6. Produce the report	Present the findings and discuss these findings in relation to the extant literature.

Step 1. Data familiarisation and preparation

While ChatGPT can be used to generate summaries of source material, which can support the rapid identification of key content and sentiments (Hitch., 2024), like Braun and Clarks' (2006) seminal work, we propose that the first step is to become familiar with the data by reading and re-reading the qualitative data, immersing yourself in the data, and making notes of interest (Bruan & Clark, 2012).

However, here, we go beyond data familiarisation and acknowledge the importance of data preparation in AI-supported qualitative data analysis. Clean data is essential when working with AI (Waardenburg & Huysman, 2022), as only when data is cleansed can AI properly analyse it (Manoharan et al., 2024).

Therefore, following data familiarisation and consistent with Jeyaraj and Zadeh (2020), we pre-processed and cleaned the data, removing various actions such as tokenisation, stemming, spell-checking, and synonyms to prepare the data set for analysis.

Step 2. Create a Prompting Frame

As LLMs are very sensitive to prompts (Drori & Te'eni, 2024), and ChatGPT is an LLM that better understands individual prompts (Chiu, 2024) using algorithms and predictive text to create new content in response to questions or prompts posted by users (Sweeney, 2023), it is important to structure and phrase prompts so that AI responses best align with human interpretations (Lo,2023).

The term prompt refers to the input users provide to guide the AI model's output, and it is the primary means of interaction between users and AI models (White et al., 2023). This textual interface can be a simple question, statement, or set of specific instructions with detailed descriptions to generate the desired responses from AI (Amatriain, 2024). As the researcher provides the prompts, these prompts are considered human intervention and feedback (Zheng et al., 2023, cited in Huang & Rocha, 2023).

To obtain high-quality outputs from AI tools, users must be able to engineer effective prompts (Eager & Brunton, 2023; Walter, 2024; Park & Choo, 2024); therefore, during this stage, the researcher acts as a 'prompt engineer' who tries to distil the meaning of something into a limited number of words (Al-Sibai, 2023).

Prompt engineering is the "art of fine-tuning the questions or commands provided to AI models to optimise their performance and guarantee that they produce the desired results" (Lo,2023, p.1). Researchers can enhance AI's functionality by fine-tuning prompts, phrased as instructions (Susarla et al., 2023), to produce new content that responds to these prompts (Benbya et al., 2023).

In practice, when ideating the prompt design, it is essential to consider not only the research question but also structure, clarity, and context (Bozkurt, 2024), as well as questions of who, what, when, where, and how (Corbin & Strauss, 2008). This involves crafting clear and logical instructions for GenAI by creating a prompting frame.

The process of crafting a good prompting frame requires researchers to reflect on the language used to express their intentions, critically evaluate their assumptions and clarify the criteria used for coding the data (Park & Choo, 2024). As ChatGPT struggles with confusingly worded questions, which can lead to incorrect answers (Dwivedi et al., 2023), the researchers followed the prompt engineering principles known as CLEAR: Concise, Logical, Explicit, Adaptive, and Reflective (Lo, 2023) to develop the prompting frame for this study. Five prompts (**see Table 2**) were developed to support data analysis.

Table 2. Prompting frame for an AI-Assisted Thematic Approach

Define Objectives	**Prompts**
Level 1 Coding (Descriptive)	Adopting an interpretive approach to inquiry generate Level 1 descriptive low inference codes from the enclosed dataset to summarise this data.
Level 2 Coding (Pattern)	Adopting an interpretive approach to inquiry generate more abstract Level 2 pattern codes, by reviewing the full data set and clustering the Level 1 codes already generated together into a smaller number of units to provide Level 2 codes.
Theme Development	Using the Level 1 & 2 codes already generated, compare these codes with the full data set and each other and further refine these codes to present overarching themes to answer the research question for this study. Provide a summation of each theme.
Quote Extraction	List all quotes from the data source which illustrate the theme (Theme Name)
Thematic Map	Generate a visual thematic map to illustrate the relationship between Level 1 Coding (Descriptive Codes), Level 2 Coding (Pattern Codes), and the Overarching Themes.

Step 3. AI Assisted Qualitative Coding

The coding process, which is at the heart of qualitative data analysis (Coffee & Atkinson, 1996), aims to help the researcher systematically organise and analyse the data, identify patterns or themes during coding, and identify relationships between different codes or themes – to move from descriptive information to patterns and abstractions (Miles & Huberman, 1994). Simply put, during the coding process, the researcher compares incidents with incidents by constantly comparing the data in order to classify them. As the researcher progresses through the analysis, each incident in the data is compared with other incidents for similarities and differences. Incidents that are found to be conceptually similar are grouped together (Strauss & Corbin, 1998; Corbin & Strauss, 2008; Braun & Clarke, 2006).

However, coding is a process which goes remarkably undocumented (Elliott, 2018), and there is no unique or generally accepted set of procedures to guide the coding process in qualitative studies, with researchers often adapting or partially using elements of different approaches (Birks et al., 2013). Coupled with this is the terminology used: "and different people use different terms to refer to the same thing and the same terms to refer to different things" (Elliot, 2018, p. 2852). As a result of this ambiguity, "there is potential for considerable confusion regarding what coding actually is so that it is doubtful whether writers who employ the term are referring to the same procedure" (Bryman & Burgess, 1994, p. 218).

For example, while leading grounded theorists (e.g. Glaser & Strauss, 1967; Corbin & Strauss, 2008) have presented clear guidelines for those adopting a Grounded Theory Methodological (GTM) approach, which also includes specific terminology and approaches to code qualitative data, Bryant and Charmaz (2007, p. 353), note that many claimed grounded theory studies, use the GTM as a coding method only, and, indeed, the term "grounded theory has almost become a blanket term for a way of coding qualitative data".

Cognisant of these criticisms and also the importance for researchers to precisely articulate the coding approach when using GenAI so that others can apply the same prompt to reproduce and verify the process adopted, this study, influenced by Braun and Clark (2006) and informed by Punch (2014) adopts a Level 1 and Level 2 coding approach to ensure that there was clear explainability (Sartori & Theodorou, 2022; Reis et al., 2020; Hansen et al.,2024) of the outcome of the responses generated by ChatGPT (Benbya et al., 2021) in a language that others can understand (Park& Choo, 2024).

According to Punch (2014), first-level coding mainly uses descriptive, low-inference codes, which are very useful in summarising segments of data and provide the basis for later higher-order coding. Second-level coding tends to focus on pattern codes, which are more abstract and involve pulling material together into a smaller number of meaningful units.

Guided by Drori and Te'eni (2024), we began this stage of data analysis by setting the system role to 'You are an interpretive researcher employing a qualitative methodological approach and the principles of thematic analysis to answer the research question 'An investigation into the everyday use of Gen AI'.

All cleansed data (Waardenburg & Huysman, 2022; Manoharan et al., 2024) was then uploaded to ChatGPT, and ChatGPT was prompted using the text in the predeveloped prompting frame to generate Level 1 Codes. (see Table 4 in the Appendix). This step was carried out multiple times to ensure that the views of different participants were captured, and it was identified where they converged and diverged in the perception of their everyday use of GenAI.

Here, we also adopted the concept of 'coding saturation'. Saturation is used in qualitative research as a criterion for discontinuing data collection and/or analysis (Saunders et al., 2019). Like theoretical saturation in Grounded Theory, wherein theoretical saturation is only reached when new data yields no new insights for the researcher (Glaser & Strauss, 1967; Birks & Mills, 2015), coding saturation occurred when, despite multiple coding cycles, further coding cycles yielded no new Level 1 codes. During this phase, several Level 1 codes were identified.

The next phase involved prompting ChatGPT using thc wording in thc prompting frame to review the Level 1 codes and the complete dataset to generate Level 2 Codes. (see Table 5 in the Appendix). This step followed the same approach as Level 1 coding and was carried out multiple times to ensure that the views of different participants were captured. Once again, we employed the concept of 'coding saturation,' which was only achieved when further Level 2 coding cycles yielded no new insights.

Step 4. AI-Assisted Theme Identification

As ChatGPT is well-suited to pattern-matching exercises (Susarla et al.,2023) and can capture a particular form of tacit knowledge (Ilk et al., 2020), using the prompt in the prompting frame, ChatGPT was prompted to cluster all the Level 1 and 2 codes together (see Table 6 in the Appendix) to generate themes to answer the research question.

Similar to the coding process, these themes were reviewed multiple times to ensure they reflected the findings. Theme saturation was only achieved when further ChatGPT coding cycles yielded no new insights.

As it is possible to request suggestions from ChatGPT for specific quotations related to data analysis results, which, according to Morgan (2023), is vital for a full-fledged analysis, similar to Hitch (2024), ChatGPT was prompted to 'List all quotes from the data source which illustrate the theme. Once ChatGPT generated this list, the researchers checked for AI hallucinations, as ChatGPT can sometimes generate quotes or data that do not exist in the original dataset (Perkins & Roe, 2024). Once confirmed, the researchers selected the quotes they considered to have adequately captured the essence of the theme.

Step 5. Humans in the loop quality check

A fundamental requirement when using GenAI is understanding, adapting, and striking the right balance between human expertise and AI capabilities (Perkins & Roe, 2024). Therefore, in the interest of a rigorous validation process to maintain research validity (Perkins & Roe, 2024), we adopted a human-in-the-loop (Jarvenpaa & Klein, 2024), multidisciplinary (Fügener et al., 2021; Dwivedi et al., 2023) two-step approach to ensure that unique human knowledge is maintained (Fügener et al., 2021).

Firstly, as interpretive researchers embrace the notion that different researchers may come to different conclusions when coding and analysing the same data (Hammersley, 2012), guided by Denzin (1989) and similar to Bano et al. (2023), we adopted the principles of investigator triangulation to correlate the findings (Denzin, 1989). In doing so, four independent reviewers from diverse disciplines who were not involved in the study were first asked to review the raw data and then review the Level 1 Codes (Descriptive Codes), Level 2 Codes (Pattern Codes), and Overarching Themes generated by Chat GPT to provide feedback on whether they considered the AI-generated data analysis reflective of the raw data.

We then asked the reviewers to consider the study in its entirety against the proposed RATED principles (Responsibility, Accountability, Transparency, Explainability, Disclosure). In doing so, we developed a quality check framework in Microsoft Forms **(Table 3)** similar to what is often employed in scientific research (Swanborn,1996) to confirm the trustworthiness of the study's findings (Lincoln & Guba, 2000), a key consideration when conducting AI-supported activities (Ma et al., 2024; Turel & Kalhan, 2023).

Table 3. Humans in the loop quality check

Quality Check	Questions
Responsibility	Is it evident that the researcher (s) have conducted the research responsibly to safeguard against potential biases or discriminatory practices embedded in AI?
Accountability	Are there any possible positive and/or negative consequences that the AI-generated research outputs may have?
Transparency	Based on the information provided, could the findings of this research study be replicated?
Explainability	Are the outcomes of the responses generated by AI explainable?
Disclosure	Have the researchers disclosed the use of AI?

Step 6. Present the findings

The final stage in the process is to present the findings and discuss these findings in relation to the existing literature.

FINDINGS AND DISCUSSION

This exploratory study aimed to investigate individuals' experiences of their everyday use of Generative AI. Following data analysis, which employed an AI-Assisted Thematic Approach, three themes were identified: i) A Journey of Discovery, ii) Workplace Partner-Friend or Foe and iii) AI's Ethical Crossroads.

The following paragraphs will discuss these themes and their relationship to the existing literature. Exemplar quotes identified by ChatGPT and selected by the researchers will be used where relevant. A series of identifiers, such as Participant 1 (P1), will be used to protect participants' privacy.

Theme 1. A Journey of Discovery

Similar to the literature (e.g., Berente et al., 2021; Chiu et al., 2024; Farrukh Moin et al., 2025), participants disclosed that GenAI has become an important tool in their daily lives, likening it to 'the biggest change to society since the World Wide Web entered our homes' (P72). They were using GenAI to assist with everything from general knowledge searches and holiday planning to content creation and coding support. For example, (P22) perceived the use of GenAI as an extension of their cognitive processes, using it to refine thoughts, generate ideas, or structure their work. They remarked, 'I use it to refine my work, do research, and act as a source of information in general. Similarly, (P21) acknowledged, 'AI has made me better at thinking about thinking as I need to write clear and logical instructions'.

In particular, participants reported using GenAI at home to make day-to-day tasks less cumbersome. They found that navigating GenAI was straightforward, increasing its perceived utility and usefulness (Venkatesh & Davis, 2000; Gupta et al., 2024). As (P45) explained, 'I am using GenAI for holiday itinerary ideas, troubleshooting problems around the home/car/garden, and image creation for kids' school projects. It is so easy to use and so useful.

However, as the literature acknowledges, participants were initially unsure about the platform and lacked trust in GenAI outputs, which resulted in delayed acceptance (e.g., Ghazizadeh et al., 2012; Hoff & Bashir, 2015; Neyazi et al.,2024). As (P14) disclosed, 'At first, I did not know what to expect. I was a little afraid, but now I use it every day!'

The findings revealed that this 'fear' (P12), which is widely acknowledged in the literature (e.g. Reis et al., 2020; Sartori & Theodorou, 2022; Yu et al., 2023) centred around the reliability and authenticity of GenAI outputs and the integrity of the data used to inform these outputs, which is a significant concern (e.g. Sweeney, 2023; Gupta et al., 2024; Wang & Liang, 2024) for citizens and organisations as documented by 63% of respondents. For instance, (P1) stated, 'I do not trust the output except for in very small quantities, and only when I have provided it with a large dataset of templates'. Similarly, (P6) remarked, 'Fast access to a starting point for enquiry is useful, but it takes time to learn what to trust and how to refine the responses'.

However, despite existing studies which have suggested that GenAI might dilute human knowledge as individuals become overly reliant on GenAI's outputs (Gupta et al., 2024; Sarker et al., 2023), over 76% of participants remarked on the knowledge-enhancing potential of GenAI, with some using it as a 'personal tutor' (P63). For example, (P46) stated, 'It effectively boosts the average worker's IQ by at least 20 points'. Similarly, (P34) noted, 'I have learned a thing or maybe 1000 things from AI, because AI knows every Python library that exists… I sure don't!' while (P41) mentioned, 'Last week I learned how to make a cake; it is so much faster than searching everything up on the internet'. Therefore, as (P8) remarked, 'It is a journey of discovery', 'We are learning as we go', (P22), and no one really 'knows what will happen next in this space' (P35).

Theme 2- Workplace Partner-Friend or Foe

In a professional environment, over 80% of participants disclosed that GenAI is reshaping their professional lives. This is confirmed in the extant literature, with numerous scholars acknowledging that GenAI will significantly reshape the future of businesses (Benbya et al., 2021; Monod et al., 2024; O'Dea,2024).

Participants reported that this ranged from enhancing productivity to introducing concerns about dependency and erosion of expertise. For instance, as the literature suggests (e.g., Dennis et al., 2023; Perkins & Roe, 2024; dos Santos & Williamson, 2024) some participants experienced GenAI as a collaborative partner that refines their work, providing structure, research insights, and efficiency in drafting communications and reports as (P5) shared, 'GenAI can be beneficial for initial document structures, summarising dense reports, debugging Excel formulas, or creating templates for social media posts based on bullet points'. Whilst others revealed that GenAI is a useful tool to streamline repetitive or time-consuming tasks, thereby freeing up time for more complex or creative thinking. For example, (P22) remarked, 'It saves time on manual insignificant tasks and allows focus on tasks requiring creativity and critical thinking'. This sentiment was also echoed by (P7), who emphasised its perceived utility (Davis, 1989) in this area, stating, 'It assists in composing emails, drafting documents, summarising dense reports, and even debugging Excel formulas or creating templates for social media posts, thereby saving time for the company and the client'.

Nevertheless, the findings show that, as the literature indicates (e.g. Benbya et al., 2021; Yu et al., 2023; Wessel et al., 2023) that the impact of GenAI on job displacement could be an area of concern as (P51) expressed, 'I am concerned that AI will, as it already has to some extent, will take over my job. I observe a lot of generated content from others in the organisation, and it is evident that they used AI-generated content'. Consequently, participants believed, in line with the literature (e.g., Yu et al., 2023; Monod et al., 2024; Mariani & Dwivedi, 2024), that substantial upskilling will be necessary to help employees adapt and collaborate effectively with AI. Otherwise, as (P19) noted, 'GenAI might be used without a full understanding of its limitations or without disclosure'.

It was also evident that there is ongoing tension in the workplace between GenAI as a productivity enhancer and the fear of overreliance. This concern is widely acknowledged in the extant literature (e.g., Chan & Hu, 2023; Sarker et al., 2023; Drori & Te'eni, 2024) and was articulated by several participants, including (P33), who remarked, 'Eighty per cent of my work involves coding, and AI helps—but I also worry about relying on it too much'. Similarly, (P18) stated, 'At some point, you might end up depending on Generative AI more, which could erode creativity and self-expression'.

Nevertheless, participants acknowledged that, in contrast to data integrity already mentioned, individuals were also concerned about workplace integrity and colleagues not disclosing their use of GenAI as a support tool, with (P34) noting, 'People could be hiding behind the technology and pretending they have skills they do not, particularly as in the workplace, the stakes are higher.' However, as (P60) acknowledges, 'It can only be beneficial if you know what you are doing'.

Theme 3- AI's Ethical Crossroads

The final theme identified is AI's Ethical Crossroads. This is unsurprising as the ethical use of GenAI is a recurring theme in the literature (Chiu, 2024; French & Shim, 2024; O'Dea, 2024), with participants expressing concerns regarding GenAI's reliability, fairness, long-term impact, propensity towards misinformation, and data privacy and security.

For example, participants expressed concerns about GenAI's inclination to generate inaccurate or misleading information, a phenomenon often referred to in the literature as "hallucination" (Retkowsky et al., 2024; Sarker et al., 2023; Susarla et al., 2023), which (P19) suggested could be 'especially problematic in academic and professional settings'. However, while GenAI's accuracy is improving, many remain wary. As (P37) remarked, 'Accuracy of content from AI is a concern, it appears to be improving, but it is still unreliable,' although 'It is a pattern you learn to spot' (P33).

Conversely, others were concerned about intellectual property theft as GenAI heightens existing threats, posing challenges for governance and compliance (Gregor, 2024), with P22 questioning, 'Using my information to train other models is a big concern. What happens to the data we input? As a result, individuals were taking 'extra precautions' (P11), particularly in educational environments, where they are advising students to verify AI-generated content. This was explained by (P2), who stated, 'I am very conscious not to include personal information and repeatedly warn students that AI-generated content needs to be verified from authentic sources'.

Participants also had 'legitimate concerns' (P24) similar to those expressed by Prem (2024) that larger organisations could 'gain a monopoly' (P10), resulting in 'the dominance of centralised AI providers, which could wield significant influence, wealth, and have access to extensive data' thus 'devolving responsibility to big tech, giving them a lot of power' (P13) which could result in individuals not knowing 'where their data would end up' (P68).

Adding to these ethical dilemmas is the environmental impact of AI. As indicated in the study's findings and acknowledged in the literature (Gupta et al., 2024; Kemp, 2024; Yu et al., 2023), participants were aware of the vast computational power required to train and run AI models, which consume enormous amounts of energy and water. This was evident in the response provided by (P27), who remarked, 'I am aware of the energy it uses compared to just searching something up; it is not good for the environment'. This sentiment was also echoed by (P45), who noted, 'Large AI data centres consume massive amounts of water to cool servers. Should we be wasting water on such simple tasks?

However, despite these socio-technical challenges (Sartori & Theodorou, 2022) participants acknowledged that this is just another phase in rapid technological advancement, akin to the advent of the 'dictionary or Google for information' (P74) with a recognition that AI is now an 'inseparable part of modern life' (P4) while the challenge ahead is 'not whether AI will remain, but how to ensure its use is ethical, responsible, and beneficial to society' (P45). Therefore, according to (P51), 'we are at a crossroads; we must weigh up the benefits and the impact and consider this when we are using Gen AI'.

CONTRIBUTION, LIMITATIONS, AND FUTURE RESEARCH DIRECTIONS

Our study aimed to investigate the everyday use of Generative AI, and in the spirit of the study, we endeavoured to use GenAI to support qualitative data analysis. With this in mind, we have drawn inspiration from the TAM (Davis, 1989) and the socio-technical perspective (Trist & Bamforth, 1951) when considering the uptake and use of this digital development.

In doing so, our findings contribute to the limited body of knowledge on citizens' everyday use of GenAI (Crusoe et al., 2024; Andersen et al., 2025). It was evident that, despite concerns, GenAI has rapidly permeated work and home life, and as predicted by Hitch (2024), GenAI has become a way of life similar to the internet.

The study also demonstrates that, despite suggestions that GenAI will lead to knowledge dilution (Gupta et al., 2024; Sarker et al., 2023), this is not the case. Instead, participants considered that GenAI improved their knowledge and also allowed them time to focus on more complex activities. Drawing on these findings, it could be argued that there is a need to revisit the rhetoric of GenAI as a tool that will dilute knowledge, and to acknowledge that, with the correct training and support in place, GenAI offers significant potential for knowledge generation both inside and outside the home.

This study also addresses the call for researchers to embrace and capitalise on this new digital development (Sarker et al., 2023; Yoo, 2024) by making a novel contribution to the literature on GenAI as a tool to support qualitative data analysis (e.g. Chang et al., 2021; Morgan,2023; Perkins & Roe, 2024; Chiu, 2024) by proposing a new AI-assisted Thematic Approach (AITA) and in doing so we present detailed guidelines for others to follow. However, this should be considered an adjunct to existing approaches and not a replacement (Hitch, 2024).

From a research tradition perspective, the positivist tradition has held prominence in the IS field (Orlikowski & Baroudi, 1991; Chen & Hirschheim, 2004; Richardson & Robinson, 2007). Nevertheless, there have been calls for IS researchers to move away from the positivist roots of its past (& Myers, 1999; Hansen et al., 2011) as positivist quantitative methods are limited in providing depth and detail in contextual situations in IS research (Conboy et al., 2012). This study addresses this gap, moving beyond the traditional positive approaches and methods commonly adopted by IS researchers.

Nevertheless, like all research, this exploratory study has several limitations. Firstly, due to the qualitative nature of this research, the small sample size means that traditional statistical measures of reliability cannot confirm the findings, as the nature of qualitative research generates outcomes that cannot be viewed as facts or objective truths; they are the result of an interaction between the researcher and the researched (Silverman, 2020). Nevertheless, generalisability to a broader population was not an objective of this research; instead, the main objective of the research was to contribute to the body of knowledge by furthering the understanding of individuals' everyday use of GenAI.

Furthermore, this investigation proposes a new approach to AI-assisted qualitative data analysis that is open to scrutiny. This is to be welcomed and embraced, and further investigations following the approach adopted are encouraged. Finally, although GenAI is a relatively new phenomenon, further research is needed to determine its long-term impact on individuals' everyday lives.

CONCLUSION

Williard McCarty (2010, p.1), presenting their Inaugural Lecturer, at King's College London, remarked, "I celebrate computing as one of our most potent speculative instruments, for its enabling of competent hands to force us all to rethink what we trusted that we knew… I celebrate learning for its own sake, but by other means than our predecessors had. "I celebrate our responsibilities to those we teach, not merely to prepare them for the world as it is, but more to open up, with the help of computing, the alternatives of human possibility." This sentiment is especially true of GenAI, which is easily accessible both inside and outside the home. While these findings suggest that some may view GenAI as a valuable tool to free up their time, others remain uncertain about its influence and implications. This complexity reflects not only the evolving nature of GenAI itself but also how society navigates a rapidly changing world.

REFERENCES

Al-Sibai, N. (2023, September 13). Companies are paying bonkers salaries for people good at ChatGPT: Expert prompt engineers can name their price. *The Byte*. https://futurism.com/the-byte/companies-paying-high-salaries-ChatGPT

Alberts, I. L., Mercolli, L., Pyka, T., Prenosil, G., Shi, K., Rominger, A., & Afshar-Oromieh, A. (2023). Large language models (LLMs) and ChatGPT: What will the impact on nuclear medicine be? *European Journal of Nuclear Medicine and Molecular Imaging*, *50*(6), 1549–1552. DOI: 10.1007/s00259-023-06172-w

Amatriain, X. (2024). Prompt design and engineering: Introduction and advanced methods. *arXiv*. https://doi.org//arXiv.2401.14423DOI: 10.48550

Andersen, J. P., Degn, L., Fishberg, R., Graversen, E. K., Horbach, S. P. J. M., Schmidt, E. K., Schneider, J. W., & Sørensen, M. P. (2025). Generative artificial intelligence (GenAI) in the research process: A survey of researchers' practices and perceptions. *Technology in Society*, *81*, 1–11. DOI: 10.1016/j.techsoc.2025.102813

Anis, S., & French, J. A. (2023). Efficient, explicatory, and equitable: Why qualitative researchers should embrace AI, but cautiously. *Business & Society*, *62*(6), 1139–1144. DOI: 10.1177/00076503231163286

Aoki, N., Tatsumi, T., Naruse, G., & Maeda, K. (2024). Explainable AI for government: Does the type of explanation matter to the accuracy, fairness, and trustworthiness of an algorithmic decision as perceived by those who are affected? *Government Information Quarterly*, *41*(4), 101965. Advance online publication. DOI: 10.1016/j.giq.2024.101965

Bailey, D. E., & Barley, S. R. (2020). Beyond design and use: How scholars should study intelligent technologies. *Information and Organization*, *30*(2), 100286. Advance online publication. DOI: 10.1016/j.infoandorg.2019.100286

Bano, M., Hoda, R., Zowghi, D., & Treude, C. (2023). Large language models for qualitative research in software engineering: Exploring opportunities and challenges. *Automated Software Engineering*, *31*(1), 8. Advance online publication. DOI: 10.1007/s10515-023-00407-8

Benbya, H., Pachidi, S., & Jarvenpaa, S. L. (2021). Special issue editorial: Artificial intelligence in organisations: Implications for information systems research. *Journal of the Association for Information Systems*, *22*(2), 281–303. DOI: 10.17705/1jais.00662

Benbya, H., Strich, F., & Tamm, T. (2023). Navigating the promises and perils of generative artificial intelligence for knowledge and creative work. *Journal of the Association for Information Systems, 25*(1), 23–36. DOI: 10.17705/1jais.00861

Berente, N., Gu, B., & Santhanam, R. (2021). Managing artificial intelligence. *Management Information Systems Quarterly, 45*(3), 1433–1450. DOI: 10.25300/MISQ/2021/16274

Bielska, B., Kalinowska, K., Męcfal, S., & Surmiak, A. (2024). Strengths and limitations of an online qualitative survey in times of social crisis: Example of the COVID-19 pandemic. *Qualitative Sociology Review, 20*(4), 78–100. Advance online publication. DOI: 10.18778/1733-8077.20.4.04

Birks, D. F., Fernandez, W., Levina, N., & Nasirin, S. (2013). Grounded theory method in information systems research: Its nature, diversity and opportunities. *European Journal of Information Systems, 22*(1), 1–8. DOI: 10.1057/ejis.2012.48

Birks, M., & Mills, J. (2015). *Grounded theory: A practical guide* (2nd ed.). Sage Publications.

Bozkurt, A. (2024). Tell me your prompts, and I will make them true: The alchemy of prompt engineering and generative AI. *Open Praxis, 16*(2), 111–118. DOI: 10.55982/openpraxis.16.2.661

Braun, V., & Clarke, V. (2006). Using thematic analysis in psychology. *Qualitative Research in Psychology, 3*(2), 77–101. DOI: 10.1191/1478088706qp063oa

Braun, V., & Clarke, V. (2013). *Successful qualitative research: A practical guide for beginners*. SAGE Publications.

Braun, V., & Clarke, V. (2021). Can I use TA? Should I use TA? Should I not use TA? Comparing reflexive thematic analysis and other pattern-based qualitative analytic approaches. *Counselling & Psychotherapy Research, 21*(1), 37–47. DOI: 10.1002/capr.12360

Braun, V., Clarke, V., Boulton, E., Davey, L., & McEvoy, C. (2021). The online survey as a qualitative research tool. *International Journal of Social Research Methodology, 24*(6), 641–654. DOI: 10.1080/13645579.2020.1805550

Bryant, A., & Charmaz, K. (2007). *The SAGE handbook of grounded theory*. Sage Publications., DOI: 10.4135/9781848607941

Bryman, A., & Bell, E. (2011). *Business research methods*. Oxford University Press.

Bryman, A., & Burgess, R. G. (1994). Reflections on qualitative data analysis. In Bryman, A., & Burgess, R. G. (Eds.), *Analysing qualitative data* (pp. 216–226). Routledge., DOI: 10.4324/9780203413081_chapter_11

Bryson, J. J., & Theodorou, A. (2019). How society can maintain human-centric artificial intelligence. In Toivonen-Noro, M., Saari, E., Melkas, H., & Hasu, M. (Eds.), *Human-centered digitalization and services* (pp. 305–323). Springer., DOI: 10.1007/978-981-13-7725-9_16

Cao, Y., Li, S., Liu, Y., Yan, Z., Dai, Y., Yu, P. S., & Sun, L. (2023). A comprehensive survey of AI-generated content (AIGC): A history of generative AI from GAN to ChatGPT. *Computer Science*. https://doi.org//arXiv.2303.04226DOI: 10.48550

Chan, C. K. Y., & Hu, W. (2023). Students' voices on generative AI: Perceptions, benefits, and challenges in higher education. *International Journal of Educational Technology in Higher Education, 20*(1), 43. Advance online publication. DOI: 10.1186/s41239-023-00411-8

Chang, T., DeJonckheere, M., Vydiswaran, V. G. V., Li, J., Buis, L. R., & Guetterman, T. C. (2021). Accelerating mixed methods research with natural language processing of big text data. *Journal of Mixed Methods Research, 15*(3), 398–412. DOI: 10.1177/15586898211021196

Cheligeer, C., Yang, L., Nandi, T., Doktorchik, C., Quan, H., Zeng, Y., & Singh, S. (2023). Natural language processing (NLP) aided qualitative methods in health research. *Journal of Integrated Design & Process Science, 27*(1), 41–58. DOI: 10.3233/JID-220013

Chen, W., & Hirschheim, R. (2004). A paradigmatic and methodological examination of information systems research from 1991 to 2001. *Information Systems Journal, 14*(3), 197–235. DOI: 10.1111/j.1365-2575.2004.00173.x

Chi, O., Denton, G., & Gursoy, D. (2020). Artificially intelligent device use in service delivery: A systematic review, synthesis, and research agenda. *Journal of Hospitality Marketing & Management, 29*(7), 757–786. DOI: 10.1080/19368623.2020.1721394

Chiu, T. K. F. (2024). Future research recommendations for transforming higher education with generative AI. *Computers and Education: Artificial Intelligence, 6*, 100197. Advance online publication. DOI: 10.1016/j.caeai.2023.100197

Coffee, A. J., & Atkinson, P. (1996). *Making sense of qualitative data: Complementary research strategies (and social thought)* (1st ed.). Sage Publications.

Cohen, L., Manion, L., & Morrison, K. (2007). *Research methods in education* (6th ed.). Routledge Falmer., DOI: 10.4324/9780203029053

Collins, C., Dennehy, D., Conboy, K., & Mikalef, P. (2021). Artificial intelligence in information systems research: A systematic literature review and research agenda. *International Journal of Information Management*, *60*, 102383. Advance online publication. DOI: 10.1016/j.ijinfomgt.2021.102383

Conboy, K., Fitzgerald, G., & Mathiassen, L. (2012). Qualitative methods research in information systems: Motivations, themes, and contributions. *European Journal of Information Systems*, *21*(2), 113–118. DOI: 10.1057/ejis.2011.57

Corbin, J., & Strauss, A. (2008). *Basics of qualitative research* (3rd ed.). Sage.

Correani, A., De Massis, A., Frattini, F., Petruzzelli, A., & Natalicchio, A. (2020). Implementing a digital strategy: Learning from the experience of three digital transformation projects. *California Management Review*, *62*(4), 37–56. DOI: 10.1177/0008125620934864

Cotton, D. R., Cotton, P. A., & Shipway, J. R. (2023). Chatting and cheating: Ensuring academic integrity in the era of ChatGPT. *Innovations in Education and Teaching International*, •••, 1–12. DOI: 10.1080/14703297.2023.2190148

Crusoe, J., Magnusson, J., & Eklund, J. (2024). Digital transformation decoupling: The impact of wilful ignorance on public sector digital transformation. *Government Information Quarterly*, *41*(3), 101958. Advance online publication. DOI: 10.1016/j.giq.2024.101958

Davis, F. D. (1989). Perceived usefulness, perceived ease of use, and user acceptance of information technology. *Management Information Systems Quarterly*, *13*(3), 319–340. DOI: 10.2307/249008

Deng,L. Yu,D. (2014) Deep *Learning: Methods and Applications*. Now Foundations and Trends.

Dengel, A., Gehrlein, R., Fernes, D., Görlich, S., Maurer, J., Pham, H. H., Großmann, G., & Eisermann, N. D. (2023). Qualitative research methods for large language models: Conducting semi-structured interviews with ChatGPT and BARD on computer science education. *Informatics (MDPI)*, *10*(4), 78. Advance online publication. DOI: 10.3390/informatics10040078

Dennis, A. R., Lakhiwal, A., & Sachdeva, A. (2023). AI agents as team members: Effects on satisfaction, conflict, trustworthiness, and willingness to work with. *Journal of Management Information Systems*, *40*(2), 307–337. DOI: 10.1080/07421222.2023.2196773

Denzin, N. K. (1989). Interpretive biography. *Sage (Atlanta, Ga.)*.

dos Santos, J. F. P., & Williamson, P. J. (2024). Beyond connectivity: Artificial intelligence and the internationalisation of digital firms. *Information and Organization*, *34*(4), 100538. Advance online publication. DOI: 10.1016/j.infoandorg.2024.100538

Drori, I., & Te'eni, D. (2024). Human-in-the-loop AI reviewing: Feasibility, opportunities, and risks. *Journal of the Association for Information Systems*, *25*(1), 98–109. Advance online publication. DOI: 10.17705/1jais.00867

Dwivedi, R., Nerur, S., & Balijepally, V. (2023). Exploring artificial intelligence and big data scholarship in information systems: A citation, bibliographic coupling, and co-word analysis. *International Journal of Information Management Data Insights*, *3*(2), 100185. Advance online publication. DOI: 10.1016/j.jjimei.2023.100185

Dwivedi, Y. K., Kshetri, N., Hughes, L., Slade, E. L., Jeyaraj, A., Kar, A. K., Baabdullah, A. M., Koohang, A., Raghavan, V., Ahuja, M., Albanna, H., Albashrawi, M. A., Al-Busaidi, A. S., Balakrishnan, J., Barlette, Y., Basu, S., Bose, I., Brooks, L., Buhalis, D., & Wright, R. (2023). Opinion Paper: "So what if ChatGPT wrote it?" Multidisciplinary perspectives on opportunities, challenges and implications of generative conversational AI for research, practice and policy. *International Journal of Information Management*, *71*, 102642. Advance online publication. DOI: 10.1016/j.ijinfomgt.2023.102642

Eager, B., & Brunton, R. (2023). Prompting higher education towards AI-augmented teaching and learning Practice. *Journal of University Teaching & Learning Practice*, *20*(5), 1–19. DOI: 10.53761/1.20.5.02

Eisenhardt, K. M. (1989). Building theories from case study research. *Academy of Management Review*, *14*(4), 532–550. DOI: 10.2307/258557

Elliott, V. (2018). Thinking about the coding process in qualitative data analysis. *The Qualitative Report*, *23*(11), 2850–2861. DOI: 10.46743/2160-3715/2018.3560

Faraj, S., Pachidi, S., & Sayegh, K. (2018). Working and organizing in the age of the learning algorithm. *Information and Organization*, *28*(1), 62–70. DOI: 10.1016/j.infoandorg.2018.02.005

Farrukh Moin, M., Behl, A., & Zhang, Z. (2025, August). AI in the Organizational Nexus: Building Trust, Cementing Commitment, and Evolving Psychological Contracts. *Information Systems Frontiers*, *27*(4), 1413–1424. Advance online publication. DOI: 10.1007/s10796-024-10561-3

Floridi, L. (2020). AI and its new winter: From myths to realities. *Philosophy & Technology*, *33*(1), 1–3. DOI: 10.1007/s13347-020-00396-6

French, A. M., & Shim, J. P. (2024). From Artificial intelligence to augmented intelligence: A shift in perspective, application, and conceptualization of AI. *Information Systems Frontiers: A Journal of Research and Innovation*, 1–22. https://doi.org/DOI: 10.1007/s10796-024-10562-2

Frey, C. B., & Osborne, M. A. (2017). The future of employment: How susceptible are jobs to computerisation? *Technological Forecasting and Social Change, 114*, 254–280. DOI: 10.1016/j.techfore.2016.08.019

Frith, H. (2000). Focusing on sex: Using focus groups in sex research. *Sexualities, 3*(3), 275–297. DOI: 10.1177/136346000003003001

Fügener, A., Grahl, J., Gupta, A., & Ketter, W. (2021). Will humans-in-the-loop become borgs? Merits and pitfalls of working with AI. *Management Information Systems Quarterly, 45*(3), 1527–1556. DOI: 10.25300/MISQ/2021/16553

Furr, N., Ozcan, P., & Eisenhardt, K. M. (2022). What is digital transformation? Core tensions facing established companies on the global stage. *Global Strategy Journal, 97*(4), 1–24. DOI: 10.1002/gsj.1442

Gerling, C., Meier, P., & Köhler, C. (2022). AI meets digital: a critical review on artificial intelligence in digital entrepreneurship. *European Conference on Information Systems.*

Ghazizadeh, M., Lee, J. D., & Boyle, L. N. (2012). Extending the technology acceptance model to assess automation. *Cognition Technology and Work, 14*(39), 49. DOI: 10.1007/s10111-011-0194-3

Gill, T. G., & Gill, T. R. (2020). What is research rigour? Lessons for a transdiscipline. *Informing Science, 23*, 47–77. DOI: 10.28945/4528

Glaser, B. G., & Strauss, A. L. (1967). *The discovery of grounded theory*. strategies for qualitative research. Aldine

Glikson, E., & Woolley, A. W. (2020). Human trust in artificial intelligence: Review of empirical research. *Academy of Management Annals,* (14:2), 627–660. https://doi.org/DOI: 10.5465/annals.2018.0057

Goodfellow, I., Bengio, Y., & Courville, A. (2016). *Deep learning*. MIT Press.

Gregor, S. (2024). Responsible artificial intelligence and journal publishing. *Journal of the Association for Information Systems, 25*(1), 48–60. DOI: 10.17705/1jais.00863

Gupta, R., Nair, K., Mishra, M., Ibrahim, B., & Bhardwaj, S. (2024). Adoption and impacts of generative artificial intelligence: Theoretical underpinnings and research agenda. *International Journal of Information Management Data Insights*, *4*(1), 100232. Advance online publication. DOI: 10.1016/j.jjimei.2024.100232

Haenlein, M., & Kaplan, A. (2019). A brief history of artificial intelligence: On the past, present, and future of artificial intelligence. *California Management Review*, *61*(4), 5–14. DOI: 10.1177/0008125619864925

Hammersley, M. (2012). *Methodological paradigms in educational research*. British Educational Research Association.

Hanelt, A., Bohnsack, R., Marz, D., & Antunes Marante, C. (2021). A systematic review of the literature on digital transformation: Insights and implications for strategy and organizational change. *Journal of Management Studies*, *58*(5), 1159–1197. DOI: 10.1111/joms.12639

Hansen, A. M., Kraemmergaard, P., & Mathiassen, L. (2011). Rapid adaptation in digital transformation: A participatory process for engaging IS and business leaders. *MIS Quarterly Executive*, *10*(4), 175–185.

Hansen, H. F., Lillesund, E., Mikalef, P., & Altwaijry, N. (2024). Understanding artificial intelligence diffusion through an ai capability maturity model. *Information Systems Frontiers: A Journal of Research and Innovation*, 26(6), 2147–2163. https://doi.org/DOI: 10.1007/s10796-024-10528-4

Hinings, B., Gegenhuber, T., & Greenwood, R. (2018). Digital innovation and transformation: An institutional perspective. *Information and Organization*, *28*(1), 52–61. DOI: 10.1016/j.infoandorg.2018.02.004

Hitch, D. (2024). Artificial intelligence augmented qualitative analysis: The way of the future? *Qualitative Health Research*, *34*(7), 595–606. DOI: 10.1177/10497323231217392

Hoff, K. A., & Bashir, M. (2015). Trust in automation: Integrating empirical evidence on factors that influence trust. *Human Factors*, *57*(3), 407–434. DOI: 10.1177/0018720814547570

Huang, Y.-M., & Rocha, T. (2023). *Innovative technologies in learning* (Vol. 14785). Springer Nature Switzerland.

Iansiti, M., & Lakhani, K. R. (2020). Competing in the age of AI. *Harvard Business Review*, •••, 12–19.

Ilk, N., Shang, G., & Goes, P. B. (2020). Improving customer routing in contact centers: An automated triage design based on text analytics. *Journal of Operations Management, 66*(5), 553–577. DOI: 10.1002/joom.1084

Jarvenpaa, S., & Klein, S. (2024). New frontiers in information systems: Theorising human-AI collaboration. *Journal of the Association for Information Systems, 25*(1), 110–121. DOI: 10.17705/1jais.00868

Jeyaraj, A., & Zadeh, A. H. (2020). Evolution of information systems research: Insights from topic modelling. *Information & Management, 57*(4), 103207. Advance online publication. DOI: 10.1016/j.im.2019.103207

Kanbach, D. K., Heiduk, L., Blueher, G., Schreiter, M., & Lahmann, A. (2023). The GenAI is out of the bottle: Generative artificial intelligence from a business model innovation perspective. *Review of Managerial Science*, •••, 1–32. DOI: 10.1007/s11846-023-00696-z

Kelly, R. (2023). Over-reliance on ChatGPT could harm worker performance. ITPro. Retrieved on 29 October 2024 from https://www.itpro.com/technology/artificial-intelligence/over-reliance-on-chatgpt-could-harm-worker-performance

Kemp, A. (2024). Competitive advantage through artificial intelligence: Toward a theory of situated AI. *Academy of Management Review, 49*(3), 618–635. DOI: 10.5465/amr.2020.0205

Kietzmann, J., & Park, A. (2024). Written by ChatGPT: AI, large language models, conversational chatbots, and their place in society and business. *Business Horizons, 67*(5), 453–459. DOI: 10.1016/j.bushor.2024.06.002

Klein, H. K., & Myers, M. D. (1999). A set of principles for conducting and evaluating interpretive field studies in information systems. *Management Information Systems Quarterly, 23*(1), 67–93. DOI: 10.2307/249410

Krauss, J. P., Kailer, N., & Weinmann, C.-B. (2021). Digital transformation: An overview of the current state of the art of research. *SAGE Open, 11*(3), 1–15. DOI: 10.1177/21582440211047576

Lennon, R. P., Fraleigh, R., Van Scoy, L. J., Keshaviah, A., Hu, X. C., Snyder, B. L., Miller, E. L., Calo, W. A., Zgierska, A. E., & Griffin, C. (2021). Developing and testing an automated qualitative assistant (AQUA) to support qualitative analysis. *Family Medicine and Community Health, 9*(Suppl 1), e001287. Advance online publication. DOI: 10.1136/fmch-2021-001287

Lincoln, Y. S., & Guba, E. G. (2000). Paradigmatic controversies, contradictions, and emerging confluences. In Denzin, N. K., & Lincoln, Y. S. (Eds.), *The handbook of qualitative research* (2nd ed., pp. 1065–1122). Sage Publications.

Lo, L. S. (2023). The CLEAR path: A framework for enhancing information literacy through prompt engineering. *Journal of Academic Librarianship*, *49*(4), 102720. Advance online publication. DOI: 10.1016/j.acalib.2023.102720

Lou, B., & Wu, L. (2021). AI on drugs: Can artificial intelligence accelerate drug development? Evidence from a large-scale examination of Bio-Pharma firms. *Management Information Systems Quarterly*, *45*(3), 1451–1482. DOI: 10.25300/MISQ/2021/16565

Ma, H. M., Huang, W., & Dennis, A. R. (2024). Unintended consequences of disclosing recommendations by artificial intelligence versus humans on true and fake news believability and engagement. *Journal of Management Information Systems*, *41*(3), 616–644. DOI: 10.1080/07421222.2024.2376381

Mahmood, A. M., Siponen, M., Straub, D. W., Rao, H. R., & Raghu, T. S. (2010). Moving toward black hat research in information systems security: An editorial introduction to the special issue. *Management Information Systems Quarterly*, *34*(3), 431–433. DOI: 10.2307/25750685

Manoharan, G., Ashtikar, S. P., & Nivedha, M. (2024). Integrating artificial intelligence in library management: an emerging trend. In Senthilkumar, K. (Ed.), *AI-assisted library reconstruction* (pp. 144–157). IGI Global. DOI: 10.4018/979-8-3693-2782-1.ch008

Mariani, M., & Dwivedi, Y. K. (2024). Generative artificial intelligence in innovation management: A preview of future research developments. *Journal of Business Research*, *175*, 114542. Advance online publication. DOI: 10.1016/j.jbusres.2024.114542

Marsland, M., Wilson, I., Abeyasekera, S., & Kleih, U. (2014). *Socioeconomic methodologies for natural resources research: Best practice guidelines combining quantitative (formal) and qualitative (informal) survey methods*. National Resources Institute.

McCarty, W. (2010). Attending from and to the machine. Inaugural lecture, King's College London, 2 February.

McTaggart, V., & McLaughlin, C. (2024). Lessons learned from working from home: Exploring factors considered critical to supporting successful working from home during the pandemic. *Management and Labour Studies*, *50*(1), 35–60. DOI: 10.1177/0258042X241283209

Mergel, I., Edelmann, N., & Haug, N. (2019). Defining digital transformation: Results from expert interviews. *Government Information Quarterly*, *36*(4), 101385. DOI: 10.1016/j.giq.2019.06.002

Miles, M. B., & Huberman, A. M. (1994). *Qualitative data analysis: an expanded sourcebook* (2nd ed.). Sage.

Miles, M. B., Huberman, A. M., & Saldaña, T. (2020). *Qualitative data analysis: A methods sourcebook* (4th ed.). Sage.

Mogaji, E., & Jain, V. (2024). How generative AI is (will) change consumer behaviour: Postulating the potential impact and implications for research, practice, and policy. *Journal of Consumer Behaviour*, *23*(5), 2379–2389. DOI: 10.1002/cb.2345

Moin, M. F., Behl, A., Zhang, J. Z., & Shankar, A. (2024). AI in the organizational nexus: Building trust, cementing commitment, and evolving psychological contracts. *Information Systems Frontiers: A Journal of Research and Innovation*, 1–12. https://doi.org/DOI: 10.1007/s10796-024-10561-3

Monod, E., Mayer, A.-S., Straub, D., Joyce, E., & Qi, J. (2024). From worker empowerment to managerial control: The devolution of AI tools' intended positive implementation to their negative consequences. *Information and Organization*, *34*(1), 100498. Advance online publication. DOI: 10.1016/j.infoandorg.2023.100498

Moravec, H. (2008). Rise of the robots. [Special Edition]. *Scientific American*, *18*(1), 12–19. DOI: 10.1038/scientificamerican0208-12sp

Morgan, D. L. (2023). Exploring the use of artificial intelligence for qualitative data analysis: The case of ChatGPT. *International Journal of Qualitative Methods*, *22*, 1–10. DOI: 10.1177/16094069231211248

Myers, M. (2020). *Qualitative research in business and management* (3rd ed.). Sage Publications. DOI: 10.4135/9781036234539

Neyazi, T. A., Nadaf, A. H., Tan, K. E., & Schroeder, R. (2024). Does trust in government moderate the perception towards deepfakes? Comparative perspectives from Asia on the risks of AI and misinformation for democracy. *Government Information Quarterly*, *41*(4), 101980. Advance online publication. DOI: 10.1016/j.giq.2024.101980

Ngulube, P. (2023). Improving the quality of reporting findings using computer data analysis applications in educational research in context. *Heliyon*, *9*(9), e19683. DOI: 10.1016/j.heliyon.2023.e19683

O'Dea, X. (2024). Generative AI: Is it a paradigm shift for higher education? *Studies in Higher Education*, *49*(5), 811–816. DOI: 10.1080/03075079.2024.2332944

Oosthuizen, R. (2022). The fourth industrial revolution - Smart technology, artificial intelligence, robotics and algorithms: Industrial psychologists in future workplaces. *Frontiers in Artificial Intelligence*, *5*, 913168. DOI: 10.3389/frai.2022.913168

Orlikowski, W. J., & Baroudi, J. J. (1991). Studying information technology in organisations: Research approaches and assumptions. *Information Systems Research*, *2*(1), 1–28. DOI: 10.1287/isre.2.1.1

Park, J., & Choo, S. (2024). Generative AI prompt engineering for educators: Practical strategies. *Journal of Special Education Technology*, •••, 1–7. DOI: 10.1177/01626434241298954

Paulus, T., Lester, J. N., & Davis, C. (2025). The construction of the role of AI in qualitative data analysis in the social sciences. *AI & Society*. Advance online publication. DOI: 10.1007/s00146-025-02488-3

Perkins, M., & Roe, J. (2024). The use of generative AI in qualitative analysis: Inductive thematic analysis with ChatGPT. *Journal of Applied Learning & Teaching*, *7*(1). Advance online publication. DOI: 10.37074/jalt.2024.7.1.22

Prem, E. (2024). Principles of digital humanism: A critical post-humanist view. *Journal of Responsible Technology*, *17*, 1–10. DOI: 10.1016/j.jrt.2024.100075

Punch, K. (2014). *Introduction to social research : quantitative and qualitative approaches* (3rd ed.). Sage Publications.

Rahwan, I., Cebrian, M., Obradovich, N., Bongard, J., Bonnefon, J.-F., Breazeal, C., Crandall, J. W., Christakis, N. A., Couzin, I. D., Jackson, M. O., Jennings, N. R., Kamar, E., Kloumann, I. M., Larochelle, H., Lazer, D., McElreath, R., Mislove, A., Parkes, D. C., Pentland, A., & Wellman, M. (2019). **Machine behaviour.** Nature, 568(7753), 477–486. *Nature*, •••, 477–486. DOI: 10.1038/s41586-019-1138-y

Redmiles, E. (2022). *Perspectives on Digital Humanism*. Springer.

Reis, L., Maier, C., Mattke, J., Creutzenberg, M., & Weitzel, T. (2020). Addressing user resistance would have prevented a healthcare AI project failure. *MIS Quarterly Executive*, *19*(4), 279–236. DOI: 10.17705/2msqe.00038

Retkowsky, J., Hafermalz, E., & Huysman, M. (2024). Managing a ChatGPT-empowered workforce: Understanding its affordances and side effects. *Business Horizons*, *67*(5), 511–523. DOI: 10.1016/j.bushor.2024.04.009

Richardson, H., & Robinson, B. (2007). The mysterious case of the missing paradigm: A review of critical information systems research, 1991–2001. *Information Systems Journal*, *17*(3), 251–270. DOI: 10.1111/j.1365-2575.2007.00230.x

Russell, S., & Norvig, P. (2010). *Artificial intelligence: A modern approach* (3rd ed.). Prentice Hall Press.

Saldaña, J. (2021). The coding manual for qualitative researchers. *Sage (Atlanta, Ga.)*.

Sandberg, J., & Alvesson, M. (2021). Meanings of theory: Clarifying theory through typification. *Journal of Management Studies*, *58*(2), 487–516. DOI: 10.1111/joms.12587

Sarker, S., Susarla, A., Gopal, R., & Thatcher, J. B. (2023). Democratizing knowledge creation through human-AI collaboration in academic peer review. *Journal of the Association for Information Systems*, *25*(1), 158–171. DOI: 10.17705/1jais.00872

Sartori, L., & Theodorou, A. (2022). A sociotechnical perspective for the future of AI: Narratives, inequalities, and human control. *Ethics and Information Technology*, *24*(1), 4. Advance online publication. DOI: 10.1007/s10676-022-09624-3

Saunders, M. N. K., Lewis, P., & Thornhill, A. (2019). *Research methods for business students* (8th ed.). Pearson.

Shaw, C., Yuan, L., Brennan, D., & Martin, S. (2023). *Generative AI in higher education: Fall 2023 update (Time for Class study)*. Tyton Partners. https://tytonpartners.com/app/uploads/2023/10/GenAI-IN-HIGHER-EDUCATION-FALL-2023-UPDATE-TIME-FOR-CLASS-STUDY.pdf

Shim, J. P., French, A., & Upreti, B. (2020). Speech analytics for actionable insights: Current status, recommendations, and guidance. *Foundations and Trends in Information Systems*, *4*(3), 213–274. DOI: 10.1561/2900000020

Silverman, D. (2020). *Interpreting qualitative data* (6th ed.). Sage Publications.

Simons, H. (2009). *Case study research in practice*. Sage Publications. DOI: 10.4135/9781446268322

Smyth, C., Dennehy, D., Fosso Wamba, S., Scott, M., & Harfouche, A. (2024). Artificial intelligence and prescriptive analytics for supply chain resilience: A systematic literature review and research agenda. *International Journal of Production Research*, *62*(23), 8537–8561. DOI: 10.1080/00207543.2024.2341415

Someh, I., Wixom, B. H., Beath, C. M., & Zutavern, A. (2022). Building an artificial intelligence explanation capability. *MIS Quarterly Executive*, *21*(2), 143–163. DOI: 10.17705/2msqe.00063

Stahl, B. C. (2014). Interpretive accounts and fairy tales: A critical polemic against the empiricist bias in interpretive IS research. *European Journal of Information Systems*, *23*(1), 1–11. DOI: 10.1057/ejis.2012.58

Strauss, A. L., & Corbin, J. (1998). *Basics of qualitative research: techniques and procedures for developing grounded theory* (2nd ed.). Sage Publications.

Susarla, A., Gopal, R., Thatcher, J. B., & Sarker, S. (2023). The Janus effect of generative AI: Charting the path for responsible conduct of scholarly activities in information systems. *Information Systems Research*, *34*(2), 399–408. DOI: 10.1287/isre.2023.ed.v34.n2

Swanborn, P. G. (1996). A common base for quality control criteria in quantitative and qualitative research. *Quality and Quantity. Quality & Quantity*, *30*(1), 19–35. DOI: 10.1007/BF00139833

Sweeney, S. (2023). Who wrote this? Essay mills and assessment – Considerations regarding contract cheating and AI in higher education. *International Journal of Management Education*, *21*(2), 100818. Advance online publication. DOI: 10.1016/j.ijme.2023.100818

Theodorou, A., & Dignum, V. (2020). Towards ethical and socio-legal governance in AI. *Nature Machine Intelligence*, *2*(1), 10–12. DOI: 10.1038/s42256-019-0136-y

Trist, E. L., & Bamforth, K. W. (1951). Social and psychological consequences of the longwall method of coal-getting. *Human Relations*, *4*(1), 3–38. DOI: 10.1177/001872675100400101

Turel, O., & Kalhan, S. (2023). Prejudiced against the machine? Implicit associations and the transience of algorithm aversion. *Management Information Systems Quarterly*, *47*(4), 1369–1394. DOI: 10.25300/MISQ/2022/17961

Vaswani, A., Shazeer, N., Parmar, N., Uszkoreit, J., Jones, L., Gomez, A. N., Kaiser, L., & Polosukhin, I. (2017). Attention is all you need. *Computer Science*. https://doi.org//arXiv.1706.03762DOI: 10.48550

Venkatesh, V., & Davis, F. D. (2000). A theoretical extension of the technology acceptance model: Four longitudinal field studies. *Management Science*, *46*(2), 186–204. DOI: 10.1287/mnsc.46.2.186.11926

Verhoef, P., Broekhuizen, T., Bart, Y., Bhattacharya, A., Qi Dong, J., Fabian, N., & Haenlein, M. (2021). Digital transformation: A multidisciplinary reflection and research agenda. *Journal of Business Research*, *122*, 889–901. DOI: 10.1016/j.jbusres.2019.09.022

Waardenburg, L., & Huysman, M. (2022). From coexistence to co-creation: Blurring boundaries in the age of AI. *Information and Organization*, *32*(4), 100432. Advance online publication. DOI: 10.1016/j.infoandorg.2022.100432

Wade, M., & Shan, J. (2020). COVID-19 has accelerated digital transformation, but it may have made it harder, not easier. *MIS Quarterly Executive*, *19*(3), 199–211. DOI: 10.17705/2msqe.00034

Walsham, G. (1995). The Emergence of Interpretivism in IS Research. *Information Systems Research*, *6*(4), 376–394. https://www.jstor.org/stable/23010981. DOI: 10.1287/isre.6.4.376

Walter, Y. (2024). Embracing the future of artificial intelligence in the classroom: The relevance of AI literacy, prompt engineering, and critical thinking in modern education. *International Journal of Educational Technology in Higher Education*, *21*(1), 15. DOI: 10.1186/s41239-024-00448-3

Wang, S., & Liang, Z. (2024). What does the public think about artificial intelligence? An investigation of technological frames in different technological contexts. *Government Information Quarterly*, *41*(2), 101939. Advance online publication. DOI: 10.1016/j.giq.2024.101939

Wessel, M., Adam, M., Benlian, A., Majchrzak, A., Thies, F. (2023). Generative AI and its transformative value for digital platforms. Call for Papers to the Special Issue—Journal *of Management Information Systems*. Vol.np pp 1-6, https://doi.org/ DOI: 10.1080/07421222.2025.2487315

White, J., Fu, Q., Hays, S., Sandborn, M., Olea, C., Gilbert, H., Elnashar, A., Spencer-Smith, J., & Schmidt, D. C. (2023). A prompt pattern catalogue to enhance prompt engineering with ChatGPT. *Computer Science*. https://doi.org//arXiv.2302.11382DOI: 10.48550

Williams, R. T. (2024). Paradigm shifts: Exploring AI's influence on qualitative inquiry and analysis. *Frontiers in Research Metrics and Analytics*, *9*, 1331589. Advance online publication. DOI: 10.3389/frma.2024.1331589

Yin, R. K. (2018). *Case study research and applications* (6th ed.). Sage.

Yoo, Y. (2024). Evolving epistemic infrastructure: The role of scientific journals in the age of generative AI. *Journal of the Association for Information Systems, 25*(1), 137–144. DOI: 10.17705/1jais.00870

Yu, X., Xu, S., & Ashton, M. (2023). Antecedents and outcomes of artificial intelligence adoption and application in the workplace: The sociotechnical system theory perspective. *Information Technology & People, 36*(1), 454–474. DOI: 10.1108/ITP-04-2021-0254

Zhang, B., & Dafoe, A. (2019). Artificial intelligence: American attitudes and trends. *Future of Humanity Institute, University of Oxford.* https://dx.doi.org/DOI: 10.2139/ssrn.3312874

Zhang, J., & Feng, S. (2021). Machine learning modeling: A new way to do quantitative research in social sciences in the era of AI. *Journal of Web Engineering, 20*(2), 281–302. DOI: 10.13052/jwe1540-9589.2023

ADDITIONAL READING

Alalaq, A. S. (2024). The history of the artificial intelligence revolution and the nature of generative AI work. *DS Journal of Artificial Intelligence and Robotics, 2*(4), 1–15. DOI: 10.59232/AIR-V2I4P101

Ali, S. M., Dick, S., Dillon, S., Jones, M. L., Penn, J., & Staley, R. (2023). Histories of artificial intelligence: A genealogy of power. *BJHS Themes, 8*, 1–18. DOI: 10.1017/bjt.2023.15

Bjola, C. (2022). AI for development: Implications for theory and practice. *Oxford Development Studies, 50*(1), 78–90. DOI: 10.1080/13600818.2021.1960960

Copeland, B. J. (2025). History of artificial intelligence. In *Encyclopaedia Britannica.* Retrieved 0n 14 October from https://www.britannica.com/science/history-of-artificial-intelligence

Delipetrev, B., Tsinaraki, C., & Kostić, U. (2020). *Historical evolution of artificial intelligence* (EUR 30221 EN). Publications Office of the European Union. https://doi.org/DOI: 10.2760/801580

Haber, E., Jemielniak, D., Kurasiński, A., & Przegalińska, A. (2025). The evolution of artificial intelligence: From winter to renaissance. In *Using AI in Academic Writing and Research* (pp. 13–26). Springer., DOI: 10.1007/978-3-031-91705-9_2

McCorduck, P. (2004). *Machines who think: A personal inquiry into the history and future of artificial intelligence* (2nd ed.). A. K. Peters. DOI: 10.1201/9780429258985

Nilsson, N. J. (2010). *The quest for artificial intelligence: A history of ideas and achievements*. Cambridge University Press.

KEY TERMS AND DEFINITIONS

AI-Assisted Thematic Approach (AITA): A structured six-step qualitative data analysis method that integrates generative AI tools with human interpretive judgment to enhance theme identification, coding, and synthesis.

ChatGPT: A conversational generative artificial intelligence model developed by OpenAI that uses large language models (LLMs) to generate human-like text responses based on user prompts, enabling tasks such as writing, summarising, coding, and interactive dialogue.

Coding Saturation: A qualitative research principle indicating the point at which additional data analysis yields no new codes or insights, suggesting that thematic categories are sufficiently developed.

Generative Artificial Intelligence (GenAI): A subset of AI technologies capable of producing original content—such as text, images, or code based on learned patterns from large datasets, often through models like ChatGPT.

Human-in-the-Loop: A collaborative approach in AI-assisted research where human researchers remain actively involved in guiding, validating, and interpreting AI-generated outputs to ensure quality and ethical integrity.

Prompt Engineering: The practice of crafting precise and effective input instructions (prompts) to guide generative AI systems in producing relevant and accurate outputs aligned with research objectives.

RATED Principles: An ethical framework for AI-supported research that emphasises Responsibility, Accountability, Transparency, Explainability, and Disclosure to ensure trustworthy and responsible AI use.

Socio-Technical Perspective: A theoretical lens that examines the interplay between social systems (people, culture, structures) and technical systems (tools, technologies) in shaping outcomes and practices.

Technology Acceptance Model (TAM): A theoretical framework that explains how users come to accept and use technology, focusing on perceived usefulness and perceived ease of use as key determinants.

Theme Saturation: The stage in qualitative analysis where no new themes emerge from the data, indicating that the thematic framework is comprehensive and well-developed.

APPENDIX

Table 4. Sample of AI assisted Level 1 Coding

Level 1 Codes Generated by Chat GPT	Descriptive Codes
Cognitive Extension	AI is used to enhance problem-solving, critical thinking, and conceptual development (e.g., brainstorming, refining arguments, strategic planning).
Information Synthesis	AI aids in summarising dense information, comparing reports, and filtering knowledge.
Decision Support	AI provides predictive insights, recommendations, and alternative perspectives for decision-making.
Content Generation	AI is leveraged for writing, designing, and composing creative work (e.g., presentations, marketing content, social media).
Idea Amplification	AI helps generate and refine ideas, acting as a thought partner.
Personalisation & Adaptability	Users tailor AI outputs to align with their own style, preferences, or needs.
Task Automation	AI is employed to streamline repetitive or mundane tasks (e.g., drafting emails, coding, summarising reports).
Speed & Convenience	Users highlight AI's ability to accelerate workflows, reduce cognitive load, and provide immediate access to structured information.
Optimising Workflows	AI supports planning, organising, and structuring professional and personal tasks.
Educational Aid	It is used for studying, researching, and skill development (e.g., homework assistance, coding practice, language learning).
Skill Augmentation	Users leverage AI to refine expertise, generate code, and enhance technical skills.
Exploratory Engagement	Some users 'play' with AI, using it experimentally to explore its capabilities.
Collaboration & Teamwork	AI is utilised in professional settings to enhance communication, generate reports, and support group discussions.
Interactive Engagement	AI functions as an interactive knowledge source, responding to queries and shaping discourse.
Shared AI Experiences –	Some users report discussing AI outputs with colleagues and integrating them into collaborative projects.
Trust & Accuracy Concerns	Users express scepticism about AI-generated content, citing risks of hallucinations and misinformation.
Bias & Fairness	Concerns arise about AI's training data, potential bias, and regulatory oversight.
Data Privacy & Security	Users question AI's handling of sensitive information and the implications of data collection.
Job Displacement Anxiety	Fears about AI replacing human roles, particularly in writing, creative industries, and technical work.

continued on following page

Table 4. Continued

Level 1 Codes Generated by Chat GPT	Descriptive Codes
Loss of Critical Thinking	Overreliance on AI is seen as a threat to human problem-solving and cognitive development.
AI as a Cultural Shift	AI is described as a transformative force, changing the way work, education, and everyday life are conducted.
Regulatory & Governance Concerns	Users highlight the role of policies (e.g., GDPR, AI Act) in shaping the accessibility and ethical use of AI.
Sustainability Considerations	Some responses mention AI's environmental cost (e.g., energy consumption, e-waste).

Table 5. Sample of AI Assisted Level 2 Coding

Level 2 Codes Generated by Chat GPT	Pattern Codes
AI as a Cognitive & Creative Augmentation Tool	(*Combines Level 1 Codes: Cognitive Extension, Information Synthesis, Decision Support, Content Generation, Idea Amplification, Personalisation & Adaptability*) • AI extends human thinking by assisting in decision-making, brainstorming, and conceptual development. • Supports content creation across various domains, enhancing creativity and productivity. • Helps tailor responses and outputs to align with individual preferences.
AI as an Efficiency & Workflow Optimiser	(*Combines Level 1 Codes: Task Automation, Speed & Convenience, Optimising Workflows*) • AI is used to automate repetitive tasks, reducing cognitive load. • Improves workflow efficiency by quickly structuring information, summarising content, and managing work processes. • Saves time by streamlining communication, planning, and execution of tasks.
AI as a Learning & Knowledge Companion	(*Combines Level 1 Codes: Educational Aid, Skill Augmentation, Exploratory Engagement*) • AI facilitates learning by offering educational support, research assistance, and skill development. • Enhances professional and personal growth by helping users acquire new knowledge and refine technical skills. • Some users engage with AI experimentally, testing its capabilities for learning purposes.
AI as a Collaborative & Interactive Partner	(*Combines Level 1 Codes: Collaboration & Teamwork, Interactive Engagement, Shared AI Experiences*) • AI is integrated into teamwork, supporting communication, brainstorming, and report generation. • Used as an interactive knowledge source, answering questions and shaping discussions. • Some users engage with AI in a shared context, exchanging insights and experiences

continued on following page

Table 5. Continued

Level 2 Codes Generated by Chat GPT	Pattern Codes
AI as an Ethical & Societal Challenge	(*Combines Level 1 Codes: Trust & Accuracy Concerns, Bias & Fairness, Data Privacy & Security, Job Displacement Anxiety, Loss of Critical Thinking*) • Users express concerns about AI's reliability, hallucinations, and misinformation. • Ethical worries include AI's bias, data privacy risks, and the potential impact on jobs. • Some believe AI overreliance could lead to diminished human critical thinking.
AI as a Societal & Regulatory Disruptor	(*Combines Level 1 Codes: AI as a Cultural Shift, Regulatory & Governance Concerns, Sustainability Considerations*) • AI is perceived as a transformative force reshaping work, education, and communication. • Regulatory challenges, including AI governance, GDPR, and the AI Act, influence AI adoption. • Environmental concerns arise due to AI's high energy consumption and resource use.

Table 6. AI Assisted Thematic Development.

Themes Generated by Chat GPT	Theme development and summation by GenAI
Theme 1 **A Journey of Discovery**	*(Related to Level 2 Codes: AI as a Cognitive & Creative Augmentation Tool, AI as an Efficiency & Workflow Optimiser, AI as a Learning & Knowledge Companion)* Generative AI has become an **extension of human cognition**, supporting **problem-solving, idea generation, and knowledge acquisition**. Users leverage AI to **enhance creativity** in writing, design, and conceptual work, as well as to automate repetitive tasks, thereby improving efficiency in both professional and personal contexts. It acts as a **learning companion**, helping individuals acquire new skills, conduct research, and refine their thinking processes. However, **integrity** in the inputs in and outputs from AI is concerning **Key Insights:** • AI assists in structuring and organising work, reducing cognitive load. • Users benefit from AI-driven insights and recommendations, improving decision-making. • AI enables creative expression and experimentation. • Automation of tasks leads to greater efficiency in work and personal life.
Theme 2 **Workplace Partner-Friend or Foe**	*(Related to Level 2 Codes: AI as a Collaborative & Interactive Partner, AI as an Ethical & Societal Challenge)* Generative AI is increasingly viewed as a **collaborative partner** rather than a standalone tool. Users engage with AI to **co-create, interact with, and refine their work, particularly in areas such as** communication, research, and teamwork. However, a growing tension exists between the benefits of AI assistance and the risk of **overreliance**, raising concerns about the decline of **critical thinking, creativity, and independent problem-solving**. **Key Insights:** • AI fosters collaboration by streamlining communication and teamwork. • Users integrate AI into daily workflows for assistance with writing, summarising, and organising. • Concerns exist over potential **loss of human agency**, as AI-generated content risks reducing independent thought. • Ethical considerations arise regarding **bias, misinformation, integrity and AI's impact on job security**.
Theme 3 **AI's Ethical Crossroads**	*(Related to Level 2 Codes: AI as a Societal & Regulatory Disruptor, AI as an Ethical & Societal Challenge)* As AI becomes embedded in daily life, **ethical and regulatory concerns** come to the forefront. Users express apprehensions about **data privacy, misinformation, bias, and regulatory oversight**, particularly in regions with evolving AI governance policies. Additionally, discussions about **AI's impact on employment, sustainability, and control over digital content** highlight the need for **responsible AI development and use**. **Key Insights:** • The **accuracy and reliability** of AI outputs remain a significant concern, requiring human oversight. • Data privacy and **ethical AI governance** are key considerations for many users. • **Job displacement and economic implications** ¬ of AI adoption are widely debated. • AI's **environmental impact** and energy consumption raise questions about sustainable AI practices.

Figure 2. Thematic Map

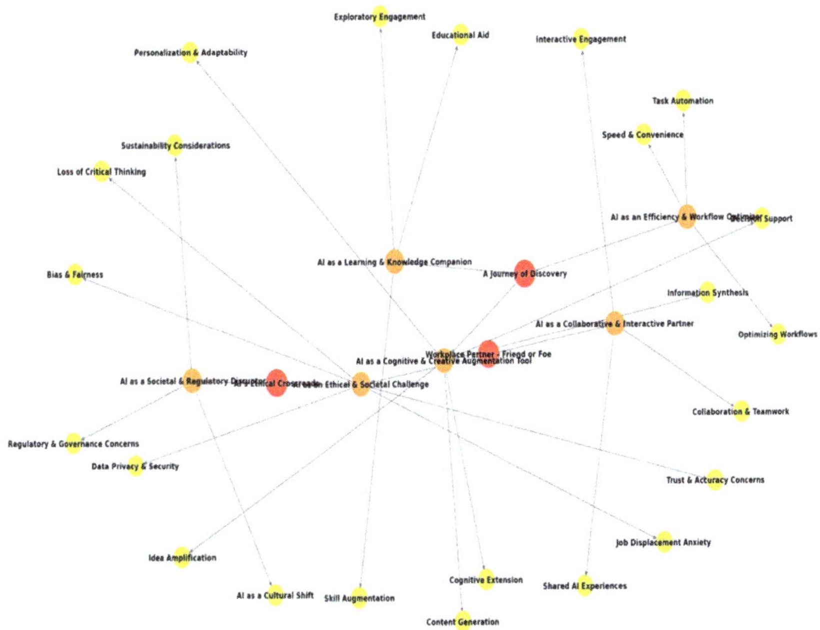

● → Overarching Themes

● → Level 2 Pattern Codes

● → Level 1 Descriptive Codes

Chapter 3

A Content Analysis Framework on AI-Driven Strategic Decision-Making in the National Basketball Association Team Management

Raymond Corona
http://orcid.org/0009-0003-3026-1762
Capitol Technology University, USA

Eugene J. Lewis
http://orcid.org/0000-0002-2956-0760
Capitol Technology University, USA

ABSTRACT

This study examines the integration of artificial intelligence (AI) into strategic decision-making in the National Basketball Association (NBA). Using Resource-Based Theory as the framework, the research explores how AI-driven analytics enhance performance optimization, player evaluation, and team strategy. By analyzing AI's role in talent management, game preparation, and financial operations, the study highlights how franchises leverage technology as a strategic asset. Findings indicate that AI contributes to competitive advantage by improving scouting accuracy, supporting in-game adjustments, and increasing operational efficiency. The research emphasizes that while AI cannot replace human expertise, it provides actionable

DOI: 10.4018/979-8-2600-0434-0.ch003

insights that strengthen organizational capabilities. This section contributes to both academic and professional discussions by showing how AI functions as a critical resource in shaping the NBA's evolving competitive landscape.

INTRODUCTION

The use of artificial intelligence in professional sports has gained substantial traction in recent years, but nowhere is its presence growing faster than in the National Basketball Association (NBA). This league, renowned for its global influence and relentless competition, is increasingly defined by the decisions made not only on the court but also behind the scenes. Draft picks, in-game adjustments, player development programs, trade negotiations, and injury prevention protocols are all shaped by decision-making frameworks that rely heavily on data. AI has become an integral part of these frameworks, enabling organisations to process more information, generate more accurate projections, and make faster, more informed strategic decisions. For a sport that demands both agility and precision, AI offers a way to align instinct with intelligence. This section was developed in response to the rapid integration of AI into basketball team operations and the growing need to understand how this integration is influencing outcomes in both individual games and across entire seasons.

The decision to study AI in the context of NBA team management arose from observing how franchises have adopted predictive models and machine learning tools to enhance their competitive decision-making. What was once considered a novelty, such as tracking player movements through SportVU cameras or using efficiency metrics to guide substitutions, has evolved into a complex system of data-supported infrastructures that influence roster construction, tactical game plans, and long-term organizational goals (Kapoor, 2021). The implementation of neural networks, computer vision, and algorithmic simulation in recent years has gone far beyond traditional analytics. Teams now utilise AI to identify patterns in opponents' play styles, predict the development trajectory of draft prospects, and simulate trade scenarios under various economic and performance-based conditions (Xu & Baghaei, 2025). These are not speculative projects. They are real systems, already in place, contributing to how modern basketball is planned and executed.

SPORTS MANAGEMENT

The era of sports management is timely because professional sports are increasingly defined by marginal gains. Unlike in earlier years, where intuition and coaching experience dominated team strategy, today's competitive edge often comes from technology that can identify opportunities and risks not visible to the naked eye. In the NBA, where every win affects playoff seeding, revenue, and long-term franchise value, teams are motivated to pursue even the smallest advantages. Artificial intelligence has emerged as one such advantage. It is being used to optimise player rotations, identify favourable matchups, and prevent injuries before they occur through real-time biometric tracking (Musat et al., 2024). In practice, these tools enable coaching staff and general managers to make decisions based on evidence rather than assumptions, often in real-time. The pace of professional basketball makes these rapid decisions not only useful but essential.

The research is relevant because the academic literature on this subject is still in development. While many studies explore sports analytics in a general sense, fewer works focus specifically on how artificial intelligence is changing the mechanics of strategic planning within NBA organizations (Choustoulakis, 2024; Nalbant & Aydın, 2022). Even fewer apply a theoretical framework to assess the long-term impact of AI on organizational effectiveness. This section aims to fill that gap by examining the broader implications of AI, beyond its technical functions, and exploring how it reshapes decision-making hierarchies, alters workflows, and redefines success.

This study focuses on multiple levels of team infrastructure. It considers the role of front office executives using AI-generated models to project salary cap outcomes and trade value; the coaching staff interpreting AI-based matchup analyses to craft defensive schemes; the medical teams assessing player fatigue with machine learning tools; and the scouts using AI to process international and college-level player data to inform draft strategy (Verma & Singh, 2024; Smith & Lee, 2025). These case applications illustrate that AI in the NBA is not confined to a single department or domain. It is woven into the entire fabric of the organization, from preseason planning to mid-season course correction.

The setting for this research is the NBA, but the relevance extends to any organization operating in high-pressure, performance-focused environments. The league's structure consists of its global audience, complex salary cap regulations, and high player turnover. This makes it an ideal framework for content analysis, enabling a deeper understanding of how AI influences human decision-making on a large scale. The NBA provides access to structured data, well-documented trends in analytics adoption, and a wide range of team strategies that reflect diverse organizational cultures. Some teams invest heavily in in-house analytics departments, while others outsource AI solutions to consulting firms or university labs. This variety makes the NBA not just a suitable but a compelling location for examining how AI functions in real-world decision-making.

STRATEGIC DECISION-MAKING

The distinction between tactical and strategic AI usage remains unclear in existing studies. Tactical use might involve game-to-game lineup adjustments or defensive matchups, while strategic use encompasses long-term roster planning and salary cap management. Anifa and Mahmoud (2024) and Choustoulakis (2024) illustrate the tactical benefits of AI in evaluating player performance and game tactics. However, these studies fail to explore the larger strategic implications and whether organizations adopt AI as a core element of team planning or merely as a support tool.

Human factors are underrepresented in the conversation. Research by Fouché (2023) and Herman and Abeza (2024) introduces the idea that legacy knowledge and historical preferences can challenge technological adoption. This dynamic becomes particularly relevant in the NBA, where front offices consist of professionals with varied levels of trust in machine learning models. Despite the growth of data science departments, coaching and scouting practices remain largely influenced by traditional, experience-based methods. The gap exists in understanding how quantitative and qualitative data are integrated or kept apart.

Organizational factors, including leadership influence, are only lightly discussed in the existing body of work. Wright and Patel (2025) suggest that leadership styles can influence AI usage in sports decision-making, but they do not empirically investigate this claim. Likewise, Gupta and Rao (2024) advocate for more robust computing infrastructure to support analytics but stop short of examining how such systems are maintained, prioritized, or even understood by top-level staff. There is also a need to differentiate between teams that have fully invested in AI integration and those that have taken a more limited or symbolic approach.

Current research provides a limited typology of how NBA franchises differ in their AI adoption rates or how these differences correlate with performance outcomes. Johnson and Brown (2024) examine how leadership style influences on-court results, offering a framework that this section adapts to analyze how leadership also governs AI implementation. Without this organizational perspective, the literature remains incomplete. Most studies present AI as a set of features or tools, without explaining the human systems through which these tools are filtered and used.

This research intends to fill the gap by centering the NBA franchise as the unit of analysis and asking how AI is institutionally adopted, interpreted, and applied in decision-making. It expands the focus from what AI can do to how teams use it. That includes analyzing organizational hierarchies, leadership communication, workflow adaptation, and the differences in how data insights are treated across contexts. The goal is to bring the conversation out of the lab and into the front office, where the real decisions are made and where the true potential or possible limitations of AI in basketball will be determined.

ARTIFICIAL INTELLIGENCE POWERED VIEWING EXPERIENCES

The introduction of AI-powered viewing experiences, such as real-time shot probability overlays and customized highlight generation, reflects a reconfiguration of how fans consume and interact with the game. These shifts require cooperation between technical developers, media executives, and league administrators, reaffirming the theory's emphasis on the mutual shaping of technology and social institutions.

Ethical considerations further deepen the relevance of this framework. AI applications that track facial expressions, emotional states, or biometric fluctuations raise concerns about player autonomy, privacy, and consent. Taylor and Smith (2025) note that while these tools offer immense analytical value, they risk dehumanizing athletes if implemented without transparent guidelines or proper safeguards. In this light, sociotechnical theory cautions against overly technocratic approaches, urging sports organizations to weigh both the utility and the ethical implications of AI deployment. The presence of human actors as gatekeepers and interpreters of AI output is not a limitation, but rather a necessary condition for the ethical and effective integration of AI.

This multidimensional framework enables a more nuanced understanding of AI's role in the NBA and not simply as a tool for efficiency, but as a catalyst for rethinking power structures, labor dynamics, and stakeholder relationships. Gupta and Rao (2024) emphasise that high-performance computing strategies are only as effective as the organisational willingness to adapt processes and roles around them. In practice, this means that a franchise's success with AI may hinge more on how they structure meetings, share data, and train staff than on the sophistication of their algorithms. Thus, sociotechnical theory helps clarify why AI's transformative power is not universal, but context dependent.

NATIONAL BASKETBALL ASSOCIATION (NBA) TEAM MANAGEMENT

NBA teams are investing in artificial intelligence to enhance decision-making in areas such as game strategy, player development, scouting, and medical forecasting; yet, a significant gap remains between the capabilities of these tools and their actual application within team operations. While AI can process large amounts of game and biometric data, produce simulations, and identify statistical patterns that are not immediately visible to the human eye, the extent to which NBA franchises are leveraging these technologies to guide management-level decisions is not well understood. In many cases, teams use AI in isolated departments without clearly defined pathways for integrating those outputs into coordinated, organization-wide strategies (Musat et al., 2024; Smith & Lee, 2025).

Teams have access to advanced systems that model player fatigue (Wesely et al., 2025), predict the success of lineups based on historical performance (Verma & Singh, 2024), and run simulations of potential trade or draft outcomes (Xu & Baghaei, 2025). However, having access to such tools does not guarantee that the insights they generate are consistently and effectively applied across decision-making structures. There is little evidence that teams share a common approach to interpreting and applying AI recommendations. In some cases, head coaches rely heavily on data analysts, while in others, intuition continues to dominate tactical choices. This inconsistency in application has led to a lack of clear standards or best practices across the league.

PROBLEM STATEMENT

The problem statement for this research is based on real data utilizing artificial intelligence strategies and competencies to determine specific outcomes. The problems statement focuses on:

(1) How does the NBA focus on AI tools and technical innovation rather than using for the purpose of data analytics for organizational use?

While studies have explored AI's capabilities in performance forecasting and predictive modelling (Choustoulakis, 2024; Mateus et al., 2025), few have addressed the operational dynamics of how AI is used by decision-makers in high-pressure, fast-paced environments. There is a missing link between what technology can do and how it is implemented in real-time strategic planning. For example, even when AI correctly identifies a heightened injury risk, coaching staff may disregard that information due to playoff pressures or subjective judgments. Similarly, AI-supported trade simulations may be undervalued by general managers if they conflict with personal scouting instincts or locker room dynamics (Fouché, 2023).

Specific Problem Statement

The specific problem is not the availability of AI technology. The specific problem is determined by:

(1) If an NBA franchise is the unit of analysis - how is Artificial Intelligence (AI) institutionally adopted, interpreted, and applied in decision-making?

SIGNIFICANCE OF THE STUDY

The significance of the study provides a viewpoint into the role of AI in correlation with human judgment and the formulation of decision-making processes. While AI can generate optimal outcomes, it cannot yet account for intangible factors such as team chemistry, player psychology, or leadership under pressure. These elements still require subjective interpretation by coaches and managers, highlighting the need for a balanced relationship between human expertise and machine-generated recommendations. As emphasized throughout this section, the most successful applications of AI in basketball occur not in isolation but as tools that enhance human decision-making without overriding other ideas (Wright & Patel, 2025; Carson & Lee, 2025).

Furthermore, evaluating the significance of standardization in how AI technologies are implemented across teams creates inconsistencies in competitive equity. Wealthier franchises with more resources may be able to afford cutting-edge systems, while smaller-market teams may struggle to keep pace. This imbalance, if left unaddressed, could lead to a widening performance gap that undermines the league's emphasis on parity. Future frameworks must consider equitable access to technological advancements, possibly through league-wide partnerships or subsidized platforms (Mateus et al., 2025; Du et al., 2024).

FINDING THE GAP IN THE RESEARCH

The most defined gap in the research is the issue of transparency and trust. Several studies have noted that front office personnel often lack a comprehensive understanding of how complex AI models generate their recommendations, which creates hesitation in relying on those models for consequential decisions (Herman & Abeza, 2024). This hesitation limits the strategic impact of AI, especially when its outputs are filtered through multiple human gatekeepers who may not share a unified interpretation of the data. Without clear lines of accountability and shared understanding, AI becomes an accessory to decision-making rather than a driving force.

This research investigates whether AI is being operationalized in ways that align with its potential. It does so by examining how NBA teams structure their use of AI across departments, how outputs are communicated, and what organizational factors affect whether AI insights are trusted and acted upon. The problem is not the lack of tools, but the absence of frameworks that link AI-generated knowledge to actual decisions that influence wins, losses, and long-term team success. This section addresses the disconnect, building on existing literature while identifying new questions about implementation, leadership structure, and the long-term integration of AI systems into team identity and strategy (Anifa & Mahmoud, 2024).

The integration of artificial intelligence into professional sports has produced a growing body of research, but much of it focuses on technological potential rather than practical implementation. Within the NBA, the use of AI tools encompasses performance analysis, injury forecasting, and player tracking; however, very few studies examine how these tools are integrated into day-to-day decision-making. While researchers such as Mohammadi et al. (2024), Musat et al. (2024), and Mateus et al. (2025) examine AI in specific contexts like marketing, diagnostics, and health management, there is little discussion of how decision-makers such as coaches, general managers, or executives actively interpret and apply these outputs. This lack of operational clarity represents a significant void in the literature.

Technological innovation has outpaced organizational analysis. Studies tend to center on the technical precision of AI models, such as the accuracy of predictive analytics or efficiency in real-time computation (Vaithianathan, 2025; Wesely et al., 2025). Xu and Baghaei (2025) consider the broader opportunities AI presents in sports, including fan engagement and strategy, but like many others, their work does not examine how internal processes within professional teams absorb or resist AI insights. As a result, the literature risks presenting AI as a neutral, universally beneficial tool, overlooking critical variables such as team culture, leadership structures, and the quality of cross-departmental communication.

RESEARCH METHODOLOGY

This research applies a qualitative, exploratory approach to understand how NBA organizations adopt and operationalize artificial intelligence within strategic decision-making processes. The complexity of AI implementation in sports, coupled with the diverse internal cultures of NBA franchises, warrants a methodology that prioritizes depth of insight over breadth of generalizability. Rather than relying solely on numerical data or predictive models, this study emphasises organisational behaviour, leadership influence, and interpretative frameworks surrounding AI usage in professional basketball settings.

A multiple-content analysis framework was selected to facilitate comparative analysis across different NBA franchises. This format enables the identification of both shared and divergent patterns in how teams integrate AI technologies into their decision workflows. It also provides a structure for analyzing contextual factors, such as the influence of executive leadership, the role of analytics departments, and the historical success or failure of AI-based initiatives within each organization. Content analysis framework methodology, as discussed by Johnson and Brown (2024), supports the nuanced exploration of organisational systems that do not conform to a single model of behaviour or adoption.

The primary unit of analysis in this study is the franchise as a holistic entity. This includes not only formal leadership, such as general managers and team presidents, but also key support structures like coaching staff, player development personnel, and data science teams. By examining each component of the decision-making ecosystem, the study aims to provide a comprehensive understanding of how AI is perceived, accepted, resisted, or misunderstood across internal boundaries.

Data sources will be drawn from a combination of document analysis, public interviews, and strategic reporting. These include statements from press conferences, published front-office commentaries, coaching clinics, and reputable insider accounts available through academic and professional sport business channels. Specific emphasis will be placed on materials where teams discuss their analytics strategy, partnership with tech firms, or internal restructuring to accommodate AI initiatives.

To ensure thematic consistency across case studies, the data will be coded using a grounded theory approach. This approach allows the research to remain open to emergent themes while remaining anchored in empirical evidence from franchise-level behaviours and statements.

Thematic coding categories will include, but are not limited to:

(1) leadership attitudes toward AI,
(2) departmental integration of analytics teams,
(3) decision-making practices involving AI tools, and
(4) resistance or friction in adopting machine learning outputs.

These codes will be refined through an iterative process as data are reviewed and cross-compared between cases. To improve validity, triangulation will be implement ed by sourcing information from multiple types of documents and public remarks. For example, if a general manager describes the use of AI in player scouting during a conference panel, that statement will be cross-validated with available internal reports, news analyses, or follow-up interviews by reputable sports journalists. This method strengthens the integrity of the findings by reducing the influence of isolated or potentially biased sources.

While the methodology prioritizes qualitative interpretation, it also recognizes the value of supportive quantitative evidence. As such, franchise performance metrics will be reviewed in the context of AI adoption timelines to identify potential correlations. However, the primary goal is not to establish causality but to explore how teams perceive and frame the role of AI in their operations. As noted by Wright and Patel (2025), qualitative exploration allows for greater sensitivity to organizational nuance, particularly when examining rapidly evolving domains like AI integration.

THEORETICAL FRAMEWORK IN CORRELATION TO THE TOPIC

The theoretical framework in correlation examines how artificial intelligence is strategically implemented within the National Basketball Association. The guiding perspectives include resource-based theory, strategic decision-making theory, and sociotechnical systems theory. These lenses help illuminate how AI acts as a competitive differentiator, how managerial decisions are shaped by technological integration, and how organizational ecosystems adapt to emerging innovations.

Resource-based theory has been adapted in recent literature to consider digital and intangible assets as key drivers of sustainable advantage. Vaithianathan (2025) frames advanced computing tools, including AI platforms, as core infrastructure for competitive capability across high-performance domains. Within the NBA, teams that integrate AI into their training systems, player health protocols, and talent acquisition pipelines leverage this technological capital to outperform rivals. As Xu and Baghaei (2025) explain, franchises that develop proprietary AI models for injury prediction or player matchup efficiency gain data leverage that cannot be easily replicated. This suggests that AI is not just a technical asset, but a resource embedded in organizational routines, difficult to imitate, and increasingly central to long-term success.

Resource-based View (RBW)

The topic is framed through Resource-Based Theory (Anifa & Mahmoud, 2024), which allows for a conceptual understanding of AI as more than a tool, instead, it is considered a resource with the potential to create sustainable advantage when deployed effectively. By applying this theoretical lens, the research will assess how AI contributes to long-term team success not merely through short-term wins but through its integration into organisational routines, decision-making systems, and competitive identity.

The goal of this section is to analyze how NBA franchises use artificial intelligence to make better strategic decisions. This includes how they select talent, allocate playing time, structure contracts, and respond to the flow of games. By highlighting both the capabilities and limitations of AI in this context, the research will offer a balanced perspective that informs future innovation and accountability in sports management. The NBA is not just adopting AI; it is being shaped by it. And as AI systems continue to grow in complexity and capability, the league will likely serve as a model for how other industries can incorporate intelligent systems into high-performance decision-making.

What makes AI especially valuable within the RBV framework is its potential for cumulative improvement. AI models trained on years of proprietary team data can evolve to reflect nuanced understandings of player tendencies, injury risks, and opponent patterns. These systems are not static; they improve with organizational use, becoming tailored to the team's evolving identity. Smith and Lee (2025) emphasize that the more a team interacts with its AI infrastructure, the more strategically irreplaceable that system becomes, reinforcing the RBV's concept of path dependency and sustained value creation through learning-based assets.

Moreover, the combination of AI with organizational culture can further protect and extend these advantages. According to Lin et al. (2020), teams that foster interdisciplinary collaboration between data scientists and coaching staff see far more actionable outcomes than those that silo their operations. The synergy between data, people, and institutional routines forms a "strategic bundle" of resources that, when coherently aligned, fulfils the RBV criteria of value, rarity, inimitability, and organisational embeddedness. This synthesis of perspectives from sociotechnical, organizational learning, and resource-based provides a comprehensive foundation for understanding how AI reshapes not only how NBA teams operate, but how they build and protect competitive edges over time.

Ultimately, integrating the resource-based view (RBV) into the theoretical framework enables a strategic understanding of how AI serves as a differentiating asset among NBA franchises. Unlike traditional physical or financial resources, AI capabilities such as proprietary algorithms, data collection infrastructures, and analytical personnel are often intangible, rare, and difficult to replicate. Wright and Patel (2025) argue that franchises that internalize these competencies can develop sustained competitive advantages, especially when AI tools are embedded into core routines like talent acquisition and performance forecasting. This perspective aligns with the RBV's focus on leveraging unique internal strengths to outperform rivals.

Strategic Decision-Making Theory

Strategic decision-making theory offers a complementary explanation of how AI influences managerial judgment and workflow processes. Rather than replacing leadership, AI augments it by narrowing uncertainty, identifying hidden patterns, and increasing the speed of information synthesis (Wright & Patel, 2025). In the NBA, this is evident in how coaching staffs now use AI-assisted systems to monitor player fatigue in real-time or assess opponents' tendencies across thousands of possessions. Mohammadi et al. (2024) argue that in dynamic environments like sports, data-driven systems enhance but do not eliminate the role of human discernment. General managers still rely on context, interpersonal chemistry, and long-term vision, but these factors are increasingly filtered through AI-enabled dashboards and statistical projections.

Sociotechnical Systems Theory

Sociotechnical systems theory offers a complementary lens by addressing the interplay between technological innovation and organizational behavior. In the context of the NBA, AI is not introduced in a vacuum; it co-evolves with team dynamics, staff workflows, and institutional values. As Carter and Nguyen (2024) argue, digital innovations are most effective when aligned with human capabilities and operational structures. Teams that have successfully implemented AI tools often restructure communication protocols between analysts, coaches, and training staff, allowing for rapid interpretation of real-time data. This alignment is essential for ensuring that AI insights are translated into actionable strategies on the court.

The integration of AI in player development pipelines further illustrates the relevance of sociotechnical perspectives. Teams now use motion tracking, biometric analysis, and predictive modelling to monitor growth and prevent overuse injuries, but the effectiveness of these systems depends on the willingness of athletes and coaching staff to trust and engage with the data. Mateus et al. (2025) describe how resistance to AI interventions can emerge when athletes perceive these tools as intrusive or when coaches feel displaced in their authority. Conversely, when properly introduced, AI serves to enhance collaboration across departments, providing a shared platform for performance evaluation and tactical adjustment.

This theory also helps account for variability in adoption across franchises. As Taylor and Nguyen (2024) explain, some teams possess the leadership infrastructure, financial resources, and data culture necessary to maximize the utility of AI, while others struggle to integrate such tools due to institutional inertia or limited expertise. Thus, successful implementation is contingent not just on the quality of the technology, but on the organizational readiness to adapt practices and values accordingly. Verma and Singh (2024) add that AI-enabled environments benefit most from leadership that promotes openness to experimentation, iterative learning, and interdepartmental trust qualities not evenly distributed across all NBA organizations. Beyond internal dynamics, sociotechnical systems theory also frames the NBA's AI adoption as part of a broader transformation in media and market-facing strategies. As Herman and Abeza (2024) explain, the league's embrace of AI is deeply intertwined with fan engagement strategies, broadcasting enhancements, and the gamification of analytics.

Organizational Learning Theory

Another central lens relevant to this study is organizational learning theory, which highlights how institutions evolve through the systematic acquisition, interpretation, and application of knowledge. In the NBA, franchises that leverage AI technologies are not merely reacting to available tools; they are actively engaged in cycles of experimentation, reflection, and refinement. Druga, Müller, and Zhai (2024) describe how AI-based coaching systems are being tested in iterative phases, with feedback loops informing both the human and machine sides of decision-making. This reciprocal adaptation underscores the NBA's shift toward knowledge-centric cultures where data becomes a core driver of competitive evolution.

Organizational learning theory also frames how NBA teams transition from ad hoc analytics usage to fully institutionalized AI strategies. As Tuyls et al. (2020) point out, the development of AI capabilities in sports involves long-term investment in infrastructure, partnerships with research institutions, and the cultivation of internal data literacy. Teams that once outsourced analytics to consultants are now building in-house departments with specialized roles, embedding AI into scouting, game preparation, and in-game decision support. This represents a structural learning process, where the organization itself evolves to become a more intelligent, responsive, and evidence-driven entity.

Furthermore, the theory provides insight into knowledge transfer within and between franchises. Nguyen and Morales (2025) argue that AI-enabled inclusivity tools such as adaptive training platforms or multilingual data visualizations facilitate broader access to strategic insights across player and staff demographics. This horizontal diffusion of knowledge enhances collaborative learning and reduces informational asymmetries between coaching, medical, and executive units. It also allows for a more participatory model of team building, where players contribute to tactical conversations grounded in shared data representations. These practices align with the core tenets of organizational learning: transparency, feedback responsiveness, and the institutionalization of reflection.

RESULTS AND FINDINGS

Focus on AI Tools and Technical Innovation Rather than Using them For the Purpose of Data Analytics for Organisational Use

The integration of artificial intelligence on AI tools and technical innovation into National Basketball Association operations has reshaped the strategic architecture of the league at multiple levels. Teams are adopting AI-enabled tools not only to support athletic performance but to generate insight across scouting, coaching, and front-office decision-making. See Table 1. How does the NBA focus on AI Tools and Technical Innovation rather than using for the purpose of data analytics for organizational use?

Table 1. How does the NBA Focus on AI Tools and Technical Innovation rather than using for the purpose of data analytics for organizational use?

Themes	Patterns	Summary of Correlation to AI Predictive Modelling
Scouting and Recruiting Processes	Anifa and Mahmoud (2024) detail how deep learning algorithms now review thousands of data points from amateur leagues, international competitions, and college programs to identify players who demonstrate statistically significant potential in decision-making, adaptability, or floor-spacing efficiency.	The algorithms account for hidden indicators of success that might be missed through traditional evaluation methods, such as microsecond reaction timing or predictive positioning. These techniques reduce human bias and expand the pool of viable candidates for recruitment.
Simulation based learning	Simulation-based learning and multi-variable performance forecasting enable franchises to construct digital models of future games, running scenario analyses on potential plays, lineups, or rotations.	According to Kapoor (2021), this approach incorporates historical trends, current player health metrics, and even psychological readiness scores to generate probabilistic outcomes of various coaching decisions.
Sequencing Models	Sequencing models trained on years of NBA data are now used to forecast likely responses by opponents in various game states, allowing coaches to pre-plan rotations, offensive counters, and defensive coverages.	These systems continuously adapt their recommendations as new data is ingested, providing teams with evolving strategic recommendations even within the flow of play. This has resulted in a measurable increase in late-game tactical success for franchises that consistently employ these tools.
Team Strategy Modelling	Through computer vision and machine learning, AI systems can tag every instance of a specific play type, defensive scheme, or matchup pattern across hundreds of game tapes (Li, 2025)	Another critical insight emerging from recent implementations of AI in the NBA is its role in augmenting coaching through decision-support interfaces. Unlike traditional analytics dashboards, which offer static historical comparisons, modern AI systems dynamically synthesize data during live play.
AI into Fan Engagement Platforms	Findings about the intersection between entertainment, personalization, and data-driven content delivery	Data interpretations may produce recommendations that conflict with human observation, raising concerns about bias embedded in training algorithms or a lack of context in behavioural metrics. These ethical complexities require ongoing institutional scrutiny and policy development to ensure responsible AI deployment (Cossich et al., 2025)
Player Surveillance Technologies	Taylor & Smith (2025) note that player surveillance technologies, particularly those involving biometric data, introduce privacy concerns about data ownership and consent.	As AI models become more sophisticated, questions arise about how much personal data can or should be used to inform performance evaluation, contract decisions, or disciplinary actions.

These findings collectively demonstrate how artificial intelligence is no longer confined to performance metrics alone. Its applications have expanded into organizational behavior, economic modeling, and even experiential design. The next pages will continue exploring findings related to ethics, coaching augmentation, and strategic talent retention.

How is AI Institutionally Adopted, Interpreted, and Applied in Decision-Making

The role of AI in the NBA as a unit analysis is multi-faceted in terms of optimization in determining adoption, interpretation, and application in decision-making. See Table 2. If an NBA franchise is the Unit of Analysis. How is Artificial Intelligence (AI) institutionally adopted, interpreted, and applied in decision-making?

Table 2. If an NBA Franchise is the Unit of Analysis – How is Artificial Intelligence (AI) institutionally adopted, interpreted, and applied in decision-making?

Themes	**Patterns**	**Summary of Correlation to AI Predictive Modelling**
Physical Monitoring	AI tools are being used to assess psychological readiness and mental fatigue	Natural processing tools applied to Team Interviews, Player Journaling Apps, and Social Media data can detect mood shifts, motivational dips or elevated stress levels (Anifa & Mahmoud, 2024)
Mental and Physical Health Conditions	Serve as AI component indicator for prompting psychology evaluations, wellness staff to intervene in subsequent physical and mental breakdowns	The integration of cognitive data into performance forecasts marks a notable evolution in what constitutes "athlete readiness."
AI Driven Scouting Systems	Change how talent evaluation is discovered and evaluated; traditionally reliant on raw data or gut feeling on a prospect	AI-Driven Scouting Systems influence and evaluate how prospects train, mental fortitude, and response to emotions under duress (Fouche, 2023)
Developmental Programs	Is another form of talent evaluation, based on metrics, potentially missed by NBA Scouts and Management Personnel to find a means of finding talent in other areas of the world (Xu and Baghaei (2025; Du et al., 2024)	The translation or correlation is the international game which has influenced the NBA to look for specific skillsets not always found within the current modelling performance metrics (Foche, 2023)
Streamlining Game Film	Through computer vision and machine learning, AI systems can tag every instance of a specific play type, defensive scheme, or matchup pattern across hundreds of game tapes (Li, 2025)	The purpose is to look for data anomalies within the evaluation of this information to determine player norms to exploit and plan for game scenarios
AI into Fan Engagement Platforms	One of the most common patterns within the NBA is the use of taglines, using music, and popular taglines, brand names, and marketing ploys to move information more rapidly and instantly, using multiple streams such as social media and sports news (Xu & Baghaei, 2025)	Each algorithm feeds into consumer behaviour and fan behaviour models (Mohammadi et al., 2024); Furthermore, tailoring digital content delivery, recommending highlight reels, merchandise, or live events to increase global reach (Lozzi et al, 2025)

From a broadcasting perspective, AI systems are being employed to generate automated commentary, adjust camera angles dynamically, and even simulate alternative endings to games. Fouché (2023) notes that networks such as CBS Sportsline, Fox Sports Network, ESPN, and NBA League Pass are already testing deep-learning models to predict pivotal moments in games and tailor highlight packages based on viewer attention spans. These capabilities are reshaping the way basketball content is consumed, especially among younger demographics accustomed to shorter, more curated content formats.

The findings also extend into global scouting and international growth. AI tools have drastically improved the ability to assess talent in non-traditional markets. Nguyen and Morales (2025) explain that deep learning systems now scan footage from semi-professional leagues, youth academies, and even informal street tournaments, using pose estimation and ball trajectory models to evaluate talent potential. These technologies help identify prospects long before they enter conventional scouting pipelines, giving teams a competitive edge and broadening the NBA's global talent footprint.

ARTIFICIAL INTELLIGENCE (AI) POLICY RECOMMENDATIONS

Artificial Intelligence (AI) policy governance within the National Basketball Association (NBA) centers on the strategic use of data analytics and AI technologies to enhance decision-making processes across the league. The NBA leverages AI-driven data collection and analysis to improve player performance, optimize team strategies, and enhance fan engagement. By implementing robust policies on data privacy, security, and ethical use, the league ensures that AI tools are used responsibly, fostering transparency and trust among stakeholders (NBA, 2023).

The league's governance framework emphasizes the integration of AI to support non-biased decision-making and operational efficiency. AI algorithms analyze vast amounts of player performance data, injury reports, and real-time game statistics to inform coaching strategies and player development programs. Additionally, AI is utilised for talent scouting and predictive modelling to identify potential rookies or undervalued players, giving teams a competitive edge. The NBA's policies specify standards for data accuracy, bias mitigation, and accountability to prevent misuse of information (NBA, 2023).

Furthermore, the NBA prioritizes ethical considerations and compliance with legal standards in its AI policy governance. This includes establishing clear guidelines on data ownership, consent, and privacy, especially for sensitive player health data. The league also continuously monitors and updates its policies to adapt to rapidly evolving AI capabilities and ethical concerns, ensuring that AI's implementation aligns with the league's values and organizational goals. Through this governance approach, the NBA exemplifies responsible leadership in integrating AI into professional sports (NBA, 2023).

FURTHER RESEARCH DIRECTIONS

Future research directions in AI continue to gain prominence within the operations of the National Basketball Association (NBA), and it becomes increasingly essential to consider areas where further inquiry could uncover new strategic, ethical, and technological opportunities. While current research has shed light on how AI enhances player performance, health monitoring, scouting, and fan engagement, there remain unexplored domains and evolving challenges that warrant further attention. One of the most pressing opportunities for future research lies in the ethical and regulatory dimensions of biometric surveillance. As NBA franchises integrate wearable technologies and AI-driven data analytics to monitor player exertion, fatigue, and injury risk in real time, questions surrounding consent, data ownership, and long-term health privacy remain largely unresolved. Addressing these concerns will require collaborative inquiry involving data scientists, legal scholars, ethicists, and professional sports organizations (Taylor & Smith, 2025; Musat et al., 2024).

Another promising research frontier involves the optimization of officiating and rule enforcement through AI-enhanced decision support systems. While current technologies such as instant replay and motion capture assist referees in making accurate calls, integrating real-time machine learning algorithms capable of assessing context, intent, and game flow introduces both technical and philosophical complexities. Future studies could investigate how such systems might reduce officiating bias or error without undermining the human element of judgment that many fans and players value (Herman & Abeza, 2024; Li, 2025). In this context, the balance between automation and tradition becomes a critical line of inquiry, particularly in a league where perception and fairness directly influence player morale and audience trust. In addition, there is growing interest in the role of AI in managing team finances and contract structures. While AI tools have shown potential in salary cap modelling and financial forecasting, future research might explore how predictive analytics can help franchises maximise long-term value from player acquisitions, trades, and extensions. Investigating how AI models can simulate career trajectories or identify undervalued talent across global markets could redefine competitive balance and talent distribution across the league (Gupta & Rao, 2024; Wright & Patel, 2025).

These directions highlight just a few of the many avenues still open for academic and industry-driven exploration. As AI technologies continue to advance and integrate more deeply into the culture and mechanics of professional basketball, the importance of multidisciplinary approaches to understanding their full implications will only grow.

Beyond team operations and officiating, a compelling area of future investigation lies in how AI can enhance the fan experience through behavioural prediction and content personalisation. With increasing amounts of data collected from ticket sales, app usage, streaming behaviour, and social media engagement, machine learning models can be employed to tailor fan experiences with high precision. Research could delve into the ethical dimensions of data usage, personalization fatigue, and the risk of algorithmic echo chambers within sports media ecosystems (Carson & Lee, 2025; Fouché, 2023). Understanding the long-term impact of hyper-personalized content delivery on audience engagement and brand loyalty remains an underexplored but potentially transformative research area.

Another significant opportunity lies in integrating AI into monitoring mental health and psychological resilience among athletes. While physical metrics are routinely captured, mental well-being often remains elusive in data-driven player development models. Future studies might examine how natural language processing, sentiment analysis, or digital behaviour tracking can provide early indicators of stress, burnout, or emotional fatigue, without breaching privacy or autonomy. Such insights could inform better support systems, coaching practices, and organizational responses to mental health concerns in high-performance environments (Choustoulakis, 2024; Smith & Lee, 2025).

Additionally, the potential for cross-league AI collaboration represents a largely untapped research frontier. Exploring shared AI infrastructures between leagues such as the NBA, NFL, and MLB could yield insights into best practices, model interoperability, and ethical standards. Research could evaluate how shared technological innovations might promote parity, elevate game quality, or even shape international sports governance. These kinds of partnerships might also stimulate comparative analyses across different sporting cultures, broadening the impact of AI tools beyond basketball alone (Johnson & Brown, 2024; Cossich et al., 2023).

As artificial intelligence (AI) becomes increasingly integrated into the operational, strategic, and entertainment aspects of the NBA, it creates a ripple effect that reveals broader societal and technological implications worth exploring through future research. One such avenue is the long-term sociocultural impact of AI-enabled decision-making in sports. For example, how might AI-driven scouting and recruitment models influence diversity and inclusion across the league? Will machine learning algorithms unintentionally reproduce historical biases present in legacy data, particularly in assessing international or underrepresented talent pools? Scholars could examine how algorithmic transparency and fairness frameworks can be embedded into these tools to mitigate such risks (Mohammadi et al., 2024; Nalbant & Aydın, 2022). Investigating algorithmic justice within a highly visible, globally consumed sport like basketball could yield insights applicable to other high-stakes domains, such as education and hiring.

Further, future research could address the evolving dynamics between AI and human labour within the context of sports journalism and analytics. As natural language generation tools and predictive modelling systems become more adept at writing post-game summaries, player profiles, or analytical breakdowns, traditional roles in sports media may be displaced or redefined. Studies may explore how human journalists and AI systems can collaborate to enhance storytelling, providing richer narratives without compromising authenticity or critical insight (Clarke & Jackson, 2025; Zhu-Tian et al., 2022). In the same vein, it will be crucial to evaluate how fan communities react to AI-generated content and whether it enhances engagement or diminishes the perceived value of human perspectives in sports discourse.

One particularly complex yet fascinating area involves investigating how AI can shape or even disrupt coaching philosophies. As coaching staff incorporate real-time data visualisation, player tracking, and tactical modelling into game preparation, the boundary between data-driven strategy and human intuition becomes increasingly blurred. Future studies could examine how coaching decisions evolve in environments saturated with data and whether they rely more on computational guidance or preserve instinctual methods refined over decades of experience (Tuyls et al., 2020; Lin et al., 2020). Furthermore, researchers could look at whether AI-supported coaching platforms can equalize performance among less experienced coaching teams by offering decision support tools typically available only to elite franchises.

Considering increasing international interest in basketball, another avenue of research lies in understanding how AI adoption varies across national leagues and what cultural or economic factors influence the speed and manner of implementation. For instance, researchers could conduct comparative studies between the NBA, EuroLeague, and Chinese Basketball Association to assess how AI-enabled player analytics, training software, or fan engagement tools are deployed across different regions. Such studies may highlight discrepancies in technological infrastructure, regulatory environments, and public sentiment that influence the global adoption of sports technologies (Hirosaki & Patel, 2025; Nguyen & Morales, 2025). These findings could inform strategies for international tech vendors, policymakers, and sports federations aiming to create equitable access to AI in sports.

Lastly, a compelling area for future research concerns the use of AI in preventing and mitigating corruption, gambling-related manipulations, or performance-enhancing drug use in professional basketball. While existing surveillance and compliance systems have traditionally relied on human oversight and whistleblower mechanisms, AI can offer more proactive monitoring solutions. For instance, anomaly detection algorithms can be trained to identify unusual player performance patterns, betting line shifts, or inconsistencies in biometric data that warrant further investigation

(Du et al., 2024; Musat et al., 2024). Integrating such tools raises important questions about transparency, due process, and technological infallibility, which are issues that intersect both ethical and legal domains. The development of accountable AI systems in this space has the potential to reinforce integrity within the league and set new precedents for other professional sports organizations.

These proposed areas not only build upon current literature but also extend the horizon of what AI can contribute to professional basketball and broader athletic ecosystems. As research continues to grow in tandem with technological innovation, it becomes vital for scholars to adopt interdisciplinary approaches that merge sports science, computer science, ethics, and cultural analysis. By doing so, future investigations can more holistically assess the trade-offs, benefits, and unintended consequences of integrating AI into one of the world's most influential sporting arenas.

ANALYZE AND VISUALIZATION TRENDS

In this section of the chapter, we analyze and visualize trends across teams and seasons. This study will utilize both descriptive and inferential statistical tools. Descriptive statistics will summarize the distribution of AI usage across NBA franchises, with variables including frequency of AI mentions, thematic diversity of usage (e.g., performance, marketing, injury prevention), and shifts in language over time. These descriptive insights will provide a foundation for more complex analysis by identifying which teams exhibit consistent patterns or deviate from league-wide norms (Xu & Baghaei, 2025; Wright & Patel, 2025).

Next, inferential methods such as correlation matrices and regression modelling will be used to investigate potential relationships between AI usage and organisational performance metrics. For example, statistical tests may be applied to explore whether an increase in AI references correlates with a reduction in average player injuries or improvements in team efficiency ratings. Although causality cannot be established through this method, the findings may suggest areas for more focused experimental research. Models will be tested for multicollinearity, heteroscedasticity, and statistical significance to ensure reliability (Li & Huang, 2023; Carson & Lee, 2025). Visualization tools will also play a key role. Platforms like Tableau and Power BI will be used to create dashboards that map AI-related activity across teams and seasons. These dashboards will include time-series plots of AI keyword frequency, heat maps of organizational investment levels, and bar graphs comparing performance indicators across AI adoption tiers. These visualizations are not only analytic tools but also communication instruments that will help present findings to academic, industry, and general audiences in an accessible format (Verma & Singh, 2024; Lin et al., 2020).

Additionally, to account for evolving terminology and inconsistent reporting standards, a terminology normalization process will be employed. This involves mapping synonymous phrases such as "predictive modelling," "machine learning optimisation," and "AI-assisted forecasting" to unified categories that reflect their underlying function. This process ensures consistency across data sources and reduces semantic ambiguity, which is critical in a field where jargon evolves rapidly. Quality assurance procedures will be employed throughout the study to validate this normalization and maintain analytic transparency (Kapoor, 2021; Tuyls et al., 2020).

Moreover, in consolidating the multi-layered methodological approach used in this study, the emphasis has been on balancing structure, depth, and adaptability. By combining qualitative content analysis framework, case-based comparisons, and statistical mapping, the research captures both the narrative and numerical dimensions of AI integration within NBA organizations. This triangulated strategy not only ensures that findings are grounded in observable patterns but also provides the flexibility to account for variances in data availability, terminology, and team-specific transparency (Hoffman & Kumar, 2024; Druga et al., 2024).

While the study is methodologically robust, several limitations must be acknowledged. The reliance on publicly accessible data introduces the potential for gaps in information, especially from franchises that maintain strategic secrecy. Additionally, the study does not access internal organizational data, which may contain richer indicators of AI use, such as proprietary algorithms or closed-door evaluations. Nonetheless, the study compensates for this by utilising reputable third-party analyses, media triangulation, and scholarly interpretation to ensure valid and reliable conclusions (Hirosaki & Patel, 2025; Smith & Lee, 2025).

Finally, the methodology described here provides the necessary foundation to examine how specific theoretical models can explain the AI phenomena observed in the NBA. The next section of ethical consideration interprets the data and guides the discussion of the study's broader implications. With a clear understanding of how data were collected, analyzed, and contextualized, the section now shifts toward establishing the intellectual lens through which findings will be assessed.

ETHICAL CONSIDERATIONS

Ethical considerations are minimal in this study, given the use of publicly available materials. No direct contact with human participants is required. However, ethical rigour will be maintained by accurately representing the context and intent of quoted materials and ensuring that all interpretations are supported by the data.

Transparency in methodological choices and documentation will further enhance the credibility of the research.

Another critical element of this study involves evaluating the degree to which AI is embedded into each NBA organization's operational rhythm. This includes more than just the presence of data tools or analytics staff; it encompasses the frequency, consistency, and context in which AI-generated insights are used in practice. Domains such as real-time decision-making during games, post-game performance breakdowns, talent scouting, injury prevention protocols, and contract negotiations all offer opportunities for AI integration. These areas have been identified by both academic researchers and industry analysts as key pressure points where the margin for success or failure is narrow and strategic precision is essential (Taylor & Smith, 2025; Wright & Patel, 2025).

Franchises like the Golden State Warriors and Toronto Raptors, both cited repeatedly in literature, are emblematic of full-cycle integration. These teams not only contract with advanced data providers like Second Spectrum, Catapult, and Synergy Sports, but they also have internal divisions where data scientists, coaches, and medical professionals collaborate to interpret and apply AI-generated insights (Carter & Nguyen, 2024; Gupta & Rao, 2024). The degree of cross-functional synthesis within these organizations allows for AI to move beyond superficial adoption into something more culturally and operationally embedded.

To measure integration levels, this study will analyze a variety of secondary data sources. Public-facing information such as conference presentations, white sections, organizational reports, and interviews with front office personnel will be reviewed. Metrics of operational alignment will include the number of staff roles dedicated to data or analytics, the public disclosure of AI usage in competitive contexts, and observable changes in decision-making that align with AI-supported strategies. Where relevant, media reports and team-authored publications will be used to cross-check organizational claims, providing both triangulation and contextual grounding (Herman & Abeza, 2024; Anifa & Mahmoud, 2024). These independent perspectives provide an important layer of triangulation, especially when internal data access is limited. This triangulation approach aligns with best practices in organisational research, offering a balance between qualitative depth and methodological rigour (Kusan & Arin, 2025; Clarke & Jackson, 2025).

Ethical considerations have also been integrated into the research design. Given the competitive nature of the NBA and the sensitivity of proprietary strategies, all data gathered will be restricted to publicly available sources. No confidential or leaked documents will be used. Furthermore, interpretations will be presented cautiously, with attention to avoiding overgeneralization or unjustified conclusions. While AI's influence is increasingly visible in professional basketball, its implementation is shaped by organizational discretion, competitive secrecy, and evolving technological standards. These contextual factors will remain foregrounded in all aspects of the analysis (Mohammadi et al., 2024; Gupta & Rao, 2024).

CONCLUSION

The integration of artificial intelligence (AI) into the operational and strategic framework of the National Basketball Association (NBA) has demonstrated both promising advancements and emerging complexities. This section provides a detailed examination of how AI-driven technologies are not only augmenting performance analytics and scouting strategies but also transforming coaching methodologies, injury prevention protocols, and fan engagement systems. Across the findings, it became evident that AI has created a multi-dimensional influence in professional basketball, redefining how decisions are made on and off the court.

One of the most significant insights from this study is the way AI has enabled more in-depth, granular player evaluations that extend beyond traditional metrics. Rather than relying solely on basic statistics, organizations are now leveraging machine learning algorithms to interpret nuanced data such as player fatigue, biometric indicators, and real-time court positioning.

These systems have empowered coaching staff and front offices with predictive tools to optimize performance, reduce risk, and adjust strategies for mid-season. Such innovations are not theoretical but are actively being applied in pilot programs by multiple NBA franchises (Kapoor, 2021; Xu & Baghaei, 2025). Equally important is the discussion surrounding how AI has altered the economic dimensions of the league. Financial modelling systems now integrate player performance data with market trends to support more efficient salary cap management, trade negotiations, and long-term planning. While not without its challenges, such as data transparency and model accuracy, the strategic use of AI in financial planning has demonstrated measurable benefits for franchises seeking sustained competitiveness (Verma & Singh, 2024; Johnson & Brown, 2024). Moreover, the integration of AI into fan experience platforms has expanded the NBA's ability to personalize content, forecast ticket demand, and even drive narrative engagement through immersive technologies. This intersection presents a compelling opportunity for future innovation, as well as regulatory scrutiny.

Despite these advances, the deployment of AI in the NBA is not without significant ethical, logistical, and operational hurdles. Chief among them is the question of data privacy. As AI tools increasingly rely on biometric and behavioural inputs, there is growing concern regarding who owns that data, how it is stored, and what consent frameworks are in place. Players, coaches, and even fans are becoming data sources, yet existing governance mechanisms within professional sports leagues have not fully caught up to these technological realities (Taylor & Smith, 2025; Kusan & Arin, 2025). The risk of misuse or unauthorized access poses a direct threat to individual autonomy and organizational integrity.

As the NBA continues to embrace AI-driven innovations, the road ahead will require a robust, multifaceted approach that incorporates not only technological capability but also organisational adaptability, stakeholder buy-in, and ethical foresight. Teams are already starting to recognize that investing in advanced algorithms is not enough. To effectively leverage AI systems, franchises must build a culture that values data literacy and strategic alignment. This includes hiring interdisciplinary staff who can translate predictive insights into actionable decisions, integrating continuous training programs for coaching staff, and promoting internal transparency so that decisions informed by AI are understood and trusted rather than blindly followed or resisted (Gupta & Rao, 2024; Wright & Patel, 2025).

Beyond internal practices, the NBA as a governing body will face increasing pressure to formalize its position on AI. While some teams move ahead with cutting-edge solutions ranging from wearable tech for performance tracking to deep learning models for draft analytics, others may fall behind due to financial constraints or institutional inertia. The NBA will need to walk a fine line between creating immersive experiences and respecting individual rights, especially as fan data becomes a valuable commodity not just for engagement but for monetization (Kusan & Arin, 2025; Taylor & Smith, 2025). A well-defined ethical infrastructure can build trust among players, fans, and business partners alike (Mohammadi et al., 2024; Herman & Abeza, 2024).

Lastly, the academic and research community must continue to contribute to this discourse. As demonstrated by this study, scholarly work plays a critical role in framing the big-picture implications of AI while also suggesting practical interventions. Interdisciplinary partnerships between sports organizations, universities, and tech firms will be essential in fostering innovation that is not only cutting-edge but also grounded in empirical evidence and real-world feasibility. As new methodologies emerge from reinforcement learning in player training to sentiment analysis in fan relations, they must be rigorously tested, ethically reviewed, and adaptively implemented. The future of AI in the NBA is not a single trajectory but a constellation of paths, each shaped by the decisions made today across technical, social, and moral dimensions.

REFERENCES

Anifa, M., & Mahmoud, A. B. (2024). Sports analytics: Data-driven sports and decision intelligence. In *Sports analytics: Data-driven sports and decision intelligence* (pp. 1–17). Springer., DOI: 10.1007/978-3-031-63573-1_1

Carson, D., & Lee, H. (2025). AI-enabled coaching platforms: Integrating strategic analytics in sports training. *Journal of Applied Cloud Computing*, *7*(3), 233–250. DOI: 10.1111/jcal.13082

Carter, L., & Nguyen, T. (2024). Integrating digital innovations and leadership to revolutionize the sports industry. In *Innovation and leadership in sports management* (Vol. 1, pp. 22–40). Springer., DOI: 10.1007/978-3-031-63573-1_3

Choustoulakis, E. (2024, November). AI-driven analysis of game tactics and player performance. In *Proceedings of the 17th International Conference of Education, Research and Innovation (ICERI 2024)* (pp. 3350–3359). IATED. https://doi.org/ DOI: 10.21125/iceri.2024.0871

Clarke, A., & Jackson, P. (2025). Data-driven message optimization in dynamic sports media: An AI approach. *Journal of Media Analytics*, *12*(1), 55–72. DOI: 10.1080/14413523.2024.2372122

Cossich, V. R. A., Carlgren, D., Holash, R. J., & Katz, L. (2023). Technological breakthroughs in sport: Current practice and future potential of artificial intelligence, virtual reality, augmented reality, and modern data visualization in performance analysis. *Applied Sciences (Basel, Switzerland)*, *13*(23), 12965. Advance online publication. DOI: 10.3390/app132312965

Druga, M., Müller, S., & Zhai, Q. (2024). AI in digital sports coaching – a systematic review. *International Journal of Sports Coaching Science*, *1*(1), 1–23. DOI: 10.1080/23750472.2024.2449016

Du, J., Mamo, Y. Z., Floyd, C., Karthikeyan, N., & James, J. D. (2024). Machine learning in sport social media research: Practical uses and opportunities. *International Journal of Sport Communication*, *1*(1), 97–106. DOI: 10.1123/ijsc.2023-0151

Fouché, R. (2023). Sport and AI: Charting the connection between historical evolution and emerging technologies. *Sport History Review*, *54*(2), 112–126. DOI: 10.1080/09523367.2023.2228118

Gupta, S., & Rao, P. (2024). High-performance computing strategies for AI in sports analytics. *Concurrency and Computation*. Advance online publication. DOI: 10.1002/cpe.70068

Herman, N., & Abeza, G. (2024). The disruptive power of AI in sport and entertainment. In *the Routledge Handbook of Sport Communication* (pp. 350–362). Routledge. https://doi.org/DOI: 10.4324/9781003430278-35

Hirosaki, M., & Patel, R. (2025). Artificial intelligence approach in biomechanics of gait and sport: A systematic literature review. *Journal of Biomedical Physics & Engineering, 13*(5), 383–402. DOI: 10.31661/jbpe.v0i0

Hoffman, J., & Kumar, R. (2024). Game theory and AI in team sports decision-making. In *Advances in leadership and digital innovations in sports* (Vol. 3, pp. 45–62). Springer., DOI: 10.1007/978-3-031-76047-1_5

Johnson, R. T., & Brown, M. P. (2024). Why do leaders matter? A revisit of player-turned-coaches from the perspective of elite basketball. *Applied Economics, 56*(34), 3759–3775. DOI: 10.1080/00036846.2023.2266606

Kapoor, S. (2021). AI-driven decision support systems in sports project management: Enhancing strategic planning. *International Journal of Artificial Intelligence, Data Science, and Machine Learning, 2*(3), 1–11. DOI: 10.63282/3050-9262.IJAIDSML-V2I3P101

Kusan, M., & Arin, A. (2025). Ethical and security issues: The impact of artificial intelligence in sports management. *Acta Scientiae et Intellectus, 10*(2), 78. Advance online publication. DOI: 10.53469/asi.2025.10.2.78

Li, A., & Huang, W. (2023). A comprehensive survey of artificial intelligence and cloud computing applications in the sports industry. *Wireless Networks, 30*(8), 6973–6984. DOI: 10.1007/s11276-023-03567-3

Li, Z. (2025). Innovation and transformation of sports event management driven by artificial intelligence: Athlete performance, referee decision support, security, and fan interaction. *Journal of Social Science Humanities and Literature, 8*(2), 3. Advance online publication. DOI: 10.53469/jsshl.2025.08(03).03

Lin, T., Yang, Y., Beyer, J., & Pfister, H. (2020). SportsXR—Immersive analytics in sports. *arXiv*. https://doi.org//arXiv.2004.08010DOI: 10.48550

Lozzi, D., Di Pompeo, I., Marcaccio, M., Alemanno, M., Krüger, M., Curcio, G., & Migliore, S. (2025). AI-powered analysis of eye tracker data in basketball games. *Sensors (Basel), 25*(11), 3572. DOI: 10.3390/s25113572

Mateus, N., Abade, E., Coutinho, D., Gómez, M.-Á., Peñas, C. L., & Sampaio, J. (2025). Empowering the sports scientist with artificial intelligence in training, performance, and health management. *Sensors (Basel), 25*(1), 139. DOI: 10.3390/s25010139

Mohammadi, S., Pashaie, S., Ghaffarisadr, S. I., & Rayner, M. (2024). Opportunities and challenges of leveraging artificial intelligence (AI) in sport marketing. In Tran, M. T. (Ed.), *Advancing the marketing technology (MarTech) revolution* (pp. 233–272). IGI Global., DOI: 10.4018/979-8-3693-4361-6.ch009

Musat, C. L., Mereuta, C., Nechita, A., Tutunaru, D., Voipan, A. E., Voipan, D., Mereuta, E., Gurau, T. V., Gurău, G., & Nechita, L. C. (2024). Diagnostic applications of AI in sports: A comprehensive review of injury risk prediction methods. *Diagnostics (Basel), 14*(22), 2516. DOI: 10.3390/diagnostics14222516

Nalbant, K. G., & Aydın, S. (2022). Literature review on the relationship between artificial intelligence technologies with digital sports marketing and sports management. *Indonesian Journal of Sport Management, 2*(2), 135–143. DOI: 10.31949/ijsm.v2i2.2876

NBA. (2023). *AI and Data Analytics in the NBA: Policy and Practice.* National Basketball Association.

Nguyen, T., & Morales, L. (2025). Accessibility to sports performance for greater inclusiveness through artificial intelligence: A systematic review. *Journal of Sports Engineering & Technology, 239*(1), 1–15. DOI: 10.1080/10255842.2025.2457118

Smith, J. A., & Lee, H. K. (2025). Big data and artificial intelligence in sports analytics. In *Big data and artificial intelligence in sports analytics* (pp. 95–112). Springer., DOI: 10.1007/978-3-662-68241-8_8-1

Taylor, J., & Smith, R. (2025). Ethical implications of AI in sports analytics. *Journal of Sport and Social Issues, 49*(2), 123–145. DOI: 10.1177/14727978251337959

Taylor, L., & Nguyen, P. (2024). Tools and open datasets to support, sustain, and secure critical digital infrastructure in sports. In *Proceedings in Sports Technology* (pp. 45–60). Springer., DOI: 10.1007/978-3-031-67256-9_3

Tuyls, K., Omidshafiei, S., Müller, P., Wang, Z., Connor, J., Hennes, D., Graham, I., Spearman, W., Waskett, T., Steele, D., Luc, P., Recasens, A., Galashov, A., Thornton, G., Elie, R., Sprechmann, P., Moreno, P., Cao, K., Garnelo, M., . . . Hassabis, D. (2020). Game plan: What AI can do for football, and what football can do for AI. *arXiv*. https://doi.org//arXiv.2011.09192DOI: 10.48550

Vaithianathan, M. (2025, January 19). The future of heterogeneous computing: Integrating CPUs, GPUs, and FPGAs for high-performance applications. International. *Journal of Emerging Trends in Computer Science and Information Technology, 6*(1), 12–23. DOI: 10.63282/3050-9246.IJETCSIT-V6I1P102

Verma, P., & Singh, A. (2024). Game tactics and strategy optimization through AI and machine learning. In *Artificial intelligence and machine learning in sport research* (pp. 150–172). IGI Global., DOI: 10.4018/978-1-978-369-385-1.ch006

Wesely, S., Hofer, E., Curth, R., Paryani, S., Mills, N., Ueberschär, O., & Westermayr, J. (2025). Artificial intelligence for objective assessment of acrobatic movements: Machine learning identification of tumbling elements in cheer sports. *Sensors (Basel)*, *25*(7), 2260. Advance online publication. DOI: 10.3390/s25072260

Wright, L. A., & Patel, M. (2025). Enhancing decision-making in team sports through AI-driven computing solutions. *International Journal of Sports Computing*, *12*(1), 45–67. DOI: 10.1002/cpe.8372

Xu, T., & Baghaei, S. (2025). Reshaping the future of sports with artificial intelligence: Challenges and opportunities in performance enhancement, fan engagement, and strategic decision-making. *Engineering Applications of Artificial Intelligence*, *142*, 109912. Advance online publication. DOI: 10.1016/j.engappai.2024.109912

Zhu-Tian, C., Yang, Q., Xie, X., Beyer, J., Xia, H., Wu, Y., & Pfister, H. (2022). Sporthesia: Augmenting sports videos using natural language. *arXiv*. https://doi.org//arXiv.2209.03434DOI: 10.48550

KEY TERMS AND DEFINITIONS

Analytics: the systematic use of data, statistical and quantitative analysis, explanatory and predictive models, and fact-based management to drive decisions and actions.

Artificial Intelligence (AI): the ability of a machine to perform cognitive functions we normally associate with human minds, such as perceiving, reasoning, learning, interacting with the environment, problem-solving, and even exercising creativity.

AI Governance: is a system of rules, practices, processes, and technological tools that are employed to ensure an organization's use of AI technologies aligns with the organization's strategies, objectives, and values; fulfills legal requirements; and meets principles of ethical AI.

Data-Driven Strategy: an approach in which decisions and strategy are explicitly guided by data analysis, metrics and insights rather than intuition or ad-hoc judgments.

Internet of Things (IoT): refers to a network of physical devices, vehicles, appliances and other physical objects that are embedded with sensors, software, and network connectivity, allowing them to collect and share data.

National Basketball Association (NBA): North America's premier professional men's basketball league, comprising franchises in the United States and Canada, functioning as the top level of commercial, organizational and competitive basketball and influencing global sports culture and business.

Predictive Modelling: consists of a collection of statistical techniques with the goal of finding a relationship between a target or response and several predictors or drivers.

Resource-Based Theory / View (RBT/RBV): contends that the possession of strategic resources provides an organization with a competitive advantage. In particular, resources that are valuable, rare, difficult to imitate and non-substitutable can enable sustained competitive advantage.

Sports Management: is the multidisciplinary academic and professional field concerned with the planning, organizing, directing, controlling, budgeting, leading and evaluating of sport-related organizations, teams, events and facilities drawing upon business, marketing, finance, law, sociology and sport-specific knowledge to support and develop the sport industry.

Strategic Decision-Making: refers to long-term, high-impact decisions made by top management that align an organization's actions with its goals and external environment, typically under conditions of uncertainty and complexity.

Chapter 4
Research Rewired AI in Computing

Otmane Azeroual
http://orcid.org/0000-0002-5225-389X
University of Hagen, Germany

ABSTRACT

This chapter explores how artificial intelligence (AI) is reshaping research methodologies in computing, moving beyond traditional paradigms of data collection, analysis, and interpretation. It examines the integration of AI-driven approaches—such as automated literature mining, intelligent hypothesis generation, simulation-based validation, and adaptive data modeling—into the research process. Particular attention is given to how these methods accelerate discovery, enhance reproducibility, and reduce human bias in computational studies. The chapter also investigates the risks of algorithmic opacity, over-automation, and ethical concerns when embedding AI into scholarly workflows. By analyzing case studies from informatics, data science, and software engineering, the chapter illustrates the transformative potential of AI-powered research pipelines while providing a critical framework for balancing innovation with methodological rigor.

INTRODUCTION

Computer science research today is confronted with an exponential increase in scientific publications, increasingly complex datasets, and growing methodological demands. Traditional research processes—from literature review and data collection to hypothesis formulation—are reaching their limits. Artificial Intelligence (AI), particularly large language models (LLMs) and explainable AI (XAI) frameworks, offers the potential to fundamentally transform these processes by optimizing work-

DOI: 10.4018/979-8-2600-0434-0.ch004

flows, automating routine tasks, and enhancing both the quality, interpretability, and scalability of scientific work (Jordan & Mitchell, 2015; Russell & Norvig, 2015; Ribeiro et al., 2016; Lundberg & Lee, 2017).

Significant progress has been achieved in systematic literature analysis through AI-based approaches. Text mining and natural language processing (NLP) technologies can accelerate the selection and evaluation of relevant literature while reducing errors (Marshall & Wallace, 2019; Ofori-Boateng et al., 2024). Modern AI pipelines also integrate semantic search, knowledge graphs, and automated clustering, enabling researchers to uncover hidden relationships between studies and detect research trends (Devlin et al., 2019; Beltagy et al., 2019; Ji et al., 2021). At the same time, challenges remain, particularly regarding methodological transparency and the "black-box" nature of modern models (Tsafnat et al., 2014; Jardim et al., 2022).

AI tools based on LLMs are increasingly extending beyond simple text analysis. They enable researchers to formulate research questions in natural language, automatically identify relevant sources, and generate concise summaries of complex topics (Doshi-Velez & Kim, 2017; Sallam, 2023). This development reflects the rise of Augmented Analytics and AI-augmented research methodology, where machine learning (ML), NLP, and generative AI function not just as auxiliary tools but as active partners in the scientific inquiry process (Davenport & Miller, 2022; Bubeck et al., 2023).

The influence of AI is also becoming evident in experimental research. AI co-scientists and autonomous research systems can generate hypotheses, design research plans, optimise experimental conditions, and even propose novel experiments autonomously (King et al., 2009; Stokes et al., 2020; Kaspar et al., 2021; Lee et al., 2024). XAI techniques, such as SHAP and LIME, allow researchers to interpret these automated outputs and validate them against domain knowledge, increasing transparency and trust.

These trends illustrate that AI in computer science is increasingly viewed not merely as a tool but as an active partner in the research process. While automation offers significant efficiency gains, it also raises questions about transparency, bias, interpretability, and ethical implications (Floridi & Cowls, 2022). Consequently, there is a need to establish comprehensive frameworks that integrate AI into the research cycle while ensuring methodological rigour, ethical compliance, and reproducibility.

This chapter provides a thorough examination of how AI is reshaping research in computing and informatics. It reviews theoretical foundations and methodological frameworks, explores AI-enhanced processes—including literature review, hypothesis generation, data analysis, XAI-supported interpretation, and experimental automation—and presents empirical case studies that highlight the advantages and challenges of AI-assisted versus traditional approaches. The chapter further addresses opportunities, benefits, challenges, ethical considerations, and policy implications associated with AI integration, before discussing potential future directions for AI-augmented research.

By doing so, this chapter demonstrates the relevance of AI not only as a technical enhancement but as a transformative agent for modern scientific inquiry. It is aimed at researchers, graduate students, and practitioners in computing and related disciplines who seek a comprehensive understanding of AI's impact on research methodology, the practical implementation of AI-driven workflows, and the ethical and epistemological considerations necessary for responsible scientific practice.

Theoretical Foundations and Background

Research in computing and informatics has traditionally relied on rigorous methodological frameworks, combining experimental design, algorithmic development, and data-driven evaluation (Wohlin et al., 2012). However, the exponential growth of data and the increasing complexity of computational systems have posed significant challenges to traditional approaches. AI, particularly ML, deep learning (DL), NLP, and LLMs, has emerged as a transformative enabler capable of enhancing, automating, and in many cases reconfiguring how research is conducted in these domains (Jordan & Mitchell, 2015; Bubeck et al., 2023).

AI technologies enable researchers to go beyond descriptive analytics and engage in predictive and prescriptive modelling (LeCun et al., 2015). XAI frameworks such as SHAP and LIME provide transparency into AI-driven decisions, allowing researchers to validate outputs, ensure reproducibility, and detect potential biases (Ribeiro et al., 2016; Lundberg & Lee, 2017). This is particularly relevant in computing, where the complexity of systems often produces non-linear, high-dimensional data patterns that cannot be efficiently captured through conventional statistical methods.

Moreover, AI is not merely a tool for data analysis but increasingly serves as an active participant in knowledge discovery, enabling systematic literature reviews, automated hypothesis generation, and adaptive methodological design (Gil et al., 2014; Bubeck et al., 2023). Such capabilities constitute what is now recognised as AI-augmented research methodology, in which human researchers collaborate with intelligent systems to advance scientific inquiry.

To provide an overview of how AI intersects with traditional research methods in computing and informatics, Table 1 compares classical methodological approaches with their AI-enhanced counterparts.

Table 1. Traditional vs. AI-Enhanced Research Methods.

Research Dimension	Traditional Approach	AI-Enhanced Approach	Key Literature
Literature Review	Manual keyword searches, bibliometric mapping	NLP-driven text mining, automated systematic reviews, semantic search, knowledge graphs	Fister & Perc, 2016; Bubeck et al., 2023; Devlin et al., 2019; Beltagy et al., 2019; Ji et al., 2021
Hypothesis Generation	Human intuition, theory-driven derivation	ML-based pattern recognition, generative modelling, predictive hypothesis frameworks	Gil et al., 2014; Jordan & Mitchell, 2015
Data Collection	Surveys, controlled experiments, and manual coding	Automated data harvesting, IoT-driven sensing, and synthetic data generation	Varian, 2014; Goodfellow et al., 2016
Data Analysis	Statistical methods (ANOVA, regression, clustering)	Deep learning, reinforcement learning, XAI-supported visualization, and advanced pattern recognition	LeCun et al., 2015; Schmidhuber, 2015; Ribeiro et al., 2016; Lundberg & Lee, 2017
Research Validation	Human peer review, replication studies	AI-assisted reproducibility checks, anomaly detection, automated benchmarking	Amershi et al., 2019; Alpaydin, 2020; Azeroual, 2025
Knowledge Dissemination	Academic writing, conferences, journals	AI-generated drafts, adaptive publishing platforms, multimodal dissemination (text, code, visualization)	Floridi & Chiriatti, 2020; van Dis et al., 2023

This comparison demonstrates that AI is increasingly integrated across the entire research lifecycle—from knowledge discovery to dissemination. While traditional methods remain foundational, the infusion of AI enables greater scalability, efficiency, interpretability, and novel insights; however, challenges such as transparency, bias mitigation, and reproducibility remain (Floridi & Cowls, 2022).

AI-ENHANCED RESEARCH PROCESSES

The integration of AI into research processes is fundamentally transforming the methodologies and practices of computing and informatics. Rather than depending solely on manual data collection, linear analysis procedures, and human intuition, AI-driven systems enable research to become systematic, scalable, adaptive, and capable of real-time optimisation. This transformation reshapes nearly every stage of scientific inquiry—from literature review and hypothesis generation to data collection, analysis, simulation, and experimental automation (Gil et al., 2014; Jordan & Mitchell, 2015; Bubeck et al., 2023).

This shift marks a departure from the traditional view of AI as a purely technical support tool. Increasingly, AI functions as a strategic co-researcher—an active participant in the research cycle that generates hypotheses, refines methods, and reveals hidden connections. However, this integration also introduces new responsibilities: ensuring interpretability, transparency, and ethical oversight, so that AI complements rather than replaces critical human reasoning (Floridi & Cowls, 2022; Azeroual, 2025).

LITERATURE REVIEW AND KNOWLEDGE DISCOVERY

One of the most visible transformations occurs in the literature review phase. With scientific knowledge expanding exponentially—doubling roughly every decade (Bornmann & Mutz, 2015)—manual review methods and keyword-based searches are no longer sufficient. Researchers now face an overwhelming volume of publications across multiple languages and disciplines, making it nearly impossible to achieve comprehensive coverage using traditional techniques.

AI, particularly through Natural Language Processing (NLP), revolutionises this process by enabling semantic search, topic modelling, and knowledge graph construction. Unlike conventional database queries that depend on exact keywords, AI-driven models interpret contextual meaning, uncovering nuanced relationships between concepts and identifying thematically related research that might otherwise remain hidden. Models such as BERT and SciBERT (Devlin et al., 2019; Beltagy et al., 2019) demonstrate how AI can detect conceptual similarities even when terminology differs, thus enabling cross-domain discovery.

Furthermore, LLMs like GPT-based architectures can summarize vast bodies of literature, identify knowledge gaps, and even suggest potential research directions or hypotheses. These tools accelerate the initial stages of inquiry and support researchers in navigating the complexity of the modern scientific landscape (Doshi-Velez & Kim, 2017; Bubeck et al., 2023). The integration of XAI ensures that such systems are not black boxes: it clarifies why certain papers or topics are prioritized, allowing for traceability and reproducibility in literature discovery (Ribeiro et al., 2016; Lundberg & Lee, 2017).

Moreover, knowledge graphs provide an additional layer of insight by visually representing relationships between concepts, methodologies, and domains. They reveal, for example, how computational methods in machine learning connect to developments in biomedicine or material science (Ji et al., 2021). These networks support interdisciplinary synthesis and feedback loops between literature review and hypothesis generation—so that insights gained from AI-driven text mining continuously inform new research directions.

HYPOTHESIS GENERATION AND TESTING

Traditionally, hypotheses have been derived from theoretical reasoning, prior findings, or intuitive insight. This human-centred approach remains foundational but is increasingly complemented by data-driven hypothesis generation powered by machine learning and probabilistic modelling. AI systems can detect correlations, anomalies, or latent structures within large datasets that humans might overlook, proposing hypotheses grounded in empirical patterns rather than theoretical expectations (Jordan & Mitchell, 2015; Beam & Kohane, 2018).

Machine learning enables researchers to move beyond static assumptions by identifying non-linear and high-dimensional relationships. In biomedicine, for example, deep learning models have revealed gene–phenotype connections that traditional statistical analyses could not detect (Stokes et al., 2020). Similarly, reinforcement learning (RL) and generative models can explore vast hypothesis spaces, suggesting new experiments or possible causal mechanisms (Carlin & Louis, 2008; Azeroual, 2025).

Crucially, these AI-generated hypotheses require interpretability and validation. XAI techniques can explain how an algorithm arrived at a particular inference, while Bayesian inference frameworks quantify uncertainty, making the process more transparent and scientifically credible. The result is a hybrid approach—one that merges machine-driven discovery with human oversight, ensuring both innovation and accountability.

DATA COLLECTION AND MANAGEMENT

The collection and management of data form another critical domain of transformation. Traditional methods often involve manual aggregation, inconsistent formats, and time-consuming preprocessing. In contrast, AI automates the ingestion of diverse data sources—from IoT devices and digital repositories to social media and scientific databases (Varian, 2014; Goodfellow et al., 2016). These systems simultaneously perform data cleaning, normalization, feature extraction, and even anomaly detection, enabling higher-quality datasets with less human intervention.

Generative AI introduces the ability to create synthetic data, expanding datasets where information is scarce or sensitive, such as in clinical trials or cybersecurity.

This preserves privacy while enhancing robustness. Furthermore, AI-driven metadata extraction and knowledge graph-based organization allow seamless integration of heterogeneous datasets, supporting interdisciplinary interoperability.

However, automated data handling raises questions of data bias and provenance. If the training data contains hidden biases or incomplete records, these can propagate through AI models. Consequently, researchers must apply diagnostic AI tools to monitor and mitigate these issues, ensuring fairness, transparency, and replicability in the resulting analyses.

DATA ANALYSIS AND VISUALIZATION

AI-driven analytics transcend the limitations of traditional statistical approaches, which are typically confined to linear relationships and small datasets. Deep learning, unsupervised clustering, and reinforcement learning techniques uncover complex, non-linear patterns within high-dimensional spaces that humans could not feasibly explore manually (LeCun et al., 2015; Aggarwal, 2018).

For instance, AI can detect subtle temporal trends in climate data, hidden user behaviors in cybersecurity logs, or novel correlations in biochemical datasets. These insights inform not only predictive modeling but also theory development. Importantly, XAI frameworks such as SHAP and LIME (Ribeiro et al., 2016; Lundberg & Lee, 2017) render these complex models interpretable, allowing researchers to trace how variables contribute to specific outcomes.

Visualization has evolved in parallel. AI-enhanced platforms combine automated analysis with interactive data exploration, enabling real-time hypothesis refinement. Rather than static charts, these tools present adaptive visualizations that evolve as models learn, allowing scientists to test assumptions dynamically and gain deeper understanding from their data (Keim et al., 2013).

CODING, SIMULATION, AND EXPERIMENTAL AUTOMATION

AI's influence also extends to the implementation and execution of research. Automated code generation models like Codex and GitHub Copilot can write, optimize, and debug code, dramatically reducing development time and human error (Chen et al., 2021). In simulation-based research, AI accelerates parameter optimization and scenario generation, enabling rapid testing of hypotheses across multiple contexts (Kumar, 2024).

In laboratory environments, Robotic Process Automation (RPA) and self-driving laboratories represent the next stage of AI-enabled experimentation. These systems autonomously conduct experiments, analyze outputs, and refine subsequent trials based on AI feedback (Kaspar et al., 2021; Lee et al., 2024). Such autonomy enhances reproducibility and efficiency while maintaining human oversight for ethical and interpretive judgment.

By connecting simulation outcomes back into hypothesis generation, AI systems create continuous feedback loops that reduce trial-and-error cycles. This represents a shift toward closed-loop research ecosystems, where empirical and computational investigations reinforce one another.

INTEGRATION, INTERDISCIPLINARY SYNERGY, AND ETHICAL OVERSIGHT

AI-enhanced research operates not as a linear pipeline but as a cyclical, adaptive system in which insights from one phase continuously inform the next. For example, findings from data analysis might suggest new literature for review or inspire modifications to experimental design. These feedback loops improve agility, responsiveness, and innovation.

Interdisciplinary collaboration amplifies the impact of AI methodologies. In computing, AI optimizes software engineering and cybersecurity; in biomedicine, it aids drug discovery and precision medicine; in material science, it accelerates the identification of compounds with desired properties (Butler et al., 2018; Esteva et al., 2019). Across all domains, AI's scalability and pattern recognition foster connections between previously disparate research areas, facilitating knowledge convergence.

Nevertheless, these advances are accompanied by ethical, epistemological, and technical challenges. The "black-box" problem threatens interpretability; biases in training data can perpetuate inequities; and reliance on proprietary AI systems may compromise reproducibility (Floridi & Cowls, 2022). Ensuring responsible AI integration requires transparent documentation, open-source collaboration, and adherence to FAIR data principles (Findable, Accessible, Interoperable, Reusable).

Furthermore, researchers must guard against over-automation—the risk that human criticality is replaced by algorithmic determinism. Ethical AI in research demands a balanced partnership between computational intelligence and human reasoning, where accountability, authorship, and decision-making remain firmly in human hands.

In summary, AI-enhanced research processes are redefining the architecture of scientific inquiry. From semantic knowledge discovery to automated experimentation, AI enables new levels of efficiency, reproducibility, and creativity. It accelerates literature analysis, enhances hypothesis formulation, automates data workflows, and facilitates deep, adaptive analytics.

Yet, AI's transformative potential comes with the obligation to preserve the core values of science—transparency, reproducibility, and ethical integrity. When used responsibly, AI serves not merely as a computational tool but as a collaborative research partner, augmenting human intelligence and expanding the boundaries of discovery.

Case Studies and Applications

AI has profoundly transformed the landscape of research methodologies, not only within computing but across a broad range of scientific disciplines. Its integration into the research workflow has redefined how hypotheses are generated, data are analyzed, and knowledge is synthesized. Traditional research methods, typically dependent on manual data collection, theory-driven hypothesis formulation, and conventional statistical analysis, face inherent limitations when applied to increasingly complex, high-dimensional, and dynamic datasets (Jordan & Mitchell, 2015; Goodfellow et al., 2016). By contrast, AI-driven methodologies—spanning ML, DL, NLP, reinforcement learning (RL), and more recently generative AI—enable researchers to approach inquiry in ways that are automated, scalable, and adaptive, offering new capabilities in knowledge discovery, experimentation, and prediction.

In computing and informatics, AI functions as both an analytical engine and a creative collaborator. In software engineering, for instance, the emergence of large language models for code generation, such as Codex and GitHub Copilot, has redefined the development process by assisting programmers in producing syntactically and functionally correct code while also suggesting optimizations based on learned contextual patterns (Vaswani et al., 2017; Chen et al., 2021). Empirical studies show that such systems can reduce debugging cycles and development time by up to 40%, marking a shift toward "augmented programming," where human creativity and AI-based automation operate in a symbiotic manner. Yet, the growing reliance on AI-generated code introduces new challenges in reproducibility, bias propagation, and accountability. Generated outputs, while efficient, may contain subtle logical or ethical flaws, underscoring the necessity of human oversight and transparent validation mechanisms.

A similar transformation is evident in cybersecurity, where AI systems have outperformed traditional signature-based intrusion detection models. Machine learning-based approaches trained on network behaviour can identify emerging threats and anomalous patterns with detection rates surpassing 95% (Sommer & Paxson, 2010; Buczak & Guven, 2015). These systems excel at recognizing previously unseen attack vectors, contributing to proactive defense mechanisms in enterprise and critical infrastructure networks. The integration of graph neural networks (GNNs) has further enhanced the interpretability and relational mapping of threat dynamics.

Nevertheless, the adversarial vulnerabilities of AI models—where malicious actors can deceive systems through adversarial examples—raise new ethical and operational questions regarding security and explainability. In this domain, XAI and responsible design principles are no longer optional but fundamental to maintaining trust and reliability in automated defense infrastructures.

Human–Computer Interaction (HCI) presents another area where AI enhances both usability and adaptability. Adaptive interfaces and predictive user models are now capable of learning from user behaviour to personalise interaction patterns and optimise workflows. Systems that combine eye-tracking and reinforcement learning techniques can adjust interface layouts in real-time, significantly improving accessibility and reducing cognitive load, especially for users with disabilities (Winograd, 2006; Zouhaier et al., 2021). However, these developments necessitate new considerations around privacy, user autonomy, and informed consent. As AI systems become increasingly integrated into user interfaces, maintaining ethical integrity through transparent design becomes crucial.

Beyond computing, AI's transformative potential extends across multiple scientific and societal domains. In biomedicine, DL models process vast quantities of genomic, imaging, and clinical data to uncover disease mechanisms, predict treatment responses, and accelerate drug discovery (Esteva et al., 2019; Topol, 2019). A landmark example is DeepMind's AlphaFold 2, which solved the long-standing problem of protein structure prediction with unprecedented precision (Jumper et al., 2021). Similar applications in predictive diagnostics, such as cancer detection or cardiovascular risk modelling, demonstrate that AI-driven systems can complement medical expertise with speed and analytical depth. Still, challenges persist regarding interpretability, generalizability across populations, and ethical compliance, particularly in relation to data privacy and bias in clinical datasets. To address these limitations, researchers are increasingly adopting hybrid frameworks that combine machine inference with human validation—a paradigm known as human-in-the-loop AI—to preserve scientific accountability.

Material science offers another vivid example of AI's accelerating impact on discovery processes. Predictive modelling now enables scientists to identify compounds and materials with targeted properties, drastically reducing the time required for experimentation (Butler et al., 2018). Machine learning models analyze molecular structures, thermodynamic variables, and performance data to suggest new candidates for synthesis. Reinforcement learning further optimizes experimental design by autonomously selecting the next set of experiments to perform, guided by real-time feedback. This dynamic, self-improving process not only enhances discovery speed but also contributes to sustainability by minimizing material waste and energy consumption. Nevertheless, the interpretability of predictive models remains a critical challenge, particularly when AI suggests materials that lack clear theoretical justification. Recent approaches in neuro-symbolic AI—combining logical reasoning with DL—offer a promising direction for bridging this gap between prediction and scientific explanation.

In climate science and environmental monitoring, AI has become indispensable for analyzing large-scale data from satellites, sensors, and simulations. Machine learning algorithms are used to predict extreme weather events, model climate change dynamics, and optimize energy consumption in smart grids. AI-assisted remote sensing enables real-time monitoring of environmental degradation, deforestation, and biodiversity loss with unprecedented precision. For example, convolutional and recurrent neural networks process petabytes of Earth observation data to generate adaptive models for regional forecasting and disaster prevention. Yet, as AI-derived predictions increasingly influence policy decisions, reproducibility and transparency become vital ethical imperatives. Researchers stress the importance of open models and auditable AI systems to ensure the accountability of climate-related predictions.

The educational domain also benefits from AI-driven innovation. Intelligent tutoring systems and adaptive learning platforms employ predictive analytics to personalize instruction, identify struggling students, and dynamically adjust learning materials. Generative AI further extends these possibilities by creating tailored educational content and assessment tools. However, this growing reliance on algorithmic decision-making raises pressing questions about data privacy, algorithmic bias, and the preservation of pedagogical agency. Effective governance frameworks must therefore ensure that AI augments rather than replaces human educators, maintaining fairness and transparency within digital learning ecosystems.

Across these diverse applications, the comparative advantages of AI-enhanced methods can be analyzed through multiple dimensions. AI-driven workflows offer scalability, handling datasets of unprecedented size and complexity; speed, enabling real-time or near-real-time analysis; precision, revealing hidden patterns and correlations; and interdisciplinarity, allowing integration of heterogeneous data from various domains. At the same time, these systems face challenges related to data bias, explainability, reproducibility, and computational cost. These dynamics are concisely illustrated in Table 2, which contrasts traditional and AI-driven methodologies across core stages of research—ranging from literature review and hypothesis generation to data analysis, visualization, and experimentation. The table highlights how AI not only accelerates individual stages but also creates new forms of interconnected research workflows where insights from one phase inform and refine subsequent processes.

Table 2. Comparative Table: AI-Enhanced vs. Traditional Methods.

Research Stage	Traditional Methods	AI-Enhanced Methods	Advantages of AI	Limitations / Challenges
Literature Review & Knowledge Discovery	Manual search, keyword-based review	NLP-based text mining, semantic search	Faster, comprehensive, pattern discovery	Data bias, quality of sources
Hypothesis Formulation & Testing	Heuristic, theory-driven	ML-driven hypothesis generation, predictive modeling	Scalability, identification of complex patterns	Model interpretability, overfitting
Data Analysis & Visualization	Statistical models, spreadsheets	ML/DL analytics, automated visualization	Handles large datasets, real-time insights	Requires computational resources, expertise
Experimentation & Simulation	Manual coding, lab experiments	Automated simulation, RL-driven experiment design	Faster iterations, adaptive experimentation	Reproducibility, model assumptions
Application Domains	Limited to domain-specific expertise	Cross-domain integration, AI-assisted workflows	Interdisciplinary scalability, efficiency	Integration complexity, validation required

Moreover, as visualized in Figure 1, modern AI-driven research workflows are inherently iterative and self-correcting. Unlike the linear logic of traditional scientific inquiry, these systems operate as adaptive cycles in which models continuously learn from new data, simulations guide experimental priorities, and feedback loops refine hypotheses. This cyclical architecture enhances both efficiency and resilience, fostering a culture of continuous improvement and knowledge co-creation between humans and machines.

Figure 1. AI-Driven Stages in Modern Research Workflows.

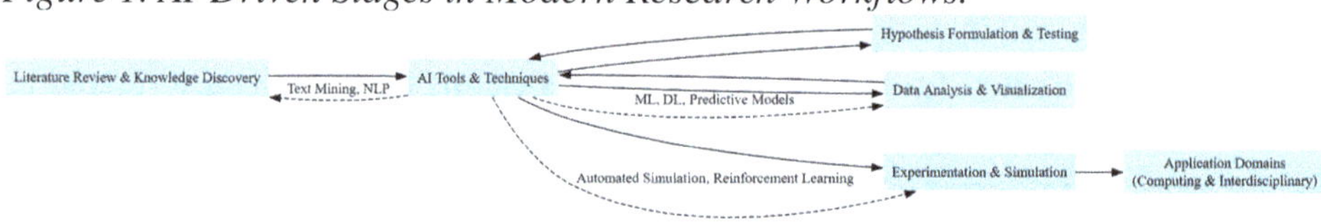

Nevertheless, the widespread integration of AI into research is not without risk. The opacity of deep learning models can obscure causal relationships, potentially undermining interpretability and scientific transparency. Biases embedded in training data may propagate social inequities or distort empirical results. Reproducibility becomes a pressing issue when AI systems depend on proprietary algorithms or datasets that cannot be independently verified. Ethical governance is therefore essential to ensure accountability, clarity of authorship, and equitable access to AI tools. To address these concerns, the scientific community is increasingly embracing frameworks grounded in Responsible AI, the FAIR principles (Findable, Accessible, Interoperable, and Reusable), and open science practices. These initiatives aim to harmonize innovation with integrity by embedding transparency, explainability, and fairness throughout the AI research pipeline.

Taken together, these empirical examples demonstrate that AI-enhanced research processes offer transformative advantages in speed, scalability, and predictive power across both computational and interdisciplinary domains. Traditional methods remain vital for conceptual grounding and human interpretation, but they are now complemented by autonomous and adaptive systems that can identify hidden relationships and accelerate discovery. The convergence of human expertise and machine intelligence defines the emerging paradigm of hybrid research ecosystems—systems that are not merely data-driven but knowledge-driven, not solely efficient but also ethically and epistemologically sound. By aligning AI's technical potential with human-centred scientific values, researchers can harness its full capabilities to confront the complexity of modern scientific challenges while preserving the rigour, transparency, and creativity at the heart of research practice.

OPPORTUNITIES AND BENEFITS

AI is fundamentally reshaping the research landscape, creating opportunities that extend far beyond technical automation. Its influence spans efficiency, scalability, reproducibility, knowledge discovery, and researcher development, offering a profound reconfiguration of how science is conducted, validated, and disseminated. These opportunities do not only transform computing disciplines but reverberate across the entire spectrum of scientific inquiry—from biomedicine and materials science to social sciences and humanities—where AI augments human creativity, accelerates discovery, and deepens methodological rigor.

A central advantage of AI-driven research methodologies lies in their capacity to radically enhance efficiency. Through automation of labor-intensive processes such as data preprocessing, algorithmic coding, and statistical analysis, AI enables researchers to devote greater attention to high-level cognitive tasks including hypothesis formulation, theoretical reasoning, and interpretation of results (Jordan & Mitchell, 2015). Tasks that traditionally required days or weeks—such as literature screening, data labeling, or code optimization—can now be performed in hours. In computational sciences, for instance, automated hyperparameter optimization and model selection allow experiments to run autonomously, significantly reducing human workload while increasing precision. This shift reflects a broader epistemological transformation from manual, sequential workflows toward continuous, adaptive research cycles supported by machine intelligence.

Equally transformative is AI's ability to scale research beyond the limits of human capacity. Modern science increasingly depends on data volume, velocity, and variety. Traditional analytical methods struggle with the complexity of high-dimensional data generated by sensors, simulations, or social platforms. AI systems, however, are designed to operate efficiently within such complexity. Deep learning architectures, distributed computing, and generative models enable the processing, analysis, and visualization of terabytes of data at speeds unattainable by manual methods (Esteva et al., 2019). In climate science, for example, AI can integrate heterogeneous satellite, environmental, and socioeconomic data to predict long-term ecological trends. In digital humanities, NLP-based models process vast corpora of historical texts, revealing thematic structures and semantic shifts that would remain invisible to conventional qualitative analysis. In both cases, AI not only handles scale but also extracts meaning, converting data abundance into actionable knowledge.

Reproducibility—one of the most persistent challenges in modern research—benefits substantially from AI's capacity for standardization and traceability. Automated data pipelines ensure that every step of data processing and model training is logged, parameterized, and documented, allowing exact replication of experimental conditions (Peng, 2011; Baker, 2016). Workflow orchestration tools such as MLflow or Data Version Control (DVC) preserve full histories of datasets, algorithms, and hyperparameters, enabling transparent comparison of results. Moreover, AI-assisted reproducibility frameworks use anomaly detection to identify inconsistencies in replication studies and flag potential errors in data handling or model interpretation. These innovations strengthen scientific integrity by embedding accountability and transparency within computational processes. However, they also underscore the importance of open data and open algorithms—without which reproducibility risks becoming confined to proprietary ecosystems.

Perhaps the most exciting promise of AI lies in its capacity to enable new forms of knowledge discovery. Generative models, including transformer architectures and diffusion networks, are capable of producing synthetic datasets, generating plausible hypotheses, and identifying hidden relationships in complex systems (Radford et al., 2019). In biomedical research, AI systems simulate experimental conditions to predict protein structures, drug-target interactions, or disease progression patterns with remarkable accuracy (Jumper et al., 2021). In materials science, predictive AI models identify molecular configurations with desired mechanical or thermal properties, guiding experimental synthesis more efficiently than random trial-and-error approaches (Butler et al., 2018). These capabilities represent a shift from descriptive to predictive and prescriptive science, where AI does not merely analyze existing phenomena but actively proposes new ones.

Crucially, these knowledge-discovery processes are iterative and interactive. AI does not simply replace the scientist but acts as an epistemic partner that amplifies human insight through computational exploration. For instance, AI can propose unconventional hypotheses derived from multi-dimensional correlations that human intuition might overlook, while researchers provide contextual judgment to evaluate their plausibility. This hybrid interplay between machine-generated inference and human reasoning enhances creativity and accelerates discovery cycles, producing more robust and innovative outcomes. Such collaborative intelligence represents the core of what scholars describe as Augmented Intelligence—a paradigm in which humans and machines co-create knowledge rather than compete for cognitive dominance (Gil et al., 2014; Davenport & Miller, 2022).

Another major opportunity arises from AI's impact on researcher training and capacity building. Intelligent tutoring systems, virtual research assistants, and AI-enabled mentoring platforms guide emerging scholars through complex methodological decisions, data analysis techniques, and literature synthesis (Madanchian & Taherdoost, 2024). By automatically recommending relevant frameworks, datasets, and analytical strategies, these systems help early-career researchers overcome steep learning curves. For instance, doctoral students using AI-based recommendation engines can identify suitable modeling approaches, statistical tests, or citation networks more efficiently than through manual exploration. This democratization of research expertise contributes to a more inclusive scientific ecosystem, where access to advanced analytical capacity is no longer limited by institutional or disciplinary boundaries.

AI also facilitates collaboration and community building. Cloud-based research environments and AI-driven collaboration platforms allow geographically dispersed teams to work synchronously, share curated datasets, and co-author documents in real time. Semantic search engines and knowledge graphs help connect researchers with complementary expertise, fostering interdisciplinary projects that bridge computing, biology, engineering, and social sciences. Such platforms exemplify the shift toward collaborative intelligence, where human networks and algorithmic systems coalesce to accelerate collective problem solving. In doing so, AI acts not only as a computational tool but as a social catalyst for research innovation.

Strategically, AI contributes to evidence-based research planning and foresight. Predictive modeling enables researchers and institutions to forecast experimental outcomes, funding trends, and emerging research areas, allowing for informed prioritization of projects with the highest potential impact (Silver ct al., 2016). Bibliometric AI tools such as OpenAlex and Semantic Scholar leverage NLP and citation network analysis to identify knowledge gaps, trace the evolution of scientific topics, and detect early signals of emerging paradigms. These insights support data-driven decision-making in research management, policy formation, and funding allocation—transforming AI into a meta-research tool that enhances the governance and coordination of science itself.

Moreover, AI supports sustainable and ethical research practices by optimizing resource use and reducing redundancy. Automated experiment design minimizes waste in materials and energy, while algorithmic efficiency reduces computational carbon footprints. In environmental science, AI models contribute to the development of sustainable solutions by optimizing experimental variables and simulating ecological interventions before physical implementation. This convergence of efficiency and responsibility aligns AI-enhanced research with broader sustainability goals and responsible innovation frameworks.

However, the full realization of these benefits depends on critical awareness and ethical integration. AI's transformative potential can only be harnessed responsibly if researchers maintain transparency, interpretability, and inclusivity in its application. Embedding XAI techniques into scientific workflows ensures that algorithmic reasoning remains interpretable, preserving the epistemic legitimacy of AI-derived insights. Similarly, implementing the FAIR principles—making data findable, accessible, interoperable, and reusable—enhances reproducibility and collaboration across the global research community. These practices transform AI from a technical utility into a cornerstone of open, transparent, and responsible science.

Finally, AI reshapes the very culture of scientific inquiry by introducing new epistemological perspectives. As machines increasingly participate in hypothesis generation, data interpretation, and even theoretical modeling, researchers are prompted to reconsider what constitutes scientific reasoning and authorship. This evolving relationship between human cognition and computational intelligence does not diminish the role of the scientist but redefines it—shifting focus from information processing to critical interpretation, from manual analysis to conceptual synthesis. The most significant opportunity, therefore, is not simply to make research faster or larger, but to make it more reflective, integrative, and anticipatory, equipped to address complex, interconnected global challenges.

In summary, the opportunities and benefits of AI-enhanced research are multilayered and far-reaching. AI enables efficiency through automation, scalability through high-dimensional computation, reproducibility through standardized workflows, and discovery through generative intelligence. It democratizes access to research expertise, fosters collaboration, informs strategic decision-making, and supports ethical and sustainable research practices. Yet, these advantages come with the responsibility to ensure that AI remains transparent, fair, and aligned with human values. If implemented thoughtfully, AI will not only accelerate science but also enrich it—transforming the pursuit of knowledge into a more dynamic, inclusive, and ethically grounded endeavor.

CHALLENGES AND LIMITATIONS

Despite the transformative potential of AI in modern research, several critical challenges and limitations persist that necessitate systematic attention to ensure the responsible and effective implementation of AI. These challenges extend across technical, methodological, ethical, and epistemological dimensions and directly influence the reliability, fairness, and transparency of AI-enhanced research processes.

One of the most persistent and widely discussed issues is the "black-box problem." Many AI systems—particularly deep neural networks—operate through multilayered, nonlinear transformations that yield powerful predictions yet provide little insight into their internal reasoning processes. While these systems achieve high levels of accuracy, their opacity poses significant difficulties for scientific validation and interpretability. When researchers cannot fully understand why an algorithm produces a certain output, the capacity for critical evaluation diminishes, potentially undermining trust in research outcomes and reproducibility.

In response, the field of XAI has emerged, aiming to make model decision-making processes more transparent. Tools such as SHAP, LIME, and attention visualization mechanisms enable partial interpretation of complex models (Lundberg & Lee, 2017). However, these methods often provide simplified or fragmented explanations that may not align with the original logic of the model. The challenge lies not only in creating explanations but in ensuring that they are meaningful, contextually accurate, and interpretable to domain experts. This tension between model complexity and interpretability underscores a broader epistemic challenge: the need to balance predictive performance with scientific understanding.

Another major limitation involves data bias and representational inequality. AI systems reflect the characteristics of their training data, meaning that any biases—whether demographic, cultural, or methodological—can propagate through research workflows and distort conclusions. For instance, biomedical AI trained predominantly on Western patient data may yield less accurate predictions for other populations, and NLP models can reproduce linguistic and gender biases present in the source texts. In scientific contexts, biased training data can skew results and reinforce systemic inequalities in global knowledge production.

To address these concerns, researchers are adopting bias auditing frameworks and fairness-aware learning algorithms, alongside transparent documentation practices such as Datasheets for Datasets. Ethical AI frameworks, including the OECD AI Principles and EU AI Act, increasingly emphasize the detection, mitigation, and reporting of bias. Nonetheless, achieving genuine fairness requires reflexivity: researchers must recognize how their own methodological choices—such as data selection, labeling, and problem framing—influence the fairness and inclusivity of their findings.

Closely related to bias is the issue of data privacy and consent, particularly in research domains involving sensitive human data such as health, behavioral, or social information. Even anonymized datasets can be vulnerable to re-identification when combined with external data sources. The use of privacy-preserving methods—such as federated learning, differential privacy, and secure multiparty computation—has gained importance in safeguarding participant autonomy and data confidentiality (Floridi & Cowls, 2022). These methods enable collective learning from distributed data without compromising privacy, aligning AI research practices with international ethical and legal standards, including the GDPR.

Beyond ethical considerations, methodological transparency and reproducibility remain central challenges. AI models often involve extensive preprocessing pipelines, hyperparameter tuning, and iterative optimization procedures that are difficult to replicate without full disclosure. The absence of standardized reporting formats can hinder peer validation and long-term reliability (Peng, 2011; Baker, 2016). To address this issue, the research community increasingly relies on open science practices, including version-controlled code repositories, shared datasets, and open-access model archives. Structured documentation tools such as Model Cards (Mitchell et al., 2019) and Dataset Statements promote reproducibility by clearly describing model behavior, data provenance, and intended use. However, widespread adoption of these practices remains inconsistent, particularly in interdisciplinary fields where standards are still evolving.

A practical constraint involves dependence on proprietary AI platforms and closed ecosystems. The most advanced AI models, such as GPT-based architectures or multimodal transformers, are developed and maintained by private corporations that restrict access through commercial licenses and APIs. This reliance limits research autonomy, as updates or policy changes by the providers can disrupt ongoing projects or invalidate reproducibility. Moreover, unequal access to such systems exacerbates disparities between well-funded institutions and smaller research entities, leading to concentration of innovation within specific geographic and economic regions. The promotion of open-source AI frameworks and publicly funded model repositories is therefore essential to democratize access and ensure scientific sovereignty.

A further challenge relates to the balance between AI-driven automation and human oversight. While automation enhances efficiency and scalability, excessive reliance on algorithmic decision-making risks diminishing the role of human interpretation. Automated tools can identify patterns but lack contextual awareness and ethical judgment. Consequently, errors or misleading correlations may go unnoticed if researchers fail to critically evaluate AI outputs. Effective human–AI collaboration requires human-in-the-loop systems, where algorithmic results are continuously reviewed, validated, and contextualized by domain experts. Human oversight ensures accountability, enabling researchers to correct biases, reinterpret findings, and maintain methodological rigor.

AI's increasing integration into research also raises epistemological challenges. Traditional scientific reasoning emphasizes causality, transparency, and theoretical grounding, whereas AI often prioritizes correlation, prediction, and optimization. This paradigm shift risks privileging empiricism over explanation—where the ability to predict outcomes may overshadow the goal of understanding underlying mechanisms. Such a transformation requires careful calibration: researchers must combine data-driven models with domain theory to ensure that AI augments rather than replaces causal reasoning. Otherwise, scientific inquiry risks becoming dependent on pattern recognition without explanatory depth.

Moreover, resource intensity and environmental impact present growing concerns. Training large-scale models requires significant computational power, specialized hardware, and substantial energy consumption. Studies estimate that the carbon footprint of training a single large AI model can rival that of hundreds of transatlantic flights. This raises critical questions regarding the sustainability of AI-driven research. The emerging paradigm of Green AI advocates for more energy-efficient algorithms, model compression, and transparent reporting of computational costs to align AI innovation with environmental responsibility. Integrating sustainability metrics into research evaluation criteria could help balance technological advancement with ecological stewardship.

At an institutional level, the skills and literacy gap remains a significant barrier to effective AI adoption. The integration of machine learning, data analytics, and computational modeling into traditional research disciplines demands technical expertise that many researchers currently lack. Without sufficient understanding of AI methodologies, researchers risk misinterpreting results or applying algorithms inappropriately. Addressing this limitation requires systemic investment in AI education, interdisciplinary training, and collaborative infrastructures that bridge technical and domain expertise. Initiatives such as AI research schools, open-access tutorials, and institutional capacity-building programs are instrumental in cultivating a scientifically literate and ethically aware research community.

Another emerging challenge concerns accountability and governance. As AI-generated results become integral to publications, questions arise about authorship, intellectual responsibility, and ethical liability. Determining who is accountable for errors in AI-assisted research—whether the developer, the operator, or the algorithm itself—remains unresolved. Moreover, discussions persist about whcther AI systems should be recognized as research contributors or remain tools under human direction. Addressing these governance questions requires clear institutional policies, standardized authorship criteria, and the inclusion of algorithmic accountability mechanisms within ethical review processes.

The governance issue also extends to peer review and evaluation frameworks. Traditional peer review processes may be ill-equipped to assess AI-based research outputs, particularly when results depend on proprietary algorithms or massive computational resources unavailable to reviewers. Developing new peer-review standards that account for AI reproducibility, data ethics, and model transparency is essential for maintaining scientific integrity in the AI era.

Finally, infrastructural and collaboration challenges persist in integrating AI within interdisciplinary research. Many institutions lack adequate computational resources, standardized data infrastructures, or collaborative frameworks to support large-scale AI projects. Fragmented data silos, incompatible metadata standards, and inadequate interoperability hinder the potential of AI to connect research across domains. Establishing shared infrastructures, such as federated data platforms and research knowledge graphs, can mitigate these limitations by enabling data sharing, model reuse, and cross-institutional collaboration.

Addressing these challenges requires more than technical solutions—it demands a systemic rethinking of research governance, ethics, and methodology. Technical progress must be matched with transparency, accountability, and inclusivity to ensure that AI enhances, rather than compromises, the credibility and trustworthiness of scientific inquiry.

ETHICAL AND EPISTEMOLOGICAL CONSIDERATIONS

The deployment of AI into research processes introduces profound ethical and epistemological questions that extend beyond technical performance or methodological efficiency. These considerations influence how research is conducted, how results are interpreted, how credit and responsibility are assigned, and how scientific knowledge itself is conceptualized. Addressing them is essential to ensure that AI-driven research remains transparent, accountable, and aligned with the foundational principles of scientific inquiry.

A central ethical dilemma concerns the division of cognitive labor between humans and AI systems. While AI can automate data collection, pattern recognition, and hypothesis generation, the human researcher remains the epistemic authority responsible for formulating research questions, interpreting findings, and maintaining methodological rigor. Striking an appropriate balance between automation and human oversight is therefore crucial. Overreliance on algorithmic outputs can erode critical reflection, reduce epistemic accountability, and risk delegating evaluative judgment to systems that lack interpretive understanding. Human expertise must remain central—not only to validate results but also to contextualize them within theoretical and ethical frameworks that AI cannot grasp.

The issue of accountability and authorship becomes increasingly complex in AI-supported research. When AI systems generate hypotheses, perform analyses, or contribute textually to manuscripts, traditional notions of authorship and intellectual responsibility are challenged. Researchers must specify the extent of AI involvement to maintain ethical transparency and avoid misrepresentation. International publishing standards such as those by COPE and the ICMJE now require authors to declare whether generative AI was used in drafting or data analysis. Beyond authorship, questions of intellectual property and data ownership arise, particularly when proprietary AI models or third-party datasets are employed. These tools often operate as black boxes with unclear data provenance, creating uncertainty around the traceability and reproducibility of findings. Transparent disclosure of data sources, model provenance, and algorithmic dependencies is therefore essential for ethical compliance and scientific integrity.

From a theoretical perspective, AI challenges established epistemological frameworks of knowledge creation. Traditional science is grounded in human reasoning, intentionality, and theoretical explanation. In contrast, AI operates through statistical optimization, correlation detection, and large-scale pattern recognition. This shift raises the question of whether AI's outputs constitute genuine knowledge or merely refined forms of information processing. If AI autonomously identifies hypotheses or generates insights beyond direct human comprehension, does it act as a co-producer of knowledge or remain a sophisticated instrument under human control?. While some scholars conceptualize AI as an "epistemic agent," it fundamentally lacks consciousness, intentionality, and moral responsibility. Its contributions therefore depend on human interpretation to transform computational outputs into meaningful, contextually grounded knowledge.

These epistemological tensions also highlight a potential shift from causal explanation to predictive modeling. AI excels at forecasting outcomes but often provides limited insight into why they occur. This predictive paradigm risks marginalizing theory-driven research in favor of empirically successful but conceptually opaque models. The challenge for contemporary science lies in integrating AI's predictive strength with explanatory depth—ensuring that machine-derived correlations are critically examined through human reasoning and theoretical reflection. Such integration requires epistemic humility: recognizing both the power and the limits of algorithmic inference within the broader landscape of human understanding.

Ethically, the deployment of AI in research extends beyond laboratory boundaries to encompass societal consequences and collective accountability. AI models are embedded with assumptions and values reflected in their training data and design choices. These latent biases can reinforce existing inequalities or generate new forms of epistemic exclusion—for example, by privileging data-rich disciplines or under-representing non-Western perspectives in global knowledge systems. Consequently, researchers bear a moral obligation to assess how their methodological decisions—such as data selection, labeling practices, or model tuning—may influence not only scientific accuracy but also social justice and inclusivity.

Ensuring ethical transparency in AI research thus involves more than technical auditing; it requires procedural and cultural reforms. Frameworks such as the FAIR principles (Findable, Accessible, Interoperable, Reusable) and Responsible Research and Innovation (RRI) approaches promote openness and accountability by embedding ethical reflection into every stage of the research lifecycle. Similarly, emerging standards for XAI contribute to interpretability and foster public trust, even when full transparency of deep models remains technically unattainable. However, ethical assurance cannot rely solely on compliance mechanisms; it must become a normative commitment shared by all participants in the research ecosystem.

The responsibility gap between human and machine agency further complicates ethical evaluation. When AI-driven systems produce erroneous analyses or biased results, attributing accountability becomes difficult. Is the responsibility borne by the developer, the researcher who applied the model, or the institution that endorsed its use? Without clear governance structures, accountability may diffuse, weakening both ethical responsibility and institutional trust. To mitigate this, several organizations advocate for the inclusion of algorithmic accountability statements in publications, detailing not only the technical performance of AI models but also their ethical implications, decision boundaries, and potential limitations.

Moreover, AI's growing participation in research forces a reconsideration of epistemic authority and autonomy. If AI-generated results increasingly shape theory development, literature reviews, or policy recommendations, the question arises: who—or what—defines the boundaries of credible knowledge? Safeguarding epistemic pluralism requires ensuring that AI does not narrow scientific inquiry to data patterns aligned with dominant paradigms or corporate datasets. Maintaining diversity in data sources, theoretical perspectives, and methodological approaches is essential to prevent the homogenization of scientific reasoning.

Finally, the ethical horizon of AI in research extends to public engagement and societal trust. Scientific legitimacy depends not only on internal rigor but also on the credibility of research in the eyes of society. When AI contributes to discoveries, the public must be confident that these results are transparent, fair, and beneficial. Researchers therefore carry a dual obligation: to advance innovation responsibly and to communicate the limits of AI with honesty and clarity. This includes acknowledging uncertainty, addressing ethical dilemmas openly, and promoting literacy about how AI participates in the research process.

In sum, ethical and epistemological reflection is not an auxiliary concern but a structural necessity for the sustainable integration of AI in research. Human judgment, transparency, and reflexivity remain indispensable in transforming algorithmic capabilities into genuine scientific insight—ensuring that the pursuit of knowledge through AI continues to serve both truth and society.

FUTURE RESEARCH DIRECTIONS

The future of AI in research promises to redefine the foundations of scientific inquiry by transforming how knowledge is generated, validated, and shared. The coming decade will witness increasing automation, deeper collaboration between human researchers and intelligent systems, and the integration of AI into educational and institutional frameworks. These developments are not merely technological but structural, reshaping the very culture and organization of research across disciplines.

A major frontier in this evolution is the rise of self-driving laboratories and autonomous research environments. These facilities, powered by AI and robotics, can independently design, execute, and refine experiments without continuous human supervision. By coupling real-time data acquisition with adaptive learning algorithms, such systems can iteratively optimize experimental conditions, explore vast parameter spaces, and even propose new research directions based on preliminary results. The outcome is a continuous feedback loop of experimentation and discovery—far faster than traditional manual workflows. Beyond efficiency, these laboratories foster a new epistemic paradigm in which discovery emerges from the dynamic interaction between machine learning systems, robotic automation, and human strategic oversight. Early implementations in materials science and synthetic biology already demonstrate substantial gains in research throughput and reproducibility, suggesting that self-driving labs may become foundational infrastructures in data-intensive disciplines.

Closely related to this development is the emergence of AI-augmented research teams, where humans and AI systems operate as integrated collaborators rather than in hierarchical roles. In these hybrid teams, AI handles data preprocessing, predictive modeling, and large-scale simulations, while human researchers focus on interpretation, theory-building, and ethical judgment. This complementary interaction combines computational precision with conceptual creativity, enabling teams to tackle complex, interdisciplinary problems that exceed the cognitive limits of individuals or purely algorithmic systems. The collaboration extends beyond efficiency: it cultivates new forms of reasoning that blend intuition with algorithmic exploration, transforming how scientific hypotheses are conceived and validated. Such configurations will likely become central to "augmented science," where human expertise and machine intelligence coevolve to expand the boundaries of discovery.

The integration of AI into education and research management represents another critical trajectory. As AI becomes embedded in all stages of the research lifecycle, academic institutions will need to prepare the next generation of scientists for AI-driven environments. Curricula in graduate and doctoral programs must increasingly emphasize computational literacy, algorithmic thinking, and data ethics, alongside traditional research design and critical reasoning. Training programs that blend theory with hands-on experience in AI-assisted research will cultivate researchers capable of designing, auditing, and interpreting AI-driven studies responsibly. At the institutional level, research management systems may evolve toward intelligent infrastructures that assist in grant evaluation, data stewardship, and performance assessment. Funding agencies are already beginning to prioritize projects that integrate AI methodologies, not only for their scientific promise but also for their potential to enhance transparency, efficiency, and interdisciplinarity in research ecosystems.

These transformations also call for a reconfiguration of policy frameworks and governance mechanisms. As AI's role in research expands, new standards must be established to regulate algorithmic transparency, data stewardship, and reproducibility. Policymakers and scientific bodies need to address the risks of dependency on proprietary systems and unequal access to computational resources, which could exacerbate disparities between institutions or nations. Open-source models, interoperable data infrastructures, and shared computational platforms can mitigate such inequalities by democratizing access to AI-driven research capabilities. At the same time, regulatory frameworks should promote accountability through algorithmic auditing, documentation of decision pipelines, and ethical review processes for AI-assisted studies.

Several open research questions remain unresolved and will shape the next phase of AI integration. How can interpretability be guaranteed in increasingly complex models without sacrificing predictive power? What forms of human–machine interaction yield the most epistemically reliable outcomes? How should responsibility and authorship be defined when AI contributes substantively to discovery? These questions require interdisciplinary inquiry, drawing insights from computer science, philosophy of science, law, and ethics. The answers will determine not only how AI reshapes the mechanics of research but also how the scientific community conceptualizes creativity, authorship, and epistemic authority in the age of intelligent systems.

Ultimately, the future of AI in research will depend on maintaining a productive balance between automation and agency, ensuring that machine intelligence amplifies rather than replaces human judgment. As AI systems assume greater roles in experimentation, data interpretation, and even hypothesis formation, the responsibility of scientists will shift toward curating, validating, and contextualizing algorithmic insights. The success of this transformation will rely on fostering cultures of transparency, collaboration, and reflexivity—where technological progress aligns with ethical integrity and the pursuit of meaningful, socially beneficial knowledge.

CONCLUSION

The examination of AI-driven methodologies in computing and informatics reveals a profound transformation in how research is conceived, conducted, and interpreted. The integration of artificial intelligence across the research lifecycle—spanning literature review, hypothesis generation, data analysis, simulation, and experimental automation—marks a decisive shift from linear, manual processes toward adaptive, data-rich, and self-optimizing research ecosystems. AI no longer functions merely as an analytical instrument; it acts as a co-evolving agent within the research process, amplifying human capabilities and reshaping epistemic practices across disciplines.

Across the domains explored, AI demonstrates an unprecedented capacity to enhance efficiency, scalability, and insight generation. Automated text mining, semantic search, and knowledge graph construction accelerate literature review and knowledge discovery, enabling researchers to navigate exponentially growing scientific outputs with greater precision. Machine learning and deep learning models detect complex, nonlinear relationships in large-scale datasets, while generative and reinforcement learning techniques support hypothesis formulation and predictive modelling beyond traditional statistical boundaries. Moreover, automation in coding, simulation, and laboratory experimentation reduces repetitive manual work, enhances reproducibility, and enables rapid iteration cycles, thereby transforming how empirical inquiry unfolds in practice.

These developments carry significant implications for the future of computing and informatics. AI expands the scope of problems that researchers can address, facilitating the exploration of multifaceted questions in software engineering, cybersecurity, human–computer interaction, and data science, as well as in interdisciplinary fields such as biomedicine and materials research. Through AI-augmented analytical pipelines and predictive frameworks, scientists can uncover previously hidden correlations, model complex systems, and test hypotheses at scales and speeds unattainable by human effort alone. This capacity to bridge disciplines positions AI as a foundational catalyst for cross-domain innovation.

Equally transformative is the shift in the role of the researcher. As AI systems increasingly handle computational and procedural complexity, human expertise becomes centered on conceptual reasoning, interpretive judgment, and ethical governance. Researchers evolve from data operators to strategic orchestrators of AI-driven workflows, responsible for ensuring transparency, reliability, and social accountability. This shift underscores the importance of methodological literacy and ethical reflection as essential competencies in future scientific education. Accordingly, universities and research institutions must integrate AI literacy, data ethics, and algorithmic accountability into curricula, preparing the next generation of scientists to engage critically with intelligent systems rather than depend passively on them.

At the same time, the integration of AI into research raises crucial epistemological and ethical considerations. Questions of interpretability, fairness, and responsibility remain central to the legitimacy of AI-assisted discoveries. Ensuring that AI systems are transparent, auditable, and aligned with scientific values is vital to preserving trust and reproducibility in research outcomes. Moreover, the democratization of AI infrastructure—through open-source tools, equitable access to computational resources, and international collaboration—will determine whether the benefits of AI-augmented research are distributed inclusively or concentrated within privileged institutions and economies.

Looking ahead, the research landscape is entering an era characterized by deep human–AI collaboration. Emerging paradigms such as self-driving laboratories, AI-augmented research teams, and intelligent research management systems illustrate the potential of autonomous yet accountable science. These environments foster continuous learning loops, where data and hypotheses evolve dynamically through feedback between human insight and machine adaptation. In this "rewired" ecosystem, AI does not replace the human intellect but magnifies it—enhancing creativity, expanding the frontiers of analysis, and accelerating the translation of knowledge into societal impact.

Ultimately, the convergence of AI and computing research heralds a paradigm grounded in three interdependent values: innovation, rigour, and responsibility. Innovation drives the exploration of new methods and domains; rigour ensures transparency, reproducibility, and epistemic integrity; and responsibility safeguards ethical alignment and societal benefit. Together, these principles define the contours of a new scientific paradigm—one that is adaptive, data-driven, and inherently interdisciplinary, positioning AI not as an auxiliary tool but as an essential partner in shaping the future of knowledge production.

REFERENCES

Aggarwal, C. C. (2018). Machine learning for text: An introduction. In *Machine learning for text* (pp. 1–16). Springer International Publishing. DOI: 10.1007/978-3-319-73531-3_1

Alpaydin, E. (2020). *Introduction to machine learning* (4th ed.). MIT Press.

Amershi, S., Weld, D., Vorvoreanu, M., Fourney, A., Nushi, B., Collisson, P., & Horvitz, E. (2019, May). Guidelines for human-AI interaction. In *Proceedings of the 2019 chi conference on human factors in computing systems* (pp. 1-13).

Azeroual, O. (2025). Beyond Black Boxes: Adaptive XAI for Dynamic Data Pipelines. In *Proceedings of the 17th International Joint Conference on Knowledge Discovery, Knowledge Engineering and Knowledge Management* - (Volume 2), pp. 428-437. DOI: 10.5220/0013736100004000

Baker, M. (2016). 1,500 scientists lift the lid on reproducibility. *Nature, 533*(7604), 452–454. DOI: 10.1038/533452a

Beam, A. L., & Kohane, I. S. (2018). Big data and machine learning in health care. *Journal of the American Medical Association, 319*(13), 1317–1318. DOI: 10.1001/jama.2017.18391

Beltagy, I., Lo, K., & Cohan, A. (2019). SciBERT: A pretrained language model for scientific text. *arXiv preprint arXiv:1903.10676*. DOI: 10.18653/v1/D19-1371

Bornmann, L., & Mutz, R. (2015). Growth rates of modern science: A bibliometric analysis based on the number of publications and cited references. *Journal of the Association for Information Science and Technology, 66*(11), 2215–2222. DOI: 10.1002/asi.23329

Bubeck, S., Chandrasekaran, V., Eldan, R., Gehrke, J., Horvitz, E., Kamar, E., . . . Zhang, Y. (2023). Sparks of artificial general intelligence: Early experiments with GPT-4. *arXiv preprint arXiv:2303.12712*.

Buczak, A. L., & Guven, E. (2015). A survey of data mining and machine learning methods for cyber security intrusion detection. *IEEE Communications Surveys and Tutorials, 18*(2), 1153–1176. DOI: 10.1109/COMST.2015.2494502

Butler, K. T., Davies, D. W., Cartwright, H., Isayev, O., & Walsh, A. (2018). Machine learning for molecular and materials science. *Nature, 559*(7715), 547–555. DOI: 10.1038/s41586-018-0337-2

Carlin, B. P., & Louis, T. A. (2008). *Bayesian methods for data analysis*. CRC press. DOI: 10.1201/b14884

Chen, M., Tworek, J., Jun, H., Yuan, Q., Pinto, H. P. D. O., Kaplan, J., . . . Zaremba, W. (2021). Evaluating large language models trained on code. *arXiv preprint arXiv:2107.03374.*

Davenport, T. H., & Miller, S. (2022). *Working with AI: Real stories of human-machine collaboration*. MIT Press. DOI: 10.7551/mitpress/14453.001.0001

Devlin, J., Chang, M. W., Lee, K., & Toutanova, K. (2019). Bert: Pre-training of deep bidirectional transformers for language understanding. In *Proceedings of the 2019 conference of the North American chapter of the association for computational linguistics: human language technologies, volume 1 (long and short papers)* (pp. 4171-4186).

Doshi-Velez, F., & Kim, B. (2017). Towards a rigorous science of interpretable machine learning. *arXiv preprint arXiv:1702.08608.*

Esteva, A., Robicquet, A., Ramsundar, B., Kuleshov, V., DePristo, M., Chou, K., Cui, C., Corrado, G., Thrun, S., & Dean, J. (2019). A guide to deep learning in healthcare. *Nature Medicine*, *25*(1), 24–29. DOI: 10.1038/s41591-018-0316-z

Fister, I., & Perc, M. (2016). Toward the discovery of citation cartels in citation networks. *Frontiers in Physics (Lausanne)*, *4*, 240569. DOI: 10.3389/fphy.2016.00049

Floridi, L., & Chiriatti, M. (2020). GPT-3: Its nature, scope, limits, and consequences. *Minds and Machines*, *30*(4), 681–694. DOI: 10.1007/s11023-020-09548-1

Floridi, L., & Cowls, J. (2022). A unified framework of five principles for AI in society. *Machine learning and the city: Applications in Architecture and Urban Design*, 535-545.

Gil, Y., Greaves, M., Hendler, J., & Hirsh, H. (2014). Amplify scientific discovery with artificial intelligence. *Science*, *346*(6206), 171–172. DOI: 10.1126/science.1259439

Goodfellow, I., Bengio, Y., Courville, A., & Bengio, Y. (2016). Deep Learning: Vol. 1. *No. 2*. MIT press.

Jardim, P. S. J., Rose, C. J., Ames, H. M., Echavez, J. F. M., Van de Velde, S., & Muller, A. E. (2022). Automating risk of bias assessment in systematic reviews: A real-time mixed methods comparison of human researchers to a machine learning system. *BMC Medical Research Methodology*, *22*(1), 167. DOI: 10.1186/s12874-022-01649-y

Ji, S., Pan, S., Cambria, E., Marttinen, P., & Yu, P. S. (2021). A survey on knowledge graphs: Representation, acquisition, and applications. *IEEE Transactions on Neural Networks and Learning Systems*, *33*(2), 494–514. DOI: 10.1109/TNNLS.2021.3070843

Jordan, M. I., & Mitchell, T. M. (2015). Machine learning: Trends, perspectives, and prospects. *Science*, *349*(6245), 255–260. DOI: 10.1126/science.aaa8415

Jumper, J., Evans, R., Pritzel, A., Green, T., Figurnov, M., Ronneberger, O., Tunyasuvunakool, K., Bates, R., Žídek, A., Potapenko, A., Bridgland, A., Meyer, C., Kohl, S. A. A., Ballard, A. J., Cowie, A., Romera-Paredes, B., Nikolov, S., Jain, R., Adler, J., & Hassabis, D. (2021). Highly accurate protein structure prediction with AlphaFold. *Nature*, *596*(7873), 583–589. DOI: 10.1038/s41586-021-03819-2

Kaspar, C., Ravoo, B. J., van der Wiel, W. G., Wegner, S. V., & Pernice, W. H. (2021). The rise of intelligent matter. *Nature*, *594*(7863), 345–355. DOI: 10.1038/s41586-021-03453-y

Keim, D., Qu, H., & Ma, K. L. (2013). Big-data visualization. *IEEE Computer Graphics and Applications*, *33*(4), 20–21. DOI: 10.1109/MCG.2013.54

King, R. D., Rowland, J., Oliver, S. G., Young, M., Aubrey, W., Byrne, E., & Whelan, K. E. (2009). The automation of science. *Science*, *324*(5923), 85–89. DOI: 10.1126/science.1165620

Kumar, A. (2024). The role of simulation and modeling in artificial intelligence: A review. *International Journal of Modeling, Simulation and Scientific Computing*, *15*(03), 2430002. DOI: 10.1142/S1793962324300024

LeCun, Y., Bengio, Y., & Hinton, G. (2015). Deep learning. *Nature*, *521*(7553), 436–444. DOI: 10.1038/nature14539

Lee, S., Kim, J., & Park, H. (2024). AI-assisted scientific discovery: From data to hypotheses. *Nature Machine Intelligence*, *6*(1), 45–55.

Lundberg, S. M., & Lee, S. I. (2017). A unified approach to interpreting model predictions. *Advances in Neural Information Processing Systems*, •••, 30.

Madanchian, M., & Taherdoost, H. (2024). AI-Powered Innovations in High-Tech Research and Development: From Theory to Practice. *Computers, Materials & Continua*, *81*(2), 2133–2159. DOI: 10.32604/cmc.2024.057094

Marshall, I. J., & Wallace, B. C. (2019). Toward systematic review automation: A practical guide to using machine learning tools in research synthesis. *Systematic Reviews*, *8*(1), 163. DOI: 10.1186/s13643-019-1074-9

Ofori-Boateng, R., Aceves-Martins, M., Wiratunga, N., & Moreno-Garcia, C. F. (2024). Towards the automation of systematic reviews using natural language processing, machine learning, and deep learning: A comprehensive review. *Artificial Intelligence Review, 57*(8), 200. DOI: 10.1007/s10462-024-10844-w

Peng, R. D. (2011). Reproducible research in computational science. *Science, 334*(6060), 1226–1227. DOI: 10.1126/science.1213847

Radford, A., Wu, J., Child, R., Luan, D., Amodei, D., & Sutskever, I. (2019). Language models are unsupervised multitask learners. *OpenAI blog, 1*(8), 9.

Ribeiro, M. T., Singh, S., & Guestrin, C. (2016). " Why should I trust you?" Explaining the predictions of any classifier. In *Proceedings of the 22nd ACM SIGKDD International Conference on Knowledge Discovery and Data Mining* (pp. 1135-1144). DOI: 10.18653/v1/N16-3020

Russell, S., & Norvig, P. (2015). *Artificial intelligence: A modern approach* (4th ed.). Pearson.

Sallam, M. (2023). The utility of ChatGPT as an example of large language models in healthcare education, research and practice: Systematic review on the future perspectives and potential limitations. MedRxiv, 2023-02.

Schmidhuber, J. (2015). Deep learning in neural networks: An overview. *Neural Networks, 61*, 85–117. DOI: 10.1016/j.neunet.2014.09.003

Silver, D., Huang, A., Maddison, C. J., Guez, A., Sifre, L., Van Den Driessche, G., Schrittwieser, J., Antonoglou, I., Panneershelvam, V., Lanctot, M., Dieleman, S., Grewe, D., Nham, J., Kalchbrenner, N., Sutskever, I., Lillicrap, T., Leach, M., Kavukcuoglu, K., Graepel, T., & Hassabis, D. (2016). Mastering the game of Go with deep neural networks and tree search. *Nature, 529*(7587), 484–489. DOI: 10.1038/nature16961

Sommer, R., & Paxson, V. (2010). Outside the closed world: On using machine learning for network intrusion detection. In *2010, IEEE Symposium on Security and Privacy* (pp. 305-316). IEEE.

Stokes, J. M., Yang, K., Swanson, K., Jin, W., Cubillos-Ruiz, A., Donghia, N. M., MacNair, C. R., French, S., Carfrae, L. A., Bloom-Ackermann, Z., Tran, V. M., Chiappino-Pepe, A., Badran, A. H., Andrews, I. W., Chory, E. J., Church, G. M., Brown, E. D., Jaakkola, T. S., Barzilay, R., & Collins, J. J. (2020). A deep learning approach to antibiotic discovery. *Cell, 181*(2), 475–483. DOI: 10.1016/j.cell.2020.04.001

Topol, E. (2019). *Deep medicine: how artificial intelligence can make healthcare human again.*

Tsafnat, G., Glasziou, P., Choong, M. K., Dunn, A., Galgani, F., & Coiera, E. (2014). Systematic review automation technologies. *Systematic Reviews, 3*(1), 74. DOI: 10.1186/2046-4053-3-74

Van Dis, E. A., Bollen, J., Zuidema, W., Van Rooij, R., & Bockting, C. L. (2023). ChatGPT: Five priorities for research. *Nature, 614*(7947), 224–226. DOI: 10.1038/d41586-023-00288-7

Varian, H. R. (2014). Big data: New tricks for econometrics. *The Journal of Economic Perspectives, 28*(2), 3–28. DOI: 10.1257/jep.28.2.3

Vaswani, A., Shazeer, N., Parmar, N., Uszkoreit, J., Jones, L., Gomez, A. N., & Polosukhin, I. (2017). Attention is all you need. *Advances in Neural Information Processing Systems*, •••, 30.

Winograd, T. (2006). Shifting viewpoints: Artificial intelligence and human–computer interaction. *Artificial Intelligence, 170*(18), 1256–1258. DOI: 10.1016/j.artint.2006.10.011

Wohlin, C., Runeson, P., Höst, M., Ohlsson, M. C., Regnell, B., & Wesslén, A. (2012). *Experimentation in software engineering* (Vol. 236). Springer. DOI: 10.1007/978-3-642-29044-2

Zouhaier, L., Hlaoui, Y. B. D., & Ayed, L. B. (2021). A reinforcement learning based approach of context-driven adaptive user interfaces. In *2021, IEEE 45th Annual Computers, Software, and Applications Conference (COMPSAC)* (pp. 1463-1468). IEEE. DOI: 10.1109/COMPSAC51774.2021.00217

Chapter 5

Research Redefined AI Tools, Trends, and Innovations in the Modern Era

Mangubek Tursunbayevich Urazmetov
http://orcid.org/0009-0004-2155-7809
Urgench State University, Uzbekistan

Shakhlo Samanova
http://orcid.org/0009-0004-2155-7809
Mamun University, Uzbekistan

Sirojbek Tursunbayevich Sultanov
Tashkent State Transport University, Uzbekistan

Munir Ahmad
http://orcid.org/0000-0003-4836-6151
Survey of Pakistan, Pakistan

ABSTRACT

This chapter delves into key AI tools and methodologies that propel advancements across various domains. ML techniques enable systems to learn from data autonomously by employing supervised learning and unsupervised learning. Deep learning, leveraging neural networks, excels in medical image analysis to enhance diagnostic accuracy. Reinforcement learning enables AI for tasks like robot navigation with Q-learning optimizing strategies. Generative AI innovates by generating new data, aiding drug discovery through synthetic protein structures. Transfer learning enhances NLP tasks like sentiment analysis by adapting models efficiently. Big data

DOI: 10.4018/979-8-2600-0434-0.ch005

analytics tools manage extensive datasets, facilitating genomic research in identifying disease markers. High-performance computing resources like GPUs and TPUs accelerate AI computations, while explainable AI frameworks ensure transparency in applications such as healthcare diagnostics. These tools drive interdisciplinary AI innovation, emphasizing ethical and impactful applications from healthcare to climate modeling.

INTRODUCTION

The discovery area of AI is now rapidly progressing more than ever before and has signaled a significant shift in almost every aspect of our existence (Bhattamisra et al., 2023; Grilli & Pedota, 2024; Hermann & Hermann, 2022; López-Chila et al., 2024; Neumann et al., 2024; Nitin & Gupta, 2024; Ruiz-Real et al., 2021). Such a rise is driven by continuous attempts in research to uncover more effective tooling and generally enhanced toolkits that have not been seen before. Nevertheless, managing an expanding number of promotions and rapidly changing the nature of the marketplace environment requires a keen view of current trends and innovations. But at the core of this imperative is research and development and ever-evolving problem-solving. In recent years, new tools and methodologies have emerged right from their buds, and on the other hand, the existing set of methodologies is also evolving and developing. Thus, there is this unending cycle that requires researchers to really do a lot of discovering and readjusting all the time. It is not just a question of informing oneself on such advances, rather it is a matter of strategic necessity to fully capitalize on the potential.

It, therefore, becomes important to explore it as a requirement of functionality in the contemporary world. As we contemplate the widespread adoption of AI in various industries, research settings, and in society, it remains imperative to keep in pace with new developments for maximum impact. The need for an enhanced understanding of the changing landscape of AI research and the progress that has already been made is imperative when it comes to optimizing business processes, developing new hypotheses for scientific investigations, or approaching social problems (Baum, 2021; Doumpos et al., 2023; Liu et al., 2021; Parashar et al., 2023; Rubinger et al., 2023; Williams et al., 2023). This chapter seeks to explore the importance of assessing the current research on AI as a guide towards its future direction. The acquisition of this knowledge would allow the researchers to make informed decisions about what area to focus on in their research and the methods to use in attaining the results that lead to the progress of this field of research.

The objective of this chapter is to provide an overview of the current landscape of AI in research. The chapter is structured to provide a clear and concise exploration of the topic. It begins by introducing the fundamental concepts of AI in research and its role in driving overall AI progress. Subsequently, the chapter explores specific tools and methodologies currently shaping the field. It then examines emerging trends and recent innovations, highlighting their potential impact. Finally, it concludes by discussing the challenges and limitations associated with these advancements and offers considerations for future research directions.

TOOLS AND METHODOLOGIES

This section explores the essential tools and methodologies that empower AI.

Machine Learning (ML) Techniques

Machine learning forms the backbone of many advancements in artificial intelligence (AI). The ability to learn from data and improve performance over time without being explicitly programmed is what makes ML so powerful. Several key ML techniques have emerged, each with unique capabilities and applications. These techniques may include supervised learning, deep learning, reinforcement learning, and generative AI, among others.

Deep Learning

Deep learning is inspired by the structure and function of the human brain, specifically the neural networks that comprise it. This technique utilizes artificial neural networks with multiple layers (hence "deep") to learn complex patterns from data. The multiple layers allow the network to learn increasingly abstract representations of the input data, making deep learning particularly effective for tasks that require understanding high-dimensional data (Goodfellow et al., 2016; Vasilev et al., 2019).

Researchers may use deep learning to develop an AI model that can analyze medical images and detect signs of disease with high accuracy. For instance, in radiology, a deep learning model can be trained on thousands of labeled images of X-rays or MRIs. Through the training process, the model learns to identify features associated with various conditions, such as tumors or fractures. By using convolutional neural networks (CNNs), which are particularly well-suited for image processing tasks, the model can achieve diagnostic performance comparable to or even exceeding that of human radiologists. This capability not only enhances diagnostic accuracy but also significantly reduces the time required to interpret medical images, thereby improving patient outcomes.

Reinforcement Learning (RL)

RL differs from other ML techniques in that it allows AI models to learn through trial and error within a simulated environment. Instead of learning from a labeled dataset, the model learns to make decisions by receiving feedback in the form of rewards or penalties (Joshi et al., 2021; Kiran et al., 2022). This technique is particularly valuable for tasks that require an agent to interact with an environment, such as robotics or game playing.

Researchers may use RL to train an AI model to control a virtual robot, allowing it to learn optimal strategies for completing specific tasks within that environment. For instance, consider training a robot to navigate complex terrains. The robot starts with no knowledge of the environment and gradually learns to optimize its movements by receiving rewards for successful navigation and penalties for obstacles encountered. Over time, through techniques such as Q-learning or policy gradients, the robot develops a policy that maximizes its cumulative reward, effectively learning the most efficient paths and maneuvers. This approach can be extended to real-world applications, such as autonomous vehicles, where the RL-trained model can help the vehicle learn to navigate safely and efficiently in varied traffic conditions.

Generative AI

Generative AI focuses on creating entirely new data, such as images, text, or even code. This branch of AI is powered by models that learn the underlying distribution of the input data and can generate new samples from this learned distribution. Generative models, including generative adversarial networks (GANs) and variational autoencoders (VAEs), are powerful tools for research and practical applications, enabling tasks like generating synthetic data for training other AI models or creating realistic simulations for testing purposes (Alqahtani et al., 2021; Kalota, 2024; Ooi et al., 2023; Ramdurai & Adhithya, 2023).

Researchers may use generative AI to create synthetic datasets of protein structures, accelerating drug discovery research by allowing them to test potential new drugs on a vast array of virtual molecules. In this context, GANs could be trained on existing protein structure data to generate new, plausible protein configurations. These synthetic proteins can then be used to simulate how new drugs might interact with them, providing valuable insights into potential efficacy and safety before any actual laboratory work is conducted. This use of generative AI not only speeds up the drug discovery process but also reduces costs and the need for extensive physical experiments.

Supervised Learning

Supervised learning involves training an AI model on labeled data, where the model learns to map input data to the desired output (Arundel et al., 2020; Debener et al., 2023). The "supervision" comes from the labeled dataset used during training, which provides the correct answers for the model to learn from. This technique is widely used in tasks such as classification and regression. Suppose researchers want to predict housing prices based on features like location, size, and number of rooms. By using a supervised learning algorithm such as linear regression or decision trees, they can train a model on historical housing data where each instance is labeled with its actual selling price. Once trained, the model can accurately predict prices for new houses based on their features, helping real estate agents and buyers make informed decisions.

Unsupervised Learning

Unsupervised learning involves training on data without labeled responses. The goal is to explore the structure of the data and extract meaningful insights or patterns (Debener et al., 2023; Usama et al., 2019). Clustering and association are common tasks within unsupervised learning. In customer segmentation, unsupervised learning techniques like k-means clustering can group customers based on similarities in purchasing behavior without any predefined labels. By analyzing transaction data from a retail business, the algorithm identifies clusters of customers who exhibit similar buying patterns. This segmentation helps marketers tailor marketing strategies and promotions to different customer segments, thereby improving customer satisfaction and maximizing revenue.

Transfer Learning

Transfer learning involves leveraging knowledge gained from solving one problem to help solve a different but related problem (Iman et al., 2023; Panigrahi et al., 2021; Zhuang et al., 2021). In the context of neural networks, transfer learning allows pre-trained models to be adapted for new tasks with less data and training time. In natural language processing (NLP), transfer learning has been revolutionized by models like BERT (Bidirectional Encoder Representations from Transformers). Researchers can fine-tune a pre-trained BERT model on a specific NLP task, such as sentiment analysis or question answering, using a relatively small dataset. This approach accelerates model development and improves performance compared to training from scratch, making it a valuable technique in areas where labeled data is limited or costly to acquire.

Semi-Supervised Learning

Semi-supervised learning combines elements of supervised and unsupervised learning. It uses a small amount of labeled data and a large amount of unlabeled data to improve learning accuracy (Song et al., 2023; van Engelen & Hoos, 2020; Yang et al., 2023). In speech recognition, semi-supervised learning techniques can enhance the accuracy of transcribing spoken language. By training a model on a small set of transcribed audio samples and a vast collection of unlabeled speech data, the model can generalize better to different accents and speech variations. This approach reduces the need for extensive manual transcription and improves the overall robustness of the speech recognition system.

Big Data Analytics Tools

The ever-growing volume of data presents both challenges and opportunities for research in AI. Big data analytics tools are essential for handling, processing, and deriving insights from large datasets. These tools enable researchers to efficiently manage and analyze massive amounts of data, leading to more robust and sophisticated AI models. Key big data analytics tools include Apache Spark, TensorFlow, Hadoop, and Apache Flink.

Apache Spark

Apache Spark is an open-source distributed processing framework designed to handle and analyze massive datasets across clusters of computers. It provides an interface for programming entire clusters with implicit data parallelism and fault tolerance. Spark's capabilities make it ideal for tasks that require processing large-scale data quickly and efficiently (Akil et al., 2017; Longkumer & Hussain Mazumder, 2024; Nga & Thuy, 2024; Ullah et al., 2024).

Researchers might use Apache Spark to process large-scale genomic data. By leveraging Spark's distributed computing capabilities, they can efficiently perform complex analyses, such as identifying genetic variants associated with diseases. For instance, analyzing millions of DNA sequences to find common genetic markers that correlate with a particular disease would be computationally intensive. Spark's in-memory processing ensures faster data retrieval and computation, enabling researchers to gain valuable insights from vast genomic datasets that would be impractical to analyze using traditional methods. This capability accelerates the discovery of genetic predispositions to diseases, aiding in the development of personalized medicine.

TensorFlow

TensorFlow is a popular open-source platform developed by Google, providing a flexible framework for building, training, and deploying ML models. It supports a wide range of tasks, from image recognition to natural language processing, and is designed to run on various platforms, including CPUs, GPUs, and TPUs. TensorFlow's versatility and powerful libraries make it a preferred tool for many researchers and developers (Joseph et al., 2021; Lee et al., 2021; Pang et al., 2020; Vasilev et al., 2019).

Researchers might use TensorFlow to develop a deep learning model for automatic speech recognition (ASR). By leveraging TensorFlow's extensive libraries and tools, they can preprocess audio data, build and train a neural network, and deploy the model to convert spoken language into text. For instance, the model can be trained on a large dataset of speech recordings annotated with their corresponding transcripts. TensorFlow's scalability allows the model to be trained on large datasets, improving its accuracy and robustness in understanding diverse speech patterns and accents. This capability is crucial for developing ASR systems that can function effectively in real-world applications, such as virtual assistants and transcription services.

Hadoop

Hadoop is an open-source framework that enables the distributed processing of large datasets across clusters of computers using simple programming models. It is designed to scale up from single servers to thousands of machines, each offering local computation and storage. Hadoop's robust architecture makes it suitable for handling vast amounts of unstructured data (Akil et al., 2017; Bekker, 2017; Guan et al., 2024; Ma et al., 2023; Ullah et al., 2024).

Researchers might use Hadoop to analyze social media data for sentiment analysis. By distributing the data processing tasks across a Hadoop cluster, they can efficiently handle and analyze vast amounts of social media posts to determine public sentiment on various topics. For instance, using Hadoop's MapReduce programming model, they can process and analyze millions of tweets or Facebook posts to gauge public reaction to a new product launch or a political event. This analysis helps companies and policymakers understand public opinion trends and make data-driven decisions.

Apache Flink

Apache Flink is an open-source stream-processing framework that performs stateful computations over data streams. It is designed for both batch and stream processing, providing high throughput and low-latency performance. Flink's real-time data processing capabilities make it ideal for applications that require immediate data analysis and response (Akil et al., 2017; Henning & Hasselbring, 2024; Ullah et al., 2024).

Researchers might use Apache Flink to monitor and analyze real-time sensor data from an IoT network. By processing the continuous stream of data with Flink, they can detect anomalies or patterns in real time, enabling prompt responses to potential issues. For instance, in a smart city, sensors deployed across infrastructure can send real-time data about traffic, air quality, or energy consumption. Flink's capability to handle both batch and stream processing allows for comprehensive analysis, integrating historical data with real-time insights for more informed decision-making. This application can help city planners optimize traffic flow, reduce pollution, and improve energy efficiency, enhancing the quality of urban life.

Figure 1 shows the comparative analysis of the key features of big data analytics tools.

Figure 1. Comparative Analysis of Big Data Analytics Tools

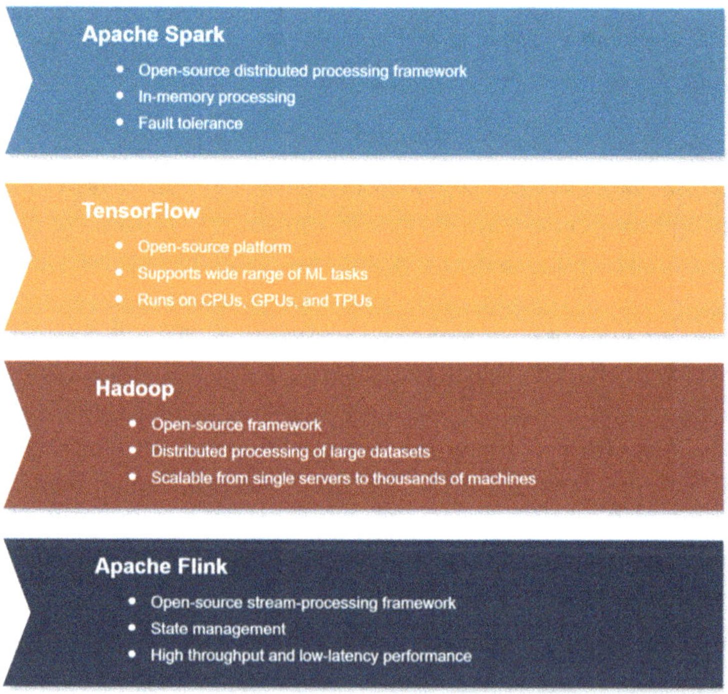

High-Performance Computing (HPC) Resources

The complex computations involved in training and running AI models often require significant processing power. High-performance computing (HPC) resources address this need by providing the necessary computational capabilities to handle these demanding tasks. HPC resources are essential for enabling the development, training, and deployment of sophisticated AI models at scale. Key HPC resources include Graphics Processing Units (GPUs), Tensor Processing Units (TPUs), Field-Programmable Gate Arrays (FPGAs), and Supercomputers.

Graphics Processing Units (GPUs)

Originally designed for graphics processing, GPUs have proven highly effective in accelerating AI computations due to their parallel processing capabilities. Unlike traditional Central Processing Units (CPUs), which are optimized for sequential processing, GPUs can handle thousands of operations simultaneously, making them ideal for the matrix and vector operations commonly used in AI algorithms (Alahmadi et al., 2020; Aqib et al., 2019).

Researchers can leverage GPUs to train complex AI models significantly faster than using traditional CPUs. For instance, in the field of deep learning, training a convolutional neural network (CNN) for image classification tasks involves numerous matrix multiplications and convolutions. Using a GPU, such as the NVIDIA Tesla V100, researchers can distribute these operations across thousands of cores, drastically reducing the training time from weeks to days or even hours. This acceleration is crucial for iterative experimentation and fine-tuning of models, enabling faster development cycles and more rapid advancements in AI research.

Tensor Processing Units (TPUs)

TPUs are specialized AI accelerators designed specifically for machine learning tasks. Developed by companies like Google, TPUs offer even greater processing power than GPUs for specific types of AI workloads. TPUs are optimized for the high-throughput requirements of machine learning, providing enhanced performance and efficiency for training and inference of deep learning models (Munanday et al., 2023; Rech Junior et al., 2022).

Researchers may use TPUs to accelerate the training of large-scale natural language processing (NLP) models. For example, training a transformer-based model like BERT (Bidirectional Encoder Representations from Transformers) requires significant computational resources due to the model's large number of parameters and the complexity of its operations. By utilizing Google's TPU Pods, which connect multiple TPUs to work together, researchers can reduce the training time from several weeks to just a few days. This capability allows for rapid experimentation and deployment of state-of-the-art NLP models in applications such as text summarization, translation, and sentiment analysis.

Field-Programmable Gate Arrays (FPGAs)

FPGAs are integrated circuits that can be configured by the user after manufacturing. They offer a customizable and reconfigurable computing solution that can be optimized for specific tasks. FPGAs provide a balance between the flexibility of CPUs and the performance of specialized accelerators like GPUs and TPUs (Choi, 2023; Podlubne & Gohringer, 2023; Trimberger & Moore, 2014). Researchers may use FPGAs for real-time data processing in autonomous vehicles. The computational requirements for processing sensor data from cameras, LiDAR, and radar in real time are substantial. FPGAs can be programmed to perform specific operations, such as object detection and sensor fusion, with high efficiency and low latency. This capability ensures that autonomous vehicles can make quick and accurate decisions based on their environment, enhancing safety and performance.

Supercomputers

Supercomputers are high-performance computing systems that consist of thousands of processors working in parallel. They are designed to perform extremely complex computations at unprecedented speeds, making them invaluable for large-scale simulations and data-intensive research. Researchers may use supercomputers to simulate climate models and predict future weather patterns. Climate modeling involves solving complex mathematical equations that describe atmospheric processes over long periods and across large geographic areas. Supercomputers, such as those at the National Center for Atmospheric Research (NCAR), can perform these calculations at a scale and speed that is impossible with conventional computing resources. By using supercomputers, researchers can improve the accuracy of climate predictions, inform policy decisions, and better understand the impacts of climate change.

High-Performance Computing Clusters

HPC clusters consist of interconnected computers that perform parallel processing tasks. These clusters can scale from a few nodes to thousands, providing a flexible and scalable solution for various scientific and engineering applications. Researchers can use HPC clusters for genomics research, such as sequencing and analyzing large sets of DNA data. By distributing the computational workload across multiple nodes in the cluster, tasks like genome assembly, variant calling, and annotation can be performed much more quickly and efficiently. This capability accelerates the pace of genomic research, leading to faster discoveries in fields such as personalized medicine and evolutionary biology. Figure 2 shows the comparative analysis of the key features of high-performance computing resources.

Figure 2. Comparative Analysis of HPC Resources

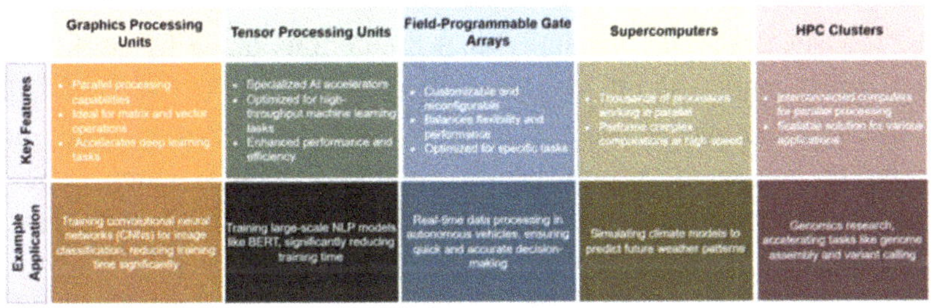

Explainable AI (XAI) Frameworks

As AI models become more sophisticated and integrated into critical decision-making processes, understanding how these models arrive at their outputs becomes crucial. Explainable AI (XAI) frameworks address this challenge by providing tools and techniques to make AI model behavior more transparent and interpretable. These frameworks are essential for building trust, ensuring accountability, and identifying potential biases in AI systems. Key XAI frameworks include SHAP (SHapley Additive exPlanations), LIME (Local Interpretable Model-agnostic Explanations), and others such as Integrated Gradients, Counterfactual Explanations, and Anchors.

SHAP (SHapley Additive exPlanations)

SHAP is a game-theoretic approach to explain the output of machine learning models. It helps researchers understand the contribution of individual features in a dataset to the model's predictions. SHAP values are based on the concept of Shapley values from cooperative game theory, which fairly distributes the “payout” (in this case, the prediction) among the “players” (features) (Lundberg, 2019; Wang et al., 2022; Younisse et al., 2022).

Suppose researchers are using a machine learning model to predict loan approval based on features such as credit score, income, employment status, and loan amount.

SHAP can be used to explain how each feature contributes to the prediction for a specific applicant. For instance, SHAP values might reveal that a high credit score and stable employment significantly contribute to a positive loan approval decision, while a high loan amount might have a negative impact. By visualizing these contributions, researchers can better understand the model's decision-making process, ensuring that it aligns with fair lending practices and regulatory requirements.

LIME (Local Interpretable Model-agnostic Explanations)

LIME focuses on explaining individual predictions made by an AI model. It works by approximating the model locally around the prediction to be explained with an interpretable model, such as a linear regression. This approach provides insights into the reasoning behind specific outputs, helping researchers build trust in the model's accuracy and reliability (Cartolano et al., 2024; Chen et al., 2023; Mehta & Passi, 2022).

Consider a healthcare application where an AI model predicts the likelihood of a patient developing a particular disease based on medical records. Using LIME, researchers can explain the model's prediction for an individual patient. For example, if the model predicts a high risk of diabetes, LIME might show that factors such as high BMI, age, and family history of diabetes are significant contributors to this prediction. This localized explanation helps healthcare providers understand and trust the AI model's output, and also provides actionable insights for patient care.

Integrated Gradients

Integrated Gradients is a method for attributing the prediction of a deep learning model to its input features. It computes the gradients of the prediction output with respect to the input features, integrating these gradients over a path from a baseline input to the actual input (Pourdarbani et al., 2023; Schwegler et al., 2023). This technique is particularly useful for interpreting complex neural networks. In image recognition tasks, researchers can use Integrated Gradients to identify which pixels in an image contributed most to the model's classification decision. For instance, if a model identifies a dog in an image, Integrated Gradients can highlight the specific regions of the image (such as the dog's ears and eyes) that were most influential in making this decision. This visualization helps researchers verify that the model is focusing on relevant features and not being misled by irrelevant background details.

Anchors

Anchors are rules that sufficiently explain a prediction and are robust to small perturbations in the input (X. Zhang et al., 2022). These rules act as "anchors" that hold the prediction stable and provide clear, interpretable conditions under which the model makes its predictions. In a sentiment analysis application, an anchor for a positive review prediction might be the presence of words like "excellent" and "highly recommend." These words serve as strong indicators that anchor the model's prediction. Researchers can use these anchors to understand which features consistently lead to specific predictions, ensuring the model's decisions are based on meaningful and stable patterns.

KEY TRENDS AND RECENT INNOVATIONS

The landscape of research AI is constantly in flux, driven by exciting trends and groundbreaking advancements.

AutoML

AutoML tools are revolutionizing AI research by automating various stages of the machine learning workflow. Traditionally, this process involved significant expertise in data preparation, model selection, hyperparameter tuning, and evaluation. AutoML tools simplify these steps, allowing researchers with less experience to leverage the power of AI effectively. By automating these complex tasks, AutoML enhances productivity, improves model performance, and reduces the time required to deploy AI solutions (Brändle et al., 2022; Ferreira et al., 2021; Mumuni & Mumuni, 2024; Paladino et al., 2023; Schwen et al., 2022; Weerts et al., 2024).

AutoML democratizes AI research, making it more accessible to a wider range of researchers from diverse scientific backgrounds. By lowering the barrier to entry, AutoML enables scientists, engineers, and analysts without extensive machine learning expertise to develop sophisticated models. This inclusivity fosters innovation and collaboration across various fields, accelerating the pace of discovery and enabling the application of AI to a broader array of problems. For example, an ecologist with limited AI expertise can leverage AutoML to develop a model that analyzes satellite imagery and predicts deforestation patterns in a specific region. The ecologist can input raw satellite images into an AutoML tool, which then handles data preprocessing, feature extraction, model selection, and hyperparameter tuning. The resulting model can identify areas at high risk of deforestation with high accuracy, providing valuable insights for conservation efforts and policy-making. Some of the key AutoML Tools are described below.

Google AutoML

Google AutoML offers a suite of machine-learning tools that enable users to build custom models with minimal effort. It supports various tasks, including image classification, object detection, natural language processing, and tabular data analysis. For example, a healthcare researcher could use Google AutoML to develop a model that predicts disease outbreaks based on environmental and epidemiological data. By uploading the relevant datasets, the researcher can utilize Google AutoML to automatically preprocess the data, select the best model architecture, and tune the hyperparameters. This results in a high-performing model that can forecast disease outbreaks, aiding in proactive public health measures.

H2O.ai

H2O.ai provides an open-source platform for automatic machine learning, offering tools like H2O AutoML and Driverless AI. These tools automate the entire machine-learning pipeline, from data ingestion to model deployment. For instance, a financial analyst can use H2O AutoML to develop a credit scoring model. By feeding historical financial data into H2O AutoML, the tool automatically handles data cleaning, feature engineering, model training, and validation. The analyst can then deploy the optimized model to assess the creditworthiness of loan applicants, improving the decision-making process in lending institutions.

Auto-Sklearn

Auto-Sklearn is an automated machine-learning library built on top of the popular Scikit-learn library. It leverages ensemble learning and Bayesian optimization to find the best machine-learning pipelines. For example, a marketing researcher could employ Auto-Sklearn to predict customer churn. By providing customer data, such as purchase history and engagement metrics, Auto-Sklearn automates the selection and tuning of algorithms to create a robust predictive model. This model helps the researcher identify at-risk customers, allowing the company to implement targeted retention strategies.

Microsoft Azure AutoML

Microsoft Azure AutoML is a cloud-based service that automates the machine learning lifecycle. It supports a range of machine learning tasks, including classification, regression, and time series forecasting. For instance, an energy sector researcher can use Azure AutoML to forecast energy consumption patterns. By inputting historical energy usage data, weather conditions, and other relevant factors, Azure AutoML automates the creation of a predictive model. This model helps utility companies optimize energy distribution, manage resources more efficiently, and reduce operational costs.

Figure 3 shows the comparative analysis of the strengths of these AutoML tools. AutoML tools democratize AI research by automating complex tasks, making it accessible to researchers with varying expertise levels. This inclusivity fosters innovation and accelerates the application of AI across diverse fields, from healthcare to finance, marketing, and energy. Each tool has its unique strengths, catering to specific needs and applications, ultimately enhancing productivity and model performance.

Figure 3. Comparative Analysis of AutoML Tools

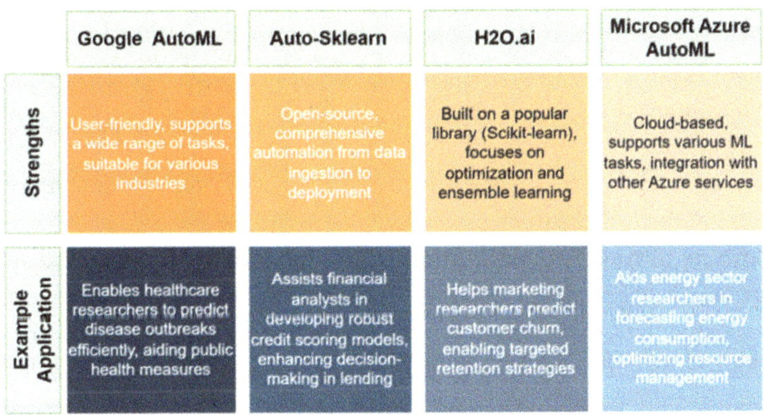

Cross-Pollination of AI with Other Fields

AI techniques are increasingly applied to diverse fields, enhancing understanding and capabilities through mutual learning and adaptation (Langley et al., 2022; Macpherson et al., 2021; Myers et al., 2024; Neupane et al., 2024; Prakash et al., 2024; Silva et al., 2024; Zhou et al., 2024).

Neuroscience and AI

AI is revolutionizing neuroscience by enabling the analysis of complex brain activity data. Machine learning algorithms can process vast amounts of neural data, extracting patterns and correlations that contribute to our understanding of brain function and behavior. Insights gained from AI-driven analyses can inform hypotheses about neural processes and aid in the development of more accurate models of brain function.

Neuroscientists might use AI techniques to analyze functional magnetic resonance imaging (fMRI) data to map brain regions involved in specific cognitive tasks, such as memory retrieval or language processing. Machine learning algorithms can identify patterns in brain activity that correspond to different mental states or behaviors, providing insights into how neural circuits operate. This knowledge not only enhances our understanding of the brain but also facilitates the development of AI systems that mimic cognitive functions or assist in diagnosing neurological disorders.

Robotics and AI

The synergy between robotics and AI has led to significant advancements in autonomous systems and human-robot interaction. AI algorithms enable robots to perceive and interpret their environment, make decisions in real-time, and adapt to dynamic conditions. In turn, insights from robotics—such as biomechanics and human factors—inform the development of AI models that enhance robot performance and interaction capabilities.

Researchers in robotics might integrate AI for computer vision and reinforcement learning to enable robots to navigate complex environments autonomously. By analyzing visual data from onboard cameras and using AI algorithms to interpret surroundings, robots can identify obstacles, plan optimal paths, and perform tasks with precision. This integration not only improves the efficiency and safety of robotic operations but also expands the applications of AI-driven automation in industries such as manufacturing, logistics, and healthcare.

Natural Language Processing (NLP) and AI

The combination of AI with NLP has transformed how we interact with and understand language. AI-powered NLP models can analyze, interpret, and generate human language at scale, facilitating applications in translation, sentiment analysis, chatbots, and content generation. Conversely, linguistic theories and insights into language structure and semantics inform the development of more sophisticated NLP algorithms.

Researchers may apply AI and NLP to analyze vast amounts of scientific literature to extract knowledge and insights. By developing AI models capable of understanding technical jargon, identifying key concepts, and summarizing complex information, researchers can accelerate the discovery of new scientific findings. For instance, AI-driven analysis of medical research papers can uncover correlations between genetic markers and disease susceptibility, guiding biomedical research and personalized medicine initiatives.

CHALLENGES AND LIMITATIONS

Despite the remarkable progress in research AI, significant challenges and limitations remain. Addressing these issues is crucial for ensuring the responsible and ethical development of AI.

Data Availability and Bias

The accessibility and quality of data are fundamental challenges in training effective AI models, impacting their performance, generalization, and potential biases (Draghi et al., 2024; Onyebuchi Nneamaka Chisom et al., 2024; Radanliev, 2024). Addressing these challenges involves strategies to ensure sufficient, representative data while mitigating biases that can perpetuate societal inequalities. Training robust AI models typically demands large volumes of high-quality data. However, acquiring such data can be resource-intensive, particularly for research groups with limited resources. Moreover, data bias poses a critical issue, where skewed or incomplete datasets may lead to AI models that exhibit unfair or discriminatory behavior, particularly against underrepresented groups.

To overcome these challenges and foster more inclusive and accurate AI applications, several strategies and solutions are employed:

Data Augmentation Techniques

Data augmentation involves generating synthetic data from existing datasets, thereby expanding the volume and diversity of training data available. Techniques such as image rotation, cropping, and color manipulation in computer vision or text paraphrasing and augmentation in natural language processing can enhance model robustness and reduce overfitting. For example, in medical imaging, researchers might use data augmentation to create variations of existing X-ray images by flipping, rotating, or adjusting contrast levels. This augmentation technique helps train AI models to recognize medical conditions more effectively, even with limited original data.

Focus on Diverse Datasets

Actively seeking and utilizing datasets that encompass diverse demographics, contexts, and scenarios is crucial for training AI models that generalize well across different populations. Incorporating data from various socioeconomic backgrounds, ethnicities, genders, and geographic regions helps mitigate biases and ensures that AI systems are fair and equitable. For example, in financial services, using diverse datasets that include information from different income brackets and regions helps develop credit scoring models that are more inclusive and less biased towards specific demographic groups. This approach supports fair lending practices and reduces the risk of systemic discrimination.

Building Fairness Metrics

Developing and integrating fairness metrics into the AI development pipeline is essential for identifying and addressing biases within models. Fairness metrics evaluate the performance of AI systems across different subgroups defined by sensitive attributes such as race, gender, or age. By quantitatively measuring fairness, researchers can detect disparities and adjust algorithms to promote equitable outcomes. For instance, in hiring practices, AI tools can be evaluated using fairness metrics to ensure that candidates from all demographic backgrounds receive equitable consideration. Metrics such as disparate impact analysis can identify biases in recruitment algorithms and prompt adjustments to mitigate discrimination and promote diversity.

Computational Bottlenecks

The advancement of AI models is often hindered by computational bottlenecks, where the complexity of computations exceeds the capabilities of current hardware and resources (Cai et al., 2022; Sufyan et al., 2023; Y. Zhang et al., 2023). Overcoming these challenges requires innovative solutions to expand computational capacity and efficiency. The complexity of training and running cutting-edge AI models demands substantial computational resources. Current hardware limitations can constrain the development of more sophisticated models, hindering progress in AI research and application. To address computational bottlenecks and enable the development of advanced AI models, researchers and engineers employ several strategic solutions:

Hardware Advancements

Continuous advancements in hardware technologies, such as next-generation Graphics Processing Units (GPUs) and Tensor Processing Units (TPUs), are crucial for expanding computational capabilities. These specialized processors are designed to accelerate specific types of AI computations, such as matrix operations in deep learning models. For example, researchers developing large-scale language models like GPT-3 rely on advanced GPUs from NVIDIA to handle massive amounts of training data and complex model architectures efficiently. These GPUs enable faster training times and support the deployment of more complex AI applications in natural language processing.

Efficient Algorithm Development

Researchers are actively exploring and developing more efficient AI algorithms that reduce the computational resources required for model training and inference. Optimization techniques, algorithmic advancements, and model compression methods contribute to making AI models more computationally economical without sacrificing performance. For instance, in computer vision, researchers have developed efficient convolutional neural network architectures like MobileNet and EfficientNet. These models are designed to achieve high accuracy with fewer parameters and computations, making them suitable for deployment on resource-constrained devices like smartphones and embedded systems.

Cloud-Based Computing Platforms

Leveraging cloud-based computing platforms offers researchers access to scalable and cost-effective high-performance computing resources. Cloud providers offer on-demand access to GPUs, TPUs, and large-scale computing clusters, enabling researchers to overcome local hardware limitations and accelerate AI model development. For example, AI startups and academic research teams use cloud platforms like Amazon Web Services (AWS), Google Cloud Platform (GCP), and Microsoft Azure for training deep learning models. These platforms provide scalable compute instances optimized for AI workloads, facilitating rapid experimentation and deployment of AI applications.

Ethical Considerations

The rapid advancement of AI research brings forth ethical challenges that require careful consideration to ensure the responsible development, deployment, and use of

AI technologies (Kleinrichert, 2024; Rajpurohit & Dobhada, 2024). Addressing these concerns involves implementing frameworks, guidelines, and technologies that uphold principles of fairness, transparency, privacy, and human oversight. As AI technologies become increasingly integrated into various facets of society, several ethical concerns emerge: AI systems often require vast amounts of data to train and operate effectively. Protecting individuals' privacy while utilizing sensitive data poses significant challenges, particularly regarding data anonymization and consent. Moreover, AI's capabilities raise concerns about potential misuse, including issues of bias, discrimination, and unintended consequences in decision-making processes. Further, ensuring AI systems make decisions fairly and transparently is crucial to avoid reinforcing biases or perpetuating inequalities in diverse societal contexts. To address these ethical challenges and promote responsible AI development, researchers and policymakers are exploring various considerations and solutions.

Privacy-Preserving AI Techniques

Implementing privacy-preserving techniques allows AI models to be trained on sensitive data while preserving individual privacy. Techniques such as federated learning, differential privacy, and encrypted computation enable data to be analyzed without exposing sensitive information to unauthorized parties. For example, healthcare researchers can use federated learning to collaborate on medical research without sharing patient data directly. Each hospital trains an AI model on its local data, and only encrypted model updates are shared, ensuring patient privacy while benefiting from collective insights.

Developing Ethical Guidelines

Establishing ethical guidelines and frameworks is essential for guiding AI research, development, and deployment. These guidelines outline principles for fairness, accountability, transparency, and the responsible use of AI technologies across industries and applications. For example, organizations like the IEEE and the European Commission have developed ethical guidelines for AI that emphasize transparency, accountability, and human oversight. These guidelines help ensure AI technologies are developed and deployed in ways that benefit society while minimizing potential risks.

Human Oversight and Control

Maintaining human oversight and control mechanisms within AI systems is crucial for mitigating risks associated with AI decision-making processes. Humans provide critical judgment, ethical reasoning, and intervention capabilities to ensure AI decisions align with ethical norms and societal values. For example, autonomous vehicle technologies incorporate fail-safe mechanisms that allow human drivers to override AI decisions in critical situations. This human-in-the-loop approach ensures safety and ethical considerations are prioritized in autonomous driving systems.

FUTURE RESEARCH DIRECTIONS

Future research in the field of AI-driven research methodologies should focus on advancing human–AI collaborative frameworks, establishing standardized evaluation and validation mechanisms for AI-generated knowledge, and developing transparent and ethically grounded models that ensure accountability and trust. Further inquiry is needed into how AI can accelerate interdisciplinary research by integrating diverse data sources and enabling new forms of knowledge discovery at scale, while also examining its cognitive and pedagogical impacts on researcher training, academic identity, and scholarly practices. Additionally, localized and context-aware AI systems must be prioritized to ensure inclusion and relevance across diverse cultural and linguistic settings, alongside exploring sustainable and energy-efficient model architectures to reduce the environmental footprint of large-scale AI applications. Finally, future work should consider how AI insights can be effectively translated into real-world problem solving and policy decision-making, particularly in domains such as governance, public health, climate adaptation, and education.

CONCLUSION

The intersection of advanced tools and methodologies in AI has ushered in a transformative era across various disciplines. From foundational machine learning techniques that power predictive analytics and image recognition to cutting-edge deep learning models revolutionizing medical diagnostics and autonomous systems, AI continues to reshape how we approach complex problems. The integration of high-performance computing resources and big data analytics tools has democratized access to computational power, enabling researchers to tackle unprecedented challenges at scale. Moreover, the development of explainable AI frameworks ensures transparency and accountability in AI decision-making, addressing critical ethical considerations. As these technologies evolve, their synergistic impact across domains underscores AI's potential to drive innovation, improve decision-making processes, and enhance societal well-being, paving the way for a future where intelligent systems play a central role in advancing human progress.

Recent trends and innovations in research AI reflect a dynamic landscape driven by advancements in AutoML and interdisciplinary applications across fields like neuroscience, robotics, and natural language processing (NLP). AutoML tools such as Google AutoML, H2O.ai, Auto-Sklearn, and Microsoft Azure AutoML are democratizing AI by automating complex tasks and enabling researchers from diverse backgrounds to develop sophisticated models effectively. These tools are pivotal in accelerating discoveries and applications, from disease outbreak prediction to credit scoring and customer churn analysis. Moreover, the cross-pollination of AI with neuroscience is enhancing our understanding of brain function, while AI-driven robotics is revolutionizing autonomous systems. In NLP, AI models are transforming language analysis and interaction, driving advances in translation, sentiment analysis, and content generation. Despite these strides, challenges like data bias and computational bottlenecks remain critical, demanding solutions such as diverse dataset incorporation and hardware advancements to ensure ethical and inclusive AI development.

REFERENCES

Akil, B., Zhou, Y., & Röhm, U. (2017). On the usability of Hadoop MapReduce, Apache Spark & Apache flink for data science. *Proceedings - 2017 IEEE International Conference on Big Data, Big Data 2017, 2018-January*. DOI: 10.1109/BigData.2017.8257938

Alahmadi, S., Mohammed, T., Albeshri, A., Katib, I., & Mehmood, R. (2020). Performance analysis of sparse matrix-vector multiplication (Spmv) on graphics processing units (gpus). *Electronics (Basel)*, *9*(10), 1675. Advance online publication. DOI: 10.3390/electronics9101675

Alqahtani, H., Kavakli-Thorne, M., & Kumar, G. (2021). Applications of Generative Adversarial Networks (GANs): An Updated Review. *Archives of Computational Methods in Engineering*, *28*(2), 525–552. Advance online publication. DOI: 10.1007/s11831-019-09388-y

Aqib, M., Mehmood, R., Alzahrani, A., Katib, I., Albeshri, A., & Altowaijri, S. M. (2019). Smarter traffic prediction using big data, in-memory computing, deep learning and gpus. *Sensors (Basel)*, *19*(9), 2206. Advance online publication. DOI: 10.3390/s19092206 PMID: 31086055

Arundel, S. T., Li, W., & Wang, S. (2020). GeoNat v1.0: A dataset for natural feature mapping with artificial intelligence and supervised learning. *Transactions in GIS*, *24*(3), 556–572. Advance online publication. DOI: 10.1111/tgis.12633

Baum, S. D. (2021). Artificial Interdisciplinarity: Artificial Intelligence for Research on Complex Societal Problems. *Philosophy & Technology*, *34*(S1), 45–63. Advance online publication. DOI: 10.1007/s13347-020-00416-5

Bekker, A. (2017). *Spark vs. Hadoop MapReduce: Which big data framework to choose*. ScienceSoft.

Bhattamisra, S. K., Banerjee, P., Gupta, P., Mayuren, J., Patra, S., & Candasamy, M. (2023). Artificial Intelligence in Pharmaceutical and Healthcare Research. In *Big Data and Cognitive Computing* (Vol. 7, Issue 1). DOI: 10.3390/bdcc7010010

Brändle, S., Hanussek, M., Blohm, M., & Kintz, M. (2022). Evaluation of Representation Models for Text Classification with AutoML Tools. *Lecture Notes in Networks and Systems, 359 LNNS*. DOI: 10.1007/978-3-030-89880-9_24

Cai, H., Lin, J., Lin, Y., Liu, Z., Tang, H., Wang, H., Zhu, L., & Han, S. (2022). Enable Deep Learning on Mobile Devices: Methods, Systems, and Applications. *ACM Transactions on Design Automation of Electronic Systems*, *27*(3), 1–50. Advance online publication. DOI: 10.1145/3486618

Cartolano, A., Cuzzocrea, A., & Pilato, G. (2024). Analyzing and assessing explainable AI models for smart agriculture environments. *Multimedia Tools and Applications*, *83*(12), 37225–37246. Advance online publication. DOI: 10.1007/s11042-023-17978-z

Chen, K. Y., Lee, H. C., Lin, T. C., Lee, C. Y., & Ho, Z. P. (2023). Deep Learning Algorithms with LIME and Similarity Distance Analysis on COVID-19 Chest X-ray Dataset. *International Journal of Environmental Research and Public Health*, *20*(5), 4330. Advance online publication. DOI: 10.3390/ijerph20054330 PMID: 36901338

Chisom, O. N., Biu, P. W., Umoh, A. A., Obaedo, B. O., Adegbite, A. O., & Abatan, A. (2024). Reviewing the role of AI in environmental monitoring and conservation: A data-driven revolution for our planet. *World Journal of Advanced Research and Reviews*, *21*(1), 161–171. Advance online publication. DOI: 10.30574/wjarr.2024.21.1.2720

Choi, H. S. (2023). Siamese Neural Network for User Authentication in Field-Programmable Gate Arrays (FPGAs) for Wearable Applications. *Electronics (Basel)*, *12*(19), 4030. Advance online publication. DOI: 10.3390/electronics12194030

Debener, J., Heinke, V., & Kriebel, J. (2023). Detecting insurance fraud using supervised and unsupervised machine learning. *The Journal of Risk and Insurance*, *90*(3), 743–768. Advance online publication. DOI: 10.1111/jori.12427

Doumpos, M., Zopounidis, C., Gounopoulos, D., Platanakis, E., & Zhang, W. (2023). Operational research and artificial intelligence methods in banking. In *European Journal of Operational Research* (Vol. 306, Issue 1). DOI: 10.1016/j.ejor.2022.04.027

Draghi, B., Wang, Z., Myles, P., & Tucker, A. (2024). Identifying and handling data bias within primary healthcare data using synthetic data generators. *Heliyon*, *10*(2), e24164. Advance online publication. DOI: 10.1016/j.heliyon.2024.e24164 PMID: 38288010

Ferreira, L., Pilastri, A., Martins, C. M., Pires, P. M., & Cortez, P. (2021). A Comparison of AutoML Tools for Machine Learning, Deep Learning and XGBoost. *Proceedings of the International Joint Conference on Neural Networks, 2021-July*. DOI: 10.1109/IJCNN52387.2021.9534091

Goodfellow, I., Bengio, Y., & Courville, A. (2016). *Deep learning*. MIT press.

Grilli, L., & Pedota, M. (2024). Creativity and artificial intelligence: A multilevel perspective. *Creativity and Innovation Management, 33*(2), 234–247. Advance online publication. DOI: 10.1111/caim.12580

Guan, S., Zhang, C., Wang, Y., & Liu, W. (2024). Hadoop-based secure storage solution for big data in cloud computing environment. *Digital Communications and Networks, 10*(1), 227–236. Advance online publication. DOI: 10.1016/j.dcan.2023.01.014

Henning, S., & Hasselbring, W. (2024). Benchmarking scalability of stream processing frameworks deployed as microservices in the cloud. *Journal of Systems and Software, 208*, 111879. Advance online publication. DOI: 10.1016/j.jss.2023.111879

Hermann, E., & Hermann, G. (2022). Artificial intelligence in research and development for sustainability: The centrality of explicability and research data management. *AI and Ethics, 2*(1), 29–33. Advance online publication. DOI: 10.1007/s43681-021-00114-8

Iman, M., Arabnia, H. R., & Rasheed, K. (2023). A Review of Deep Transfer Learning and Recent Advancements. In *Technologies* (Vol. 11, Issue 2). DOI: 10.3390/technologies11020040

Joseph, F. J. J., Nonsiri, S., & Monsakul, A. (2021). Keras and TensorFlow: A Hands-On Experience. In *EAI/Springer Innovations in Communication and Computing*. DOI: 10.1007/978-3-030-66519-7_4

Joshi, D. J., Kale, I., Gandewar, S., Korate, O., Patwari, D., & Patil, S. (2021). Reinforcement Learning: A Survey. *Advances in Intelligent Systems and Computing, 1311 AISC*. DOI: 10.1007/978-981-33-4859-2_29

Kalota, F. (2024). A Primer on Generative Artificial Intelligence. In *Education Sciences* (Vol. 14, Issue 2). DOI: 10.3390/educsci14020172

Kiran, B. R., Sobh, I., Talpaert, V., Mannion, P., Sallab, A. A. A., Yogamani, S., & Perez, P. (2022). Deep Reinforcement Learning for Autonomous Driving: A Survey. *IEEE Transactions on Intelligent Transportation Systems, 23*(6), 4909–4926. Advance online publication. DOI: 10.1109/TITS.2021.3054625

Kleinrichert, D. (2024). Empathy: An ethical consideration of AI & others in the workplace. *AI & Society, 39*(6), 2743–2757. Advance online publication. DOI: 10.1007/s00146-023-01831-w

Langley, C., Cirstea, B. I., Cuzzolin, F., & Sahakian, B. J. (2022). Theory of Mind and Preference Learning at the Interface of Cognitive Science, Neuroscience, and AI: A Review. In *Frontiers in Artificial Intelligence* (Vol. 5). DOI: 10.3389/frai.2022.778852

Lee, T., Singh, V. P., & Cho, K. H. (2021). *Tensorflow and Keras Programming for Deep Learning*. DOI: 10.1007/978-3-030-64777-3_11

Liu, N., Shapira, P., & Yue, X. (2021). Tracking developments in artificial intelligence research: Constructing and applying a new search strategy. *Scientometrics*, *126*(4), 3153–3192. Advance online publication. DOI: 10.1007/s11192-021-03868-4 PMID: 34720254

Longkumer, I., & Hussain Mazumder, D. (2024). Improving cancer prediction using feature selection in spark environment. *Concurrency and Computation*, *36*(2), e7903. Advance online publication. DOI: 10.1002/cpe.7903

López-Chila, R., Llerena-Izquierdo, J., Sumba-Nacipucha, N., & Cueva-Estrada, J. (2024). Artificial Intelligence in Higher Education: An Analysis of Existing Bibliometrics. In *Education Sciences* (Vol. 14, Issue 1). DOI: 10.3390/educsci14010047

Lundberg, S. (2019). *SHAP (SHapley Additive exPlanations)*. Github Repository.

Ma, C., Zhao, M., & Zhao, Y. (2023). An overview of Hadoop applications in transportation big data. In *Journal of Traffic and Transportation Engineering (English Edition)* (Vol. 10, Issue 5). DOI: 10.1016/j.jtte.2023.05.003

Macpherson, T., Churchland, A., Sejnowski, T., DiCarlo, J., Kamitani, Y., Takahashi, H., & Hikida, T. (2021). Natural and Artificial Intelligence: A brief introduction to the interplay between AI and neuroscience research. *Neural Networks*, *144*, 603–613. Advance online publication. DOI: 10.1016/j.neunet.2021.09.018 PMID: 34649035

Mehta, H., & Passi, K. (2022). Social Media Hate Speech Detection Using Explainable Artificial Intelligence (XAI). *Algorithms*, *15*(8), 291. Advance online publication. DOI: 10.3390/a15080291

Mumuni, A., & Mumuni, F. (2024). Automated data processing and feature engineering for deep learning and big data applications: A survey. *Journal of Information and Intelligence*. DOI: 10.1016/j.jiixd.2024.01.002

Munanday, A. P., Sazali, N., Harun, W. S. W., Kadirgama, K., & Jamaludin, A. S. (2023). Analysis of Convolutional Neural Networks for Facial Expression Recognition on GPU, TPU and CPU. *Journal of Advanced Research in Applied Sciences and Engineering Technology*, *31*(3), 50–67. Advance online publication. DOI: 10.37934/araset.31.3.5067

Myers, D., Mohawesh, R., Chellaboina, V. I., Sathvik, A. L., Venkatesh, P., Ho, Y. H., Henshaw, H., Alhawawreh, M., Berdik, D., & Jararweh, Y. (2024). Foundation and large language models: fundamentals, challenges, opportunities, and social impacts. In *Cluster Computing* (Vol. 27, Issue 1). DOI: 10.1007/s10586-023-04203-7

Neumann, O., Guirguis, K., & Steiner, R. (2024). Exploring artificial intelligence adoption in public organizations: A comparative case study. *Public Management Review, 26*(1), 114–141. Advance online publication. DOI: 10.1080/14719037.2022.2048685

Neupane, S., Mitra, S., Fernandez, I. A., Saha, S., Mittal, S., Chen, J., Pillai, N., & Rahimi, S. (2024). Security Considerations in AI-Robotics: A Survey of Current Methods, Challenges, and Opportunities. *IEEE Access : Practical Innovations, Open Solutions*, *12*, 22072–22097. Advance online publication. DOI: 10.1109/ACCESS.2024.3363657

Nga, H. T. T., & Thuy, A. N. T. (2024). Recommender System with Apache Spark. *Lecture Notes in Networks and Systems*, *786*, 487–497. Advance online publication. DOI: 10.1007/978-981-99-6547-2_37

Nitin, & Gupta, S. B. (2024). Artificial Intelligence in Smart Agriculture: Applications and Challenges. In *Current Applied Science and Technology* (Vol. 24, Issue 2). DOI: 10.55003/cast.2023.254427

Ooi, K. B., Tan, G. W. H., Al-Emran, M., Al-Sharafi, M. A., Capatina, A., Chakraborty, A., Dwivedi, Y. K., Huang, T. L., Kar, A. K., Lee, V. H., Loh, X. M., Micu, A., Mikalef, P., Mogaji, E., Pandey, N., Raman, R., Rana, N. P., Sarker, P., Sharma, A., & Wong, L. W. (2023). The Potential of Generative Artificial Intelligence Across Disciplines: Perspectives and Future Directions. *Journal of Computer Information Systems*. Advance online publication. DOI: 10.1080/08874417.2023.2261010

Paladino, L. M., Hughes, A., Perera, A., Topsakal, O., & Akinci, T. C. (2023). Evaluating the Performance of Automated Machine Learning (AutoML) Tools for Heart Disease Diagnosis and Prediction. *AI*, *4*(4), 1036–1058. Advance online publication. DOI: 10.3390/ai4040053

Pang, B., Nijkamp, E., & Wu, Y. N. (2020). Deep Learning With TensorFlow: A Review. In *Journal of Educational and Behavioral Statistics* (Vol. 45, Issue 2). DOI: 10.3102/1076998619872761

Panigrahi, S., Nanda, A., & Swarnkar, T. (2021). A Survey on Transfer Learning. *Smart Innovation. Systems and Technologies*, *194*, 781–789. Advance online publication. DOI: 10.1007/978-981-15-5971-6_83

Parashar, M., Deblanc-Knowles, T., Gianchandani, E., Parker, L. E., & Lin, H. Y. (2023). Strengthening and Democratizing Artificial Intelligence Research and Development. *Computer*, *56*(11), 85–90. Advance online publication. DOI: 10.1109/MC.2023.3284568

Podlubne, A., & Gohringer, D. (2023). A Survey on Adaptive Computing in Robotics: Modelling, Methods and Applications. *IEEE Access : Practical Innovations, Open Solutions*, *11*, 53830–53849. Advance online publication. DOI: 10.1109/ACCESS.2023.3281190

Pourdarbani, R., Sabzi, S., Nadimi, M., & Paliwal, J. (2023). Interpretation of Hyperspectral Images Using Integrated Gradients to Detect Bruising in Lemons. *Horticulturae*, *9*(7), 750. Advance online publication. DOI: 10.3390/horticulturae9070750

Prakash, N., Atiq, A., Shahid, M., Rani, J., & Dikshit, S. (2024). Merging Minds and Machines: The Role of Advancing AI in Robotics. *EAI Endorsed Transactions on Internet of Things*, *10*. Advance online publication. DOI: 10.4108/eetiot.4658

Radanliev, P. (2024). Artificial intelligence: Reflecting on the past and looking towards the next paradigm shift. *Journal of Experimental & Theoretical Artificial Intelligence*. Advance online publication. DOI: 10.1080/0952813X.2024.2323042

Rajpurohit, L., & Dobhada, S. (2024). Scientific writing and its ethical considerations using AI tools. In *Oral Oncology Reports* (Vol. 9). DOI: 10.1016/j.oor.2024.100196

Ramdurai, B., & Adhithya, P. (2023). The Impact, Advancements and Applications of Generative AI. *International Journal on Computer Science and Engineering*, *10*(6), 1–8. Advance online publication. DOI: 10.14445/23488387/IJCSE-V10I6P101

Rech, R. L.Junior, Malde, S., Cazzaniga, C., Kastriotou, M., Letiche, M., Frost, C., & Rech, P. (2022). High Energy and Thermal Neutron Sensitivity of Google Tensor Processing Units. *IEEE Transactions on Nuclear Science*, *69*(3), 567–575. Advance online publication. DOI: 10.1109/TNS.2022.3142092

Rubinger, L., Gazendam, A., Ekhtiari, S., & Bhandari, M. (2023). Machine learning and artificial intelligence in research and healthcare. *Injury*, *54*, S69–S73. Advance online publication. DOI: 10.1016/j.injury.2022.01.046 PMID: 35135685

Ruiz-Real, J. L., Uribe-Toril, J., Torres, J. A., & Pablo, J. D. E. (2021). Artificial intelligence in business and economics research: Trends and future. *Journal of Business Economics and Management*, *22*(1), 98–117. Advance online publication. DOI: 10.3846/jbem.2020.13641

Schwegler, M., Müller, C., & Reiterer, A. (2023). Integrated Gradients for Feature Assessment in Point Cloud-Based Data Sets. *Algorithms, 16*(7), 316. Advance online publication. DOI: 10.3390/a16070316

Schwen, L. O., Schacherer, D., Geißler, C., & Homeyer, A. (2022). Evaluating generic AutoML tools for computational pathology. *Informatics in Medicine Unlocked, 29*, 100853. Advance online publication. DOI: 10.1016/j.imu.2022.100853

Silva, B. M. F., Todt, E., Nascimento, T. P., Curvelo, C., & Gonçalves, L. M. G. (2024). Robotics Meets AI and Vision in South America (Topical Collection). *Journal of Intelligent & Robotic Systems, 110*(1), 11. Advance online publication. DOI: 10.1007/s10846-023-02039-2

Song, Z., Yang, X., Xu, Z., & King, I. (2023). Graph-Based Semi-Supervised Learning: A Comprehensive Review. *IEEE Transactions on Neural Networks and Learning Systems, 34*(11), 8174–8194. Advance online publication. DOI: 10.1109/TNNLS.2022.3155478 PMID: 35302941

Sufyan, M., Shokat, Z., & Ashfaq, U. A. (2023). Artificial intelligence in cancer diagnosis and therapy: Current status and future perspective. In *Computers in Biology and Medicine* (Vol. 165). DOI: 10.1016/j.compbiomed.2023.107356

Trimberger, S. M., & Moore, J. J. (2014). FPGA security: Motivations, features, and applications. *Proceedings of the IEEE, 102*(8), 1248–1265. Advance online publication. DOI: 10.1109/JPROC.2014.2331672

Ullah, F., Dhingra, S., Xia, X., & Babar, M. A. (2024). Evaluation of distributed data processing frameworks in hybrid clouds. *Journal of Network and Computer Applications, 224*, 103837. Advance online publication. DOI: 10.1016/j.jnca.2024.103837

Usama, M., Qadir, J., Raza, A., Arif, H., Yau, K. L. A., Elkhatib, Y., Hussain, A., & Al-Fuqaha, A. (2019). Unsupervised Machine Learning for Networking: Techniques, Applications and Research Challenges. *IEEE Access : Practical Innovations, Open Solutions, 7*, 65579–65615. Advance online publication. DOI: 10.1109/ACCESS.2019.2916648

van Engelen, J. E., & Hoos, H. H. (2020). A survey on semi-supervised learning. *Machine Learning, 109*(2), 373–440. Advance online publication. DOI: 10.1007/s10994-019-05855-6

Vasilev, I., Slater, D., Spacagna, G., Roelants, P., & Zocca, V. (2019). *Python Deep Learning: Exploring deep learning techniques and neural network architectures with Pytorch, Keras, and TensorFlow*. Packt Publishing Ltd.

Wang, K., Yan, L. Z., Li, W. Z., Jiang, C., Wang, N. N., Zheng, Q., Dong, N. G., & Shi, J. W. (2022). Comparison of Four Machine Learning Techniques for Prediction of Intensive Care Unit Length of Stay in Heart Transplantation Patients. *Frontiers in Cardiovascular Medicine*, *9*, 863642. Advance online publication. DOI: 10.3389/fcvm.2022.863642 PMID: 35800164

Weerts, H., Pfisterer, F., Feurer, M., Eggensperger, K., Bergman, E., Awad, N., Vanschoren, J., Pechenizkiy, M., Bischl, B., & Hutter, F. (2024). Can Fairness be Automated? Guidelines and Opportunities for Fairness-aware AutoML. *Journal of Artificial Intelligence Research*, *79*, 639–677. Advance online publication. DOI: 10.1613/jair.1.14747

Williams, K., Berman, G., & Michalska, S. (2023). Investigating hybridity in artificial intelligence research. *Big Data & Society*, *10*(2), 20539517231180577. Advance online publication. DOI: 10.1177/20539517231180577

Yang, X., Song, Z., King, I., & Xu, Z. (2023). A Survey on Deep Semi-Supervised Learning. *IEEE Transactions on Knowledge and Data Engineering*, *35*(9), 8934–8954. Advance online publication. DOI: 10.1109/TKDE.2022.3220219

Younisse, R., Ahmad, A., & Abu Al-Haija, Q. (2022). Explaining Intrusion Detection-Based Convolutional Neural Networks Using Shapley Additive Explanations (SHAP). *Big Data and Cognitive Computing*, *6*(4), 126. Advance online publication. DOI: 10.3390/bdcc6040126

Zhang, X., Wan, F., Liu, C., Ji, X., & Ye, Q. (2022). Learning to Match Anchors for Visual Object Detection. *IEEE Transactions on Pattern Analysis and Machine Intelligence*, *44*(6), 3096–3109. Advance online publication. DOI: 10.1109/TPAMI.2021.3050494 PMID: 33434120

Zhang, Y., He, G., Ma, L., Liu, X., Hjorth, J. J. J., Kozlov, A., He, Y., Zhang, S., Kotaleski, J. H., Tian, Y., Grillner, S., Du, K., & Huang, T. (2023). A GPU-based computational framework that bridges neuron simulation and artificial intelligence. *Nature Communications*, *14*(1), 5798. Advance online publication. DOI: 10.1038/s41467-023-41553-7 PMID: 37723170

Zhou, B., Yang, G., Shi, Z., & Ma, S. (2024). Natural Language Processing for Smart Healthcare. *IEEE Reviews in Biomedical Engineering*, *17*, 4–18. Advance online publication. DOI: 10.1109/RBME.2022.3210270 PMID: 36170385

Zhuang, F., Qi, Z., Duan, K., Xi, D., Zhu, Y., Zhu, H., Xiong, H., & He, Q. (2021). A Comprehensive Survey on Transfer Learning. In *Proceedings of the IEEE* (Vol. 109, Issue 1). DOI: 10.1109/JPROC.2020.3004555

ADDITIONAL READING

Ali, I., Warraich, N. F., & Butt, K. (2025). Acceptance and use of artificial intelligence and AI-based applications in education: A meta-analysis and future direction. *Information Development, 41*(3 Special Issue: "Artificial Intelligence initiatives"), 859–874. DOI: 10.1177/02666669241257206

Ezzaim, A., Dahbi, A., Aqqal, A., & Haidine, A. (2024). AI-based learning style detection in adaptive learning systems: a systematic literature review. *Journal of Computers in Education 2024 12:3, 12*(3), 731–769. DOI: 10.1007/s40692-024-00328-9

Fuellen, G., Kulaga, A., Lobentanzer, S., Unfried, M., Avelar, R. A., Palmer, D., & Kennedy, B. K. (2025). Validation requirements for AI-based intervention-evaluation in aging and longevity research and practice. *Ageing Research Reviews, 104*, 102617. DOI: 10.1016/j.arr.2024.102617 PMID: 39643211

Luo, J., Zheng, C., Yin, J., & Teo, H. H. (2025). Design and assessment of AI-based learning tools in higher education: a systematic review. *International Journal of Educational Technology in Higher Education 2025 22:1, 22*(1), 1–27. DOI: 10.1186/s41239-025-00540-2

Ortega-Bolaños, R., Bernal-Salcedo, J., Germán Ortiz, M., Galeano Sarmiento, J., Ruz, G. A., & Tabares-Soto, R. (2024). Applying the ethics of AI: a systematic review of tools for developing and assessing AI-based systems. *Artificial Intelligence Review 2024 57:5, 57*(5), 1–30. DOI: 10.1007/s10462-024-10740-3

Shi, H., & Aryadoust, V. (2024). A systematic review of AI-based automated written feedback research. *ReCALL, 36*(2), 187–209. DOI: 10.1017/S0958344023000265

KEY TERMS AND DEFINITIONS

Big Data Analytics: Big Data Analytics involves collecting, processing, and analyzing extremely large and complex datasets to uncover trends, patterns, relationships, and actionable insights. It enables organizations to make data-driven decisions by using advanced statistical methods, machine learning techniques, and scalable computational tools to handle high-volume, high-velocity, and high-variety data.

Deep Learning: Deep Learning is a subset of machine learning that uses multi-layered artificial neural networks to automatically learn patterns from large amounts of data. These networks can model highly complex relationships and are particularly effective in tasks such as image recognition, natural language processing, and speech analysis.

Explainable AI (XAI): Explainable AI refers to methods and techniques that make the internal workings of artificial intelligence systems understandable to humans. XAI aims to provide transparency about how decisions, predictions, or outputs are generated by complex models, ensuring that users can interpret, trust, and validate the behavior of AI systems.

Generative AI: Generative AI refers to models and algorithms capable of creating new and original content, such as text, images, audio, code, and simulations, based on the data they have been trained on. Examples include ChatGPT, DALL·E, and Midjourney.

High-Performance Computing (HPC): High-Performance Computing refers to the use of powerful computational systems, such as supercomputers and clustered computing environments, to perform large-scale calculations and data processing tasks at very high speeds.

Natural Language Processing (NLP): Natural Language Processing is a field of artificial intelligence that focuses on enabling computers to understand, interpret, generate, and respond to human language. NLP combines linguistics, computer science, and machine learning to support applications such as chatbots, sentiment analysis, speech recognition, translation systems, and text summarization.

Chapter 6
Ethical Usage of Artificial Intelligence Tools in Information Science Research

Bolaji David Oladokun
http://orcid.org/0000-0002-7826-9187
Federal University of Technology, Ikot Abasi, Nigeria

Adeyinka Tella
http://orcid.org/0000-0002-5382-4471
University of Ilorin, Nigeria & University of South Africa, Pretoria, South Africa

ABSTRACT

Artificial intelligence (AI) tools are rapidly transforming research practices in information science, enabling new approaches to literature review, hypothesis development, academic writing, referencing, and data analysis. This study provides a critical examination of the ethical issues surrounding the use of AI applications such as ChatGPT, Elicit, Scite, Grammarly, and Semantic Scholar in scholarly work. Using a conceptual research design, the paper synthesizes evidence from recent literature, policy documents, and grey sources to explore the opportunities and risks of AI adoption. The analysis identifies five major ethical concerns: transparency and disclosure, authorship and intellectual ownership, data privacy and confidentiality, bias and inclusivity, and institutional responsibility. To address these challenges, the study proposes ethical guidelines that include mandatory AI usage statements, human-in-command authorship principles, responsible data practices, equity-driven AI literacy, and institutional oversight mechanisms.

DOI: 10.4018/979-8-2600-0434-0.ch006

INTRODUCTION

The growing proliferation of artificial intelligence (AI) in academia has fundamentally reshaped how research is conceptualized, conducted, and communicated. Generative AI tools such as ChatGPT, Elicit, Scite, Grammarly, and Semantic Scholar are increasingly being integrated into scholarly workflows, supporting activities ranging from literature synthesis and hypothesis formulation to writing assistance and data analysis. These tools hold significant promise in enhancing research efficiency and productivity, particularly in fields like information science, where data-intensive methods, rapid knowledge discovery, and technological adaptability are essential. For instance, AI has been described as a transformative force in academic writing, offering opportunities to streamline scholarly tasks that traditionally required extensive human effort (Khalifa & Albadawy, 2024; Lund et al., 2023).

In information science research, generative AI introduces new paradigms in knowledge creation and dissemination (Singh et al., 2025; Zhang et al., 2025). Its ability to process vast data, generate coherent text, and identify hidden patterns provides researchers with unprecedented support in managing complex research processes. Studies suggest that AI-driven tools can contribute to more inclusive and accessible scholarship by lowering technical barriers for early-career researchers and non-native English speakers (Amirjalili et al., 2024; Raheem et al., 2023). This highlights AI's potential as a valuable partner in promoting equity and innovation within the discipline.

Beyond efficiency and productivity, several scholars view AI as a democratizing force in global research. According to Bolanos et al. (2024) and Khalifa and Albadawy (2024), AI-driven platforms expand access to advanced analytical capabilities previously limited to well-funded institutions, thereby empowering scholars in resource-constrained environments to engage competitively in global scholarship. Similarly, Raheem, Anjum, and Ghafar (2023) and Nelson, Santamaría, and Javens (2024) highlight that AI tools help bridge linguistic and cognitive divides by supporting non-native English speakers and those with limited methodological training. These perspectives underscore the transformative potential of AI as a tool for inclusion and epistemic justice, promoting equitable participation in knowledge production across geographies and disciplines.

However, alongside these opportunities are profound ethical concerns that challenge the foundations of research integrity. The reliance on AI raises questions about authorship, originality, and academic honesty. Scholars warn that excessive dependence on generative AI risks diluting genuine intellectual contribution and may foster new forms of plagiarism and ghostwriting (Eke, 2023; Dergaa et al., 2023). Similarly, issues of transparency and disclosure have become urgent, as debates persist over whether and how to acknowledge AI's involvement in scholarly outputs

(Hosseini et al., 2023; Ganjavi et al., 2024). Beyond authorship, ethical dilemmas extend to data privacy, bias, and inclusivity, since AI systems are often trained on datasets that reflect existing inequities or lack cultural sensitivity (Dinçer, 2024; Schlagwein & Willcocks, 2023). Such concerns have led academic publishers, institutions, and policymakers to issue evolving guidance on the responsible use of AI in research (Chetwynd, 2024; *Nature* Editorial, 2023).

Against this backdrop, this chapter seeks to critically interrogate the ethical usage of AI tools within information science research. Its objectives are threefold: first, to examine the practical implications of using AI tools for core research activities such as literature review, hypothesis generation, academic writing, referencing, and data analysis; second, to explore the unintended consequences of these tools through an ethical lens, particularly in relation to research integrity, authorship, data privacy, bias, and inclusivity; and third, to develop actionable guidelines and principles that researchers, educators, institutions, and policymakers can adopt to ensure responsible AI usage throughout the research lifecycle. By situating the discussion within the context of information science, this chapter aims to provide a reflective and practice-oriented framework for balancing innovation with ethical responsibility in the age of intelligent machines. While many studies identify discrete ethical issues like authorship or bias, and international bodies propose high-level principles, there is a lack of synthesized, actionable guidelines tailored specifically for the workflows and values of information science researchers. This chapter aims to bridge that gap

CONCEPTUAL AND THEORETICAL BACKGROUND

Artificial Intelligence (AI) has become a defining technological advancement in academia, bringing new possibilities for research practices while simultaneously provoking ethical debates. Conceptually, AI refers to the development of computational systems capable of performing tasks that typically require human intelligence, such as language processing, pattern recognition, problem-solving, and decision-making (Dinçer, 2024). In academic research, AI is broadly categorized into analytical AI which is focused on processing data and identifying patterns, and generative AI, which emphasizes the creation of novel text, images, and other outputs that mimic human creativity. Generative AI in particular has gained prominence because of its ability to produce coherent narratives, synthesize literature, and assist in drafting research outputs, activities traditionally seen as central to scholarly expertise (Lund et al., 2023; Amirjalili, Neysani, & Nikbakht, 2024).

The ethical discussion surrounding AI in research has evolved in tandem with the technology's development. In the early automation phase of the 1950s–1980s, AI in academia was limited to expert systems and data indexing tools, raising concerns about data reliability and intellectual ownership (Rahm, 2021). The machine learning era (1990s–2010s) introduced algorithmic decision-making and predictive analytics, which brought questions of transparency, bias, and accountability to the forefront (Yeung & Fernandes, 2022). In the current generative AI phase (2020s onward), technologies like ChatGPT and Gemini have continued to blur the boundaries between human creativity and machine output.

The scope of AI in research extends well beyond text generation. Tools such as ChatGPT, DeepSeek, Gemini, Elicit, Scite, and Semantic Scholar are now being used for literature discovery, summarization, and hypothesis exploration. Grammarly and Quillbot assist in refining academic writing, ensuring linguistic clarity, and reducing barriers for non-native English speakers (Raheem et al., 2023). Meanwhile, AI-driven bibliometric platforms and reference managers support citation accuracy and streamline research communication. These tools enhance efficiency by automating repetitive tasks, enabling researchers to devote more time to critical analysis and original contribution (Khalifa & Albadawy, 2024). Yet, as scholars such as Harati (2024) and Moffatt and Hall (2024) observe, the seamless integration of these tools into academic workflows has blurred the boundaries between human and machine contributions, raising pressing questions about ownership, accountability, and the authenticity of scholarship.

While the adoption of AI promises significant productivity gains, it has also exposed structural weaknesses in the ethical frameworks that underpin academic integrity. Research indicates that uncritical dependence on AI tools can foster risks of plagiarism, dilute intellectual authorship, and compromise transparency (Eke, 2023; Dergaa et al., 2023). Hosseini et al. (2023) argue that disclosure of AI involvement is critical to preserving scholarly trust, while Gulumbe et al. (2025) highlight the inconsistencies across universities and publishers in setting clear policies for AI usage. This lack of harmonization leaves researchers in a vulnerable space, particularly in developing contexts where institutional policies are either absent or underdeveloped.

To address these challenges, international organizations have proposed theoretical frameworks that advocate for a more responsible deployment of AI in research. UNESCO's Recommendation on the Ethics of Artificial Intelligence (2021) provides a global governance model centered on human rights, inclusivity, and sustainability. As Van Norren (2023) explains, this framework is rooted in cross-cultural perspectives, including the African Ubuntu philosophy, which stresses collective well-being and shared responsibility. Such an approach is particularly relevant to information science, a field that values information equity and community access to knowledge. In parallel, the Organisation for Economic Co-operation and Development (OECD) developed its Recommendation of the Council on Artificial Intelligence, which emphasizes inclusivity, fairness, robustness, and accountability in AI systems (Yeung, 2020). These principles advocate for AI to complement human capacities rather than displace them, a perspective echoed by Weldon et al. (2024), who caution that ethical regulation should balance protection against harm with the need to sustain innovation.

To move beyond descriptive accounts, it is crucial to ground the ethical challenges of AI within established theories of information ethics. Floridi (2013) provides a foundational lens by conceptualizing information as a moral entity in itself, suggesting that every act of data manipulation, generation, or omission carries moral weight. From this standpoint, AI systems are not neutral tools but moral agents embedded in informational ecosystems that demand stewardship and accountability. Similarly, Mason et al.'s (1986) framework, which includes Privacy, Accuracy, Property, and Accessibility, remains relevant to evaluating AI's impact on scholarly communication. These four principles correspond closely with the five themes explored in this chapter: privacy relates to data protection, accuracy to authorship integrity, property to intellectual ownership, and accessibility to inclusivity and institutional responsibility. Capurro (2006) further contributes a socio-cultural dimension, arguing that information systems must be interpreted through the lived experiences and values of their users. Applying this to AI suggests that ethical evaluation must consider not only universal principles but also contextual factors, such as linguistic diversity, cultural knowledge systems, and power asymmetries between Global North and South researchers.

Theoretical pathways for operationalizing these frameworks have been outlined by Stix (2021), who identifies compliance-based governance, risk-adaptive governance, and ethics-driven governance as practical models for regulating AI. Within academia, these models can be translated into policies that enforce disclosure of AI-generated content, safeguard against data manipulation, and mandate AI literacy training for scholars. Similarly, Truby (2020) stresses that AI governance must align with the United Nations Sustainable Development Goals (SDGs), particularly in preventing the misuse of AI to spread misinformation and in ensuring equitable access to digital technologies in resource-limited regions. UNESCO's guidelines extend this by recommending capacity-building initiatives to close the digital divide, a priority for African institutions that often lag in AI adoption and ethical preparedness (Van Norren, 2023). The integration of these international frameworks into academic practice underscores the imperative for local adaptation.

METHODOLOGY

This study adopts a conceptual research design to interrogate the ethical issues surrounding the use of artificial intelligence (AI) tools in information science research. Conceptual research is particularly suitable for examining evolving domains such as AI ethics, as it emphasizes the synthesis of theories, principles, and debates rather than generating new empirical data (Tammaro, 2011; Tucker et al., 2014). The data for this study were derived exclusively from secondary sources, drawing from a wide range of peer-reviewed literature, policy documents, and grey literature. Peer-reviewed articles and conference proceedings published between 2020 and 2025 were retrieved through systematic searches of databases such as Scopus, Web of Science, and Google Scholar using terms including "AI in academic writing," "generative AI ethics," "AI tools in research," and "AI and information science." Policy documents and international guidelines on Artificial Intelligence were also reviewed to provide a global perspective on responsible AI deployment.

To ensure the relevance and rigor of the sources analyzed, selection was guided by clear criteria. Included materials were those that explicitly discussed the role of AI tools in research and academic writing, highlighted ethical concerns such as transparency, authorship, privacy, or governance, and were relevant to information science or closely related disciplines such as library science, education, or digital scholarship. Sources were excluded if they focused narrowly on technical aspects of AI without ethical reflection or if they lacked analytical depth and scholarly rigor. This process allowed for the collection of a rich body of material that aligned closely with the study's objectives. The analysis was carried out through thematic synthesis, which involved identifying recurring arguments and concepts in the reviewed sources and coding them under key categories derived from the research focus. Five thematic areas were identified and refined during the analysis: transparency and disclosure, authorship and academic integrity, data privacy and confidentiality, bias and inclusivity, and ethical competence and institutional responsibility. This process ensured both structure and flexibility, as it enabled the study to capture dominant ethical concerns while also accommodating emerging issues highlighted in the literature.

The rigor and replicability of the study were enhanced through several strategies. A triangulation of sources was employed by combining scholarly literature, international policies, and grey literature, thereby reducing the risk of bias that might result from reliance on a single source type. Transparency was maintained by carefully documenting the search strategies, databases used, and selection criteria, ensuring that other scholars could replicate the approach with updated datasets or adapt it to other disciplines.

ETHICAL CONSIDERATIONS IN THE USE OF AI TOOLS IN RESEARCH

The growing use of artificial intelligence in research has introduced significant ethical issues alongside its many benefits. These issues are no longer abstract; they are evident in journal policies, institutional debates, and researchers' everyday practices. To capture this complexity, the ethical considerations in this study are organized into five themes: transparency and disclosure, authorship and academic integrity, data privacy and confidentiality, bias and inclusivity, and ethical competence and institutional responsibility.

Transparency and Disclosure

Transparency and disclosure are fundamental ethical concerns in the use of AI tools for academic research. Scholars emphasize that failing to disclose when AI tools contribute to a manuscript undermines trust and accountability in scholarship. Hosseini et al. (2023) argue that researchers must explicitly acknowledge AI use, detailing the tool, version, and extent of its involvement, while stressing that AI systems should not be listed as co-authors since they lack moral and legal responsibility. Similar concerns are echoed in the *Nature* (2023) editorial, which establishes ground rules for AI use in research, including mandatory disclosure to maintain transparent science. Ganjavi et al. (2024) provide empirical evidence that many journals now include AI-related instructions to authors, but these policies are inconsistent and often insufficiently clear. Aslam and Nisar (2024) add that without harmonized disclosure standards, researchers may inadvertently engage in unethical practices, thereby eroding scholarly integrity. As transparency defines the credibility of scholarly outputs, it inevitably intersects with authorship, since undisclosed AI contributions raise questions about who truly deserves intellectual credit in the academic record.

Authorship and Academic Integrity

Authorship and academic integrity represent another pressing theme. Generative AI blurs the boundary between genuine human contribution and machine assistance, raising questions about originality and intellectual ownership. Amirjalili et al. (2024) show in their comparative analysis that while AI-generated texts can meet structural and linguistic standards of academic writing, they lack a distinctive authorial voice and critical reasoning, making authorship attribution problematic. Dergaa et al. (2023) similarly caution that ChatGPT-generated texts, though polished, can compromise authenticity and create doubts about originality. Eke (2023) warns that the rise of generative AI threatens academic integrity by facilitating plagiarism and ghostwriting, particularly if scholars present AI-generated content as their own. Moffatt and Hall (2024) frame this as a deeper ethical dilemma, asking whether AI should ever be considered a co-author, given that authorship is traditionally linked to accountability and responsibility. Studies by Gustilo, Ong, and Lapinid (2024) further demonstrate how educators grapple with balancing AI's potential for supporting learning with its potential to enable academic dishonesty. Just as questions of authorship challenge the integrity of the final written product, the data fed into these systems during the research process raise equally critical concerns about privacy and confidentiality, especially regarding how sensitive research materials are handled by AI tools.

Data Privacy and Confidentiality

Data privacy and confidentiality form a third critical concern. Many AI systems require users to input sensitive or unpublished data, raising risks around storage, reuse, and exploitation by third-party providers. Carobene et al. (2024) highlight these risks in the context of scientific publishing, especially in peer review, where confidentiality is central but could be compromised if reviewers or authors employ AI without safeguards. Farhi et al. (2023), examining students' perceptions of ChatGPT, found that a significant proportion of respondents were uneasy about how the system manages user data, with concerns about privacy violations and lack of clarity regarding ownership of submitted content. Dinçer (2024) adds that insufficient transparency from AI developers about how data is processed intensifies the risk, making it imperative for researchers to exercise caution when engaging with such tools. Concerns about data privacy extend beyond technical protection; they shape how inclusive and equitable AI systems can be. Indeed, the biases that emerge from data practices directly influence the fairness and inclusivity of AI-generated outputs.

Bias, Fairness, and Inclusivity

Bias, fairness, and inclusivity are equally salient issues. Because AI models are trained on large-scale datasets that are often skewed or unrepresentative, their outputs can perpetuate or even amplify biases. Ngulube and Mosha (2025) and Schlagwein and Willcocks (2023) argue that generative AI systems inevitably reflect the biases of their training data, producing outputs that may marginalize underrepresented groups or reinforce stereotypes. Dergaa et al. (2023) caution that even when AI-generated text appears objective, hidden biases can distort interpretations and limit inclusivity in scholarship. Lund et al. (2023) point out that generative AI presents a "new academic reality" where fluency and coherence may mask underlying biases, particularly disadvantaging scholars from non-dominant linguistic or cultural backgrounds. Raheem, Anjum, and Ghafar (2023) reinforce this by noting that while AI tools like Grammarly and Quillbot help with linguistic precision, they often impose Western-centric writing norms that may dilute cultural nuance and researcher voice. Addressing bias and promoting inclusivity requires not only technical solutions but also the cultivation of ethical competence and institutional accountability, ensuring that researchers and organizations alike uphold shared values of fairness and integrity in AI adoption.

Ethical Competence and Institutional Responsibility

Finally, ethical competence and institutional responsibility are essential in mitigating the risks associated with AI adoption. Aslam and Nisar (2024) reveal through their web content analysis that academic publishers vary widely in their AI policies, leaving scholars uncertain about acceptable practices. Gulumbe et al. (2025) further demonstrate that universities and publishers occupy shifting positions, with some banning AI use outright while others permit limited applications, resulting in a lack of standardization. Chetwynd (2024) stresses that clear institutional guidelines are urgently needed, as researchers cannot individually navigate the complex ethical terrain of AI without collective norms. Mirbabaie et al. (2022) argue for integrating information ethics into AI governance frameworks to ensure that researchers develop the competence needed to use these tools responsibly. Subaveerapandiyan et al. (2025) add empirical weight by showing that PhD scholars in India often perceive AI tools as effective but express uncertainty about ethical boundaries, underscoring the need for formal training and guidance. As a result, these findings suggest that institutional codes of conduct, capacity-building initiatives, and continuous professional education are necessary to safeguard research integrity in the AI era. These ethical competencies and institutional roles provide the groundwork for translating theory into practice, as illustrated in the following case studies that examine real-world manifestations of transparency, authorship, and bias dilemmas in AI-assisted research.

REAL-WORLD ILLUSTRATIONS OF ETHICAL DILEMMAS IN AI-ASSISTED RESEARCH

The ethical complexities surrounding artificial intelligence in academic research become more tangible when examined through real-world cases that demonstrate how universities, publishers, and research organizations are grappling with challenges of transparency, authorship, and bias. Across diverse academic settings, these experiences reveal the evolving intersection between innovation and accountability, showing how policy, practice, and ethics must continually adapt to technological change. They also underscore the fact that ethical principles are not abstract ideals, but operational realities shaped by institutional responses and disciplinary cultures.

A prominent example of the shift toward transparency in scholarly communication emerged in early 2023, when the *Nature* journal group detected AI-generated abstracts in manuscript submissions that failed to disclose the use of ChatGPT. In response, *Nature* swiftly revised its editorial policy, mandating that authors must explicitly declare the name, version, and purpose of any AI tool used in manuscript preparation (Nature Editorial, 2023). The *BMJ* adopted a similar stance, introducing mandatory AI disclosure statements in all submissions (Ganjavi et al., 2024). These policy reforms represent a turning point in academic publishing, where transparency has become a core ethical requirement rather than an optional courtesy. By institutionalising disclosure norms, publishers are signalling that responsible AI use must be traceable and accountable, ensuring that technological assistance does not obscure intellectual ownership or undermine the authenticity of scholarly communication.

The implications of authorship and integrity extend beyond journal publishing into university research settings. In one widely discussed incident in 2023, a graduate student at a UK university incorporated AI-generated text into a thesis draft without acknowledgement (AI World Journal, 2024). When discovered, the institution's ethics committee ruled that the student had violated academic integrity policies by misrepresenting AI-generated content as original work. This led to the revision of the university's academic honesty code, which was expanded to include explicit provisions for AI use in writing and research. Students are now required to document any AI assistance, including the name of the tool and the nature of its contribution, aligning institutional practice with emerging publishing standards. This case underscores a growing recognition that academic honesty in the age of AI demands clear disclosure, reflexive self-reporting, and explicit accountability mechanisms. It also illustrates how institutions are transitioning from reactive to proactive stances, embedding AI-related ethics into existing educational and research integrity frameworks.

Beyond issues of transparency and authorship, questions of bias and inclusivity reveal how the ethical challenges of AI are entangled with broader epistemic inequalities in global knowledge production. Schlagwein and Willcocks (2023) observed that AI-generated literature summaries often disproportionately cite Western research, inadvertently reinforcing existing hierarchies in scholarly visibility and representation. Such algorithmic bias marginalizes perspectives from underrepresented regions, particularly from Africa and Asia, thereby perpetuating structural inequities in academic discourse. In contrast, the African Open Science Platform has taken a different approach by piloting AI tools trained on regional and multilingual datasets, aiming to enhance epistemic inclusivity and local knowledge visibility (Van Norren, 2023). These examples highlight the dual potential of AI to either amplify or mitigate bias, depending on how datasets are curated and ethical frameworks operationalized. They suggest that inclusivity in AI ethics must extend beyond data fairness to encompass cultural representation and the equitable distribution of scholarly voice.

Equally instructive are the institutional and corporate efforts to develop coherent oversight mechanisms for AI use in research and education. Universities such as Stanford, Harvard, and the University of Cambridge have each issued interim policies outlining acceptable uses of generative AI in academic writing, teaching, and research supervision. These policies converge on a few key principles: responsible disclosure of AI use, human accountability for final outputs, and the explicit prohibition of listing AI systems as co-authors. Similarly, Elsevier's (2024) official statement on "AI in Scholarly Communication" reaffirms that while AI can serve as a supportive tool for improving productivity and precision, it cannot replace human authorship or ethical judgment. Together, these institutional responses demonstrate the global momentum toward operationalizing ethical AI guidelines in concrete, enforceable terms. They also reveal that ethical governance in AI is increasingly multi-layered, shaped not only by individual researcher responsibility but also by organizational culture, professional norms, and legal accountability.

ETHICAL GUIDELINES FOR RESPONSIBLE AI USAGE IN INFORMATION SCIENCE RESEARCH

The adoption of AI tools in information science research must be guided by principles that go beyond identifying risks; it requires concrete practices that researchers can implement to uphold integrity and accountability. Drawing on the literature and aligning with global frameworks such as UNESCO's Recommendation on the Ethics of Artificial Intelligence (2021) and the OECD AI Principles, the following guidelines propose practical pathways for embedding ethical responsibility into the research process (Table 1).

Table 1: Guidelines for Responsible AI Usage in Information Science Research

Guideline	Description
AI Usage Statement	Researchers should include a standardized statement in their outputs, documenting how, when, and why AI tools were used. This ensures transparency and traceability.
Human-in-Command Authorship	AI should never replace human authorship. Researchers must curate, interpret, and remain fully accountable for all outputs generated with AI assistance.
Ethical Data Handling	Treat all AI interactions as data-sensitive. Confidential, unpublished, or proprietary materials should not be uploaded into unsecured or commercial AI systems.
Equity-Oriented AI Literacy	Researchers should be trained to detect and correct cultural, linguistic, or epistemic biases in AI outputs, ensuring inclusivity and fairness in scholarship.
Institutional AI Ethics Registries	Universities and research bodies should establish registries or oversight boards to track AI use in research, requiring declarations of intended AI engagement.

A first guideline is the creation of an AI usage statement in research outputs, modeled after conflict-of-interest or funding disclosures. This statement should document not only that AI was used, but also how, at what stage of the research, and with what purpose. Unlike generic disclosure requirements, this approach standardizes reporting and allows readers and reviewers to trace AI's contribution with clarity (Hosseini et al., 2023; Ganjavi et al., 2024). For information science, where reproducibility and traceability are core values, this practice ensures that the role of AI in shaping knowledge claims is visible and subject to scrutiny.

Second, researchers should adopt a "human-in-command" principle of authorship. This means that AI can never replace intellectual ownership but can serve as an aid whose contributions are curated and interpreted by the researcher. To operationalize this, institutions could require that manuscripts include a reflexive note where authors explain their own interpretive role in shaping AI-assisted outputs. Such reflexivity preserves originality and ensures that, as argued by Amirjalili, Neysani, and Nikbakht (2024) and Moffatt and Hall (2024), the human author remains accountable for the integrity of the final product.

A third guideline calls for ethical data handling protocols for AI engagement. Researchers must treat prompts and inputs to AI systems as data transactions requiring the same ethical scrutiny as human subject data. This means not uploading unpublished manuscripts, sensitive bibliometric data, or confidential institutional records into AI systems without assurance of privacy protections (Carobene et al., 2024; Dinçer, 2024). Institutions in turn should develop secure, locally hosted AI systems for scholarly use, minimizing reliance on commercial platforms where data policies remain opaque. Fourth, information science researchers should cultivate AI literacy with an equity lens. Rather than treating AI as a neutral tool, training should focus on understanding the cultural and epistemic biases embedded in its outputs. Scholars must be trained to detect, question, and correct biased content, especially where minority perspectives or culturally nuanced contexts are at risk of being flattened by dominant linguistic or cultural norms (Schlagwein & Willcocks, 2023; Raheem, Anjum, & Ghafar, 2023). To institutionalize this, curricula and professional development programs should integrate modules on auditing AI outputs for inclusivity, thus aligning with UNESCO's emphasis on fairness and diversity.

Finally, a novel guideline for responsible AI usage is the establishment of institutional AI ethics registries within universities and research bodies. Much like institutional review boards (IRBs), these registries would provide oversight of AI use in scholarly work. Researchers would submit short declarations of intended AI use at the start of their projects, enabling institutions to track patterns, identify risks, and develop context-sensitive guidelines. This echoes Aslam and Nisar's (2024) call for harmonized policies and responds to the uncertainty documented by Subaveerapandiyan et al. (2025) among doctoral scholars regarding what constitutes acceptable AI use.

EMERGING TECHNOLOGICAL SOLUTIONS FOR STRENGTHENING AI ETHICS

Beyond traditional policy frameworks and ethical education initiatives, recent technological innovations are playing a pivotal role in mitigating the risks associated with AI-assisted research. The increasing growth of generative AI tools has prompted researchers, publishers, and technology developers to pursue technical solutions that reinforce transparency, accountability, and integrity in scholarly communication (AK-Kfairy et al., 2024). These emerging technologies reflect an important shift in AI ethics, from reactive regulation to proactive system design, where ethical safeguards are embedded within the technology itself. By combining algorithmic detection, provenance tracking, and fairness-oriented design, the research community is beginning to operationalize ethics not merely as a philosophical ideal but as a measurable and enforceable standard of practice.

A major advancement in this domain is the development of AI output detection tools designed to identify whether a text has been generated or heavily modified by artificial intelligence. Platforms such as GPTZero and Turnitin's AI Writing Detection Suite employ sophisticated linguistic pattern recognition, probability modeling, and metadata analysis to flag AI-generated content in academic submissions. According to Resnik and Hosseini (2025), these detection systems provide crucial support to journal editors, educators, and peer reviewers in differentiating between human-authored and machine-generated writing, thereby safeguarding academic integrity. However, while detection tools enhance accountability, they also present limitations. Their accuracy varies across models, and false positives may unfairly penalize writers who legitimately use AI for minor editing or grammar correction. This indicates the need for transparency on both sides; that is, authors must disclose legitimate AI use, and detection tools must maintain openness regarding their methodologies and confidence thresholds to avoid reinforcing distrust in scholarly ecosystems.

Complementing these detection tools are advances in digital watermarking and provenance tracking, which aim to establish traceability and verifiable authorship of AI-generated content. Digital watermarking involves embedding imperceptible identifiers, such as cryptographic markers or metadata tags, into AI-generated text, images, or datasets to authenticate their origin. Kasani et al. (2024) explain that watermarking not only aids in identifying the source of AI-generated material but also helps deter plagiarism, unauthorized reuse, and data manipulation. At the forefront of these developments is the Coalition for Content Provenance and Authenticity (C2PA), a global consortium working to standardize technical protocols that enable users to verify the origin and modification history of digital content (Collomosse & Parsons, 2024). The implementation of such systems is particularly relevant in scholarly publishing, where the capacity to verify AI involvement could enhance transparency, foster trust in digital outputs, and protect intellectual property rights. Nevertheless, challenges remain; watermarking can be removed or altered through reformatting, and provenance tracking requires global interoperability and institutional buy-in to become effective across publishing platforms.

Equally transformative is the emerging paradigm of *ethical-by-design* artificial intelligence—a principle that advocates embedding ethical values into the technological architecture of AI systems from their inception. Instead of addressing ethical failures post-deployment, this approach seeks to integrate fairness, accountability, and transparency into the design, training, and operation of AI tools. Truby (2020) and Weldon et al. (2024) argue that implementing *explainable AI (XAI)* models, systems that can articulate the rationale behind their outputs, can significantly enhance human oversight and reduce opacity in decision-making. Similarly, *bias auditing protocols* and *fairness-aware data sampling* ensure that datasets used for AI training are representative, inclusive, and sensitive to cultural and linguistic diversity. When aligned with UNESCO's (2021) human-rights-based approach to AI ethics, ethical-by-design frameworks can transform AI from a potential ethical risk into a vehicle for equity and justice in research. Such integration requires interdisciplinary collaboration between ethicists, data scientists, and policymakers to ensure that ethical reflection is not an afterthought but a guiding design parameter.

These technological interventions collectively demonstrate that the future of AI ethics in research will rely not only on moral reasoning and regulatory enforcement but also on innovation within the technology itself. The convergence of detection systems, watermarking, and ethical-by-design principles marks a new phase in responsible AI governance, one that emphasizes prevention over correction. Yet, these advances also raise new questions: Who controls the standards of detection? How can watermarking protocols respect privacy and data sovereignty? And how might explainable AI balance technical transparency with proprietary constraints? Addressing these questions will determine whether such technologies can fulfill their ethical promise or inadvertently introduce new forms of surveillance and inequality. Ultimately, technological solutions must work in tandem with ethical judgment, institutional accountability, and cultural awareness to create a sustainable and trustworthy AI ecosystem in scholarly research.

IMPLICATIONS OF THE CHAPTER

This paper has implications for policy and practice in information science research. The findings indicate the urgent need for clear, standardized policies that govern the use of AI tools in research. Policymakers within academic institutions and professional bodies must prioritize the development of harmonized disclosure norms, mandating that scholars explicitly state the type, purpose, and extent of AI involvement in their research. Such policies will enhance transparency, uphold academic integrity, and reduce the risk of unethical practices such as ghostwriting or hidden AI contributions. Moreover, institutions should establish AI ethics registries or oversight committees, modeled after research ethics boards, to monitor and regulate the responsible use of AI across different stages of the research lifecycle. This step will provide accountability mechanisms and help align institutional practices with global ethical frameworks, such as those proposed by UNESCO and the OECD.

For practice, the findings highlight the necessity of equipping researchers and educators with adequate AI literacy, particularly with a focus on equity and inclusivity. Training programs and curricula in information science must integrate modules on detecting and mitigating AI-driven biases, managing data privacy risks, and balancing the benefits of AI efficiency with critical human interpretation. This will ensure that scholars do not merely adopt AI tools as neutral aids but engage with them reflexively, aware of their cultural, epistemic, and ethical limitations. Additionally, researchers are encouraged to adopt the "human-in-command" principle, where AI is positioned as a supportive tool rather than a substitute for intellectual authorship, thereby preserving originality and accountability in scholarly outputs.

At a broader professional level, the study's findings imply that journals, publishers, and scholarly associations must revisit their editorial policies to incorporate clear guidelines on AI usage. This includes establishing uniform disclosure requirements and rejecting practices that allow AI to be recognized as a co-author. Such measures will not only safeguard academic integrity but also build consistency in how AI-related scholarship is evaluated, disseminated, and archived. In sum, the study's implications point toward the dual responsibility of policy and practice to balance innovation with ethical responsibility. For policy, the focus must be on creating robust governance structures and harmonized guidelines. For practice, the emphasis should be on cultivating researcher competence, accountability, and inclusivity in AI adoption. Together, these measures will ensure that the integration of AI tools into information science research enhances scholarly productivity and innovation without undermining ethical standards or the authenticity of intellectual contributions.

FURTHER RESEARCH DIRECTIONS

Although this chapter has provided a comprehensive conceptual analysis of the ethical use of artificial intelligence (AI) tools in information science research, several areas remain open for deeper empirical investigation. Future research should move beyond conceptual synthesis to include empirical studies that examine how researchers in different regions and disciplines actually engage with AI tools in their daily scholarly practices. Such studies could explore the extent to which ethical guidelines are understood, adopted, or enforced within academic environments.

Further studies are also needed to evaluate the effectiveness of institutional and publisher-level AI policies. As the study revealed, there are significant inconsistencies across universities and journals regarding the disclosure of AI use and the recognition of human authorship. Comparative research across institutions, countries, or disciplines would help determine which policy models most effectively uphold academic integrity and promote responsible AI use. This could also include longitudinal studies to track how these policies evolve as AI technologies continue to advance.

In addition, future research should focus on developing and testing frameworks for AI literacy in higher education and research institutions. While this chapter highlighted the importance of equity-oriented AI literacy, more evidence is needed to understand the most effective pedagogical strategies for training researchers to detect bias, ensure inclusivity, and manage ethical dilemmas in AI-assisted scholarship. Experimental or action research could be used to design and evaluate training interventions that foster ethical competence among academics and students.

There is also a pressing need for cross-cultural and global South perspectives in AI ethics research. Current debates are often dominated by frameworks and standards originating in the Global North. Future studies should, therefore, examine how cultural values, socio-economic conditions, and infrastructural realities shape ethical perceptions of AI in Africa, Asia, and Latin America. Such research would contribute to building more inclusive and context-sensitive ethical frameworks that reflect diverse epistemological traditions and local research realities. Finally, further research could investigate the technological dimension of AI ethics, particularly the development and adoption of ethical-by-design AI systems, digital watermarking, and provenance tracking in scholarly publishing. Evaluating these technologies' practicality, limitations, and adoption barriers within academic environments would help bridge the gap between ethical theory and technological implementation.

CONCLUSION

This study has explored five interrelated ethical domains, including transparency, authorship, data privacy, bias, and institutional responsibility, to provide an integrated perspective on the responsible use of AI in information science research. While tools such as ChatGPT, DeepSeek, Gemini, Elicit, Scite, Grammarly, and Semantic Scholar are reshaping how research is conducted by improving efficiency, accessibility, and knowledge discovery, they also introduce pressing concerns around authorship, transparency, privacy, bias, and academic integrity. Without clear ethical guidelines, the benefits of AI could be overshadowed by risks that compromise the trustworthiness and authenticity of scholarly work.

The authors affirm that the way forward lies in adopting a balanced approach that values both innovation and ethical responsibility. Researchers must retain ownership of their intellectual contributions, treating AI as a supportive resource rather than a replacement for human creativity and accountability. At the same time, institutions and policymakers should establish robust frameworks that include disclosure standards, responsible data handling practices, AI literacy programs, and institutional oversight mechanisms. These measures will not only safeguard academic integrity but also promote fairness and inclusivity in the application of AI across diverse research contexts. In sum, the integration of AI into information science research is both inevitable and transformative. However, its success depends on the collective commitment of researchers, educators, institutions, and policymakers to embed transparency, accountability, and equity into its use. By doing so, the discipline can harness AI as a powerful tool for advancing knowledge while preserving the values of integrity, inclusivity, and trust that are central to scholarly practice. The coherence of these five themes demonstrates that addressing one ethical concern in isolation is insufficient; only by integrating transparency, accountability, and inclusivity across all stages of research can AI truly serve as a trustworthy partner in knowledge production.

REFERENCES

Al-Kfairy, M., Mustafa, D., Kshetri, N., Insiew, M., & Alfandi, O. (2024). Ethical challenges and solutions of generative AI: An interdisciplinary perspective. *Informatics (MDPI)*, *11*(3), 58. DOI: 10.3390/informatics11030058

Amirjalili, F., Neysani, M., & Nikbakht, A. (2024, March). Exploring the boundaries of authorship: A comparative analysis of AI-generated text and human academic writing in English literature. *Frontiers in Education*, *9*(1), 1347421. Advance online publication. DOI: 10.3389/feduc.2024.1347421

Aslam, M. S., & Nisar, S. (2024). Ethical Considerations for artificial intelligence tools in academic research and manuscript preparation: A web content analysis. In *Digital Transformation in Higher Education, Part B: Cases, Examples and Good Practices* (pp. 155–196). Emerald Publishing Limited. DOI: 10.1108/978-1-83608-424-220241007

Bolanos, F., Salatino, A., Osborne, F., & Motta, E. (2024). Artificial intelligence for literature reviews: Opportunities and challenges. *Artificial Intelligence Review*, *57*(10), 259. DOI: 10.1007/s10462-024-10902-3

Capurro, R. (2006). Towards an ontological foundation of information ethics. *Ethics and Information Technology*, *8*(4), 175–186. DOI: 10.1007/s10676-006-9108-0

Carobene, A., Padoan, A., Cabitza, F., Banfi, G., & Plebani, M. (2024). Rising adoption of artificial intelligence in scientific publishing: Evaluating the role, risks, and ethical implications in paper drafting and review process. [CCLM]. *Clinical Chemistry and Laboratory Medicine*, *62*(5), 835–843. DOI: 10.1515/cclm-2023-1136 PMID: 38019961

Chetwynd, E. (2024). Ethical use of artificial intelligence for scientific writing: Current trends. *Journal of Human Lactation*, *40*(2), 211–215. DOI: 10.1177/08903344241235160 PMID: 38482810

Collomosse, J., & Parsons, A. (2024). To authenticity, and beyond! Building safe and fair generative AI upon the three pillars of provenance. *IEEE Computer Graphics and Applications*, *44*(3), 82–90. DOI: 10.1109/MCG.2024.3380168 PMID: 38905025

Dergaa, I., Chamari, K., Zmijewski, P., & Saad, H. B. (2023). From human writing to artificial intelligence generated text: Examining the prospects and potential threats of ChatGPT in academic writing. *Biology of Sport*, *40*(2), 615–622. DOI: 10.5114/biolsport.2023.125623 PMID: 37077800

Dinçer, S. (2024). The use and ethical implications of artificial intelligence in scientific research and academic writing. *Educational Research & Implementation, 1*(2), 139–144. DOI: 10.14527/edure.2024.10

Editorial, N. (2023). Tools such as ChatGPT threaten transparent science; here are our ground rules for their use. *Nature, 613*(7945), 612. DOI: 10.1038/d41586-023-00191-1 PMID: 36694020

Eke, D. O. (2023). ChatGPT and the rise of generative AI: Threat to academic integrity? *Journal of Responsible Technology, 13*, 100060. DOI: 10.1016/j.jrt.2023.100060

Farhi, F., Jeljeli, R., Aburezeq, I., Dweikat, F. F., Al-shami, S. A., & Slamene, R. (2023). Analyzing the students' views, concerns, and perceived ethics about ChatGPT usage. *Computers and Education: Artificial Intelligence, 5*, 100180. DOI: 10.1016/j.caeai.2023.100180

Floridi, L. (2013). *The ethics of information*. Oxford University Press. DOI: 10.1093/acprof:oso/9780199641321.001.0001

Ganjavi, C., Eppler, M. B., Pekcan, A., Biedermann, B., Abreu, A., Collins, G. S., Gill, I. S., & Cacciamani, G. E. (2024). Publishers' and journals' instructions to authors on use of generative artificial intelligence in academic and scientific publishing: Bibliometric analysis. *BMJ (Clinical Research Ed.), 384*, e077192. Advance online publication. DOI: 10.1136/bmj-2023-077192 PMID: 38296328

Gulumbe, B. H., Audu, S. M., & Hashim, A. M. (2025). Balancing AI and academic integrity: What are the positions of academic publishers and universities? *AI & Society, 40*(3), 1775–1784. DOI: 10.1007/s00146-024-01946-8

Gustilo, L., Ong, E., & Lapinid, M. R. (2024). Algorithmically-driven writing and academic integrity: Exploring educators' practices, perceptions, and policies in AI era. *International Journal for Educational Integrity, 20*(1), 3. DOI: 10.1007/s40979-024-00153-8

Harati, K. (2024). ChatGPT and AI-powered writing tools: Unveiling risks and ethical challenges in scientific writing. *Journal of Reviews in Medical Sciences, 4*(1), 1-6. https://.doi.org.DOI: 10.22034/jrms.2024.493520.1031

Hosseini, M., Resnik, D. B., & Holmes, K. (2023). The ethics of disclosing the use of artificial intelligence tools in writing scholarly manuscripts. *Research Ethics, 19*(4), 449–465. DOI: 10.1177/17470161231180449 PMID: 39749232

Kasani, P. H., Cho, K. H., Jang, J. W., & Yun, C. H. (2024). Influence of artificial intelligence and chatbots on research integrity and publication ethics. *Science Editing, 11*(1), 12–25. DOI: 10.6087/kcse.323

Khalifa, M., & Albadawy, M. (2024). Using artificial intelligence in academic writing and research: An essential productivity tool. *Computer Methods and Programs in Biomedicine Update*, *100145*, 100145. Advance online publication. DOI: 10.1016/j.cmpbup.2024.100145

Lund, B. D., Wang, T., Mannuru, N. R., Nie, B., Shimray, S., & Wang, Z. (2023). ChatGPT and a new academic reality: Artificial intelligence-written research papers and the ethics of the large language models in scholarly publishing. *Journal of the Association for Information Science and Technology*, *74*(5), 570–581. DOI: 10.1002/asi.24750

Mason, R. O., Collins, C. P., & Cox, E. L. (1986). Four ethical issues of the information age. *Management Information Systems Quarterly*, *10*(1), 5–12. DOI: 10.2307/248873

Mirbabaie, M., Brendel, A. B., & Hofeditz, L. (2022). Ethics and AI in information systems research. *Communications of the Association for Information Systems*, *50*(1), 38. DOI: 10.17705/1CAIS.05034

Moffatt, B., & Hall, A. (2024). Is AI my co-author? The ethics of using artificial intelligence in scientific publishing. *Accountability in Research*, •••, 1–17. DOI: 10.1080/08989621.2024.2386285 PMID: 39109816

Nelson, A. S., Santamaría, P. V., & Javens, J. S. (2024, November). Students' Perceptions of Generative AI Use in Academic Writing. In *Conference Proceedings. Innovation in Language Learning 2024.*

Ngulube, P., & Mosha, N. F. V. (2025). Integrating artificial intelligence-based technologies 'safely' in academic libraries: An overview through a scoping review. *Technical Services Quarterly*, *42*(1), 46–67. DOI: 10.1080/07317131.2024.2432093

Raheem, B. R., Anjum, F., & Ghafar, Z. N. (2023). Exploring the profound impact of artificial intelligence applications (Quillbot, Grammarly and ChatGPT) on English academic writing: A systematic review. [IJIR]. *International Journal of Integrative Research*, *1*(10), 599–622. DOI: 10.59890/ijir.v1i10.366

Rahm, L. (2021). Education, automation and AI: A genealogy of alternative futures. *Learning, Media and Technology*, *48*(1), 6–24. DOI: 10.1080/17439884.2021.1977948

Resnik, D. B., & Hosseini, M. (2025). The ethics of using artificial intelligence in scientific research: New guidance needed for a new tool. *AI and Ethics*, *5*(2), 1499–1521. DOI: 10.1007/s43681-024-00493-8 PMID: 40337745

Schlagwein, D., & Willcocks, L. (2023). 'ChatGPT et al.': The ethics of using (generative) artificial intelligence in research and science. *Journal of Information Technology, 38*(3), 232–238. DOI: 10.1177/02683962231200411

Singh, R., Kim, J. Y., Glassy, E. F., Dash, R. C., Brodsky, V., Seheult, J., de Baca, M. E., Gu, Q., Hoekstra, S., & Pritt, B. S. (2025). Introduction to generative artificial intelligence: Contextualizing the future. *Archives of Pathology & Laboratory Medicine, 149*(2), 112–122. DOI: 10.5858/arpa.2024-0221-RA PMID: 39631430

Stix, C. (2021). Actionable principles for artificial intelligence policy: Three pathways. *Science and Engineering Ethics, 27*(1), 15. DOI: 10.1007/s11948-020-00277-3 PMID: 33608756

Subaveerapandiyan, A., Kalbande, D., & Ahmad, N. (2025). Perceptions of effectiveness and ethical use of AI tools in academic writing: A study among PhD scholars in India. *Information Development, 41*(3), 728–746. DOI: 10.1177/02666669251314840

Tammaro, A. M. (2011). Library and Information Science (LIS) education: A conceptual framework towards "Europeisation". *Journal of the Bangladesh Association of Young Researchers, 1*(1), 1–13. DOI: 10.3329/jbayr.v1i1.6950

Truby, J. (2020). Governing artificial intelligence to benefit the UN sustainable development goals. *Sustainable Development (Bradford), 28*(4), 946–959. DOI: 10.1002/sd.2048

Tucker, V. M., Weedman, J., Bruce, C. S., & Edwards, S. L. (2014). Learning portals: Analyzing threshold concept theory for LIS education. *Journal of Education for Library and Information Science*, •••, 150–165.

Van Norren, D. E. (2023). The ethics of artificial intelligence, UNESCO and the African Ubuntu perspective. *Journal of Information. Communication and Ethics in Society, 21*(1), 112–128. DOI: 10.1108/JICES-04-2022-0037

Weldon, M. N., Thomas, G., & Skidmore, L. (2024). Establishing a future-proof framework for AI regulation: Balancing ethics, transparency, and innovation. *Transactions: The Tennessee Journal of Business Law, 25*(2), 2.

World Journal, A. I. (2024, January 10). *Challenge between professors and university students: Using AI for theses.* Retrieved on 12 October 2025 from https://aiworldjournal.com/challenge-between-professors-and-university-students-using-ai-for-theses/#:~:text=In%202023%2C%20a%20graduate%20student,the%20field%20of%20social%20sciences

Yeung, K. (2020). Recommendation of the council on artificial intelligence (OECD). *International legal materials, 59*(1), 27-34. DOI: 10.1017/ilm.2020.5

Yeung, R. C., & Fernandes, M. A. (2022). Machine learning to detect invalid text responses: Validation and comparison to existing detection methods. *Behavior Research Methods*, *54*(6), 3055–3070. DOI: 10.3758/s13428-022-01801-y PMID: 35175566

Zhang, Q., Zuo, J., & Yang, S. (2025). Research on the impact of generative artificial intelligence (GenAI) on enterprise innovation performance: A knowledge management perspective. *Journal of Knowledge Management*, *29*(7), 2238–2257. DOI: 10.1108/JKM-10-2024-1198

ADDITIONAL READING

Kim, S. J. (2024). Research ethics and issues regarding the use of ChatGPT-like artificial intelligence platforms by authors and reviewers: A narrative review. *Science Editing*, *11*(2), 96–106. DOI: 10.6087/kcse.343

Lim, B., Seth, I., Cevik, J., Mu, X., Sofiadellis, F., Cuomo, R., & Rozen, W. M. (2025). Artificial intelligence tools in surgical research: A narrative review of current applications and ethical challenges. *Surgeries*, *6*(3), 55. DOI: 10.3390/surgeries6030055

Mumtaz, S., Carmichael, J., Weiss, M., & Nimon-Peters, A. (2025). Ethical use of artificial intelligence-based tools in higher education: Are future business leaders ready? *Education and Information Technologies*, *30*(6), 7293–7319. DOI: 10.1007/s10639-024-13099-8

KEY TERMS AND DEFINITIONS

Academic Integrity: Academic integrity refers to honesty, originality, and ethical behavior in academic work, requiring transparency in how AI tools are used to avoid plagiarism or misrepresentation.

AI Ethics: AI ethics examines the moral and societal implications of artificial intelligence, focusing on transparency, accountability, fairness, privacy, and respect for human rights.

AI Limitations: AI limitations refer to the inherent risks in AI tools, including misinformation, factual inaccuracy, and the absence of verifiable sources, which may affect research validity.

AI Reliability: AI reliability relates to the factual accuracy, consistency, and reproducibility of AI-generated outputs used in research.

Artificial Intelligence (AI): Artificial Intelligence refers to computer systems designed to perform cognitive tasks such as reasoning, problem-solving, and text generation. In research, it includes generative tools like ChatGPT that assist in writing, data analysis, and knowledge organization.

Authorship: Authorship represents the recognition of intellectual contribution and responsibility for scholarly work. In AI-supported writing, it requires clarifying human input versus machine-generated content.

Bias: Bias denotes systematic distortion or unfairness in AI systems resulting from non-representative datasets, flawed algorithms, or human design errors.

Confidentiality: Confidentiality is the ethical obligation to ensure that information shared within research contexts is safeguarded and disclosed only with proper authorization.

Data Privacy: Data privacy concerns protecting personal and sensitive data from unauthorized access, misuse, or exposure when processed by AI systems.

Digital Watermarking: Digital watermarking involves embedding invisible identifiers in AI-generated content to trace its origin and prevent unauthorized use.

Disclosure: Disclosure means formally acknowledging the specific AI tools, versions, and their roles in manuscript preparation or data processing to promote accountability and authenticity.

Ethical-by-Design: Ethical-by-design means embedding fairness, accountability, and transparency into AI systems from the design phase to prevent ethical lapses during use.

Ethics: Ethics refers to the moral principles and professional standards that guide human conduct, ensuring fairness, honesty, and responsibility in research and scholarly communication.

Inclusivity: Inclusivity ensures that AI systems and research practices promote diversity, equity, and fair representation across regions, cultures, and languages.

Institutional Responsibility: Institutional responsibility refers to the duty of universities, publishers, and research bodies to establish and enforce ethical policies that guide AI use in scholarship.

Provenance Tracking: Provenance tracking refers to documenting the history and modifications of digital content to ensure transparency and authenticity in AI-assisted outputs.

Transparency: Transparency involves clearly communicating how AI tools are used in the research process, allowing others to understand their influence on findings and outputs.

Chapter 7
Artificial Intelligence Enhanced Taguchi–RSM Framework for Intelligent and Sustainable Optimization

Shalom Akhai
http://orcid.org/0000-0002-7533-457X
Maharishi Markandeshwar Engineering College, Maharishi Markandeshwar University, Mullana, India

Mahapara Abbass
http://orcid.org/0000-0002-0838-2598
Maharishi Markandeshwar Engineering College, Maharishi Markandeshwar University, Mullana, India

ABSTRACT

This chapter presents an AI-enhanced Taguchi–Response Surface Methodology (RSM) framework that merges statistical design with computational intelligence for intelligent and sustainable optimization. The Taguchi method provides robust parameter screening using orthogonal arrays, while RSM offers precise regression modeling and response visualization. Artificial intelligence tools, including neural networks and genetic algorithms, refine the hybrid process by capturing nonlinear relationships and adaptive patterns. The framework minimizes experimental effort, enhances prediction accuracy, and supports multi-objective optimization. Applications across manufacturing, civil engineering, materials processing, and

DOI: 10.4018/979-8-2600-0434-0.ch007

energy systems demonstrate its versatility and sustainability. The proposed hybrid model bridges traditional experimentation with AI-driven automation, enabling data-driven innovation and efficient decision-making in the Industry 4.0 landscape.

INTRODUCTION

Optimisation underpins progress in science, technology, and industry, enabling the pursuit of higher efficiency, quality, and performance (Lipski & Zaleski, 2019; Sadollah et al., 2020). Early optimisation relied on trial and error, adjusting one variable at a time while keeping others constant. Although simple, this approach demanded considerable time and resources and failed to capture complex variable interdependencies, emphasising the need for systematic and data-driven methods (Brownlee et al., 2022; Chhajer & Roy, 2024; Rogers et al., 2024).

The emergence of the Design of Experiments (DoE) framework in the mid-20th century revolutionisedoptimisation by allowing simultaneous analysis of multiple factors and their interactions. Within DoE, the Taguchi Method and response surface methodology (RSM) became key approaches that integrate mathematical rigour with practical experimentation (Botha et al., 2021; "Optimisation Approach to Evaluate the Solar, UV and LED Visible Light Fenton Processes for Pollutant Degradation in Landfill Leachate," 2022; Popović, 2020).

The Taguchi Method, developed by Genichi Taguchi, focuses on robust design by minimising the influence of uncontrollable "noise" factors. Using orthogonal arrays, it identifies significant parameters with fewer experiments, improving quality and reliability in manufacturing and process design (Hisam et al., 2024; Kumar & Lavanya, 2024; Shie et al., 2019). In contrast, RSM models continuous variable systems using polynomial regression and designs such as Central Composite Design (CCD) and Box–Behnken Design (BBD). It predicts optimal conditions in nonlinear systems and complements Taguchi by refining results with mathematical precision (Bheel et al., 2024; Chen, 2023).

With the rise of digital transformation, optimisation has expanded beyond traditional experimentation. The fusion of computational intelligence and data analytics has produced hybrid statistical–computational frameworks that integrate Taguchi and RSM with artificial intelligence (AI), machine learning (ML), genetic algorithms (GA), and multi-objective optimisation (Arteaga-Cabrera et al., 2025; Mairpady et al., 2021; Paw et al., 2023). These combinations enable adaptive modelling, non-linear trend detection, and autonomous optimisation through self-learning systems (Lai et al., 2023; Zimmer & Debelak, 2023).

Modern industries face complex, multi-response optimisation challenges—from balancing cost, energy, and quality in manufacturing to optimising formulations in healthcare. Hybrid Taguchi–RSM frameworks, enhanced with computational algorithms such as neural networks and fuzzy logic, address these challenges by combining robustness, adaptability, and predictive accuracy (Kushwah et al., 2024; Wichapa et al., 2024; Bubeník et al., 2025; Hakim et al., 2025).

The increasing accessibility of computational tools such as Minitab, Design Expert, and MATLAB has further accelerated the adoption of these frameworks. Minitab simplifies experimental design, Design Expert provides advanced modelling and visualisation, and MATLAB enables integration with AI and simulation environments—bridging theoretical models and practical applications (Saad et al., 2023; Sung et al., 2025; Zimmer & Debelak, 2023).

Overall, this chapter establishes an AI-enhanced Taguchi–RSM hybrid framework that unites Taguchi's robustness with RSM's analytical precision and computational intelligence. Its novelty lies in consolidating research, clarifying methodological interlinkages, and demonstrating how hybrid AI-based models improve accuracy, adaptability, and sustainability across diverse scientific and engineering fields—advancing the vision of Industry 4.0-driven intelligent optimisation.

Broader Relevance Across Disciplines

While the Taguchi and RSMtechniques were originally developed for applications in manufacturing and quality control, their adaptability and conceptual strength have allowed them to become essential tools across a wide range of disciplines. In manufacturing and mechanical engineering, these methods are extensively employed to optimise machining parameters, reduce tool wear, and improve surface finish quality. They help engineers design experiments that identify the most influential factors affecting product performance and operational efficiency (Koli et al., 2024; Kumar et al., 2021). In civil and environmental engineering, Taguchi and RSM approaches are used to refine concrete mix designs, predict compressive strength, and evaluate sustainability parameters such as material substitution and durability under variable conditions (Aziz et al., 2024; Bheel et al., 2024; Rosa et al., 2023). In chemical and process engineering, these statistical frameworks assist in maximising reaction yields, enhancing catalyst efficiency, and minimising waste through systematic experimentation (Aldulaijan et al., 2023; Pahija et al., 2024). Their applications extend further into healthcare and biotechnology, where they play a critical role in optimising drug formulations, controlling bio-process parameters, and improving the precision of diagnostic systems (Akhai & Khang, 2024; Bellonzi et al., 2024; Kuo & Shieh, 2020).

Similarly, in the energy and environmental sector, these methods are used to enhance renewable energy efficiency, analyze combustion processes, and control emission parameters for cleaner and more sustainable technologies (Khan et al.,

2023; Sadeq et al., 2025). Beyond traditional engineering fields, the relevance of Taguchi and RSM has expanded into data science and machine learning, where they are employed to fine-tune model hyperparameters, enhance prediction accuracy, and streamline algorithmic performance (Hisam et al., 2024; Mairpady et al., 2021; Paw et al., 2023). The ability of these methodologies to integrate experimental data with computational analysis has made them indispensable for cross-disciplinary innovation. Their flexibility, structured experimentation, and compatibility with modern computational tools allow researchers across domains to design efficient, data-driven studies that lead to robust and optimised solutions (Hariharasakthisudhan et al., 2024; Hope et al., 2023; Xue et al., 2022).

This chapter synthesises the theoretical foundations and practical applications of the Taguchi Method and RSM, emphasising their integration within intelligent optimisation frameworks. It highlights how these methodologies enhance experimental precision, foster innovation, and support the transition toward data-driven, sustainable, and high-performance systems. The conceptual framework for the proposed Hybrid Taguchi–RSM Intelligent Optimisation Model is outlined in Appendix I.

FUNDAMENTALS OF INTELLIGENT OPTIMISATION

The Taguchi Method represents a landmark development in quality engineering and experimental design. Developed by Genichi Taguchi, it introduced a robust framework for achieving reliability and cost-effectiveness through systematic parameter optimisation. Unlike traditional approaches focused solely on improving average performance, the Taguchi philosophy minimises variability caused by uncontrollable "noise" factors, ensuring consistent outcomes under varying conditions (Goodarzian et al., 2022; Hisam et al., 2024; Moussa et al., 2022).

Philosophy of Robust Design

Taguchi's philosophy centres on robust design, which seeks to create products and processes that perform consistently despite external disturbances. He argued that quality should be built into systems rather than inspected afterwards, viewing deviations from target values as losses due to inefficiency and waste. The quality loss function links variation to cost, showing that even minor deviations can lead to significant long-term performance losses (Lee et al., 2024; Li et al., 2019). This focus on performance consistency reflects a shift in quality management and aligns with intelligent optimisation principles, where systems adapt to uncertainty (Fallahi et al., 2024; Juul-Nyholm, 2024).

Orthogonal Arrays and Experimental Efficiency

A defining element of the Taguchi framework is the use of orthogonal arrays (OAs)—predefined, statistically balanced matrices that enable simultaneous testing of multiple parameters with minimal experimentation. OAs maintain factor independence and ensure unbiased results. For instance, four factors at three levels require 81 full factorial experiments, but only nine trials using an L9 OA (Banashooshtari et al., 2024; Mensah et al., 2019). Standardized OAs such as L4, L9, L16, and L25 suit different levels and factor combinations (Hisam et al., 2024; Karazi et al., 2019; Tsai, 2023). This structured design exemplifies intelligent experimentation by maximising information gain with minimal effort (Janković et al., 2021; Miikkulainen et al., 2020).

Signal-to-Noise (S/N) Ratio

Taguchi's major analytical innovation is the Signal-to-Noise (S/N) ratio, which integrates mean performance (signal) and variability (noise) into a single robustness metric. Maximising the S/N ratio ensures stability and reliability across operating conditions (Wadhwa et al., 2024). Three types of S/N ratios are employed:

- Larger-the-better for maximising outcomes such as tensile strength,
- Smaller-the-better for minimising defects or roughness, and
- Nominal-the-best when an exact target value is desired.

This transformation onto a logarithmic scale minimises outlier influence, allowing experimenters to identify optimal conditions resistant to noise (Abad & Momayezi, 2024; Kushwah et al., 2024; Manning et al., 2022).

Implementation Steps

Implementation of the Taguchi Method follows a concise structure: selecting key parameters and levels, designing the experiment using an appropriate OA (e.g., L9, L16, L25), conducting trials, and analyzing results using S/N ratios and Analysis of Variance (ANOVA). The analysis identifies critical factors influencing performance, while confirmation runs validate predicted optimal combinations (Hisam et al., 2024; Kumar & Lavanya, 2024).

Illustrative Example

In a machining optimisation study, four parameters—cutting speed, feed rate, depth of cut, and coolant flow—each at three levels, were analyzed using an L9 OA. Only nine trials were required instead of eighty-one. After measuring surface roughness, the "smaller-the-better" S/N ratio was applied. Main effect plots and ANOVA identified the most influential parameters, leading to an optimised combination that significantly reduced surface roughness and variability (Akhai et al., 2021; Akhai et al., 2016; Kumar & Lavanya, 2024). This demonstrates Taguchi's efficiency and precision in deriving practical engineering solutions.

Advantages and Limitations

The Taguchi Method offers exceptional simplicity, reduced experimentation, and robustness, but is less effective in highly nonlinear or multi-response systems where strong variable interactions exist. It assumes independence among factors, which limits its ability to capture complex behaviours. Therefore, integration with RSM, artificial neural networks (ANNs), or genetic algorithms (GAs) is often recommended to improve modelling accuracy and handle nonlinearities (Bheel et al., 2024; Paw et al., 2023; Wichapa et al., 2024).

Taguchi in the Era of Intelligent Optimisation

With the advent of computational intelligence, the Taguchi Method now functions as a critical component of hybrid intelligent optimisation frameworks. It serves as a preliminary screening tool to identify dominant parameters and refine the search space for advanced algorithms such as RSM, neural networks, and fuzzy systems. This integration enhances predictive power, reduces experimental costs, and supports data-driven learning (Dikshit et al., 2022; Kumar et al., 2022; Kumar et al., 2025; Batra et al., 2020). The structured, high-quality datasets generated through Taguchi designs also enable machine learning models to achieve higher accuracy and generalisation (Hisam et al., 2024; Nikzad et al., 2024; Zubair et al., 2024).

Thus, the Taguchi Method has transcended its traditional role as a quality improvement tool to become a core element of intelligent optimisation, applicable across manufacturing, materials science, energy systems, and healthcare technologies (Hisam et al., 2024; Wichapa et al., 2024).

RESPONSE SURFACE METHODOLOGY (RSM): CONCEPT AND DEVELOPMENT

RSM is a powerful statistical and mathematical approach used for modelling and optimising processes where several independent variables influence a desired response. RSM provides a structured framework to explore relationships among these factors, build predictive models, and determine optimal operating conditions. Its strength lies in its ability to represent complex interactions through mathematical equations and graphical surfaces, enabling researchers to visualise, interpret, and control system behaviour effectively. The method evolved from the early principles of regression analysis and the design of experiments, and has now become a fundamental tool in engineering, industrial design, and scientific research for multi-variable optimisation (Bell et al., 2021; Janković et al., 2021; Kumar & Reji, 2023).

Concept and Philosophy Behind RSM

RSM models the relationship between process outputs and multiple input variables using empirical data from designed experiments. As the true functional relationship is often unknown, RSM employs regression-based polynomial equations to approximate system behaviour, predict responses, and optimise performance (Eckert et al., 2021; Ghaebi & Soleymani, 2024; Pasha et al., 2023). Viewing the process as a continuous response surface across the experimental domain, RSM enables identification of optimal or target conditions while visually revealing factor interactions (Kumar & Reji, 2023; Nourmohammadzadeh et al., 2023; Onyelowe et al., 2024). In contrast to the Taguchi Method's focus on robustness, RSM emphasises precision and is frequently applied after Taguchi screening to refine and model key influencing parameters (Sai et al., 2023; Kumar & Reji, 2023).

Mathematical Foundation

Mathematically, the relationship between the response Y and the independent variables x1,x2,…,xnx can be represented as:

$$Y = f(x1,x2,\ldots,xn) + \varepsilon$$

where ε\varepsilonε represents the random error term. The true functional relationship fff is usually unknown, but within a localised experimental region, it can be approximated using a low-order polynomial function.

For most engineering applications, either a first-order (linear) or a second-order (quadratic) polynomial model is sufficient.

(a) First-Order Model

$$Y = \beta 0 + i = 1\sum n\beta ixi + \varepsilon$$

This model is suitable when the response varies linearly with the input factors. It helps identify the general direction in which the response changes but may not capture curvature effects (Yan et al., 2025).

(b) Second-Order Model

$$Y = \beta 0 + i = 1\sum n\beta ixi + i = 1\sum n\beta iixi2 + i < j\sum \beta ijxixj + \varepsilon$$

This quadratic model accounts for curvature and interaction effects, providing a more accurate representation of real-world systems. The inclusion of squared and interaction terms allows for mapping the response surface in three dimensions, which can be visualised through contour and 3D plots.

The coefficients βi, βii, and βij are estimated using regression analysis, typically by the method of least squares. Once the model is built, statistical measures such as R^2 (coefficient of determination), adjusted R^2, and p-values are used to assess its adequacy and predictive power (Mohammed et al., 2024).

Experimental Design in RSM

RSM relies on structured designs to efficiently generate data for modelling. Central Composite Design (CCD) combines factorial, axial, and centre points to capture curvature effects (Bayuo et al., 2020; Brito et al., 2024).Box–Behnken Design (BBD) avoids extreme conditions, ensuring economical and safe experimentation within bounded domains.Both are supported by software such as Design Expert, Minitab, and MATLAB (Bayuo et al., 2020; Brito et al., 2024).

Visualisation and Interpretation

Visualisation is RSM's key strength.Contour plots display two-factor interactions via equal-response curves.3D surface plots highlight peaks, valleys, and saddle points.

These tools simplify the communication of complex data and guide optimisation (Kepdieu et al., 2023; Paw et al., 2023; Hafid et al., 2023).

Model Adequacy and Validation

Reliability is confirmed through ANOVA, R^2/Adjusted R^2, residual plots, and lack-of-fit tests. A valid model shows statistical significance, minimal residual bias, and high predictive accuracy (Mihalovits & Keméný, 2022; Raposo & Barceló, 2021; Strobl et al., 2022).

Illustrative Example

In a chemical process optimisation, CCD with five levels of temperature and concentration was used to model yield. The response surface revealed a curvature trend—yield increasing to an optimum before declining. Numerical optimisation identified optimal conditions validated experimentally, demonstrating RSM's precision and efficiency (Peng et al., 2020; Victoria et al., 2023).

Advantages and Limitations

RSM enables predictive polynomial modelling, interaction visualisation, and efficient single/multi-response optimisation through desirability functions. Its limitations include dependency on polynomial assumptions, data sensitivity, and reduced accuracy for highly nonlinear systems. These are mitigated by hybridization with Artificial Neural Networks (ANNs), Genetic Algorithms (GAs), or Fuzzy Logic (Kumari & & Lavanya, 2024; Mairpady et al., 2021; Onyelowe et al., 2024).

Role in Intelligent Optimisation

RSM provides the analytical backbone of hybrid intelligent systems, defining interpretable models that guide AI-based refinements. It defines the search domain for neural or evolutionary algorithms, which iteratively adjust parameters for nonlinear, multi-objective optimisation (Singh & Akhai, 2024; Ghaheh et al., 2024; Nourmohammadzadeh et al., 2023; Pasha et al., 2023). This synergy bridges experimental design with intelligent computation, positioning RSM as an essential component of modern data-driven optimisation.

COMPARATIVE ANALYSIS: TAGUCHI VS. RSM

The Taguchi Method and RSM are pivotal optimisation tools that differ in philosophy yet complement each other in practical application. Taguchi emphasises robustness and simplicity, while RSM prioritises precision and predictive modelling, together forming a balanced framework for intelligent experimentation and process optimisation (Rudolph et al., 2024; Stegherr et al., 2020).Table 1 presents an analytical comparison of the Taguchi method and RSM frameworks as applied in intelligent optimisation studies.

Philosophical Foundation

Taguchi focuses on minimising variation and achieving consistent performance under noise using orthogonal arrays, emphasising robustness and reliability (Kumar & Lavanya, 2024; Yang et al., 2022). In contrast, RSM employs regression-based polynomial modeling to quantify and optimize system behavior with analytical accuracy (Pasha et al., 2023; Paw et al., 2023).

Experimental and Analytical Framework

Taguchi utilises discrete orthogonal arrays for efficient screening of parameters, ideal for moderate system complexity (Elsayed et al., 2022; Hisam et al., 2024). RSM applies continuous designs such as CCD and BBD to capture curvature and interactions, enabling deeper analysis at the cost of more trials (Bheel et al., 2024; Kumar & Reji, 2023). Analytically, Taguchi relies on Signal-to-Noise ratios, ANOVA, and main effect plots, while RSM uses regression analysis, contour plots, and statistical validation metrics like F-tests, p-values, and R^2 for high-precision optimisation (Rahman et al., 2022; Triller et al., 2023).

Focus and Application

Taguchi is best suited for early-stage screening, discrete parameters, and cost-sensitive industrial applications that demand robustness (Fernández-García et al., 2020; Hisam et al., 2024; Karazi et al., 2019; Moussa et al., 2022). RSM, by contrast, excels in detailed modelling, continuous systems, and multi-response optimisation, making it ideal for research and advanced product development (Hisam et al., 2024; Karazi et al., 2019; Kumar & Lavanya, 2024).

Strengths and Limitations

Taguchi's advantages include simplicity, low cost, and minimal experimentation, but it is limited in modelling complex nonlinear interactions and multi-response systems. RSM offers superior visualisation and modelling capability but requires more data, assumes polynomial relationships, and can be noise-sensitive (Chelladurai et al., 2020; Ishola & McKenna, 2021; Zhang & Wu, 2021).

Integration and Complementarity

The hybrid Taguchi–RSM approach merges experimental economy with analytical rigour—Taguchi identifies critical parameters, while RSM refines their relationships for precise optimisation. This synergy strengthens modern intelligent optimisation systems by linking experimental design with computational modelling, advancing data-driven, sustainable, and high-performance engineering (Kumar & Lavanya, 2024; Kushwah et al., 2024; Hakim et al., 2025; Wegener et al., 2021; Yarasu & Podgornik, 2025). Collectively, these frameworks represent complementary paradigms that bridge practical robustness and mathematical sophistication, aligning with the goals of intelligent and sustainable industrial transformation (Bubeník et al., 2025; Ghadoui et al., 2023; Huang et al., 2024).

Table 1. Comparison between Taguchi and RSM Methods

Aspect	Taguchi Method	Response Surface Methodology (RSM)
Primary Focus	Robust design and reduction of process variability	Precise modelling and prediction of system behaviour
Experimental Design	Uses orthogonal arrays (discrete levels) for efficient screening	Uses Central Composite or Box–Behnken designs (continuous variables)
Computation & Analysis	Based on S/N ratio and ANOVA, simple and practical	Regression modelling with polynomial equations; mathematically intensive
Best Use Stage	Early screening and identification of key factors	Fine-tuning and detailed optimisation after variable selection
Strengths	Reduces the number of experiments; improves robustness; cost-effective	Captures curvature and interactions; provides predictive and graphical insight
Limitations	Less effective for nonlinear or multi-response systems	Requires more runs; sensitive to data quality and assumes polynomial behaviour
Integration Potential	Often used as a preliminary step in hybrid models	Serves as the refinement and modelling stage in hybrid frameworks

HYBRID TAGUCHI–RSM FRAMEWORKS

The hybrid Taguchi–RSM framework integrates Taguchi's efficient experimental design with the analytical precision of RSM to address nonlinear, multi-variable systems across manufacturing, materials, and process engineering. This hybridisation enhances optimisation accuracy while minimising experimental cost and time, offering a balance between robustness and mathematical depth (Bheel et al., 2024; Hisam et al., 2024; Kumar & Lavanya, 2024; Modi & Gopal, 2020). Taguchi efficiently identifies significant factors but lacks predictive modelling, whereas RSM provides quantitative insight at the expense of more experimentation; their combination merges Taguchi's economy with RSM's precision to create adaptive, data-driven optimisation frameworks applicable in aerospace, biotechnology, and other complex industries (Kumar & Lavanya, 2024; Shie et al., 2019; Wichapa et al., 2024; Aje & Adie, 2020; Moussa et al., 2022). Typically, hybrid approaches follow a two-stage process—Taguchi screens key factors using orthogonal arrays, followed by RSM's Central Composite or Box–Behnken designs to model curvature and interactions—thereby achieving high optimisation accuracy with fewer trials (Cuesta et al., 2022; López et al., 2019; Mahapatra et al., 2020; Carabajal et al., 2019; Fischer et al., 2019; Serani & Diez, 2024). Such integration has proven effective in applications like precision machining, polymer blending, additive manufacturing, and thermal processing, leading to improved product quality, reduced cost, and enhanced reliability (Chandra et al., 2023; Elshaer et al., 2024; Kumar et al., 2022; Rao et al., 2023; Rudrapati et al., 2020; Castéran et al., 2020; Sun et al., 2019).

Moreover, incorporating computational intelligence—such as Artificial Neural Networks (ANNs), Fuzzy Logic, and Evolutionary Algorithms like Genetic Algorithms (GAs) and Particle Swarm Optimisation (PSO)—further enhances predictive capability, uncertainty management, and global optimisation (Ao et al., 2022; Darabkh & Al-Akhras, 2025; Harudin et al., 2020; Ojha et al., 2019; Danesh & Danesh, 2023; Roy et al., 2021; Yu & Liu, 2024). Collectively, these integrations yield self-learning, adaptive optimisation systems that continuously refine accuracy and robustness, aligning with the goals of intelligent, data-driven, and sustainable industrial innovation (Hardian et al., 2020; Nikzad et al., 2024).

COMPUTATIONAL TOOLS AND SOFTWARE IMPLEMENTATION

The effective application of Taguchi, RSM, and hybrid frameworks relies on robust computational tools that facilitate experimental design, statistical analysis, and visualisation with accuracy and reproducibility. As research and industries embrace automation and AI integration, tools like Minitab, Design Expert, MATLAB, and Python-based frameworks have become essential for large-scale, multi-objective optimisation (Aje & Adie, 2020; Hisam et al., 2024; Simone et al., 2025).

Minitab offers a user-friendly interface for Design of Experiments (DoE) and Taguchi–RSM applications. It provides templates for orthogonal arrays, automatic computation of Signal-to-Noise (S/N) ratios, ANOVA, and interaction plots, enabling users to quickly identify optimal conditions. Its regression and visualisation capabilities support RSM, generating contour and surface plots for predictive modelling. Minitab's automated optimisation and result interpretation make it ideal for industrial practitioners and educators (Aje & Adie, 2020; Hisam et al., 2024).

Design Expert, developed by Stat-Ease Inc., is the leading platform for RSM and advanced DoE. It supports CCD, BBD, and D-optimal designs, offering powerful 3D visualisation and desirability function tools for multi-response optimisation. It's integrated regression, ANOVA, and diagnostic features ensure statistical validity and practical relevance. The software's clarity and publication-ready graphics have made it a top choice in research and industrial innovation (Czyrski & Jarzębski, 2020; Victoria et al., 2023; Bachorz et al., 2024; Mohammed et al., 2024; Hejazi & Moradpour, 2019; Zăvoianu et al., 2020).

MATLAB serves as a versatile computational environment for customised hybrid modelling. It enables researchers to design Taguchi orthogonal arrays, compute S/N ratios, perform ANOVA, and conduct RSM modelling using its Optimisation and Curve Fitting Toolboxes. Unlike predefined software, MATLAB supports direct integration with AI and machine learning algorithms—such as ANNs, Fuzzy Logic, and GAs—through its specialised toolboxes. This makes it highly suited for intelligent optimisation in fields like smart manufacturing and predictive maintenance (Dang et al., 2019; Goranov et al., 2023; Shojaei et al., 2021; Karazi et al., 2019; Melaku & Bogale, 2023; Onokwai et al., 2024; Heirendt et al., 2019; Salgotra et al., 2023).

Collectively, these computational tools bridge traditional statistical design and modern intelligent computation, supporting automation, adaptability, and reproducibility in Taguchi–RSM-based research and applications.

Python-based frameworks

The widespread adoption of open-source platforms like Python has revolutionisedoptimisation by providing accessible, powerful alternatives to proprietary software for implementing Taguchi and RSM methodologies. Python libraries support everything from experimental design to advanced hybrid modelling:

- pyDOE enables easy creation of factorial and Central Composite Designs (CCD), generating Taguchi-style orthogonal arrays for structured experimentation (Jones, 2019; Simone et al., 2025; Xu et al., 2024).
- Statsmodels offers comprehensive statistical tools—regression, ANOVA, and residual diagnostics—essential for RSM modelling and validation (Pham et al., 2020; Popović, 2020; Rahman et al., 2022).
- Scikit-Opt and SciPy support single and multi-objective optimisation, integrating smoothly with algorithms like PSO and GAs (Aldeghi et al., 2021; Tsoulos et al., 2025; Venkatesh et al., 2024).

- Scikit-Learn complements RSM by providing machine learning capabilities for regression, classification, and cross-validation, enhancing predictive modelling (Ghader, 2025; Zhang & Wu, 2021).

Python's flexibility allows the creation of custom hybrid workflows, merging the statistical rigour of Taguchi–RSM with AI's adaptive intelligence. Its open-source nature fosters reproducibility, scalability, and collaboration, aligning with modern data-driven and IoT-integrated research paradigms (Ranjan et al., 2023; Talati, 2021).

Overall, Python democratizes optimisation, enabling academia and industry to conduct high-level analyses without cost barriers. Alongside tools like Minitab, Design Expert, and MATLAB, it ensures that the Taguchi–RSM hybrid framework remains robust, reproducible, and future-ready—bridging traditional experimentation with intelligent, automated computational systems (Botha et al., 2021; Zubair et al., 2024).

Table 2. Comparative Summary of Computational Tools for Taguchi–RSM and Hybrid Optimisation

Software / Platform	Key Features	Primary Advantages	Limitations / Challenges	Typical Use Case
Minitab	GUI-based statistical software for Design of Experiments (DoE); supports Taguchi arrays, S/N ratios, ANOVA, and regression modelling.	Simple, automated workflow; ideal for quick industrial analysis; minimal coding required.	Limited flexibility; unsuitable for advanced AI integration.	Industrial quality control, process optimisation, training labs.
Design Expert	Specialised for RSM (CCD, BBD, D-optimal designs); offers desirability functions and model validation.	Strong visualisation (3D plots, contour maps); built-in multi-response optimisation.	Proprietary; less adaptable to custom algorithms.	Advanced design modelling in chemical, pharmaceutical, and materials research.
MATLAB	Script-based numerical computing environment; supports RSM and Taguchi scripting, regression, and AI toolboxes (ANN, GA, Fuzzy Logic).	Highly flexible; allows AI integration and custom hybrid model development.	Requires programming expertise; licensed software.	Intelligent optimisation, predictive maintenance, and hybrid AI-RSM studies.
Python Frameworks (pyDOE, Statsmodels, Scikit-Opt, Scikit-Learn)	Open-source libraries enabling Taguchi/RSM designs, regression, optimisation, and machine learning integration.	Free, modular, reproducible; supports cloud and IoT data integration.	Needs coding skills and library management; validation must be manually ensured.	Scalable research projects, AI-driven optimisation, and academic model development.

APPLICATIONS ACROSS MULTIPLE DOMAINS

The combined use of the Taguchi Method and RSM—along with their hybrid extensions—has demonstrated remarkable versatility across scientific and industrial domains. By balancing experimental efficiency with analytical precision, these methods enable systematic modelling, analysis, and optimisation of complex multi-variable systems that would otherwise require exhaustive experimentation (Dudek, 2025; Pasupuleti et al., 2024).

In manufacturing and materials processing, hybrid Taguchi–RSM models optimise machining, welding, casting, and additive manufacturing by identifying key factors such as cutting speed, feed rate, and tool geometry through Taguchi's orthogonal arrays, while RSM quantitatively models their interactions. This synergy enhances surface finish, tool life, and dimensional accuracy, accelerates product development, and supports sustainable, energy-efficient manufacturing (Elahi et al., 2023; Tănase et al., 2024).

In civil and environmental engineering, these frameworks guide the development of low-carbon, high-performance concretes and sustainable materials by optimising mix proportions, curing regimes, and activator concentrations. Hybrid Taguchi–RSM applications in geopolymers, self-compacting concretes, and fibre-reinforced systems improve strength, durability, and eco-efficiency (Faridmehr et al., 2021; Mashaly et al., 2023; Tipu et al., 2025; Dhiman et al., 2025). Similarly, in environmental engineering, they refine wastewater treatment, bioreactor design, and adsorption processes by quantifying the combined effects of parameters such as pH, dosage, and contact time (Mohanadevi & Dhanabalan, 2024; Sheikhmohammadi et al., 2024).

In healthcare and biotechnology, Taguchi–RSM frameworks are used for drug formulation, bioprocess optimisation, and diagnostic system calibration. They enable systematic control of excipient ratios, reaction conditions, and process parameters, improving bioavailability, yield, and stability. In nanobiotechnology, these models regulate particle size and dispersity for drug delivery and imaging applications, while ensuring robustness under variable clinical conditions (Bakhaidar et al., 2022; Hisam et al., 2024; Sims et al., 2020; Yoo et al., 2023; Kaziz et al., 2022; Kok et al., 2024; Mala et al., 2025).

In the energy and thermal sectors, RSM predicts heat-transfer behaviour and Taguchi ensures operational robustness, jointly optimising parameters in heat exchangers, combustion systems, and cooling devices (Lohith et al., 2024; Venkatesh et al., 2024). These techniques also enhance renewable energy and storage applications—optimising solar collector geometry, biofuel feedstock composition, and electrode materials to improve energy conversion efficiency and sustainability (Prasertpong et al., 2019; Sajdak et al., 2025).

Within artificial intelligence (AI) and machine learning (ML), hybrid Taguchi–RSM frameworks are increasingly used for efficient hyperparameter tuning. Taguchi's orthogonal arrays provide systematic sampling, while RSM models nonlinear relationships among learning rate, network depth, and batch size, accelerating convergence and improving generalization compared with traditional grid or random search (Nguyen et al., 2024; Yu & Zhu, 2020; Ali et al., 2023; Ishfaq et al., 2023; Pal, 2023; Säfken & Rügamer, 2024; Sipper, 2022).

Collectively, across all these sectors, the Taguchi–RSM hybrid paradigm serves as a universal, sustainable optimisation framework—merging statistical design, predictive modelling, and adaptive intelligence to deliver precision, efficiency, and resilience in complex real-world systems (Paw et al., 2023; Schwaninger & Ott, 2023; Wei et al., 2024).

HYBRID STATISTICAL–COMPUTATIONAL RESEARCH APPROACHES

The rise of hybrid statistical–computational optimisation reflects the convergence of structured experimental design and computational intelligence. By uniting deterministic experimentation (Taguchi, RSM) with adaptive learning (AI, ML), researchers achieve interpretable, self-evolving systems capable of continuous performance refinement. This integration transforms static models into dynamic, data-driven frameworks for real-time, sustainable, and predictive optimisation—bridging traditional statistics with intelligent computation (Fózer et al., 2023; Zubair et al., 2024). Table 3 highlights representative case studies of hybrid Taguchi–RSM applications across disciplines.

Table 3. Sustainable and Energy-Efficient Optimisation Case Studies Using Taguchi–RSM or Hybrid Frameworks

Application / Case Study	Description of Sustainable or Energy-Efficient Outcome	Reference
Internal Gear Pump Optimisation	Taguchi + RSM + ANOVA used to reduce energy consumption of internal gear pumps from 156.1 Wh/m³ → 92.0 Wh/m³, improving hydraulic efficiency and lowering operational cost.	(Öztürk et al., 2022)
Residential Solar Energy Optimisation (Jordan)	Taguchi and Box–Behnken (RSM) applied to PV systems to minimise annual electricity use and enhance sustainable energy management in residential buildings.	(Al-Husban et al., 2023)
Hydraulic Ram Pump Performance	Comparative study of Taguchi and RSM showing how design optimisation improves energy transfer efficiency in low-energy hydraulic water pumps for rural/off-grid systems.	(Romelin et al., 2024)
Polymer Composite Manufacturing Optimisation	Integration of machine learning with RSM to optimise composite processing parameters, reducing waste and energy use while enhancing material performance.	(Raza et al., 2025)

Integration with Artificial Intelligence and Machine Learning

Integrating AI and ML with Taguchi–RSM transforms traditional statistical optimisation into intelligent, data-driven modelling. Machine learning, especially ANNs, captures nonlinear and complex relationships that Taguchi or RSM alone cannot (Feng et al., 2024; Jaddi et al., 2024; Onyelowe et al., 2024). Data from experimental designs train ANNs to predict untested outcomes, while GA and PSO further refine global optimisation (Jong et al., 2020; Kumari & & Lavanya, 2024; Zubair et al., 2024). This synergy creates adaptive systems that self-learn and enhance prediction accuracy, widely applied in machining, chemical, aerospace, and energy optimisation (Boukredera et al., 2025; Mishra et al., 2022; Elsayed et al., 2022).

Multi-Objective Optimisation

Real-world engineering problems often involve conflicting goals—like maximising strength while minimising weight—making multi-objective optimisation (MOO) essential (Mavi et al., 2023; Wang et al., 2024). Traditional RSM focuses on single responses, but hybrid models using desirability functions and Pareto-based algorithms handle multiple objectives simultaneously (Paw et al., 2023; Teng et al., 2019). Desirability methods convert responses into a 0–1 scale, while Pareto optimisation, through algorithms like MOGA, NSGA-II, and DE, identifies non-dominated trade-off solutions (Heddar et al., 2023; Nouri-Moghaddam et al., 2020; Zong et al., 2022). These approaches are applied across materials, manufacturing, biomedical, and environmental fields to balance performance, efficiency, and sustainability (Bangi & Kwon, 2019; Fózer et al., 2023; Teh et al., 2019).

Decision Support Systems

The integration of hybrid statistical–computational research into Decision Support Systems (DSS) enables data-driven, adaptive, and sustainable decision-making across industries (Oppioli et al., 2023; Zubair et al., 2024). Built on the Taguchi–RSM–AI framework, DSS platforms combine experimental modelling with real-time data monitoring and predictive feedback, creating closed-loop optimisation systems (Christensen et al., 2021; Pithan et al., 2023).Applications include adaptive control in smart manufacturing, energy management, and pharmaceutical production, where systems autonomously adjust parameters for efficiency and quality (Kodumuru et al., 2025; Niazi, 2023). These hybrid DSS models also enhance collaboration through visual dashboards and simulations, turning static analyses into dynamic optimisation environments (Pandi et al., 2022; Weidele et al., 2023). Such systems mark a shift toward autonomous, self-learning, and sustainable technologies defining Industry 4.0 and beyond (Fózer et al., 2023; Tangjitsitcharoen et al., 2025).

FUTURE SCOPE, EMERGING TRENDS, AND RESEARCH DIRECTIONS

The future of optimisation lies in intelligent, adaptive, and sustainable systems shaped by the convergence of digitalisation, automation, and artificial intelligence (AI). The Taguchi–RSM hybrid framework anchors this transformation, supporting continuous learning, real-time decision-making, and cross-domain adaptability (Kalach et al., 2024; Polo et al., 2025; Zeng, 2024).

Emerging Trends

In the Industry 4.0 era, Taguchi–RSM hybrids form the analytical foundation of digital twins and cyber-physical systems, ensuring predictive maintenance, real-time control, and reliability (Akhai & Wadhwa, 2024; Singh & Akhai, 2024; Friederich et al., 2021; Kumar & Misra, 2020; Kronberger et al., 2020; Kumar et al., 2022; Belim et al., 2024; Soori et al., 2023). Their integration with IoT and edge computing enables localised, sensor-driven optimisation in manufacturing, energy, and agriculture (Elias et al., 2022; Msakni et al., 2023; Artiushenko et al., 2024; Li et al., 2022; Awais et al., 2025; Chen & Huang, 2025). In materials and nanotechnology, the framework aids in optimising nanocomposite synthesis and energy material processing for green, cost-efficient production (Lai et al., 2023; Lei et al., 2021; Martínez & Shadman, 2024; Drakopoulos et al., 2021; Lee et al., 2025).

The integration of deep learning with Taguchi–RSM enables self-updating, explainable models capable of handling nonlinear, high-dimensional data, while reinforcement learning and explainable AI (XAI) improve adaptability and transparency (Onyelowe et al., 2024; Zubair et al., 2024; Dörfler et al., 2024; Treloar et al., 2022; Agrawal et al., 2025; Sudjianto & Zhang, 2021). Meanwhile, cloud and quantum computing are revolutionising scalability and computational speed, allowing distributed and large-scale optimisation (Huang et al., 2024; Yang et al., 2023).

Future Research Directions

Future work should focus on:

1. Integrating advanced AI models—deep learning, reinforcement learning, and XAI—for adaptive and transparent optimisation.
2. Linking digital twins and IoT ecosystems to enable real-time, data-driven Taguchi–RSM optimisation.
3. Embedding sustainability metrics, including life-cycle assessment (LCA) and circular-economy analysis, into hybrid frameworks.

4. Advancing multi-objective optimisation through evolutionary algorithms, fuzzy decision-making, and desirability-based approaches.
5. Expanding cross-disciplinary applications to healthcare, energy storage, climate modeling, and socio-technical innovation.
6. Developing automated, open-source platforms to promote accessibility, scalability, and reproducibility.

Collectively, these directions signify a transition toward Industry 5.0, where human–AI collaboration, real-time intelligence, and sustainability converge. The Taguchi–RSM hybrid framework will continue to serve as a vital link between statistical experimentation and intelligent computation, driving the evolution of autonomous, adaptive, and environmentally responsible systems (Sharma et al., 2020; Zubair et al., 2024).

CONCLUSION

Optimisation is central to scientific and industrial progress, evolving from trial-and-error experimentation to modern, data-driven strategies. Among the influential frameworks are the Taguchi Method and RSM, each embodying distinct philosophies—Taguchi focusing on robustness and practicality, while RSM emphasises precision and predictive modelling. Their integration has led to hybrid Taguchi-RSM frameworks, enhancing experimental accuracy and broadening the applicability of statistical optimisation to complex nonlinear and multi-response systems.

This chapter examined the theoretical basis, design philosophy, and adaptability of these methods across various fields, including manufacturing, engineering, healthcare, and artificial intelligence. Taguchi-RSM hybrids act as powerful decision-making tools, identifying optimal conditions, improving robustness, reducing costs, and supporting sustainable innovation. They integrate seamlessly with digital tools and machine learning algorithms, solidifying their role in modern optimisation.

As computational intelligence advances, Taguchi and RSM are evolving into crucial components of hybrid statistical-computational ecosystems. The combination of methodologies like Artificial Neural Networks, Fuzzy Logic, and Genetic Algorithms enables adaptive systems capable of self-learning. Looking forward, these hybrid systems will be vital in Industry 4.0 and IoT ecosystems, facilitating autonomous optimisation and accelerated discovery in advanced materials.

Ultimately, the shift toward AI-driven statistical modelling—using deep learning on Taguchi-RSM datasets—will redefine optimisation, making it a self-correcting and sustainable process. The synergy between Taguchi and RSM exemplifies the convergence of classical statistical reasoning with computational intelligence, empowering researchers to conduct smarter experiments and drive innovation in efficiency and sustainability.

REFERENCES

Abad, Z. A. N., & Momayezi, F. (2024). Optimizing modular hub location in air and road transportation systems. *Journal of Project Management*, *9*(3), 277–300. DOI: 10.5267/j.jpm.2024.3.001

Agrawal, R., Gupta, T., Gupta, S., Chauhan, S. S., Patel, P., & Hamdare, S. (2025). Fostering trust and interpretability: Integrating explainable AI (XAI) with machine learning for enhanced disease prediction and decision transparency. *Diagnostic Pathology*, *20*(1), 105. Advance online publication. DOI: 10.1186/s13000-025-01686-3

Aje, O. F., & Adie, A. J. (2020). Response surface methodology and Taguchi method based applications – A Review. Zenodo *(CERN European Organization for Nuclear Research)*. European Organization for Nuclear Research. DOI: 10.5281/zenodo.4467686

Akhai, S., & Khang, A. (2024). Innovations in medical manufacturing: A review of 3D printing, robotics, and internet of things (IoT). *The Quantum Evolution*, 226-241. DOI: 10.1201/9781032642079-10

Akhai, S., Singh, V. P., & John, S. (2016). Human performance in industrial design centers with small unit air conditioning systems. *Journal of Advanced Research in Production Industrial Engineering*, *3*(2), 5–11.

Akhai, S., Srivastava, P., Sharma, V., & Bhatia, A. (2021, August). Investigating weld strength of AA8011-6062 alloys joined via friction-stir welding using the RSM approach. In *Journal of Physics: Conference Series*, 1950(1), 012016. IOP Publishing. DOI: 10.1088/1742-6596/1950/1/012016

Akhai, S., & Wadhwa, A. S. (2024). Wireless networking solutions for climate control and comfort in modern air conditioning systems. In *Intelligent Networks* (pp. 165–181). CRC Press., DOI: 10.1201/9781003541363-11

Al-Husban, Y., Al-Ghriybah, M., Gaeid, K. S., Takialddin, A. S., Handam, A., & Alkhazaleh, A. H. (2023). Optimisation of the residential solar energy consumption using the taguchi technique and box-behnken design: A case study for Jordan. *International Journal on Energy Conversion*, *11*(1), 25. Advance online publication. DOI: 10.15866/irecon.v11i1.22672

Aldeghi, M., Häse, F., Hickman, R. J., Tamblyn, I., & Aspuru-Guzik, A. (2021). Golem: An algorithm for robust experiment and process optimisation. *Chemical Science (Cambridge)*, *12*(44), 14792–14807. DOI: 10.1039/D1SC01545A

Aldulaijan, N., Marsden, J. A., Manson, J. A., & Clayton, A. D. (2023). Adaptive mixed variable bayesian self-optimisation of catalytic reactions. *Reaction Chemistry & Engineering*, *9*(2), 308–316. DOI: 10.1039/D3RE00476G

Ali, H., Muthudoss, P., Chauhan, C., Ilango, K., Kumar, D., Paudel, A., & Ramasamy, G. (2023). Machine learning-enabled NIR Spectroscopy. Part 3: Hyperparameter by Design (HyD) Based ANN-MLP Optimisation, Model Generalizability, and Model Transferability. *AAPS PharmSciTech*, *24*(8), 254. Advance online publication. DOI: 10.1208/s12249-023-02697-3

Ao, S.-I., Gelman, L., Karimi, H. R., & Tiboni, M. (2022). Advances in machine learning for sensing and condition monitoring. *Applied Sciences (Basel, Switzerland)*, *12*(23), 12392. DOI: 10.3390/app122312392

Arteaga-Cabrera, E., Ramírez-Márquez, C., Sánchez-Ramírez, E., Segovia-Hernández, J. G., Osorio, O., & Gómez-Salazar, J. A. (2025). Advancing optimisation strategies in the food industry: From traditional approaches to multi-objective and technology-integrated solutions. *Applied Sciences (Basel, Switzerland)*, *15*(7), 3846. DOI: 10.3390/app15073846

Artiushenko, V., Lang, S., Lerez, C., Reggelin, T., & Hackert-Oschätzchen, M. (2024). Resource-efficient Edge AI solution for predictive maintenance. *Procedia Computer Science*, *232*, 348–357. DOI: 10.1016/j.procs.2024.01.034

Awais, M., Wang, X., Hussain, S., Aziz, F., & Mahmood, M. Q. (2025). Advancing precision agriculture through digital twins and smart farming technologies: A Review [Review of *Advancing Precision Agriculture Through Digital Twins and Smart Farming Technologies: A Review*]. *AgriEngineering*, *7*(5), 137. Multidisciplinary Digital Publishing Institute. DOI: 10.3390/agriengineering7050137

Aziz, A., Mehboob, S. S., Tayyab, A., Khan, D., Hayyat, K., Ali, A., & Qureshi, Q. B. I. L. (2024). Enhancing sustainability in self-compacting concrete by optimizing blended supplementary cementitious materials. *Scientific Reports*, *14*(1), 12326. Advance online publication. DOI: 10.1038/s41598-024-62499-w

Bachorz, R. A., Nowak, D., & Ratajewski, M. (2024). QSPRmodeler - An open source application for molecular predictive analytics. *Frontiers in Bioinformatics*, *4*, 1441024. Advance online publication. DOI: 10.3389/fbinf.2024.1441024

Bakhaidar, R. B., Naveen, N. R., Basim, P., Murshid, S. S. A., Kurakula, M., Alamoudi, A. J., Bukhary, D. M., Jali, A. M., Majrashi, M., Alshehri, S., Alissa, M., & Ahmed, R. A. (2022). Response surface methodology (RSM) powered formulation development, optimisation and evaluation of thiolated based mucoadhesive nanocrystals for local delivery of simvastatin. *Polymers*, *14*(23), 5184. DOI: 10.3390/polym14235184

Banashooshtari, H., Khamehchi, E., Rashidi, F., & Dargi, M. (2024). An experimental study of epoxy-based nanocomposite for chemical consolidation in a sandstone reservoir with high clay content. *Journal of Petroleum Exploration and Production Technology, 14*(10), 2769–2784. DOI: 10.1007/s13202-024-01853-0

Bangi, M. S. F., & Kwon, J. S. (2019). Deep hybrid modeling of chemical process: Application to hydraulic fracturing. *Computers & Chemical Engineering, 134*, 106696. DOI: 10.1016/j.compchemeng.2019.106696

Batra, N., Kaur, J., & Batra, N. K. (2020). SmartCOdetect: An automated car window opening system on detection of carbon monoxide. *International Journal of Vehicle Structures & Systems, 12*(4), 398–404. DOI: 10.4273/ijvss.12.4.09

Bayuo, J., Abukari, M. A., & Pelig-Ba, K. B. (2020). Optimisation using central composite design (CCD) of response surface methodology (RSM) for biosorption of hexavalent chromium from aqueous media. *Applied Water Science, 10*(6), 135. Advance online publication. DOI: 10.1007/s13201-020-01213-3

Belim, M., Meireles, T., Gonçalves, G., & Pinto, R. (2024). Forecasting models analysis for predictive maintenance. *Frontiers in Manufacturing Technology, 4*, 1475078. Advance online publication. DOI: 10.3389/fmtec.2024.1475078

Bell, N. L., Kupper, M., & Cronin, L. (2021). Design of experiments for optimisation of polyoxometalate syntheses. *Chemistry of Materials, 33*(18), 7263–7271. DOI: 10.1021/acs.chemmater.1c01401

Bellonzi, N., Kunitsa, A. A., Cantin, J. T., Campos-Gonzalez-Angulo, J. A., Radin, M. D., Zhou, Y., Johnson, P. D., Martínez-Martínez, L. Á., Jangrouei, M. R., Brahmachari, A. S., Wang, L., Patel, S., Kodrycka, M., Loaiza, I., Lang, R. A., Aspuru-Guzik, A., Izmaylov, A. F., Fontalvo, J. R., & Cao, Y. (2024). Feasibility of accelerating homogeneous catalyst discovery with fault-tolerant quantum computers. *arXiv (Cornell University)*. https://doi.org//arxiv.2406.06335DOI: 10.48550

Bheel, N., Shams, M. A., Sohu, S., Buller, A. S., Najeh, T., Ismail, F. I., & Benjeddou, O. (2024). A comprehensive study on the impact of human hair fiber and millet husk ash on concrete properties: Response surface modeling and optimisation. *Scientific Reports, 14*(1), 13569. Advance online publication. DOI: 10.1038/s41598-024-63050-7

Botha, N., Inglis, H. M., Coetzer, R., & Labuschagne, F. J. (2021). Statistical design of experiments: An introductory case study for polymer composites manufacturing applications. *MATEC Web of Conferences, 347*, 28. DOI: 10.1051/matecconf/202134700028

Boukredera, F. S., Hadjadj, A., Youcefi, M. R., & Ouadi, H. (2025). AI-driven optimisation of drilling performance through torque management using machine learning and differential evolution. *Processes (Basel, Switzerland)*, *13*(5), 1472. DOI: 10.3390/pr13051472

Brownlee, A. E. I., Epitropakis, M. G., Mulder, J., Paelinck, M., & Burke, E. (2022). A systematic approach to parameter optimisation and its application to flight schedule simulation software. *Journal of Heuristics*, *28*(4), 509–538. DOI: 10.1007/s10732-022-09501-8

Bubeník, P., Rakyta, M., Buzalka, M., Biňasová, V., & Kovaríková, Z. (2025). Optimisation of business processes using artificial intelligence. *Electronics (Basel)*, *14*(11), 2105. DOI: 10.3390/electronics14112105

Carabajal, M., Teglia, C. M., Cerutti, S., Culzoni, M. J., & Goicoechea, H. C. (2019). Applications of liquid-phase microextraction procedures to complex samples assisted by response surface methodology for optimisation. *Microchemical Journal*, *152*, 104436. DOI: 10.1016/j.microc.2019.104436

Castéran, F., Ibáñez, R., Argerich, C., Delage, K., Chinesta, F., & Cassagnau, P. (2020). Application of machine learning tools for the improvement of reactive extrusion simulation. *Macromolecular Materials and Engineering*, *305*(12), 2000375. Advance online publication. DOI: 10.1002/mame.202000375

Chandra, A., Yadav, A., Singh, S., & Arora, P. K.Pawan Kumar Arora. (2023). Optimisation of machining parameters for CNC milling of fibre reinforced polymers. *Evergreen*, *10*(2), 765–773. DOI: 10.5109/6792826

Chelladurai, S. J. S., Murugan, K., Ray, A. P., Upadhyaya, M., Narasimharaj, V., & Gnanasekaran, S. (2020). Optimisation of process parameters using response surface methodology: Review of *Optimisation of process parameters using response surface methodology: A review. Materials Today Proceedings, 37*, 1301. Elsevier BV. DOI: 10.1016/j.matpr.2020.06.466

Chen, D., & Huang, Y. (2025). Integrating reinforcement learning and large language models for crop production process management optimisation and control through a new knowledge-based deep learning paradigm. *Computers and Electronics in Agriculture*, *232*, 110028. DOI: 10.1016/j.compag.2025.110028

Chen, Q. (2023). Response surface methodology in the optimisation of natural product extraction and separation techniques. *Academic Journal of Science and Technology*, *5*(3), 149–151. DOI: 10.54097/ajst.v5i3.7902

Chhajer, H., & Roy, R. (2024). Rationalised experiment design for parameter estimation with sensitivity clustering. *Scientific Reports, 14*(1), 25864. Advance online publication. DOI: 10.1038/s41598-024-75539-2

Christensen, M., Yunker, L. P. E., Adedeji, F., Häse, F., Roch, L. M., Gensch, T., Gomes, G. dos P., Zepel, T., Sigman, M. S., Aspuru-Guzik, A., & Hein, J. E. (2021). Data-science driven autonomous process optimisation. *Communications Chemistry, 4*(1), 112. Advance online publication. DOI: 10.1038/s42004-021-00550-x

Cuesta, I. I., Díaz, A., Rojo, M. A., Peral, L. B., Martínez, J., & Alegre, J. M. (2022). Parameter optimisation in selective laser melting on C300 steel. *Applied Sciences (Basel, Switzerland), 12*(19), 9786. DOI: 10.3390/app12199786

Czyrski, A., & Jarzębski, H. (2020). Response Surface Methodology as a Useful Tool for Evaluation of the Recovery of the Fluoroquinolones from Plasma—The Study on Applicability of Box-Behnken Design, Central Composite Design and Doehlert Design. *Processes (Basel, Switzerland), 8*(4), 473. DOI: 10.3390/pr8040473

Danesh, M., & Danesh, S. (2023). Optimal design of adaptive neuro-fuzzy inference system using PSO and ant colony optimisation for estimation of uncertain observed values. *Soft Computing, 28*(1), 135–152. DOI: 10.1007/s00500-023-09194-6

Dang, M. P., Dao, T.-P., Chau, N. L., & Le, H. G. (2019). Effective hybrid algorithm of Taguchi Method, FEM, RSM, and Teaching Learning-Based optimisation for multiobjective optimisation design of a compliant rotary positioning stage for nanoindentation tester. *Mathematical Problems in Engineering, 2019*(1). DOI: 10.1155/2019/4191924

Darabkh, K. A., & Al-Akhras, M. (2025). Evolutionary cost analysis and computational intelligence for energy efficiency in internet of things-enabled smart cities: Multi-sensor data fusion and resilience to link and device failures. *Smart Cities, 8*(2), 64. DOI: 10.3390/smartcities8020064

de Brito, T. G., Nunes, T. C., de Paiva, E. J., de Lacerda, J. C., Brito, R. F., & Brito, R. P. (2024). Optimisation of iron ore flotation parameters using response surface methodology. *Research Square (Research Square)*. DOI: 10.21203/rs.3.rs-5183292/v1

Dhiman, S., Seema, S., & Akhai, S. (2025). Performance evaluation of OPC 43 concrete incorporating micro silica and recycled ceramic tile waste as partial cement and coarse aggregate replacements. *Asian Journal of Civil Engineering*, 1-19. DOI: 10.1007/s42107-025-01504-7

Dikshit, I., Bhushan, G., Singh, R. P., & Batra, N. K. (2022). Statistical modeling and optimisation of density in a novel carbon/jute fiber reinforced hybrid composite. *Materials Today: Proceedings, 63*, 259-263. https://doi.org/ rights and contentDOI: 10.1016/j.matpr.2022.03.052Get

Dörfler, F., He, Z., Belgioioso, G., Bolognani, S., Lygeros, J., & Muehlebach, M. (2024). Toward a Systems Theory of Algorithms. *IEEE Control Systems Letters, 8*, 1198–1210. DOI: 10.1109/LCSYS.2024.3406943

Drakopoulos, S. X., Gholamipour-Shirazi, A., Macdonald, P. M., Parini, R., Reynolds, C. D., Burnett, D. L., Pye, B., O'Regan, K., Wang, G., Whitehead, T. M., Conduit, G. J., Cazacu, A., & Kendrick, E. (2021). Formulation and manufacturing optimisation of lithium-ion graphite-based electrodes via machine learning. *Cell Reports. Physical Science, 2*(12), 100683. DOI: 10.1016/j.xcrp.2021.100683

Dudek, G. (2025). Computational intelligence and machine learning: Advances in models and applications. *Electronics (Basel), 14*(8), 1530. DOI: 10.3390/electronics14081530

Eckert, T., Klein, F. C., Frieler, P., Thunich, O., & Abetz, V. (2021). Experimental design in polymer chemistry—A guide towards true optimisation of a raft polymerization using design of experiments (DoE). *Polymers, 13*(18), 3147. DOI: 10.3390/polym13183147

Elahi, M., Afolaranmi, S. O., Lastra, J. L. M., & García, J. A. P. (2023). A comprehensive literature review of the applications of AI techniques through the lifecycle of industrial equipment. *Discover Artificial Intelligence, 3*(1), 43. Advance online publication. DOI: 10.1007/s44163-023-00089-x

Elias, J. R., Chard, R., Levental, M., Liu, Z., Foster, I., & Chaudhuri, S. (2022). Real-time streaming and event-driven control of scientific experiments. *arXiv (Cornell University)*. https://doi.org//arxiv.2205.01476DOI: 10.48550

Elsayed, M. E. A., Abdo, S., Attia, A. A. A., Attia, E.-A., & ElRahman, M. A. (2022). Parametric optimisation for the design of gravity energy storage system using Taguchi method. *Scientific Reports, 12*(1), 19648. Advance online publication. DOI: 10.1038/s41598-022-20514-y

Elshaer, R. N., El-Aty, A. A., Sayed, E. M., Barakat, A. F., & Sobh, A. S. (2024). Optimisation of machining parameters for turning operation of heat-treated Ti-6Al-3Mo-2Nb-2Sn-2Zr-1.5Cr alloy by Taguchi method. *Scientific Reports, 14*(1), 16494. Advance online publication. DOI: 10.1038/s41598-024-65786-8

Fallahi, A., Tehrani, Y., & Mokhtari, H. (2024). Optimisation of a coordinated sustainable multi-vendor multi-livestock multi-rancher supply chain for growing products. *Scientia Iranica*, *0*(0), 0. Advance online publication. DOI: 10.24200/sci.2024.62208.7714

Faridmehr, I., Nehdi, M. L., Huseien, G. F., Baghban, M. H., Sam, A. R. M., & Algaifi, H. A. (2021). Experimental and informational modeling study of sustainable self-compacting geopolymer concrete. *Sustainability (Basel)*, *13*(13), 7444. DOI: 10.3390/su13137444

Feng, F., Ghorbani, H., & Radwan, A. E. (2024). Predicting groundwater level using traditional and deep machine learning algorithms. *Frontiers in Environmental Science*, *12*, 1291327. Advance online publication. DOI: 10.3389/fenvs.2024.1291327

Fernández-García, R., Lalatsa, A., Statts, L., Bolás-Fernández, F., Ballesteros, M. P., & Serrano, D. R. (2020). Transferosomes as nanocarriers for drugs across the skin: Quality by design from lab to industrial scale. *International Journal of Pharmaceutics*, *573*, 118817. DOI: 10.1016/j.ijpharm.2019.118817

Fischer, V., Kucia-Tran, R., Lewis, W. J., & Velayudhan, A. (2019). Hybrid optimisation of preparative chromatography for a ternary monoclonal antibody mixture. *Biotechnology Progress*, *35*(5), e2849. Advance online publication. DOI: 10.1002/btpr.2849

Fózer, D., Nimmegeers, P., Tóth, A. J., Varbanov, P. S., Klemeš, J. J., Mizsey, P., Hauschild, M. Z., & Owsianiak, M. (2023). Hybrid prediction-driven high-throughput sustainability screening for advancing waste-to-dimethyl ether valorization. *Environmental Science & Technology*, *57*(36), 13449–13462. DOI: 10.1021/acs.est.3c01892

Friederich, J., Francis, D. P., Lazarova-Molnar, S., & Mohamed, N. (2021). A framework for data-driven digital twins of smart manufacturing systems. *Computers in Industry*, *136*, 103586. DOI: 10.1016/j.compind.2021.103586

Ghader, S. (2025, October). AI-powered geotechnics: Enhancing rock mass classification for safer engineering practices. *Rock Mechanics and Rock Engineering*, *58*(10), 11319–11349. Advance online publication. DOI: 10.1007/s00603-024-04189-7

Ghadoui, M. E., Mouchtachi, A., & Majdoul, R. (2023). A hybrid optimisation approach for intelligent manufacturing in plastic injection molding by using artificial neural network and genetic algorithm. *Scientific Reports*, *13*(1), 21817. Advance online publication. DOI: 10.1038/s41598-023-48679-0

Ghaebi, H., & Soleymani, E. (2024). Investigating the effect of parameters in the thermodynamic analysis of the solid oxide fuel cell cycle using response surface methodology. *Research Square (Research Square)*. DOI: 10.21203/rs.3.rs-5053080/v1

Ghaheh, F. S., Razbin, M., Tehrani, M., Vayghan, L. Z. A., & Sadrjahani, M. (2024). Modeling and optimisation of dyeing process of polyamide 6 and woolen fabrics with plum-tree leaves using artificial intelligence. *Scientific Reports*, *14*(1), 15067. Advance online publication. DOI: 10.1038/s41598-024-64761-7

Goodarzian, F., Kumar, V., & Ghasemi, P. (2022). Investigating a citrus fruit supply chain network considering CO2 emissions using meta-heuristic algorithms. *Annals of Operations Research*. Advance online publication. DOI: 10.1007/s10479-022-05005-7

Goranov, A. I., Sleighter, R. L., Yordanov, D. A., & Hatcher, P. G. (2023). TEnvR: MATLAB-based toolbox for environmental research. *Analytical Methods*, *15*(40), 5390–5400. DOI: 10.1039/D3AY00750B

Hafid, M., Istiyanto, J., & Nasruddin, N. (2023). Multiobjective optimisation of dimension and position of elliptical crush initiator on crashworthiness performance of square tube using response surface methodology. *Frontiers of Mechanical Engineering*, *9*, 1273447. Advance online publication. DOI: 10.3389/fmech.2023.1273447

Hakim, B. A. E., Abdel-Goad, M., Awad, M. E., & Shoaib, A. M. (2025). AI enhanced model predictive control for optimizing LPG recovery through integrated computational modeling design of experiments and multivariate regression. *Scientific Reports*, *15*(1), 29249. Advance online publication. DOI: 10.1038/s41598-025-13899-z

Hardian, R., Liang, Z., Zhang, X., & Székely, G. (2020). Artificial intelligence: The silver bullet for sustainable materials development. *Green Chemistry*, *22*(21), 7521–7528. DOI: 10.1039/D0GC02956D

Hariharasakthisudhan, P., Rajendran, S., Khalkar, V., Devarajan, G., & Logesh, K. (2024). Data-driven healthcare: The role of computational methods in medical innovation. *Computer Modeling in Engineering & Sciences*, *1*. Advance online publication. DOI: 10.32604/cmes.2024.056605

Harudin, N., Jamaludin, K. R., Ramlie, F., Muhtazaruddin, M. N., Razali, C. M. C., & Muhamad, W. Z. A. W. (2020). Binary particle swarm optimisation for variables selection optimisation in Taguchi's T-Method. *MATEMATIKA*, *69*, 69–84. Advance online publication. DOI: 10.11113/matematika.v36.n1.1181

Heddar, M. I. E., Mehdi, B., Matougui, N., Tahan, A., & Jahazi, M. (2023). Adaptable multi-objective optimisation framework: Application to metal additive manufacturing. *Research Square (Research Square)*. DOI: 10.21203/rs.3.rs-3751025/v1

Heirendt, L., Arreckx, S., Pfau, T., Mendoza, S. N., Richelle, A., Heinken, A., Haraldsdóttir, H. S., Wachowiak, J., Keating, S., Vlasov, V., Magnúsdóttir, S., Ng, C. Y., Preciat, G., Žagare, A., Chan, S. H. J., Aurich, M. K., Clancy, C. M., Modamio, J., Sauls, J. T., & Fleming, R. M. T. (2019). Creation and analysis of biochemical constraint-based models using the COBRA Toolbox v.3.0. *Nature Protocols, 14*(3), 639–702. DOI: 10.1038/s41596-018-0098-2

Hejazi, T., & Moradpour, M. (2019). Developing a flexible methodology for modeling and solving multiple response optimisation problems. *Stochastics and Quality Control, 34*(2), 103–113. DOI: 10.1515/eqc-2018-0024

Hisam, M. W., Dar, A. A., Elrasheed, M. O., Khan, M. S., Gera, R., & Azad, I. (2024). The versatility of the Taguchi method: Optimizing experiments across diverse disciplines. *Journal of Statistical Theory and Applications : JSTA, 23*(4), 365–389. Advance online publication. DOI: 10.1007/s44199-024-00093-9

Hope, T., Downey, D., Weld, D. S., Etzioni, O., & Horvitz, E. (2023). A computational inflection for scientific discovery. *Communications of the ACM, 66*(8), 62–73. DOI: 10.1145/3576896

Huang, S., Yang, K., Qi, S., & Wang, R. (2024). When large language model meets optimisation. *Swarm and Evolutionary Computation, 90*, 101663. DOI: 10.1016/j.swevo.2024.101663

Ishfaq, K., Sana, M., & Ashraf, W. M. (2023). Artificial intelligence–built analysis framework for the manufacturing sector: Performance optimisation of wire electric discharge machining system. *International Journal of Advanced Manufacturing Technology, 128*(11-12), 5025–5039. DOI: 10.1007/s00170-023-12191-6

Ishola, N. B., & McKenna, T. F. L. (2021). Influence of process parameters on the gas phase polymerization of ethylene: RSM or ANN statistical methods? *Macromolecular Theory and Simulations, 30*(6), 2100059. Advance online publication. DOI: 10.1002/mats.202100059

Jaddi, H., Hmaidi, A. E., Ousmana, H., Berrada, M., Aouragh, M. H., Iallamen, Z., Kasse, Z., Ouali, A. E., Boufala, M., Ragragui, H., & Saouita, J. (2024). Predicting Nitrate levels in the saïss water table: A comparative study of machine learning methods. *BIO Web of Conferences, 115*, 3001. DOI: 10.1051/bioconf/202411503001

Janković, A., Chaudhary, G., & Goia, F. (2021). Designing the design of experiments (DOE) – An investigation on the influence of different factorial designs on the characterization of complex systems. *Energy and Building*, *250*, 111298. DOI: 10.1016/j.enbuild.2021.111298

Jones, M. (2019). Design and optimisation of oleochemical processes. *Research Portal Denmark*, 150. Retrieved on 10 October 2025 from https://local.forskningsportal.dk/local/dki-cgi/ws/cris-link?src=dtu&id=dtu-f63e262f-6567-467d-bcdb-c97fbe7bde66&ti=Design%20and%20Optimisation%20of%20Oleochemical%20Processes

Jong, W., Huang, Y.-M., Lin, Y.-Z., Chen, S., & Chen, Y. (2020). Integrating Taguchi method and artificial neural network to explore machine learning of computer aided engineering. *Zhongguo Gongcheng Xuekan*, *43*(4), 346–356. DOI: 10.1080/02533839.2019.1708804

Juul-Nyholm, H. K. (2024). *Multi-objective design indicators: A quantitative paradigm for early engineering design of deeply integrated products* [Doctoral dissertation, Technical University of Denmark]. Technical University of Denmark. Retrieved on 10 October 2025 from https://local.forskningsportal.dk/local/dki-cgi/ws/cris-link?src=dtu&id=dtu-b7c12d31-0ef3-47a2-803d-a407053c22ac&ti=Multi-Objective%20Design%20Indicators%20%3A%20A%20Quantitative%20Paradigm%20for%20Early%20Engineering%20Design%20of%20Deeply%20Integrated%20Products

Kalach, F. E., Yousif, I., Wuest, T., Sheth, A., & Harik, R. (2024). Cognitive manufacturing: Definition and current trends. *Journal of Intelligent Manufacturing*, *36*(6), 3695–3715. DOI: 10.1007/s10845-024-02429-9

Karazi, S., Moradi, M., & Benyounis, K. Y. (2019). Statistical and numerical approaches for modeling and optimizing laser micromachining process-review. In *Elsevier eBooks*. Elsevier BV., DOI: 10.1016/B978-0-12-803581-8.11650-9

Kaziz, S., Mariem, I. B., Echouchene, F., Belkhiria, M., & Belmabrouk, H. (2022). Taguchi optimisation of integrated flow microfluidic biosensor for COVID-19 detection. *The European Physical Journal Plus*, *137*(11), 1235. Advance online publication. DOI: 10.1140/epjp/s13360-022-03457-1

Kepdieu, J. M., Djangang, C. N., Njimou, J. R., Tchanang, G., Ekani, C. J., Maicaneanu, S. A., & Tizaoui, C. (2023). Mathematical modelling using full factorial design in response surface methodology applied in the adsorptive removal of dye basic blue 9 from synthetic aqueous solutions onto oryza sativa husk-derived nano-silica doped smectic clay. *Research Square (Research Square)*. DOI: 10.21203/rs.3.rs-3776135/v1

Khan, M., Chuenchart, W., Surendra, K. C., & Khanal, S. K. (2023). Applications of artificial intelligence in anaerobic co-digestion: Recent advances and prospects. *Bioresource Technology*, *370*, 128501. DOI: 10.1016/j.biortech.2022.128501

Kodumuru, R., Sarkar, S., Parepally, V., & Chandarana, J.Publishing Institute. (2025). Artificial intelligence and internet of things integration in pharmaceutical manufacturing: A smart synergy. *Pharmaceutics*, *17*(3), 290. DOI: 10.3390/pharmaceutics17030290

Kok, C. L., Ho, C. K., Lee, T. K., Loo, Z. Y., Koh, Y. Y., & Chai, J. P. (2024). A novel and low-cost cloud-enabled iot integration for sustainable remote intravenous therapy management. *Electronics (Basel)*, *13*(10), 1801. DOI: 10.3390/electronics13101801

Koli, T. A., Arajpure, V. G., & Patil, V. (2024). "Optimizing lathe process capability using non-metallic gears: A comparative analysis of machining performance." *Research Square (Research Square)*. DOI: 10.21203/rs.3.rs-5288850/v1

Kronberger, G., Bachinger, F., & Affenzeller, M. (2020). Smart manufacturing and continuous improvement and adaptation of predictive models. *Procedia Manufacturing*, *42*, 528–531. DOI: 10.1016/j.promfg.2020.02.037

Kumar, G., Goyal, K. K., Batra, N. K., & Mehdi, H. (2025). Optimizing reconfigurable manufacturing system configuration selection with multi-objective grey wolf optimisation. [IJIDeM]. *International Journal on Interactive Design and Manufacturing*, *19*(8), 5567–5582. DOI: 10.1007/s12008-024-02150-0

Kumar, G., Goyal, K. K., Batra, N. K., & Rani, D. (2022). Single part reconfigurable flow line design using fuzzy best worst method. *Opsearch*, *59*(2), 603–631. DOI: 10.1007/s12597-021-00550-4

Kumar, G., Tomer, A., Kumar, M., Goel, P., & Singh, O. P. (2021). Optimisation of turning process variables for surface integrity using taguchi technique. *IOP Conference Series. Materials Science and Engineering*, *1149*(1), 12029. DOI: 10.1088/1757-899X/1149/1/012029

Kumar, P., & Misra, J. P. (2020). Process modeling and optimisation using ANN and RSM during dry turning of titanium alloy used in automotive industry. *Proceedings of the Institution of Mechanical Engineers. Part D, Journal of Automobile Engineering*, *235*(7), 2040–2050. DOI: 10.1177/0954407020969255

Kumar, R., & Reji, M. (2023). Response surface methodology (RSM): An overview to analyze multivariate data. *Indian Journal of Microbiology Research*, *9*(4), 241–248. DOI: 10.18231/j.ijmr.2022.042

Kumari, P., & Lavanya, M. (2024). Optimization strategies for corrosion management in industries with artificial neural network and response surface technology: A comprehensive review. *Journal of Bio- and Tribo-Corrosion*, *10*(3), 59. DOI: 10.1007/s40735-024-00863-z

Kuo, C., & Shieh, C. (2020). Biocatalytic process optimisation. *Catalysts*, *10*(11), 1303. DOI: 10.3390/catal10111303

Kushwah, A., Sharma, P. K., Kushwaha, H. L., Sharma, B. B., Shrivastava, A. K., Nag, R. H., Chowdhury, M., Carpenter, G., & Yadav, R. (2024). Hybrid taguchi-grey relational analysis approach for optimizing cutter operational parameters in selective cauliflower harvesting. *Scientific Reports*, *14*(1), 29709. Advance online publication. DOI: 10.1038/s41598-024-81323-z

Lai, N. S., Tew, Y. S., Zhong, X., Yin, J., Li, J., Yan, B., & Wang, X. (2023). Artificial Intelligence (AI) workflow for catalyst design and optimisation. *Industrial & Engineering Chemistry Research*, *62*(43), 17835–17848. DOI: 10.1021/acs.iecr.3c02520

Lee, C., Lin, K.-X., & Chou, C.-W. (2024). Surrogate-based optimisation framework for enhancing smt process quality and productivity in electronics manufacturing services. *Research Square (Research Square)*. DOI: 10.21203/rs.3.rs-4770565/v1

Lee, S. H., Ye, I., Lee, C., Kim, J., Choi, G., Nam, S. C., & Park, I. (2025). accelerating battery material optimization through iterative machine learning. *arXiv preprint arXiv:2505.18162*. DOI: 10.48550/ARXIV.2505.18162

Lei, B., Kirk, T., Bhattacharya, A., Pati, D., Qian, X., Arróyave, R., & Mallick, B. K. (2021). Bayesian optimisation with adaptive surrogate models for automated experimental design. *npj Computational Materials*, *7*(1), 194. Advance online publication. DOI: 10.1038/s41524-021-00662-x

Li, H., Wang, S., & Tang, R. (2019). Robust optimal design of zero/low energy buildings considering uncertainties and the impacts of objective functions. *Applied Energy*, *254*, 113683. DOI: 10.1016/j.apenergy.2019.113683

Li, Z., Fei, F., & Zhang, G. (2022). Edge-to-Cloud IIoT for condition monitoring in manufacturing systems with ubiquitous smart sensors. *Sensors (Basel)*, *22*(15), 5901. DOI: 10.3390/s22155901

Lipski, J., & Zaleski, K. (2019). Implementation of artificial intelligence in optimisation of technological processes. *MATEC Web of Conferences, 252*, 3008. DOI: 10.1051/matecconf/201925203008

Lohith, N., Bhaskar, H. B., & Yogendrakumar, S. (2024). Optimisation of condensation heat transfer coefficients using Taguchi method: Influence of temperature, surface type, and coolant flow rate. *Research Square (Research Square)*. DOI: 10.21203/rs.3.rs-5421143/v1

López, G., Yate, C., Ramos, F. A., Cala, M. P., Restrepo, S., & Baena, S. (2019). Production of polyunsaturated fatty acids and lipids from autotrophic, mixotrophic and heterotrophic cultivation of galdieria sp. strain USBA-GBX-832. *Scientific Reports*, *9*(1), 10791. Advance online publication. DOI: 10.1038/s41598-019-46645-3

Mahapatra, A. P. K., Saraswat, R., Botre, M., Paul, B., & Prasad, N. (2020). Application of response surface methodology (RSM) in statistical optimisation and pharmaceutical characterization of a patient compliance effervescent tablet formulation of an antiepileptic drug levetiracetam. *Future Journal of Pharmaceutical Sciences*, *6*(1), 82. Advance online publication. DOI: 10.1186/s43094-020-00096-0

Mairpady, A., Mourad, A. I., & Mozumder, M. S. (2021). Statistical and machine learning-driven optimisation of mechanical properties in designing durable HDPE Nanobiocomposites. *Polymers*, *13*(18), 3100. DOI: 10.3390/polym13183100

Mala, S., Deep, U., Mala, E., & Akhai, S. (2025). Digital health technologies for preventive cardiology. In *Impact of Digital Solutions for Improved Healthcare Delivery* (pp. 379–400). IGI Global., DOI: 10.4018/979-8-3693-5237-3.ch014

Manning, M. S., Hassanein, M., Partridge, M. A., Jawa, V., Mora, J., Ryman, J., Barker, B., Braithwaite, C., Carleton, K., Hay, L. S., Hottenstein, C., Kubiak, R., & Devanarayan, V. (2022). Comparison of titer and signal to noise (S/N) for determination of anti-drug antibody magnitude using clinical data from an industry consortium. *The AAPS Journal*, *24*(4), 81. Advance online publication. DOI: 10.1208/s12248-022-00728-8

Martínez, V., & Shadman, F. (2024). Mathematical Modeling for Optimizing the structural properties of catalysts used for series multiple reactions. *Zeitschrift für Angewandte Mathematik und Physik*, *12*(11), 3923–3940. DOI: 10.4236/jamp.2024.1211239

Mashaly, A. A., Mahdy, M. G., & Elemam, W. E. (2023). Optimal design and characteristics of sustainable eco-friendly ultra-high-performance concrete. *Innovative Infrastructure Solutions*, *8*(12), 326. Advance online publication. DOI: 10.1007/s41062-023-01277-5

Mavi, R. K., Shekarabi, S. A. H., Mavi, N. K., Arisian, S., & Moghdani, R. (2023). Multi-objective optimisation of sustainable closed-loop supply chain networks in the tire industry. *Engineering Applications of Artificial Intelligence*, *126*, 107116. DOI: 10.1016/j.engappai.2023.107116

Melaku, N. S., & Bogale, T. M. (2023). Parameters optimisation in plasma arc cutting of AISI 1020 mild steel plate using hybrid genetic algorithm and artificial neural network. *International Journal for Simulation and Multidisciplinary Design Optimisation*, *14*, 20. DOI: 10.1051/smdo/2023012

Mensah, R. A., Kirton, S. B., Cook, M. T., Styliari, I. D., Hutter, V., & Chau, D. (2019). Optimising poly(lactic-co-glycolic acid) microparticle fabrication using a Taguchi orthogonal array design-of-experiment approach. *PLoS One*, *14*(9), e0222858. Advance online publication. DOI: 10.1371/journal.pone.0222858

Mihalovits, M., & Kemény, S. (2022). Testing the fit of regression models estimated with extremely small samples: Application in pharmaceutical stability studies. *Periodica Polytechnica. Chemical Engineering*, *66*(2), 325–339. DOI: 10.3311/PPch.18793

Miikkulainen, R., Brundage, M., Epstein, J., Foster, T., Hodjat, B., Iscoe, N., Jiang, J., Legrand, D., Nazari, S., Qiu, X., Scharff, M., Schoolland, C., Severn, R., & Shagrin, A. (2020). Ascend by evolv: Artificial intelligence-based massively multivariate conversion rate optimisation. *AI Magazine*, *41*(1), 44–60. DOI: 10.1609/aimag.v41i1.5256

Mishra, A., Sefene, E. M., Nidigonda, G., & Tsegaw, A. A. (2022). Performance evaluation of machine learning-based algorithm and Taguchi algorithm for the determination of the hardness value of the friction stir welded AA 6262 joints at a nugget zone. *arXiv (Cornell University)*. https://doi.org//arxiv.2203.11649DOI: 10.48550

Modi, M., & Gopal, A. (2020). Modelling, Investigation of process responses, surface assessment and parametric optimisation in powder mixed electrical discharge diamond grinding of TI6AL4V utilizing grey-based Taguchi approach. *Transactions of FAMENA*, *44*(3), 93–112. DOI: 10.21278/TOF.44308

Mohammed, M., Alqahtani, N. K., & Ali, S. (2024). Development, RSM-based modeling, and process optimisation of an ultrasonic coating system for extending the storage life of fresh fruits. *Frontiers in Sustainable Food Systems*, *8*, 1403164. Advance online publication. DOI: 10.3389/fsufs.2024.1403164

Mohanadevi, M., & Dhanabalan, K. (2024). Enhanced Chromium removal from water using ball milled nano-biochar derived from Eichhornia crassipes (Water hyacinth). *Research Square (Research Square)*. DOI: 10.21203/rs.3.rs-5312229/v1

Moussa, Z., Ghoniem, A. A., Elsayed, A., Alotaibi, A. S., Alenzi, A. M., Hamed, S. E., Elattar, K. M., & Saber, W. I. A. (2022). Innovative binary sorption of Cobalt(II) and methylene blue by Sargassum latifolium using Taguchi and hybrid artificial neural network paradigms. *Scientific Reports*, *12*(1), 18291. Advance online publication. DOI: 10.1038/s41598-022-22662-7

Msakni, M. K., Risan, A., & Schütz, P. (2023). Using machine learning prediction models for quality control: A case study from the automotive industry. *Computational Management Science*, *20*(1), 14. Advance online publication. DOI: 10.1007/s10287-023-00448-0

Nguyen, H. T., Jeong, H.-Y., Kim, K.-D., Xuan-Mung, N., Dao, N., Tran, H.-N., Hoang, V.-K., Anh, N. N., & Vu, M. T. (2024). A new hyperparameter tuning framework for regression tasks in deep neural network: Combined-sampling algorithm to search the optimized hyperparameters. *Mathematics*, *12*(24), 3892. DOI: 10.3390/math12243892

Niazi, S. K. (2023). The coming of age of AI/ML in drug discovery, development, clinical testing, and manufacturing: The FDA perspectives. *Drug Design, Development and Therapy*, *2691*, 2691–2725. Advance online publication. DOI: 10.2147/DDDT.S424991

Nikzad, M. H., Heidari-Rarani, M., & Mirkhalaf, M. (2024). A novel Taguchi-based approach for optimizing neural network architectures: Application to elastic short fiber composites. *Composites Science and Technology*, *259*, 110951. DOI: 10.1016/j.compscitech.2024.110951

Nouri-Moghaddam, B., Ghazanfari, M., & Fathian, M. (2020). A novel filter-wrapper hybrid gene selection approach for microarray data based on multi-objective forest optimisation algorithm. *Decision Science Letters*, *271*, 271–290. Advance online publication. DOI: 10.5267/j.dsl.2020.5.006

Nourmohammadzadeh, A., Sarhani, M., & Voß, S. (2023). A matheuristic approach for the family traveling salesman problem. *Journal of Heuristics*, *29*(4-6), 435–460. DOI: 10.1007/s10732-023-09516-9

Ojha, V., Abraham, A., & Snášel, V. (2019). Heuristic design of fuzzy inference systems: A review of three decades of research. *Engineering Applications of Artificial Intelligence*, *85*, 845–864. DOI: 10.1016/j.engappai.2019.08.010

Onokwai, A. O., Olusanya, O. O., Onifade, M. K., Anyaegbuna, B. E., Onoriode, A., & Agholor, D. (2024). Comparative analysis of response surface methodology and adaptive neuro-fuzzy inference system for predictive fault detection and optimisation in beverage industry. *Frontiers of Mechanical Engineering*, *10*, 1428717. Advance online publication. DOI: 10.3389/fmech.2024.1428717

Onyelowe, K. C., Moghal, A. A. B., Ebid, A. M., Rehman, A. U., Hanandeh, S., & Priyan, V. V. (2024). Estimating the strength of soil stabilized with cement and lime at optimal compaction using ensemble-based multiple machine learning. *Scientific Reports*, *14*(1), 15308. Advance online publication. DOI: 10.1038/s41598-024-66295-4

Oppioli, M., Sousa, M. J., Sousa, M., & de Nuccio, E. (2023). The role of artificial intelligence for management decision: A structured literature review. *Management Decision*. Advance online publication. DOI: 10.1108/MD-08-2023-1331

Öztürk, B., Kaymak, Ş., & Küçük, Ö. (2022). Taguchi and RSM based optimiation of energy consumption on internal gear pumps. *International Journal of 3D Printing Technologies and Digital Industry, 6*(1), 164-175. DOI: 10.46519/ij3dptdi.1001245

Pahija, E., Hwangbo, S., Saulnier-Bellemare, T., & Patience, G. S. (2024). Experimental methods in chemical engineering: Monte Carlo. *Canadian Journal of Chemical Engineering*, *102*(10), 3308–3321. DOI: 10.1002/cjce.25374

Pal, S. (2023). A Paradigm Shift in Research: Exploring the intersection of artificial intelligence and research methodology. *International Journal of Innovative Research in Engineering & Multidisciplinary Physical Sciences*, *11*(3), 230125. Advance online publication. DOI: 10.37082/IJIRMPS.v11.i3.230125

Pandi, A., Diehl, C., Kharrazi, A. Y., Scholz, S. A., Bobkova, E., Faure, L., Nattermann, M., Adam, D., Chapin, N., Foroughijabbari, Y., Moritz, C., Paczia, N., Cortina, N. S., Faulon, J., & Erb, T. J. (2022). A versatile active learning workflow for optimisation of genetic and metabolic networks. *Nature Communications*, *13*(1), 3876. Advance online publication. DOI: 10.1038/s41467-022-31245-z

Pasha, N., Mahdiraji, H. A., Hajiagha, S. H. R., Garza-Reyes, J. A., & Joshi, R. (2023). A multi-objective flexible manufacturing system design optimisation using a hybrid response surface methodology. *Operations Management Research : Advancing Practice Through Research*, *17*(1), 135–151. DOI: 10.1007/s12063-023-00412-w

Pasupuleti, V., Thuraka, B., Kodete, C. S., & Malisetty, S. (2024). Enhancing supply chain agility and sustainability through machine learning: Optimisation techniques for logistics and inventory management. *Logistics (Basel)*, *8*(3), 73. DOI: 10.3390/logistics8030073

Paw, J. K. S., Kiong, T. S., Kamarulzaman, M. K., Adam, A., Hisham, S., Kadirgama, K., & Benedict, F. (2023). Advancing renewable fuel integration: A comprehensive response surface methodology approach for internal combustion engine performance and emissions optimization. *Heliyon, 9*(11). Advance online publication. DOI: 10.1016/j.heliyon.2023.e22238

Peng, Y., Khaled, U., Al-Rashed, A. A. A. A., van der Meer, R. L. J., Goodarzi, M., & Sarafraz, M. M. (2020). Potential application of response surface methodology (RSM) for the prediction and optimisation of thermal conductivity of aqueous CuO (II) nanofluid: A statistical approach and experimental validation. *Physica A, 554*, 124353. DOI: 10.1016/j.physa.2020.124353

Pham, T.-D., Nguyen, N., Tran, C.-M., & Vuong, M. P. (2020). Constructing response surface designs with orthogonal quadratic effects using cyclic generators. *Chemometrics and Intelligent Laboratory Systems, 198*, 103918. DOI: 10.1016/j.chemolab.2019.103918

Pithan, L., Starostin, V., Mareček, D., Petersdorf, L., Völter, C., Munteanu, V., Jankowski, M., Konovalov, O., Gerlach, A., Hinderhofer, A., Murphy, B. M., Kowarik, S., & Schreiber, F. (2023). Closing the loop: Autonomous experiments enabled by machine-learning-based online data analysis in synchrotron beamline environments. *Journal of Synchrotron Radiation, 30*(6), 1064–1075. DOI: 10.1107/S160057752300749X

Polo, S., Rubio, E. M., Ayllón, J., & de Agustina, B. (2025). Emerging advances in sustainable manufacturing. *Processes (Basel, Switzerland), 13*(5), 1549. DOI: 10.3390/pr13051549

Popović, B. (2020). Planning, analyzing and optimizing experiments. *Journal of Engineering Management and Competitiveness, 10*(1), 15–30. DOI: 10.5937/jemc2001015P

Prasertpong, P., Jaroenkhasemmeesuk, C., Regalbuto, J. R., Lipp, J., & Tippayawong, N. (2019). Optimisation of process variables for esterification of bio-oil model compounds by a heteropolyacid catalyst. *Energy Reports, 6*, 1–9. DOI: 10.1016/j.egyr.2019.11.026

Rahman, M., Taib, N.-A. A. B., Matin, M. M., Rahman, M. M., Bakri, M. K. B., Alexanrovich, T. P., Vladimirovich, S. V., Sanaullah, K., Tazeddinova, D., & Khan, A. (2022). Optimisation of tensile strength and Young's modulus of CNT–CF/Epoxy composites using response surface methodology (RSM). *Materials (Basel), 15*(19), 6746. DOI: 10.3390/ma15196746

Ranjan, M. Kr., Barot, K., Khairnar, V., Rawal, V., Pimpalgaonkar, A., Saxena, S., & Sattar, A. Md. (2023). Python: Empowering data science applications and research. *Journal of Operating Systems Development & Trends*. DOI: 10.37591/joosdt.v10i1.576

Rao, B. A., Tak, M., Rao, R. N., & Bathe, R. (2023). Developing laser-assisted machining process for nickel based superalloy IN625 using experimental and statistical analysis. *Lasers in Manufacturing and Materials Processing, 10*(4), 681–701. DOI: 10.1007/s40516-023-00230-9

Raposo, F., & Barceló, D. (2021). Assessment of goodness-of-fit for the main analytical calibration models: Guidelines and case studies. *Trends in Analytical Chemistry, 143*, 116373. DOI: 10.1016/j.trac.2021.116373

Raza, Y., Raza, H., Ahmed, A., Quazi, M. M., Jamshaid, M., Anwar, M. T., Bashir, M. N., Younas, T., Jafry, A. T., & Soudagar, M. E. M. (2025). Integration of response surface methodology (RSM), machine learning (ML), and artificial intelligence (AI) for enhancing properties of polymeric nanocomposites-A review. *Polymer Composites, 46*(15), 13591–13627. Advance online publication. DOI: 10.1002/pc.30011

Rogers, A. W., Lane, A., Mendoza, Ć., Watson, S., Kowalski, A., Martin, P. A., & Zhang, D. (2024). Integrating knowledge-guided symbolic regression and model-based design of experiments to automate process flow diagram development. *Chemical Engineering Science, 300*, 120580. DOI: 10.1016/j.ces.2024.120580

Romelin, C. Zahedi, & Nusantara, B. C. (2024, September). Comparative analysis of response surface methodology (rsm) and taguchi method: optimisation hydraulic ram pump performance. In *Operations Research Forum* (Vol. 5, No. 4, p. 85). Cham: Springer International Publishing. DOI: 10.1007/s43069-024-00359-z

Rosa, A. C., Hammad, A. W. A., Boer, D., & Haddad, A. (2023). Use of operational research techniques for concrete mix design: A systematic review [Review of *Use of operational research techniques for concrete mix design: A systematic review*]. *Heliyon, 9*(4). Elsevier BV. DOI: 10.1016/j.heliyon.2023.e15362

Roy, N., Beauthier, C., Carletti, T., & Mayer, A. (2021). Setup of fuzzy hybrid particle swarms. *Proceedings of the Genetic and Evolutionary Computation Conference Companion*, 207. DOI: 10.1145/3449726.3459418

Rudolph, M., Kurz, S., & Rakitsch, B. (2024). Hybrid modeling design patterns. *Journal of Mathematics in Industry, 14*(1), 3. Advance online publication. DOI: 10.1186/s13362-024-00141-0

Rudrapati, R., Bandyopadhyay, A., Pal, P. K., & Rathod, L. (2020). Analysis, modeling and optimisation of surface roughness in cylindrical traverse cut grinding using factorial design, RSM and simulated annealing algorithm. *IOP Conference Series. Materials Science and Engineering*, *814*(1), 12016. DOI: 10.1088/1757-899X/814/1/012016

Saad, A., Su, B., & Bjørnson, F. O. (2023). A web-based platform for efficient and robust simulation of aquaculture systems using integrated intelligent agents. *Procedia Computer Science*, *225*, 4560–4569. DOI: 10.1016/j.procs.2023.10.454

Sadeq, A. M., Homod, R. Z., Hasan, H. A., Alhasnawi, B. N., Hussein, A. K., Jahangiri, A., & Abbas, S. (2025). Advancements in combustion technologies: A review of innovations, methodologies, and practical applications. *Energy Conversion and Management*, *X*, 100964. DOI: 10.1016/j.ecmx.2025.100964

Sadollah, A., Nasir, M., & Geem, Z. W. (2020). Sustainability and optimisation: From conceptual fundamentals to applications. *Sustainability (Basel)*, *12*(5), 2027. DOI: 10.3390/su12052027

Säfken, B., & Rügamer, D. (2024). Editorial special issue: Bridging the gap between AI and Statistics. *AStA. Advances in Statistical Analysis*, *108*(2), 225–229. DOI: 10.1007/s10182-024-00503-4

Sai, D. A., Nataraj, K. S., & Lakshmana, R. A. (2023). Response surface methodology: A statistical tool for the optimisation of responses. *Global Journal of Addiction & Rehabilitation Medicine*, *7*(1). Advance online publication. DOI: 10.19080/GJARM.2023.07.555705

Sajdak, M., Majewski, A. J., Sobek, S., Gałko, G., & Ouadi, M. (2025). Design of experiments method into upgrading pyrolytic oil for sustainable aviation fuel additives by hydrotreating and hydrocracking. *Waste Management (New York, N.Y.)*, *194*, 258–269. DOI: 10.1016/j.wasman.2025.01.012

Salgotra, R., Sharma, P., Saravanakumar, R., & Gandomi, A. H. (2023). A contemporary systematic review on meta-heuristic optimisation algorithms with their MATLAB and Python code reference. *Archives of Computational Methods in Engineering*, *31*(3), 1749–1822. DOI: 10.1007/s11831-023-10030-1

Schwaninger, M., & Ott, S. (2023). What is variety engineering and why do we need it? *Systems Research and Behavioral Science*, *41*(2), 235–246. DOI: 10.1002/sres.2964

Serani, A., & Diez, M. (2024). A survey on design-space dimensionality reduction methods for shape optimisation. *arXiv (Cornell University)*. https://doi.org//arxiv.2405.13944DOI: 10.48550

Sharma, A., Singh, Y., Tyagi, A., Singh, N. K., & Singla, A. (2020). Application of Taguchi and response surface methodology approach to a sustainable model developed for a compression-ignition engine using polanga biodiesel/diesel blends. *Proceedings of the Institution of Mechanical Engineers. Part I, Journal of Systems and Control Engineering*, *235*(6), 854–868. DOI: 10.1177/0959651820965301

Sheikhmohammadi, A., Alamgholiloo, H., Golaki, M., Khakzad, P., Asgari, E., & Rahimlu, F. (2024). Cefixime removal via WO3/Co-ZIF nanocomposite using machine learning methods. *Scientific Reports*, *14*(1), 13840. Advance online publication. DOI: 10.1038/s41598-024-64790-2

Shie, A., Lo, K.-H., Lin, W., Juan, C., & Jou, Y. (2019). An integrated model using the Taguchi method and artificial neural network to improve artificial kidney solidification parameters. *Biomedical Engineering Online*, *18*(1), 78. Advance online publication. DOI: 10.1186/s12938-019-0696-4

Shojaei, S., Shojaei, S., Band, S. S., Farizhandi, A. A. K., Ghoroqi, M., & Mosavi, A. (2021). Application of Taguchi method and response surface methodology into the removal of malachite green and auramine-O by NaX nanozeolites. *Scientific Reports*, *11*(1), 16054. Advance online publication. DOI: 10.1038/s41598-021-95649-5

Simone, M. D., Veneziano, S., Porcaro, A., & Guida, D. (2025). Design of experiments approach for structural optimisation of urban air mobility vehicles. *Actuators*, *14*(4), 176. DOI: 10.3390/act14040176

Sims, K. R., Maceren, J. P., Strand, A. I., He, B., Overby, C., & Benoit, D. S. W. (2020). Rigor and reproducibility in polymer nanoparticle synthesis and characterization. *RSC Advances*, *10*(5), 2513–2518. DOI: 10.1039/C9RA10091A

Singh, S., & Akhai, S. (2024, July). Electronics in precision control for mechanical engineering: A brief overview. In. AIP Conference Proceedings: Vol. 3121. *No. 1* (p. 020007). AIP Publishing LLC., DOI: 10.1063/5.0221464

Sipper, M. (2022). Combining deep learning with good old-fashioned machine learning. *SN Computer Science*, *4*(1), 85. Advance online publication. DOI: 10.1007/s42979-022-01505-2

Soori, M., Arezoo, B., & Dastres, R. (2023). Virtual manufacturing in Industry 4.0: A review [Review of Virtual manufacturing in Industry 4.0: A review]. *Data Science and Management*, *7*(1), 47–63. DOI: 10.1016/j.dsm.2023.10.006

Stegherr, H., Heider, M., & Hähner, J. (2020). Classifying metaheuristics: Towards a unified multi-level classification system. *Natural Computing*, *21*(2), 155–171. DOI: 10.1007/s11047-020-09824-0

Strobl, P., Schermer, E., Groetsch, D., Pointner-Gabriel, L., Voelkel, K., Pflaum, H., & Stahl, K. (2022). Identification and validation of linear friction models using ANOVA and stepwise regression. *Lubricants (Basel, Switzerland)*, *10*(11), 286. DOI: 10.3390/lubricants10110286

Sudjianto, A., & Zhang, A. (2021). Designing inherently interpretable machine learning models. *arXiv (Cornell University)*. https://doi.org//arxiv.2111.01743DOI: 10.48550

Sun, B., Yang, C., Wang, Y., Gui, W., Craig, I. K., & Olivier, L. E. (2019). A comprehensive hybrid first principles/machine learning modeling framework for complex industrial processes. *Journal of Process Control*, *86*, 30–43. DOI: 10.1016/j.jprocont.2019.11.012

Sung, J.-H., Ju, S.-H., Cho, S.-W., Joo, H. W., Seo, K.-M., & Kang, B. G. (2025). Modeling and exploratory analysis of discrete event simulations for optimizing overhead hoist transport systems and logistics in semiconductor manufacturing. *Mathematics*, *13*(7), 1167. DOI: 10.3390/math13071167

Talati, D. V. (2021). Python: The alchemist behind AI's intelligent evolution. *International Journal of Science and Research Archive*, *3*(1), 235. DOI: 10.30574/ijsra.2021.3.1.0169

Tănase, M., Portoacă, A. I., Dini ă, A., Brănoiu, G., Zamfir, F., S rbu, E.-E., & Călin, C. (2024). Optimizing mechanical properties of recycled 3D-printed pla parts for sustainable packaging solutions using experimental analysis and machine learning. *Polymers*, *16*(23), 3268. DOI: 10.3390/polym16233268

Tangjitsitcharoen, S., Tangjitsitcharoen, S., & Faccia, A. (2025). Leveraging intelligent machines for sustainable and intelligent manufacturing systems. *Journal of Manufacturing and Materials Processing*, *9*(5), 153. DOI: 10.3390/jmmp9050153

Teh, S. Y., Chua, K. B., Hong, B. H., Ling, A. J. W., Andiappan, V., Foo, D. C. Y., Hassim, M. H., & Ng, D. K. S. (2019). A hybrid multi-objective optimisation framework for preliminary process design based on health, safety and environmental impact. *Processes (Basel, Switzerland)*, *7*(4), 200. DOI: 10.3390/pr7040200

Teng, S. Y., How, B. S., Leong, W. D., Teoh, J. H., Cheah, A. C. S., Motavasel, Z., & Lam, H. L. (2019). Principal component analysis-aided statistical process optimisation (PASPO) for process improvement in industrial refineries. *Journal of Cleaner Production, 225*, 359–375. DOI: 10.1016/j.jclepro.2019.03.272

Tipu, R. K., Rathi, P., Pandya, K. S., & Panchal, V. R. (2025). Optimizing sustainable blended concrete mixes using deep learning and multi-objective optimisation. *Scientific Reports, 15*(1), 16356. Advance online publication. DOI: 10.1038/s41598-025-00943-1

Treloar, N. J., Braniff, N., Ingalls, B., & Barnes, C. (2022). Deep reinforcement learning for optimal experimental design in biology. *PLoS Computational Biology, 18*(11), e1010695. Advance online publication. DOI: 10.1371/journal.pcbi.1010695

Triller, J., Lopez, M. L., Nossek, M., & Frenzel, M. (2023). Multidisciplinary optimisation of automotive mega-castings merging classical structural optimisation with response-surface-based optimisation enhanced by machine learning. *Scientific Reports, 13*(1), 21678. Advance online publication. DOI: 10.1038/s41598-023-47937-5

Tsai, S. (2023). Dual-orthogonal arrays for order-of-addition two-level factorial experiments. *Technometrics, 65*(3), 388–395. DOI: 10.1080/00401706.2023.2173303

Tsoulos, I. G., Charilogis, V., Kyrou, G., Stavrou, V. N., & Tzallas, A. T. (2025). OPTIMUS: A multidimensional global optimisation package. *Journal of Open Source Software, 10*(108), 7584. DOI: 10.21105/joss.07584

Venkatesh, B., Kiran, A., Khan, M., Rahmani, M. K. I., Upadhyay, L., Babu, J. C., & Narayana, T. L. (2024). Performance optimisation for an optimal operating condition for a shell and heat exchanger using a multi-objective genetic algorithm approach. *PLoS One, 19*(6), e0304097. Advance online publication. DOI: 10.1371/journal.pone.0304097

Victoria, A., Hine, P. J., Ward, K., & Ries, M. E. (2023). Design of experiments in the optimisation of all-cellulose composites. *Cellulose (London, England), 30*(17), 11013–11039. DOI: 10.1007/s10570-023-05535-8

Wadhwa, S. A., Mahapara, A., Akhai, S., Kumar, D., & Kumar, P. (2024). Integrating Taguchi optimisation for multi-criteria decision making in engineering applications. In *Recent Theories and Applications for Multi-Criteria Decision-Making* (Vol. 1, No. 1, pp. 125–150). IGI Global. DOI: 10.4018/979-8-3693-6502-1.ch005

Wang, C., Zhou, H., Chang, K., Li, B., Mu, Y., Xiao, T., Liu, T., & Zhu, J. (2024). Hybrid alignment training for large language models. *arXiv (Cornell University)*. / arxiv.2406.15178DOI: 10.18653/v1/2024.findings-acl.676

Wegener, K., Weikert, S., Mayr, J., Maier, M., Akbari, V. O. A., & Postel, M. (2021). Operator integrated – concept for manufacturing intelligence. *Journal of Machine Engineering*, *21*(4), 5–28. DOI: 10.36897/jme/144359

Wei, Y., Peng, B., Xie, R., Chen, Y., Yu, Q., Wen, P., Bauer, S., Tung, P., & Raabe, D. (2024). Deep active learning for complex systems. *Research Square (Research Square)*. DOI: 10.21203/rs.3.rs-5434645/v1

Weidele, D. K. I., Afzal, S., Valente, A. N., Makuch, C., Cornec, O., Vu, L., & Haehn, D. (2023, March). AutoDOViz: Human-centered automation for decision optimization. In *Proceedings of the 28th international conference on intelligent user interfaces* (pp. 664-680). DOI: 10.1145/3581641.3584094

Wichapa, N., Pawaree, N., Nasawat, P., Chourwong, P., Sriburum, A., & Khanthirat, W. (2024). Process of solving multi-response optimisation problems using a novel data envelopment analysis variant-taguchi method. *International Journal of Technology*, *15*(6), 2038. DOI: 10.14716/ijtech.v15i6.7134

Xu, X., Li, D., Bi, J., & Moeckel, M. (2024). AutoML based workflow for design of experiments (DOE) selection and benchmarking data acquisition strategies with simulation models. *Scientific Reports*, *14*(1), 32170. Advance online publication. DOI: 10.1038/s41598-024-83581-3

Xue, X., Yu, X.-N., Zhou, D.-Y., Wang, X., Zhou, Z.-B., & Wang, F. (2022). Computational experiments: Past, present and future. *arXiv (Cornell University)*. https://doi.org//arXiv.2202.13690DOI: 10.48550

Yan, X., Guo, X., He, N., Shi, J., & Zhao, D. (2025). Model reduction of structural mechanical response in the time domain. *Scientific Reports*, *15*(1), 10203. Advance online publication. DOI: 10.1038/s41598-025-94681-z

Yang, C., Wang, X., Lu, Y., Liu, H., Le, Q. V., Zhou, D., & Chen, X. (2023). Large language models as optimizers. *arXiv (Cornell University)*. https://doi.org//arxiv.2309.03409DOI: 10.48550

Yang, J., Fan, K., Ma, H., & Zhan, D. (2022). RID-Noise: Towards robust inverse design under noisy environments. *Proceedings of the AAAI Conference on Artificial Intelligence*, *36*(4), 4654–4661. DOI: 10.1609/aaai.v36i4.20390

Yarasu, V., & Podgornik, B. (2025). An optimized machine-learning tool to predict heat treatment response of hot-work tool steels. *Results in Engineering*, *26*, 105260. DOI: 10.1016/j.rineng.2025.105260

Yoo, H. J., Kim, N., Lee, H., Kim, D., Tiong, L. C. O., Nam, H., Kim, C., Lee, S. Y., Lee, K., Kim, D.-H., & Han, S. S. (2023). Bespoke nanoparticle synthesis and chemical knowledge discovery via autonomous experimentations. *arXiv (Cornell University)*. https://doi.org//arxiv.2309.00349DOI: 10.48550

Yu, H., & Liu, J. (2024). *Deep insights into automated optimisation with large language models and evolutionary algorithms*. DOI: 10.48550/ARXIV.2410.20848

Yu, T., & Zhu, H. (2020). Hyper-parameter optimization: A review of algorithms and applications. *arXiv preprint arXiv:2003.05689*. https://doi.org//arxiv.2003.05689DOI: 10.48550

Zăvoianu, A.-C., Lughofer, E., Pollak, R. A., Eitzinger, C., & Radauer, T. (2020). A soft-computing framework for automated optimisation of multiple product quality criteria with application to micro-fluidic chip production. *Applied Soft Computing*, *98*, 106827. DOI: 10.1016/j.asoc.2020.106827

Zeng, X. (2024). A review on design of sustainable advanced materials by using artificial intelligence. *Advanced Materials & Sustainable Manufacturing*, *1*(1), 10006. DOI: 10.35534/amsm.2024.10006

Zhang, Y., & Wu, Y. (2021). Introducing machine learning models to response surface methodologies. In *IntechOpen eBooks*. IntechOpen., DOI: 10.5772/intechopen.98191

Zimmer, F., & Debelak, R. (2023). Simulation-based design optimisation for statistical power: Utilizing machine learning. *Psychological Methods*. Advance online publication. DOI: 10.1037/met0000611

Zong, C., Margesin, M., Staudt, J., Deghim, F., & Lang, W. (2022). Decision-making under uncertainty in the early phase of building façade design based on multi-objective stochastic optimisation. *Building and Environment*, *226*, 109729. DOI: 10.1016/j.buildenv.2022.109729

Zubair, S. W. H., Arafat, S. M., Khan, S. A., Niazi, S. G., Rehan, M., Arshad, M. U., Hayat, N., Aized, T., Uddin, G. M., & Riaz, F. (2024). Coupling Taguchi experimental designs with deep adaptive learning enhanced AI process models for experimental cost savings in manufacturing process development. *Scientific Reports*, *14*(1), 23428. Advance online publication. DOI: 10.1038/s41598-024-73669-1

ADDITIONAL READING

Çırak, M. (2018). Improvement of RSM Prediction and Optimisation by Using Box-Cox Transformation: Separation of Colloidal Contaminants From Mineral Processing Effluents via Electrocoagulation. In *Handbook of Research on Predictive Modeling and Optimisation Methods in Science and Engineering* (pp. 156–191). IGI Global. DOI: 10.4018/978-1-5225-4766-2.ch008

Kırbaş, İ., Peker, M., Basmacı, G., & Ay, M. (2018). Predictive modeling and optimisation of cutting forces through RSM and taguchi techniques in the turning of ASTM B574 (Hastelloy C-22). In *Handbook of Research on Predictive Modeling and Optimisation Methods in Science and Engineering* (pp. 398–417). IGI Global Scientific Publishing. DOI: 10.4018/978-1-5225-4766-2.ch018

Mukhopadhyay, A., Duari, S., Barman, T. K., & Sahoo, P. (2017). Tribology of Electroless Ni-P Coating Under Lubricated Condition: An RSM and GA Approach. [IJSEIMS]. *International Journal of Surface Engineering and Interdisciplinary Materials Science*, *5*(1), 37–57. DOI: 10.4018/IJSEIMS.2017010103

Mukhopadhyay, P., Bose, G. K., & Pain, P. (2021). MCDM-Based Optimisation of Performance Characteristics During µEDMing of SS 304. In *Machine Learning Applications in Non-Conventional Machining Processes* (pp. 18–32). IGI Global Scientific Publishing. DOI: 10.4018/978-1-7998-3624-7.ch002

Rao, S. R., & Padmanabhan, G. (2013). Multi-response optimisation of electrochemical machining of Al-Si/B4C composites using RSM. [IJMMME]. *International Journal of Manufacturing, Materials, and Mechanical Engineering*, *3*(3), 42–56. DOI: 10.4018/ijmmme.2013070103

Santhanakumar, M., & Adalarasan, R. (2015). Application of grey Taguchi based response surface methodology (GT-RSM) in injection moulding of polypropylene/E-glass composite. [IJMMME]. *International Journal of Manufacturing, Materials, and Mechanical Engineering*, *5*(1), 35–48. DOI: 10.4018/ijmmme.2015010103

APPENDIX I: AI-ENABLED HYBRID OPTIMISATION FRAMEWORK USING TAGUCHI–RSM

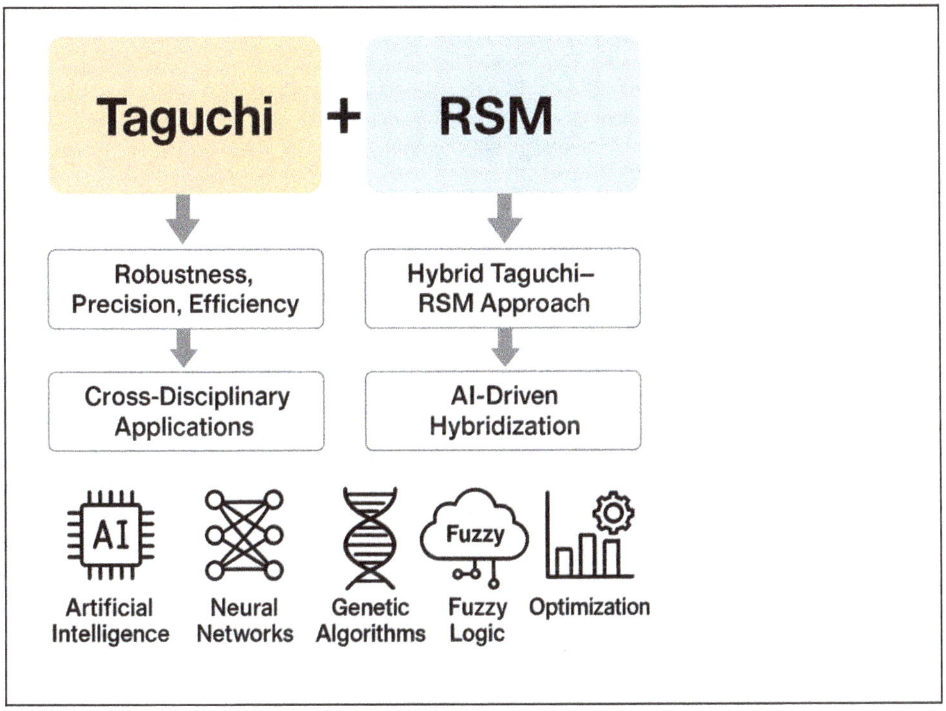

Chapter 8
Advancing Child-Centred Research Methodologies in the School Library Context

Nina Olivia Rugambwa
http://orcid.org/0000-0002-5127-965X
Kyambogo University, Uganda

Martha Lyaka
http://orcid.org/0000-0002-7676-3931
University of South Africa, South Africa

ABSTRACT

Research involving children in the field of Library and Information Science (LIS) has mainly relied on traditional data collection methods rooted in social research paradigms. Common techniques include interviews, documentary analyses, and surveys. This chapter emphasizes the importance of child-centered research methods in LIS, especially for understanding children's information-seeking behaviors in school libraries. A study conducted in two primary schools in Kampala and Mukono involved 32 children, using storytelling and flower maps, to share their views on libraries. Children perceive good libraries as safe spaces, with children's pets, knowledge hubs, places for adventure and relaxation. Conversely, children also described negative aspects of libraries, such as noise, disorganization, dirtiness, unfriendly staff, non-functional information tools, and intimidating content, which characterized bad libraries. The chapter advocates adopting child-centered methodologies to foster a more inclusive understanding of children's informational experiences and needs.

DOI: 10.4018/979-8-2600-0434-0.ch008

INTRODUCTION

Following the adoption of Article 12 of the United Nations Convention on the Rights of the Child (1989), researchers are making an effort to amplify the voices of children in research (Freire et al., 2022). The Child Rights approach, rooted in the United Nations Convention on the Rights of the Child [UNCRC] (UN, 1989), emphasises children's rights to participation, protection, and provision. In child-centred research, this framework promotes children's active involvement in studies affecting them, as outlined in Article 12, which guarantees children's right to express their views freely and have them given due weight based on their age and maturity (UN, 1989). Mackenzie (2025) argues that this approach shifts research from being conducted on children to being conducted with them, ensuring their voices shape the research design, data collection, and dissemination.

The UNCRC Committee emphasises that information must be age-appropriate and accompanied by feedback mechanisms to ensure children's views are respected, enhancing transparency and empowerment (UN Committee on the Rights of the Child, 2009).

Lake and Jamieson (2016) further argue that a child rights approach is vital for preventing violence against children, which impacts their cognitive, emotional, and psychosocial development, leading to outcomes like anxiety or aggression. By integrating Articles 19 (protection from violence) and 12 (participation), this approach ensures that child-centred research respects autonomy, addresses power imbalances, and creates safe, empowering spaces for children to engage as rights-bearing individuals with evolving capacities (Lake & Jamieson, 2016).

The children-centred approach or framework places children at the centre of the research process as active participants, enabling them to communicate their life stories (Ebrahim, 2010; Smith, 2011; Mackenzie, 2025). Additionally, it challenges adult-centric research by advocating for child-friendly methods, such as storytelling, drawing, or photo-elicitation, to elicit children's narratives directly (Dockett & Perry, 2011). By valuing children's perspectives, this framework promotes participatory and empowering research, validating their agency and fostering ownership over the research process.

Martens (2022) and Levy and Wasmuth (2025) support this framework by emphasising that providing children with accessible information enhances their ability to articulate perspectives, particularly on sensitive topics. Lake and Jamieson (2016) suggest that the children-centred framework can enable children to voice their experiences of violence or safety concerns through narrative methods, such as storytelling, to inform prevention strategies, thereby reinforcing their role as active contributors. This framework ensures that child-centred research captures authentic, child-driven insights, promoting empowerment and social justice (Lake & Jamieson, 2016).

The child-centred framework also advocates for the use of child-friendly methods, such as Play-Based and Expressive Arts, as a means of promoting safe spaces for children in research. Play-based and Expressive Arts Frameworks leverage children's natural modes of expression, including play, art, music, and drama, as research tools to access their thoughts and feelings (Koller & San Juan, 2015). These methods are particularly effective for engaging young children or those with limited verbal skills, providing non-threatening, creative avenues for participation. For instance, studies that use drawing or role-play have successfully captured children's perspectives on sensitive topics, such as family dynamics or trauma (Driessnack & Furukawa, 2012). These approaches align with child-centred principles by allowing children to communicate in safe, natural ways, enriching qualitative data (Carter & Ford, 2013).

Sanders and Farago (2018) link these frameworks to Developmentally Appropriate Practices (DAP), noting that play-based methods support children's active knowledge building and align with Piagetian and Vygotskian principles. For example, a child expressing emotions through art shows both developmental readiness and sociocultural learning (Sanders & Farago, 2018). Lake and Jamieson (2016) highlight that play-based approaches can also help promote discussions about violence prevention, enabling children to explore safety concepts creatively, which supports their emotional well-being and participation rights. By combining play and the expressive arts, researchers can foster inclusive and engaging settings that respect children's voices and provide meaningful insights into their social and emotional worlds (Carter & Ford, 2013; Lake & Jamieson, 2016). Participatory visual methods are also effective tools for gathering children's perspectives on various issues, using child-led visual data. Children are asked to draw and add text to explain their drawings. This technique has been successfully used in studies on violence against children (Ritterbusch et al., 2020). These participatory methods can also be applied to explore different research topics or themes in the field of Library and Information Science, where research involving children remains limited, especially in school libraries in Africa.

Freire et al (2022) in a systematic review of 26 studies aimed at establishing the participatory methods and approaches that can facilitate adolescents, children, and families to engage in research to inform the development of health interventions and resources for children and adolescents, reported that a variety of methods were used to do research with children, but few studies provided their participatory approach. Child participatory research in Uganda, particularly in Library and Information Science, remains limited. This study positions itself in school settings, focusing on the information behaviour of children in school libraries in primary schools.

It is also important to note that several studies have shown that professionals across various fields, such as caregiving, research, education, and support roles, often lack the skills necessary to facilitate child participation in research involving children (Freire et al., 2022). This study provides evidence on how researchers can utilise child participatory methods to understand the information-seeking behaviour of children in Ugandan school libraries.

STATUS UPDATE ON NATIONAL SCHOOL LIBRARY POLICY IN UGANDA

Quality library services contribute to educational quality and access to information, which are essential for national growth. The 1999 IFLA and UNESCO school library manifesto, as well as the 2015 IFLA school library revised guidelines (IFLA School Libraries Section Standing Committee, 2015), emphasise that well-functioning school libraries support literacy, lifelong learning, innovation, cultural development, and equitable access to information, which are critical for development. School libraries could help advance Uganda's National Development Plan (NDP) IV by promoting social and economic transformation and contributing to the goal of becoming a middle-income country under Vision 2040, through supporting access to quality education and relevant, high-quality information. However, gaps remain in Uganda's national library policy and school library development (Buwule & Ponelis, 2017). The lack of a national school library policy means there are no established guidelines for operating school libraries. Scholars (Okello-Obura & Kigongo, 2011) argued that this has led to scattered and unstandardized library operations. They noted that individual schools create their own initiatives, resulting in inadequate library facilities and a lack of nationwide coordination. The absence of coordinated government support hampers the provision of quality library services and contributes to ongoing dissatisfaction among users.

Although library professionals have held several consultative meetings and discussions with the government about a national school library policy, no concrete framework has yet been established to guide school library operations (Magara & Nyumba, 2004; Magara & Batambuze, 2009; Okello-Obura & Bukenya, 2011; Chance, Okello-Obura, & Luyombya, 2025). In Uganda's higher education sector, digital technologies are gradually being adopted in academic libraries; however, many libraries still struggle with inadequately trained staff managing them (Nakaziba & Ngulube, 2024). Therefore, challenges to providing quality library services in Uganda persist due to the lack of comprehensive national legal frameworks governing library operations. Functional academic libraries are a prerequisite for the effective integration of the new competence-based curriculum (Ernst, 2023), which has been rolled out in all educational institutions in Uganda.

PROBLEM STATEMENT

The gap in understanding young children's information-seeking behaviour remains a significant issue within Library and Information Science research. Despite a growing acknowledgement of child-focused research philosophies and methodologies, which empower children to engage in research alongside adults in a way that is both child-friendly and prioritises their well-being (Water, 2024; Faldet & Nes, 2024), there is still a scarcity of studies specifically targeting the information-seeking behaviour of young people. Some studies have been conducted in public libraries in Africa, specifically in Nigeria and Benue, on children's information needs (Ronke, Chinweike & Olokode, 2022; Tergenge, 2021). The studies revealed that it remains challenging to understand the information needs of children at different ages, which hinders the provision of information tailored to their specific needs. Consequently, the children leave the library disappointed when their information needs are not met. This finding confirms other studies that have argued that children and adults have unique information needs and seeking patterns and require innovative ways to investigate and address these needs (Oyediran, 2019; Hussain & Begun, 2021). Few studies have explored children's information needs and seeking patterns in school libraries in Uganda.

Recent studies in Uganda have examined the services provided by community libraries to children, concluding that there are limited information services and resources specifically tailored to children's information needs (Nalumansi, Kaddu & Okello-Obura, 2024; Rugambwa, Kawalya, Mutibwa, 2025). Another study in Uganda examined user satisfaction with school library services in government-aided secondary schools in Kabale District (Chance, Okello–Obura, & Luyombya, 2025). No study has yet focused on the information behaviour of children in primary schools using child-focused methodologies.

This study filled this methodological and contextual gap. Barriage (2021) noted that there is also limited research conducted with children under the age of 8, pointing out that participatory methods, which are essential for engaging children meaningfully, are often inadequately presented in existing literature. The same view had been presented in earlier studies that echoed that many studies continue to rely on traditional observation methods (Agarwal, 2014), parental reports on children's activities (Barriage, 2016), and video recordings (Given et al., 2016), which often exclude children from the research process, as emphasised by Barker and Weller (2003). There are also existing gaps regarding how young children can be supported to participate in research because adults do not fully understand different ways young children communicate (Murray, 2019; Hunleth et al., 2022; Sevon, Mustola, Sippainen, & Vlasov, 2023; Olsen, 2025).

What emerges from the discussion is that less attention has been given to how young children can participate in research, allowing their perceptions of information-seeking behaviour to be documented. This study filled the methodological gap by using flower maps, observation, and children's stories as lenses to gain insights into children's perspectives regarding their school library. The study also targeted children aged 6 to 12 years to contribute to the limited literature on children's information-seeking behaviours in school libraries in Uganda. Additionally, this study advances the ecological systems theory in the information-seeking behaviour of children in resource-constrained environments of Africa.

CONTRIBUTION OF THE STUDY

This study advances the philosophy of child-centred research methodologies supported by many researchers worldwide who are committed to promoting children's interests. It offers a methodological contribution in Uganda, particularly in the field of Library and Information Science, where the child-centred research approach and methods are rarely used in studies involving children. The research focuses on school libraries, highlighting these institutions as a foundation for developing literacy skills and achieving quality educational outcomes. This research is timely in Uganda, as it highlights the existing gaps in school libraries that may hinder the effective implementation of the recently introduced competency-based curriculum. The study advocates for the development of a school library policy to address the uncoordinated nature of library services in the country as a steppingstone for learners to become lifelong learners and independent digital citizens in the 21st century. This study highlights that the establishment of standardised library services in schools would help to achieve the Sustainable Development Goals (SDGs) SDG 4, which promotes quality education and lifelong learning, along with SDGs 5 and 10, which focus on inclusion, equity, and reducing inequalities, especially for marginalised groups like rural schools in Uganda, by enhancing access to information and ICTs. NDP IV reinforces this effort by highlighting the importance of strengthening strategic infrastructure, including the development of ICT infrastructure to foster socio-economic transformation.

NEW TRENDS: HOW CHILD PARTICIPATORY METHODS MUST EVOLVE IN A DIGITALLY SATURATED LANDSCAPE

In a digitally saturated landscape, child participatory methods must evolve to address several key issues, including the datafication of children in educational institutions and research, information literacy, digital citizenship, and inclusive digital access in schools.

Datafication of Children in Educational Institutions and Research

The interaction of children with digital information systems has increased in the 21st century, where children are increasingly relying on digital devices for their educational needs, well-being, and leisure (Bradbury, 2021; Mascheroni, 2018). There are concerns that children in educational institutions are exposed to data monitoring systems, and this data is influencing educational practices as well as the lives of children and their families (Roberts-Holmes & Bradbury, 2016; Wang, Zhao, Kleek, & Shadbolt, 2022; Mertala, Fagerlund, & Palsa, 2025). The same studies have raised concerns about knowing who is managing the data for children and what it is being used for. There is also concern that their data is shared through various platforms and can be exploited for digital engineering as well as other practices that may be harmful to children (Livingstone & Third, 2017; Wang, Zhao, Kleek, & Shadbolt, 2022). These studies have highlighted the need for more awareness, monitoring by parents of their children's online activities to safeguard children from exploitation and harm while using online information platforms. The studies also recommend design improvement and urgent policies to protect children during their online interactions.

Information literacy, Digital citizenship and Children in schools

For children in schools to access and utilise information resources, they need to be information literate. Information literacy skills will empower them to identify the information they need from reliable sources and to use it in an ethical manner (Jones, Mitchell, & Beseler, 2024; Karsgaard, 2024; Morales-Álvarez, Valdés-Cuervo, & Parra-Pérez, 2025). Without information literacy skills, children cannot exploit reliable digital platforms and resources in the digital world. This results in exclusion from digital citizenship and its varied resources. Children of the digital world must be taught how to navigate digital platforms with awareness of safety concepts, including issues such as their own privacy, how to avoid cyberbullying, and how to identify misinformation (Karsgaard, 2024). Children also need to be equipped with skills to participate in digital citizenship, including critical thinking skills to identify reliable information sources, the ability to create information from various sources, and to communicate effectively in digital environments (Fraillon, Schulz, Klinger, & Schmid, 2024). In the context of this study, libraries require the necessary infrastructure and trained staff to empower children to develop these skills. Without information literacy skills, children are vulnerable to dangers in the digital environment and cannot act responsibly while navigating the digital world.

Inclusive digital access in schools

The rights to information (Article 19), education (Article 26), and freedom of expression (Article 19) are considered essential human rights in the Universal Declaration of Human Rights. Access to information and digital technologies allows children in school libraries to instantly access reliable educational resources in multimedia formats, promoting online learning. Digital technologies also enable multiple learners to access information simultaneously in schools, thereby improving learning outcomes. Additionally, access to digital technology addresses physical space challenges common in resource-limited environments and can bridge the gap between rural and urban schools. However, it is important to recognise that many benefits of ICTs cannot be realised without an enabling environment, which includes necessary ICT infrastructure in schools (such as power sources, internet access, hardware, and software), teacher and learner training to develop digital skills, and supportive ICT policies for schools.

Inclusive digital access involves addressing the needs of marginalised groups, including children with disabilities, where many experience discrimination (Word Health Organization, 2024; United Nations, 2024), and rural populations that face numerous barriers to digital connectivity. Schools must ensure that children with disabilities have access to assistive technologies to aid them in accessing educational resources and opportunities. In Uganda, the government has made progress in facilitating access to digital resources by establishing a legal framework for the use of ICTs. This framework includes policies, regulations, and laws that support the use of ICTs, such as the Computer Misuse Act of 2011, the Electronic Digital Signature Act of 2011, the Electronic Transactions Act of 2011, the Data Protection and Privacy Act of 2019, Uganda Disability Act, 2019 among others. This legal framework aims to facilitate the equitable and responsible use of digital technologies in Uganda.

THEORETICAL FRAMEWORK OF THE STUDY

The discipline of Library and Information Science encompasses various models and theories of information-seeking behaviour that help explain how individuals seek information in diverse environments. This study focuses on Kuhlthau's information search process model (Kuhlthau, 2008), which was specifically developed in the context of school libraries to guide students in effectively seeking information for research purposes. The model emphasises that the information-seeking process involves cognitive, affective, and behavioural aspects of knowledge construction, while also highlighting the personal dimensions of information seeking. It has been instrumental in analysing information behaviour in different contexts and enhancing interventions across various information settings (e.g., Kuhlthau, 1988a; 1988b; 1996). However, this model has not addressed the concepts of communalism and the interaction of different school ecosystems as critical components influencing information-seeking behaviour, particularly in resource-constrained school library environments in Africa, and specifically in Uganda. In such contexts, individual efforts to seek information may not guarantee successful information searching.

In relation to this gap, this study explored the ecological systems theory, guided by constructs including microsystem, mesosystem, exosystem, and chronosystem, to investigate information seeking in resource-constrained school library contexts in Africa. According to Bronfenbrenner's Ecological Systems Theory (2005), individual growth and development are influenced by multiple interconnected systems. In the context of information-seeking behaviour, the theory asserts that individual information-seeking behaviour is influenced by many subsystems in the school.

The school's physical environment (micro-system- a system they interact with most frequently) is one of the systems that influences the children's perceptions about the library. In environments where the school has established a physical library, children will be introduced to the concept of a library as an important subsystem of the school, and vice versa is also true.

The school policies shape the school environment in terms of the curriculum and the time schedules for library time on the school timetable. This facilitates and enables children to utilise the library and its resources. Thus, schools without a policy on library time miss the opportunity to introduce children to the library and its role in supporting education. The schoolteachers (micro systems) in the school environment are critical stakeholders in supporting children's use of the library, as they interact with the children regularly and for many hours. These have a lot of influence on what the children know about the library. If they discuss a library, the children will understand its usefulness and be encouraged to visit the library. The parents(micro-system) also interact with the school and their children frequently, more so in this study, where children are day scholars; consequently, they influence how the children view the library. They can reinforce either their children's negative or positive perceptions of the library.

The children have peers and neighbours (Microsystem) at school and in their home environment, with whom they interact. These systems are connected to the school environment and also affect the information needs and seeking patterns of children. If children are exposed to neighbours who know about the library and an environment that values it, they are likely to be encouraged to use it, and vice versa. Therefore, the perceptions that children have about the library are influenced by their peers, teachers, parents, school administrators, and their physical school environment.

The library, as a mesosystem (interaction between microsystems), is influenced by school regulations, facilities provided in terms of physical infrastructure, information resources, and human resources, which in turn shape how the library operates and how patrons are served. The manner in which the library staff manage the school library resources, as well as their interactions with staff and patrons, impacts how the children perceive their library. Finally, the school and how the library is run are influenced by the national regulations and legal framework guiding (Chrono–systems) school libraries in Uganda. Thus, for children to effectively utilise information resources in the library, the government needs to regulate school libraries, librarians need to provide children with tailored library services, and both teachers and parents need to encourage children to utilise the library.

BROAD RESEARCH QUESTION

What are the perceptions of children aged 6-12 about a library?

Specific research questions:

- What is a library to you as a child?
- What do you like in your library?
- What don't you like in your library?

METHODOLOGY

Research Philosophy

This study was guided by the constructivist research paradigm, which posits that reality is context-specific and is based on the stories of those who have experienced it within their own environment. This study is also guided by the child-centred research approach that emphasises that children are co-creators of knowledge. Researchers (Clark & Moss, 2011; Ritterbusch et al., 2020) argue that this approach shifts research from being conducted on children to being conducted with them, ensuring their voices shape the research design, data collection, and dissemination.

Research Design

A qualitative case study research design was adopted in this study to facilitate an in-depth understanding of the children's perspectives of their school library. The design also enabled the researchers to interact with the children in their natural setting and supported the use of multiple data collection methods.

Study Population

Table 1: Demographics of study participants

Variable	Variable Types	Frequency
Gender	Male Female	16 16
Age	6 -12	32
Class	Primary Seven	9
	Primary six	8
	Primary three	9
	Primary two	6
School Type	Urban Public Primary School (Kitante Hill School) Rural Church Founded School (Kyetume Church of Uganda)	

Participants included children aged 6 to 12 years old. A total of 32 children participated. Age ranged from 6 to 12. This study was conducted in Kitante Hill Primary School, a public government school in Kampala district and Kyetume Church of Uganda Primary School in Mukono District. Convenience sampling was used to recruit the children in the study. Children in the library who were willing to participate were approached to participate in the study from primary 2 to primary 7. In the two schools, these are the only children found in the library who assented to participate in the study.

Data Collection Methods

Child-centred participatory research methods, including flower maps, drawings, and child stories, were used to engage children and gather their ideas and perspectives about their library. Drawing from personal experiences as mothers, we realised that children love bright colours and enjoy drawing. Recent studies also support this observation (Gerasymenko, 2024; Singh et al., 2024). Based on this evidence, we provided each child with three flower maps. Each map represented a research question they were asked to answer. The children wrote their responses in the petals of the flower for each of the three questions. They received the maps in three intervals, allowing them to focus on each question individually rather than all at once. Additionally, they were asked to tell stories about their library and use drawings to illustrate their stories. This was an exciting experience for the children, and they responded to the research questions with enthusiasm and calmness.

Illustrations of flower maps complemented with children's drawings are in Figures 1 to 3.

Figure 1. Flower map complemented with children's drawings (Designed by Authors)

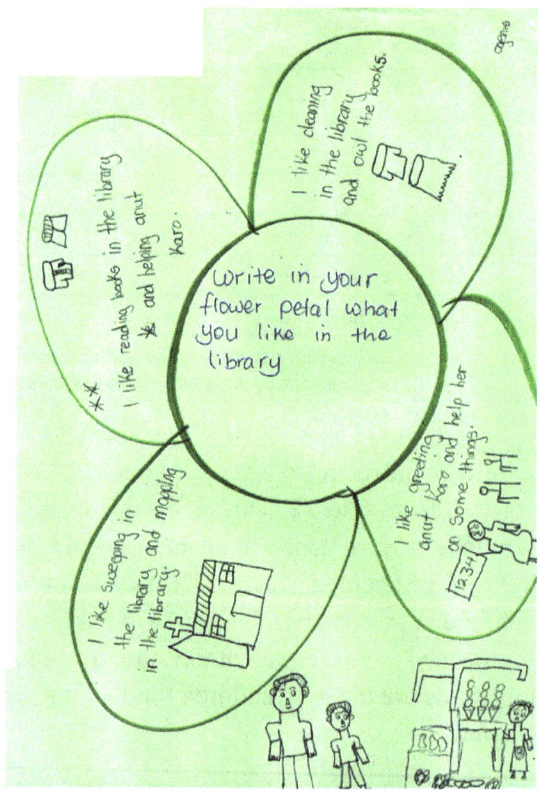

Figure 2. Flower map complemented with children's drawings (Designed by Authors)

Figure 3. Flower map complemented with children's drawings (Designed by Authors)

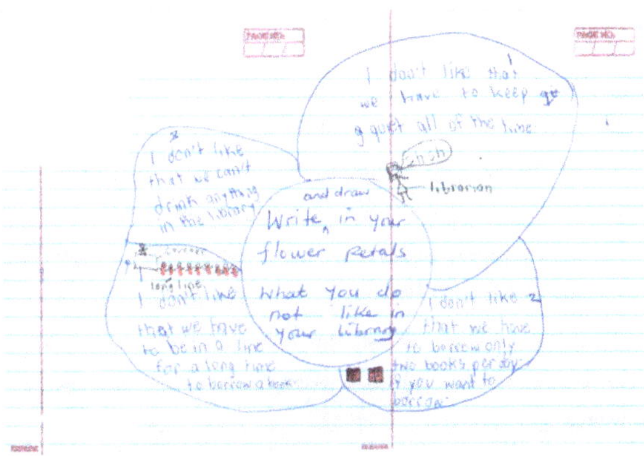

DATA ANALYSIS

Data were analysed at two levels, namely, thematic and theoretical framework analysis. Braun and Clarke's (2006) six-phase framework was employed in this study to provide a structured and rigorous approach for identifying, analysing, and reporting themes. The analysis started with familiarisation, where all data collected from stories, flower maps were examined repeatedly to gain a comprehensive understanding, supplemented by reviewing field notes and observational data. Initial observations, including recurring patterns and potential codes, were documented. The third phase involved searching for themes, during which codes were organised into potential themes based on similarities and relationships, culminating in a thematic map to visualise these connections. The fourth phase involved a thorough review of the themes at two levels: verifying that the coded data extracts aligned with the identified themes and confirming that the themes accurately represented the entire dataset, which led to the refinement of the themes. In the final phase of defining and naming themes, each theme was clearly articulated and labelled to capture its essence, with detailed descriptions provided for each theme and sub-theme to ensure clarity and consistency in interpretation. The theoretical framework analysis was also applied to the data to gain a deeper understanding of the data by linking it with the constructs that guided the study (Ngulube, 2023; Creswell & Creswell, 2017). This led to the development of new knowledge and theories about children's perceptions of their school library.

Limitations of the Study

The scope of this study focused on understanding how children perceive their library, identifying what makes it good and what makes it bad. This study did not explore how they seek and use information systems and resources. While the researchers are aware that the traditional physical libraries are outdated in the knowledge society, it was clear from the literature reviewed and observations of the libraries studied that the ICT infrastructure was still underdeveloped. Similarly, the researchers are aware of advancements in the use of digital tools in child-centred studies; this study did not use digital media in collecting the data. While the researchers employed the flower maps and stories, the pictures drawn by some children in the flower maps were not analysed. The study relied on the stories the children talked about their library, as well as their responses on the flower maps and observations, to interpret the study's findings. The findings would have been more enriched if the drawings were also interpreted.

ETHICAL ISSUES IN RESEARCH WITH CHILDREN

This section is divided into two parts. The first section presents the latest discussions on ethical practices in research involving children. This is followed by an illustration of how ethics were enforced in this study. Ethics in research with children includes seeking consent from their parents. Since children are vulnerable and cannot make their independent decision on whether to participate in research. Their parents need to consent on their behalf, then the child assents before they are allowed to engage in research (Christensen & James, 2020; Faldet & Nes, 2024).

Consent from children is crucial to ensure that they are not forced to participate in research that they are not interested in, even if a guardian or parent has given their consent. In addition, during the process of seeking the child's assent, a researcher should take responsibility to provide age-appropriate information about what the research is about, the research processes that the child will be involved in, as well as the risks and the benefits (Sharkey & Sharkey, 2020; Christensen & James, 2020).

Researchers have an obligation to protect children from harm by providing safe spaces for them to express themselves freely without fear, and to safeguard them from technologies that may affect their psychological well-being and lead to disillusionment through misinformation (Sharkey & Sharkey, 2020). The researchers caution scientists in robotics to be cautious in advancing robotics in childcare and replacing human child caregivers with machines, as this could lead to attachment disorders among children.

It is also important that a researcher employs participatory research methodologies and participatory data collection methods, such as flower maps, stories, drawings with children's explanations, environmental maps, providing children with electronic devices such as cameras and recorders for children to record their own experiences, or using cameras to capture information of interest to the researchers, are some commonly used methods. These data collection methods are essential because they empower children to participate in research by working alongside researchers, and this facilitates their voices being heard.

These methodologies are safe and child-friendly because they reduce the stress children encounter while relating with adults, and the methods are known for countering the power relations existing between children and adults that intimidate them from providing information that may not be authentic, rather influenced by the dispositions of the adult they are interacting with (Water, 2024; Loveridge, Wood, Davis-Rae, & McRae, 2024). These methodologies provide children with the independence to freely express their feelings.

Special care and caution should be taken when conducting research with children with special needs because they can be exploited because of their condition, as well as those from marginalised communities. While all children are vulnerable, those with special needs are considered more vulnerable. Researchers have an obligation to ensure that the voices of these children are heard and taken into account in research (Faldet & Nes, 2024). Additionally, researchers must provide all necessary evidence to demonstrate how their welfare and safety would be guaranteed in a study.

Furthermore, researchers should be culturally sensitive when working with children from diverse communities and must show respect for existing cultures by engaging with cultural leaders to promote ethical research (Faldet & Nes, 2024). The EECERA code for early childhood researchers provides several guidelines that must be followed for ethical research with children (Bertram et al., 2025), including maintaining the privacy and confidentiality of data provided by children, and providing clear and transparent information on how the data will be used. All data generated during the research should be secured and anonymised to protect the privacy and individual identities of children (Bertram, 2025). Additionally, respecting the child, family, and community, and ensuring respectful interactions throughout the research process are essential, along with upholding human rights, justice, equity, and democratic values at all times. The guidelines also emphasise that children should be involved in evaluating the research process after data collection, which is crucial for allowing children to raise concerns so they can be addressed. Having a feedback mechanism helps children feel valued and engaged in the research process, providing insights for future researchers on how to improve the conduct of research with children. Finally, researchers should prioritise objectivity and integrity in all research involving children to ensure the quality and credibility of the research.

How was ethics enforced in this study?

Given that the study involved minors who could not provide their own consent, the researchers obtained consent from the children's parents or guardians for their participation. A consent form was developed that clearly explained the purpose of the research and the questions the children would be asked. This form was given to the parents, and only those whose parents consented were allowed to participate in the study. The children were provided with information about the study, including potential risks, the nature of their participation, and the time commitment involved. The researchers also obtained the children's assent, and only those who gave their assent were included in the study. Children aged 6 to 12 participated, and the questions asked were simple and developed with consideration of their developmental levels. Child-friendly participatory methodologies, including flower maps and storytelling, were used to empower children to express their views on their school library without any influence from adults. The children's privacy and confidentiality were maintained by anonymising their data to safeguard their individual identities.

To reduce individual bias and promote reflexivity, the authors used verbatim quotes from the children as evidence of their perspectives.

FINDINGS

This section presents the study's findings based on the three research questions: What is a library to you as a child? What do you like in your library? What don't you like in your library? The findings have been presented systematically in the subsequent sections. This section presents findings for the first research question.

Findings on the Perceptions of Children of a School Library

- *Source of knowledge, wisdom and understanding* Many children indicated that a library is a place where they learn new knowledge, gain wisdom, and develop understanding.
- *A place where books are kept safely and borrowed* All the children indicated that a library is where books are kept and borrowed. Others added that it is a collection of storybooks and textbooks, as well as a building where books are stored safely.
- *A place where I learn to behave well* One child indicated that a library is a place where he learns how to behave. He said: "It is a place where I am happy because I learn how to behave." (Lower primary, male aged 9 years, School R1).
- *A place of relaxation, comfort* Some children defined a library as a quiet place where they can find comfort and relax their minds. One of the children said: "It is a place where I relax my mind." (Male lower primary, aged 9, School R1)
- *Research centre* The children explained that a library is a research centre since it has almost every book they need. Others added that they use the library for revision and to discover new things. Another child added that a library makes them read ahead of their teacher.

Findings on what Children like in their Library

This section presents findings on what children like in their library, presented in themes.

Library Environment: Quiet Versus Noisy Library

When children were asked about what they liked in their library, they expressed contradictory views about the library's ambience. The eldest children indicated that they liked their library because it is quiet. One of the female pupils said that:

"The library is peaceful when it is quiet.'' (Female pupil aged 12 years, school R1 05.06.2025)

Another child added that:

"We free our mind from the noise. I like the peacefulness in the library." (Male pupil aged 12 years, school R1, 05.06.2025).

On the other hand, younger children indicated that they preferred noisy environments in the library, where they could interact with their peers. Other children emphasised that they did not like it when their librarian told them to keep quiet. The younger children indicated that they would like to play and talk with other children when in the library. This view was noted especially among the younger children.

One of the pupils emphasised that:

"I do not like that we have to keep quiet all the time in the library. Sssssssh librarian." (a female pupil aged 6 years old, school R2, 05.6.2025).

Another child said:

"They tell you to keep quiet. I need to be allowed to speak to my friends and not so quiet until I leave the library" (Female pupil lower primary aged 7, school R2, 05.06.2025).

Another child emphasised that:

"They should allow us to talk quietly." (Female, lower primary aged 8, school R1).

It is worth noting that all female children preferred a noisy environment where they could interact with their peers.

A Place of Discoveries and Knowledge

Most children said they like the library because it is a place of many discoveries, such as learning new words and sounds, and learning about their country and topics not yet covered by their teachers. Other children pointed out that they visit the library to acquire knowledge and wisdom by borrowing books.

One of the pupils said that:

"I can read ahead of my teachers and I get to know what I never knew" (Female pupil upper primary aged 12, school R1, 05.06.2025).

A Place of Enjoyment and Relaxation

Some of the pupils indicated that their library is a place of enjoyment where they can read interesting story books. All children in the study reported loving the storybooks in their libraries. This emphasises storybooks as an important information need for children in this age range.

Sitting on a big mat in the library

Some pupils indicated that their librarian provides a big mat for them to sit on in their library. There was agreement among the children's responses that they loved the mat, which attracted them to the library. This finding was recorded among the female children only.

One of the female children said that:

"I like the mat in the library because it is big and we sit on it and enjoy." (Female aged 8 years, school R1, 05.06.2025)

The finding indicates that the children associated the mat with a good library and enjoyment.

Librarian's pet dog

Both male and female children in one of the schools excitedly explained that their library has a beautiful dog that the librarian keeps as a pet, which they go to see.

One of the children said that:

"I like to see the dog of the librarian in our library." (Male aged 6, school R2, 05.06.2025).

Through observation, we observed that the dog was housed near the library and the children were able to see it while entering the library.

Decorations in the library

One of the female children explained that they liked the library because it has decorations and beautiful colours on the walls and drawings. Others said they like pictures in the library.

We observed that in one of the libraries, it was decorated with bright colours, flowers, and drawings. These made the library an attractive place for the children.

Human Relationships

Positive and Negative Consequences of Human Relationships

Most of the children emphasized that they go to the library because they would love to have a good relationship with the librarian. Some children indicated that they visit the library to see if the librarian is present so that they can greet her. Male children indicated that if they have a good relationship with the librarian, it is easy for them to borrow books from the library. Two female children indicated that they assist the librarian in dusting, organizing the books, and mopping the floor because she is kind and a friend. Other children expressed concern about having a bad relationship with the librarian, as it would hinder their access to library resources.

One of the children said:

"I like greeting the librarian. I go to see her, she is kind." (Female aged 8, school R1, 05.06.2025).

Another child added that:

"I like sweeping and mopping the library to help my friend the librarian." (A female aged 6 years, school R1).

A younger child said:

"I check on the librarian to see if she is okay." (A female aged 6 years, school R1 05.062025).

However a female child in lower primary said that:

"What I do not like in the library is when you are the librarian's enemy you cannot get any book." (Female aged 10 years, School R2,05062025).

Another child added that:

"What I do not like in the library is that when you shout, the librarian chases you out." (Female aged 10 years, School R2, 05.06.2025)

The findings reveal that children greatly value having a good relationship with library staff. They put in effort to ensure that they build a close relationship with them to have access to library resources. The findings also suggest that having a bad relationship with library staff may lead to exclusion from the library and hinder access to information resources in their library.

FINDINGS ON WHAT CHILDREN DISLIKE IN THEIR LIBRARY

This section presents findings on what children dislike in their library, presented in themes and sub-themes.

Barriers to Information Access

Mistreatment by Prefects Supporting Librarians

Many children reported that they did not like the way library prefects treated them when they visited the library. Some children revealed that the prefects sometimes did not allow younger children in lower primary to access the library and mistreated them. Yet, these were their younger siblings, and they also needed to use the library. Other children indicated that the librarian gives the prefects a lot of power to mistreat them.

One of the male pupils said that:

"I don't like the children working in the library. They chase away their friends and do not allow children in lower primary to enter the library, yet these are our brothers and sisters." (Male aged 10 years, School R1, 05062025).

Another pupil said:

"I don't like the way the children who work in the library talk to us." (Male aged 9 years, School R2, 05062025).

The finding suggests that library prefects are viewed by other children as rude and a barrier to accessing information resources because they are unfriendly and prevent younger children from visiting the library.

Library Material Deficits

Some children revealed that they felt frustrated when they did not find the essential books they required in the library. Others added that their library lacked some necessary textbooks and storybooks, yet their parents could not afford to buy them. This, they indicated, disappointed them whenever they came to the library.

A male child from upper primary said that:

"The library sometimes does not have the exact book I am looking for. It makes me sad, and I leave the library, yet my parents cannot buy it. I am in primary seven, a candidate class" (Male aged, 12 years, School R2, 0506.2025).

The finding highlights the frustration children in the school library experience when their information needs are not met due to limited resources in the library.

Dysfunctional Information Tools

One of the female children revealed that some information tools in their library were not functional, but were still kept in the library. She explained that book jackets indicated that the books would make noise if they touched them, but this did not happen. The child suggested that the library needed to buy batteries to ensure the book makes noise instead of lying to the children about its' functionality. She emphasized that:

"There is a book that is written on: press me and hear a sound, but no sound is heard. They should put batteries in the book so that it can speak. Why does the librarian lie to the children?" (Female, Lower primary, 8 years old, School, R1 05.06.2025).

Congestion and Long Lines in the Library

Some children indicated that their library is often congested during break and lunch time when they are allowed to go and borrow books. As a result, they are made to stand for long in queues to borrow books, and it is difficult for younger children in the lower primary to use the library.

The finding suggests challenges in terms of physical space affecting library services in these school libraries.

Limited Human and Physical Resources in the Library

Many children at one of the schools revealed that they have no access to the library whenever their librarian falls sick. They have to wait for her to recover before they can use the library.

One of the male children in upper primary emphasized that:

" When the librarian falls sick, we have to wait until the librarian comes back to use the library."(Male, upper primary, 12 years old, School, R2 05.06.2025).

Another female child said:

"The school does not care for us in the candidate class. If the librarian is not there, the library is closed. It can be closed for a long time, and we miss a lot."(Female, upper primary 12 years old, School, R2 05.06.2025).

Absence of Light in the Library

In one of the schools, the children complained that the library lacked power and this hindered them from reading because the library was dark.

The finding highlights that the school has only one library staff member, and if she is absent, children cannot access library resources. Additionally, the school's infrastructure in terms of ICT is non-existent, as they do not even have a power source for lighting the library.

Disorganized and Dirty Library

Some children revealed that the books in the shelves were not organized in an orderly manner. This made it hard for them to identify the books they were looking for. Others added that the library has many books, which makes it difficult to choose the book they need. Another child emphasized that the library should be well-organized. One of the female children said that:

"I don't like disorganization of books in the library." (Female aged 12 years, School R2 05062025).

Some children revealed that their library was always untidy, with disorganized chairs. Others added that the library should be swept and mopped regularly, and the color of the wall in the library was not clean.

Limited Support to Young Children by Library Staff

One of the children indicated that the library staff do not assist young children in nursery to read and this was explained as one of the reasons why she did not like the library. She said:

"The librarian and her people should help the children in nursery to read. The children are left to suffer with books in the library." (Female, aged 10 years old, School R1,05062025).

Scary Books

Many children revealed that they disliked storybooks with scary stories, including stories about HIV/AIDS. They preferred reading interesting books with adventure.

Discussion of Findings on the perceptions of children of a library

The findings regarding children's perceptions of libraries reveal that they view these spaces in various ways. In this study, children identified their library as a source of knowledge, wisdom, and understanding; a place for moral instruction; a space for relaxation; and a research center. These results align with several studies that have explored children's information behavior in school libraries (Huang & Shieh, 2021; Willis et al., 2019; Kavak & Yıldız, 2025; Sansbury, 2025), all of which reported similar outcomes.

One study highlighted that school libraries promote reading for pleasure, improve children's confidence, enhance social skills, foster a sense of belonging, and provide a safe, supportive environment that encourages independent learning (Fodale &

Bates, 2011). Recent research has also indicated that libraries play a critical role in preparing children for school readiness and lifelong learning (Cahill et al., 2024).

These findings align with the perspective that libraries serve as vital centers for research and sources of knowledge. One study highlighted that school libraries promote reading for pleasure, improve children's confidence, enhance social skills, foster a sense of belonging, and provide a safe, supportive environment that encourages independent learning (Fodale & Bates, 2011). Recent research has also indicated that libraries play a critical role in preparing children for school readiness and lifelong learning (Cahill et al., 2024). These findings align with the perspective that libraries serve as vital centers for research and sources of knowledge.

The findings extend Bronfenbrenner's ecological theory, which articulates that the physical environment of libraries, along with the interactions between children, library staff, peers, school administrators, and parents, significantly impacts children's perceptions of these institutions. Consequently, their perspectives are influenced by the various subsystems within the educational framework. However, it is noteworthy that the notion of libraries as environments for technological development and access to digital resources is conspicuously absent from the children's descriptions. This absence suggests a limited exposure to Information and Communication Technologies (ICTs) and digital resources within their library experiences. Therefore, it is imperative to invest in ICT infrastructure to facilitate the integration of children into the realm of digital citizenship, thereby enabling them to reap the benefits associated with enhanced access to information.

DISCUSSION OF FINDINGS ON WHAT CHILDREN LIKE IN THEIR LIBRARY

Findings on what children like in their library revealed that older children preferred a quiet library environment to concentrate on reading, while younger children in lower primary preferred a noisy environment to interact with their peers. The findings highlight that children of different age groups have distinct information needs, which must be addressed for them to have an enjoyable experience in the library. The finding is supported by a recent study (Kavak & Yıldız, 2025) on perceptions of children aged 6-12 about the library, using drawings that emphasized that children view libraries not only as a space for accessing information but also for social interaction and play. The finding agrees with an earlier study (Huang & Shieh, 2021) on Elementary school students' perceptions of the school library and expectations of the library that emphasised that library staff need to recognise children's social and emotional needs in order to design child-friendly libraries.

For school librarians, the finding highlights the importance of being sensitive to users' needs to facilitate inclusive library services. This involves creating separate zones for noisy patrons and a quiet area for older children. As emphasised by Venny and Adli (2024), designing child-friendly spaces in libraries must prioritise the psychological needs of children to enhance the utilisation of library resources. Shoaib, Shamer, and Iqbal (2025), in their study, which included a systematic review of academic library spaces as facilitators of student engagement in higher education, found that dynamic library spaces facilitate the acquisition of various skills among students in higher education institutions, such as communication skills, teamwork, critical thinking, and technology integration, all of which are essential in the twenty-first century.

Other children indicated that they liked their library because it is a place where they make discoveries and acquire knowledge. This finding is in line with various recent studies (Kavak & Yildiz, 2025; Huang & Shieh, 2021) that have confirmed that children viewed their library as a space for accessing knowledge, technological integration, and a personal space.

Some children emphasized that they enjoy their library because they read story books, and their librarian has a pet dog which they come to see. They also come to their library because it has a large, beautiful mat on which they sit while they are there. This finding is unique few studies have highlighted pet dogs and unique mats as some of the attributes of a good library. The finding reveals innovation created by the librarians in these libraries to make the children like the library. The finding agrees with recent studies (Kavak & Yıldız, 2025; Huang & Shieh, 2021) that emphasize that children love libraries that prioritize their security, safety, and happiness. This finding also highlights the importance of child-focused research in voicing children's perspectives on an ideal library. For school librarians, the finding underscores the importance of identifying children's needs to better serve them in school libraries.

Findings also showed that children love a library where they have a good relationship with their librarian. Some children said they visit the library to assist the librarian in cleaning the library, to greet the librarian, while others said they visit because the librarian is kind. Other children perceived a bad library as one with a librarian who did not have a good relationship with them. They explained that they feared having a bad relationship with the librarian because she would not allow them to access information materials to use. This finding suggests that children perceive having a good relationship with their librarian as a significant motivator for information seeking in the school library. The finding in line with Nisa(2025), who established that creating child-centered school libraries that involved parents, children, and school administrators and promoted community cultural values strengthened reading literacy in Islamic elementary education. The study also emphasized the importance of involving all school stakeholders for successful library services in schools. This finding extends Bronfenbrenner's ecological theory in information seeking, emphasizing that the success of library services in school libraries in Uganda is dependent on all school ecosystems.

The finding highlights the need for school librarians to develop strategies to build and foster positive relationships with their patrons, encouraging access to and use of library services in their schools. Haynes' (2014) study, which focused on user satisfaction with library services among children in public libraries in sub-Saharan Africa, confirmed that libraries with trained, child-sensitive staff and child-friendly spaces reported higher levels of child engagement and supported early literacy and positive library experiences more than those without. Paton-Ash and Wilmot (2015) also noted that positive perceptions of helpful library staff by children were linked to increased library visits and usage.

Children also revealed that they liked their library because it is well decorated with beautiful drawings and colors, and a unique mat where they sit. In another school, one of the younger children emphasized that she came to the library because their librarian had a pet dog, which attracted her to come to the library. This finding highlights the importance of librarians working with children in school libraries to be sensitive to children's needs and to make deliberate efforts to create child-friendly spaces in libraries, thereby encouraging children to use the library. Sansburg(2025) documented many strategies that school librarians employed in Georgia's public libraries to promote a reading culture in children, such as inviting library spaces, curating diverse collections, and engaging with the school community.Thus, school librarians have a crucial role to play in promoting reading skills, a love of reading, and information literacy in their schools.

A recent study compliments that a child's earlier experience with language and books in the library, as well as their interaction with library staff, shapes their future academic success. This finding emphasizes the importance of children being exposed to the library and reading resources at an early age and the importance of library staff fostering good relationships to strengthen their literacy skills in preparation for easier learning at higher levels of learning. The study further suggests that fostering good relationships with children and ensuring they have a positive library experience will positively influence their love for the library. Consequently, a negative library experience would discourage them from visiting the library.

The findings expand Brofenbrenner's ecological social learning theory by emphasizing that the children's perceptions about what a good library is or is not are influenced by various sub-systems in the child's school environment. This includes how the child relates to the librarian while in the library. If the child's experience in the library is good, they will likely love and frequent their library. Another important sub-system in influencing the children's perception is the information needs of children in a school library setting, including the physical appearance of the library, and effort must be put in beautifying the library to appeal to the children. The physical environment also entails the physical infrastructure supporting library services in the school such as, electricity, computers and internet that facilitate efficient provision of library services in the schools. This theory argues that all these resources must be available to provide a good experience to the children to love their library.

DISCUSSION OF FINDINGS ON WHAT CHILDREN DISLIKE IN THEIR LIBRARY

Findings revealed that some children are mistreated by library prefects who are harsh and unkind to them and their younger siblings in nursery school, often discouraging them from visiting the library. The finding further reveals that in both schools, younger children found in the library were seen as disruptive and noisy by older children. Library staff and library prefects (children leaders elected to support library staff in schools), therefore, try to discourage and deter them from using the library. The findings further revealed that younger children, especially those in the nursery, were neglected and not supported by the library staff in reading while in the library. This finding was expressed by both young and older children in both schools. Findings revealed that older children were more supported by library staff in using the library than younger ones, yet literacy skills and a love for reading are cultivated in children at a younger age (Caldwell, 2023). The studies also revealed that children who learn literacy skills earlier are better prepared for a successful academic career than their counterparts who are not. The findings may suggest that some library staff are ignorant of their role in providing inclusive library services and their role in developing literacy skills among children in their school. Thus, it is important to conduct further investigation on whether library staff employed in school libraries are professionals in Library and Information Science. The finding also highlights the need for more training for library staff, on early childhood development and how they can promote the acquisition of literacy skills among younger children in their libraries.

The finding is in line with Haynes (2014) and Book Aid (2013), who confirmed that in sub-Saharan Africa, especially in Tanzania, Malawi, Cameroon, Zambia, Zimbabwe and Uganda, children have been discouraged by library staff from using public libraries because they are perceived as messy, noisy and disruptive. The studies argue that this has contributed to limited exposure of children to libraries in Sub-Saharan Africa.

This finding suggests that the library staff may not be sensitive to children's needs and may lack skills in implementing child-friendly strategies for managing children in school libraries. The finding may also point to a lack of awareness from the school administrators, teachers, and children on the role of school libraries in children's personal development. There is a need for sensitization for the school community on the role school libraries play in shaping pupils' personal development and acquisition of literacy skills. Evidence suggests that children who use the library are found to attain social skills, personal confidence, learning autonomy, and literary skills than their counterparts who do not. Parents also argued that use of school libraries by their children in the same study contributed positively to their reading and development of literary skills. Consequently, the need to accommodate young children in school libraries is paramount to building an inclusive school, fostering a reading culture, and acquiring the earlier-mentioned skills, especially in the competency-based curriculum that has been rolled out in primary schools in Uganda. A recent study conducted in China on creating child-friendly environments for children in public libraries supplements that such library services promote an inclusive environment for all children and support children's growth and academic development. Another related study that focused on the effective implementation of child-friendly library management (Sholikha & Wahyuni, 2023) revealed that such services contributed to increased students' reading habits in libraries.

In both schools, the children argued that they lacked some essential books on their school curriculum, yet their parents could not afford them. This finding confirms studies on school libraries in Uganda that have consistently echoed the limited funding of school libraries in Uganda for many years, resulting in inadequate information resources in the libraries (Magara & Batambuze, 2009; Kawalya, 2021; Chance, Okello-Obura, & Luyombya, 2025). This finding implies that the children in schools cannot receive quality library services and education without relevant, up-to-date information resources. The Government of Uganda needs to invest in funding libraries for improved service delivery in these institutions.

Also, one female child indicated that they had one audiobook in their library, but she was disappointed that it was not functional due to a lack of batteries. She wondered why such a book was still kept in the library, yet it was not functional. In another school, children decried the fact that they lacked electricity and could not efficiently utilize their library because it was dark. The finding illustrates the glaring digital gap that still exists in developing countries that needs to be overcome to facilitate equitable access to information to support participation in digital citizenship. While digital resources are a means of expanding access to a diversity of timely multimedia information to many users simultaneously, in these schools, none of the children identified their library as a source of digital information and technological advancement. In comparison to other studies conducted where the

ICT infrastructures are developed (Kavak & Yildiz, 2025; Huang & Shieh, 2021), children described their school libraries as a hub of technological development.

Additionally, the findings revealed that children perceive a poor library as one that is crowded with many patrons, making it difficult to access library resources. The finding highlights the need for school libraries to be expanded and upgraded to electronic library management systems so that children in these schools can access electronic resources remotely to overcome congestion in the library. The findings further highlight challenges in infrastructural development that hinder the effective provision of library services in school libraries in Uganda, which need to be addressed to improve the quality of education attained by learners. These findings confirm studies by researchers on user satisfaction of library services in school libraries in government secondary schools in Uganda (Chance, Okello-Obura & Luyombya,2025) that reported that patrons were dissatisfied with ICT infrastructure and inadequate digital literacy skills by library staff. This study also confirms findings from an earlier study (Magara, 2016) that reported that school libraries were neglected by school administrators and inadequately funded, thereby affecting the quality of education. This points to the fact that without basic school infrastructure, such as electricity, internet, and computers, the library cannot function effectively to meet its mandate. This finding also points to the need for the Government of Uganda to prioritise budgets towards equipping libraries with relevant textbooks and enhancing ICT infrastructures to facilitate more efficient and inclusive library services, especially in the competency-based curriculum, where information literacy skills are a key component to learning.

Another finding reveals that children dislike their library because the information materials are disorganised. There was consensus among the children that in their libraries, sometimes the information resources are not properly shelved and arranged. This hindered the retrieval of relevant books for their classwork and caused time wastage. The finding also emphasizes the importance of information processing and organization in facilitating information retrieval. Thus, without proper processing of information materials, the children will not access and use information resources in their libraries to meet their information needs. The finding also highlights the need for professional librarians to manage school libraries because they appreciate the importance of information processing in facilitating timely and relevant information resources for the users. It would be important to conduct studies to investigate whether school libraries are managed by professional staff.

The finding also revealed that children do not like scary stories. They prefer exciting stories that challenge their imagination and arouse their curiosity to learn. This is another information needed for children in school libraries. Thus, library staff should endeavour to acquire such information materials in the library to encourage the children to access and utilise library resources.

This finding is also in line with the ecological learning theory that emphasizes that a child's school environment and the people s/he interact with regularly influence their perception of a library. The theory also argues that the physical environment at school, in the form of infrastructure, may influence the perception of the child regarding their school library as either good or bad. Thus, all school sub-systems need to work in unison to support children to use and have a good library experience.

FURTHER RESEARCH DIRECTIONS

Innovating Digital and mobile library services to facilitate inclusive and equitable development

Researchers should investigate strategies to integrate digital technologies and mobile library services to reduce the digital gap between the rural and urban schools and to cater to underserved areas in Uganda. This will contribute to Uganda's Vision 2040, focusing on technological development and ensuring inclusive growth. This research area can also focus on how libraries can promote social equity among marginalized groups, such as children with disabilities, to advance SDG4 on quality education and the slogan leave no one behind within Uganda's development agenda.

Building the capacity of library staff and educators to support quality education under the rolled-out competence-based curriculum

Studies should be conducted on professional development models that can be used for capacity building for both librarians and teachers to facilitate foundational skills, including numeracy and literacy skills, among children. This is in line with Uganda's National Development Plan [NDP] IV that prioritises human capital development as the cornerstone for social and economic development.

CONCLUSION

This study concludes that understanding children's perspectives on library services in school libraries is crucial for enhancing service delivery. School administrators, teachers, librarians, and parents must collaborate to support these library services, as they play a vital role in developing diverse competencies in learners and complementing classroom activities. The collective efforts of all stakeholders are essential for nurturing a well-rounded child equipped with the knowledge and skills needed for the newly introduced competency-based curriculum and for succeeding in the knowledge society. The schools do not have adequate ICT infrastructure to support access and use of electronic resources that are dynamic, in multimedia.

RECOMMENDATIONS

Using Brofenner's ecological social learning theory alongside the perspectives of participants, this study makes the following recommendations:

a) Younger children were harassed and viewed as noisy and disruptive in the libraries. This has contributed to negative attitudes toward younger children using school libraries. Consequently, prefects supporting the library were often aggressive towards them and older children. This may indicate that some libraries are being managed by unprofessional staff. Therefore, the study recommends that school librarians in Uganda receive training to provide child-friendly services that cater to the diverse needs of children and foster an inclusive environment. Library spaces also need to be modified to include zones for discussion and silent reading, accommodating the different needs of children in schools. Designing flexible library spaces will also help children develop various skills, including communication, technology use, teamwork, critical thinking, and problem-solving—key skills in the competency-based curriculum.
b) The findings revealed that children did not have sufficient and relevant information resources in their library. This frustrated them from coming to the library. In line with this finding, the Ministry of Education and Sports needs to invest more in school libraries by allocating additional funds to equip them with relevant and up-to-date information resources. This will enable the children to find relevant information resources to meet their information needs and to close the gap between the rich and underprivileged children in rural areas, who are solely dependent on school information materials for their learning because their parents cannot afford to buy textbooks.
c) The findings also revealed that the rural school lacked light in the library, and this hindered the use of the library. In the urban school, the electronic audio books were not functional. This finding revealed the underdeveloped ICT infrastructure that exists in the schools, which are unable to support access to digital information resources and digital literacy skills. School libraries ought to be equipped with essential ICT infrastructure and upgraded from manual systems to integrated library management systems. This transition is vital in the information age, as it exposes learners to a wide range of fast and high-quality information resources that are essential for enhancing educational outcomes,

particularly within the newly implemented competency-based curriculum. This aligns with Uganda's NDP IV, which supports the integration of ICT in education to improve learning outcomes and prepare learners for a digital economy.

d) Finally, this study emphasizes the need for a school library policy to regulate library services in primary and secondary schools in Uganda, which are currently operating in an uncoordinated manner. There is currently no library policy in Ugandan schools to standardise library operations across sectors. This hinders the quality provision of information services in schools. Policy frameworks are important for effective library service delivery in schools. School libraries urgently need a library policy to standardize their operations in schools. Uganda's Vision 2040 emphasises the need for identification and enforcement of comprehensive policy reforms in education services, including library services, as a foundation for coordinated growth and development across sectors.

FUNDING

The authors received no financial support for this publication.

CONFLICT OF INTEREST

The authors declared no potential conflicts of interest with respect to the research, authorship, and/or publication of this chapter.

ACKNOWLEDGEMENT

I express my gratitude to AfriCHILD, the Center for the African Child Uganda, for their invaluable mentorship and for equipping me with essential knowledge and skills in child-focused research methodology. Their sponsorship of this course over the past two years has greatly influenced my emerging research focus on child-centered studies within school libraries in Uganda—an area that remains relatively unexplored in the field of library and Information science in the country.

Additionally, I extend my appreciation to Daisy Nsangi, an Assistant Lecturer in the Department of Library and Information Science at Kyambogo University, and Tyson Akankwasa, an undergraduate student in the same Department who provided crucial assistance during the data collection process.

REFERENCES

Book Aid International. (2013). Talking to users: a review of our work in five countries. Retrieved on 10 October from https://www.bookaid.org/our-work/publications-2/talking-to-users-2/

Bronfenbrenner, U. (2005). Ecological systems theory (1992). In Bronfenbrenner, U. (Ed.), *Making human beings human: Bioecological perspectives on human development* (pp. 106–173). Sage Publications Ltd.

Buwule, R. S., & Ponelis, S. R. (2017). Perspectives on university library automation and national development in uganda. *IFLA Journal*, *43*(3), 256–265. DOI: 10.1177/0340035217710539

Cahill, M., Cole, A., Joo, S., LeFebvre, L., & Garcia, A. (2024). Public library virtual storytimes: The whys, why nots, and measures of success. *Journal of Library Administration*, *64*(2), 224–242. DOI: 10.1080/01930826.2024.2305074

Caldwell, K. (2023). A New Librarian's take on ECRR: Best practices for evidence-based early literacy storytimes. *Children & Libraries*, *21*(1), 23–25. DOI: 10.5860/cal.21.1.23

Carter, B., & Ford, K. (2013). Researching children's health experiences: The place for participatory, child-centered, arts-based approaches. *Research in Nursing & Health*, *36*(1), 95–107. DOI: 10.1002/nur.21517

Chance, R., Okello-Obura, C., & Luyombya, D. (2025). User's Satisfaction with School Library Services: A Case of Selected Government-Aided Secondary Schools in Uganda. *East African Journal of Education and Social Sciences*, *6*(2), 59–70. DOI: 10.46606/eajess2025v06i02.0436

Christensen, P., & James, A. (2020). *Research with children: Perspectives and Practices* (Vol. 3, Issue 5).

Clark, A., & Moss, P. (2011). *Listening to young children: The mosaic approach*. Retrieved on 10 October from https://books.google.com/books?hl=en&lr=&id=K1eHBAAAQBAJ&oi=fnd&pg=PR3&dq=Clark,+A.,+%26+Moss,+P.+(2011).+Listening+to+young+children:+The+mosaic+approach.+National+Children%E2%80%99s+Bureau.&ots=BDNBxTYwhF&sig=RTvPaqPUKPYjvccTrF7F-MCaL6E

Creswell, J. W., & Creswell, J. D. (2017). *Research design: Qualitative, quantitative, and mixed methods approaches*. Sage Publications.

Dockett, S., & Perry, B. (2011). Researching with young children: Seeking assent. *Child Indicators Research*, *4*(2), 231–247. DOI: 10.1007/s12187-010-9084-0

Driessnack, M., & Furukawa, R. (2012). Arts-based data collection techniques used in child research. *Journal for Specialists in Pediatric Nursing*, *17*(1), 3–9. DOI: 10.1111/j.1744-6155.2011.00304.x

Ernst, M. I.Michèle I. Ernst. (2023). The Crucial Role of School Libraries in Influencing Children's Literacy and Learning. *Journal of Childhood Literacy and Societal Issues*, *2*(1), 32–39. DOI: 10.71085/joclsi.02.01.26

Faldet, A. C., & Nes, K. (2024). Valuing vulnerable children's voices in educational research. *International Journal of Inclusive Education*, *28*(5), 657–672. DOI: 10.1080/13603116.2021.1956602

Fodale, F., & Bates, J. (2011). *What is the impact of the school library on pupils' personal development? A case study of a secondary school in Northern Ireland.* University of Ulster. https://www.researchgate.net/publication/254758802

Freire, K., Pope, R., Jeffrey, K., Andrews, K., Nott, M., & Bowman, T. (2022). Engaging with children and adolescents: A systematic review of participatory methods and approaches in research informing the development of health resources and interventions. *Adolescent Research Review*, *7*(3), 335–354. DOI: 10.1007/s40894-022-00181-w

Gerasymenko, O. (2024). Exploring children's picture book design: analysis of composition and color application. In Conferința tehnico-științifică a studenților, masteranzilor și doctoranzilor (Vol. 4, pp. 2151-2154).

Haynes, J. (2014). SEEN and HEARD: Challenging the perceptions of children in libraries in sub-Saharan Africa.

Huang, L. J., & Shieh, J. C. (2021). Elementary school students' perceptions of school library and expectations of library space. In *IASL Annual Conference Proceedings*. DOI: 10.29173/iasl7977

IFLA School Libraries Section Standing Committee. (2015). IFLA school Library guidelines, 2nd Edition, International Federation of Library Associations and Institutions(IFLA).

Jones, L. M., Mitchell, K. J., & Beseler, C. L. (2024). The impact of youth digital citizenship education: Insights from a cluster randomized controlled trial outcome evaluation of the be internet awesome (BIA) curriculum. *Contemporary School Psychology*, *28*(4), 509–523. DOI: 10.1007/s40688-023-00465-5

Karsgaard, C. (2024). Democracy and planetary politics: Achille Mbembe and futures of digital citizenship education for life. *Critical Studies in Education*, *65*(5), 460–475. DOI: 10.1080/17508487.2024.2307573

Kavak, A., & Yıldız, R. (2025). Library Image in Children's Imagination: A Qualitative Analysis Based on Drawings in Türkiye. *Public Library Quarterly*, •••, 1–22. DOI: 10.1080/01616846.2025.2473164

Kawalya, J. (2021). The national library of Uganda: challenges faced in performing its institutional practices. *Library Philosophy and Practice (e-journal), 5073.*

Koller, D., & San Juan, V. (2015). Play-based interview methods for exploring young children's perspectives on inclusion. *International Journal of Qualitative Studies in Education : QSE*, *28*(5), 610–631. DOI: 10.1080/09518398.2014.916434

Kuhlthau, C. (1988a). Perceptions of the information search process in libraries: A study of changes from high school through college. *Information Processing & Management, 24*(4), 419–427. DOI: 10.1016/0306-4573(88)90045-3

Kuhlthau, C. (1988b). Longitudinal case studies of the information search process of users in libraries. *Library & Information Science Research, 10*(3), 257–304.

Kuhlthau, C. (1996). The concept of a Zone of Intervention for identifying the role of intermediaries in the information search process. *Proceedings of the American Society for Information Science Annual Meeting*, 33, 367-376.

Lake, L., & Jamieson, L. (2016). Using a child rights approach to strengthen prevention of violence against children. *South African Medical Journal, 106*(12), 1168–1172. DOI: 10.7196/SAMJ.2017.v106i12.12128

Levy, R. V., & Wasmuth, H. (2025). Developmentally Appropriate Practice and its Impact on the Field of Early Child Education and Care in the United States1. *International Perspectives on the History of Early Childhood Education: Institution–Family–Profession*, 109.

Loveridge, J., Wood, B. E., Davis-Rae, E., & McRae, H. (2024). Ethical challenges in participatory research with children and youth. *Qualitative Research, 24*(2), 391–411. DOI: 10.1177/14687941221149594

MacKenzie, A. (2025). Credible, competent contributors: Children and young people as postdigital citizen social scientists. *Postdigital Science and Education, 7*(1), 224–245. DOI: 10.1007/s42438-024-00517-w

Magara, E., & Batambuze, C. (2009). Reading promotion programmes in primary schools: A study of school library management practices in Pallisa District in Uganda. *Mousaion, 27*(si-1), 108–127.

Magara, E., & Bukirwa Nyumba, J. (2004). Towards a school library development policy for Uganda. *Library Review, 53*(6), 313–322. DOI: 10.1108/00242530410544402

Morales-Álvarez, A., Valdés-Cuervo, A. A., & Parra-Pérez, L. G. (2025). Supportive parenting and adolescents digital citizenship behaviors: The mediating role of self-regulation. *Cyberpsychology (Brno), 19*(1). Advance online publication. DOI: 10.5817/CP2025-1-2

Nakaziba, S., & Ngulube, P. (2024). Harnessing digital power for relevance: Status of digital transformation in selected university libraries in Uganda. *Collection and Curation, 43*(2), 33–44. DOI: 10.1108/CC-11-2023-0034

Ngulube, P. (2023). Improving the quality of reporting findings using computer data analysis applications in educational research in context. *Heliyon*, *9*(9), e19683. DOI: 10.1016/j.heliyon.2023.e19683

Nisa, C. (2025, November). Child-friendly library management as a strategy to strengthen reading literacy in Islamic Elementary Education. In *Proceedings of the International Conference On Education, Society and Humanity* (Vol. 3, No. 1, pp. 1006-1018).

Okello-Obura, C., & Kigongo-Bukenya, I. (2011). Library and information science education and training in Uganda: Trends, challenges, and the way forward. *Education Research International*, *2011*, 1–9. DOI: 10.1155/2011/705372

Paton-Ash, M., & Wilmot, D. (2015). Issues and challenges facing school libraries in selected primary schools in Gauteng Province, South Africa. *South African Journal of Education*, *35*(1), 1–10. DOI: 10.15700/201503062342

Ritterbusch, A. E., Boothby, N., Mugumya, F., Wanican, J., Bangirana, C., Nyende, N., Ampumuza, D., Apota, J., Mbabazi, C., Nabukenya, C., Kayongo, A., Ssembatya, F., & Meyer, S. R. (2020). Pushing the limits of child participation in research: Reflections from a youth-driven participatory action research (YPAR) initiative in Uganda. *International Journal of Qualitative Methods*, *19*, 1609406920958962. DOI: 10.1177/1609406920958962

Sansbury, S. (2025). The Role of School Librarians in Fostering School-Wide Positive Reading Culture. *School Libraries Worldwide*, *30*(2), 17–39. DOI: 10.29173/slw8811

Sharkey, N., & Sharkey, A. (2020). The crying shame of robot nannies: an ethical appraisal. In *Machine ethics and Robot Ethics* (pp. 155–184). Routledge. DOI: 10.4324/9781003074991-16

Shoaib, M., Shamer, A., & Iqbal, S. (2025). A Systematic Review of Academic Library Spaces as Facilitators of Student Engagement in Higher Education Learning. *Knowledge (Beverly Hills, Calif.)*, *4*(1), 123–134. DOI: 10.63062/tk/2k25a.41045

Sholikha, W. L., & Wahyuni, A. (2023). Child Friendly Library Management in Fostering Students' Interest in Reading. *Academia Open*, *8*(1), 10-21070.Sanders, K., & Farago, F. (2018). Developmentally Appropriate Practice in the Twenty-First Century. In *Springer International Handbooks of Education: Vol. Part F1626* (pp. 1379–1400). Springer Nature. https://doi.org/DOI: 10.1007/978-94-024-0927-7_71

Singh, M., Singari, R. M., & Bholey, M. (2024). The Influence of Color on Visual Psychology and Cognitive Behavior: A study in paediatrics hospital environment. *Educational Administration: Theory and Practice*, *30*(5), 13164–13177.

United Nations. (1989). *Convention on the Rights of the Child*. Geneva: Office of High Commissioner for Human Rights (OHCHR). Retrieved on 10 October from https://www.ohchr .org/EN/Profe ssion alInt erest /Pages/CRC.aspx

Water, T. (2024). Ethical issues in participatory research with children and young people. In *Being Participatory: Researching with children and young people: co-constructing knowledge using creative, digital and innovative techniques* (pp. 39–64). Springer International Publishing. DOI: 10.1007/978-3-031-47787-4_3

Willis, J., Hughes, H., & Bland, D. (2019). Students reimagining school libraries as spaces of learning and wellbeing. In Hughes, H., Franz, J., & Willis, J. (Eds.), *School spaces for student well-being and learning: Insights from research and practice* (pp. 121–137). Springer. DOI: 10.1007/978-981-13-6092-3_7

KEY TERMS AND DEFINITIONS

Child-Focused Participatory Methods: Child participatory methods refer to data collection methods that are child-friendly and empower children to participate in knowledge generation during the research process. These methodologies are anchored in the fact that children are accurate informants of their own experiences.

School libraries: A library is found in primary and secondary schools. It is a learned institution equipped with treasures of knowledge, organized and managed by trained personnel with the responsibility of educating children, students, men, and women, and further assisting them to develop the ability of solving their information problems (Chekweeke & Onuoha, 2018).

Chapter 9
Advancing Indigenous Knowledge Research Epistemologies and Research Methodologies in LIS Research

Nina Olivia Rugambwa
http://orcid.org/0000-0002-5127-965X
Kyambogo University, Uganda

Martha Lyaka
http://orcid.org/0000-0002-7993-613X
University of South Africa, South Africa

ABSTRACT

This chapter explores the integration and advancement of Indigenous knowledge systems within Library and Information Science (LIS) research, emphasizing the importance of Indigenous epistemologies and research methodologies. It critically examines how traditional LIS paradigms can be expanded to incorporate Indigenous ways of knowing, valuing cultural context, oral traditions, and community-based approaches. Special attention is given to the cultural context, oral traditions, and community-centered approaches that underpin these epistemologies. The chapter advocates for the development of inclusive research frameworks that respect Indigenous epistemologies, fostering more equitable and culturally responsive LIS scholarship. By advancing these paradigms, the chapter aims to promote greater recognition,

DOI: 10.4018/979-8-2600-0434-0.ch009

validation, and utilization of Indigenous knowledge in information practices and research through providing guidance on how researches biased towards indigenous knowledge systems can be effectively conducted in the LIS discipline.

UNDERSTANDING INDIGENOUS EPISTEMOLOGIES

Indigenous epistemology offers a foundational framework for understanding the knowledge systems that shape the values, beliefs, and lived experiences of indigenous peoples. These epistemologies present distinctive worldviews and methodological orientations that differ markedly from Western scientific paradigms. Recognising IK systems, therefore, means acknowledging their unique intellectual traditions, especially in fields such as environmental science, health, education, and LIS, where indigenous perspectives contribute significantly to sustainability and community wellbeing (Bansal et al., 2024; Bawack et al., 2025; Choudhury & Kalita, 2025).

A core feature of indigenous epistemology is its emphasis on relationality, ethical relationships with land, waters, animals, ancestors, and spiritual beings. Knowledge is generated through these interconnections, producing a holistic understanding of existence that integrates ecological, cultural, and spiritual dimensions. Research indicates that indigenous ecological knowledge addresses environmental issues through interdependence and stewardship practices that are often overlooked in Western paradigms (Acharibasam & McVittie, 2022; López-Quiñones et al., 2023).

This relational thinking aligns closely with emerging LIS scholarship, which highlights how IK is deeply embedded in context, community, and oral tradition. Indigenous communities possess generations of place-based expertise shaped through storywork, ceremony, and land-based learning (Williams & Shipley, 2023). Western LIS systems, by contrast, tend to compartmentalise knowledge through Eurocentric classification schemes that marginalise indigenous worldviews (Belarde-Lewis et al., 2024; Palmer et al., 2024). Recognising indigenous epistemologies in LIS, therefore, requires culturally grounded metadata, community-governed archives, and relational approaches that honour indigenous meanings and sovereignty.

RECOGNIZING INDIGENOUS WAYS OF KNOWING

Recognition of Indigenous knowledge is vital for promoting social justice and enriching fields such as education, LIS, conflict resolution, environmental science, and public health. As shown in Table 1, indigenous epistemologies differ markedly from Western models, embracing holistic, relational worldviews that emphasise interdependence among people, land, and community, shaping cultural and spiritual identity (Datta & Starlight, 2024; Ngaka, 2025).

IK systems offer essential frameworks for understanding environmental governance and sustainability. Although historically marginalised, they are increasingly recognised for their scientific value in environmental management (DePuy et al., 2021). Recognising indigenous epistemologies acknowledges that these are not merely complementary to Western science but are sophisticated intellectual traditions with unique methodologies developed over millennia (Littletree et al., 2020; Tarisayi, 2024). This recognition challenges hierarchical knowledge structures and supports horizontal integration, an approach highly relevant in LIS, where inclusive knowledge organisation, metadata design, and classification must reflect diverse epistemologies.

Acceptance of indigenous knowledge frameworks requires a transformative shift in academic and professional settings that have historically privileged Western paradigms. Community-based participatory research underscores the importance of methodologies that align with indigenous forms of knowledge production and validation (Arriola-Pacheco et al., 2025; Hatfield & Ahn, 2025). Such approaches promote ethical practice and strengthen relationships between researchers and indigenous communities, enabling more balanced knowledge exchange. Within LIS, this transformation appears in collaborative archival appraisal, indigenous-defined access protocols, and the implementation of Indigenous data governance frameworks such as the CARE (Collective Benefit, Authority to Control, Responsibility, and Ethics) Principles, which emphasise collective benefit, authority, responsibility, and ethics (Carroll et al., 2020; Chigwada & Ngulube, 2025a).

Engaging with indigenous ways of knowing requires confronting colonial legacies that produced epistemic violence and systematically devalued Indigenous knowledge (McMullin, 2025). Meaningful recognition demands decolonising education and research practices (Melo, 2025; Thambinathan & Kinsella, 2021). In LIS, this includes integrating indigenous worldviews into metadata, preservation, ethics, and community engagement, and ensuring professional respect for cultural protocols, sacred knowledge, and data sovereignty. Recognition of indigenous knowledge requires dismantling historical inequities and fostering respectful dialogue across knowledge systems. This commitment supports decolonised LIS practice and strengthens sustainability, resilience, and ethical engagement with the complex, relational epistemologies that underpin Indigenous cultural and intellectual traditions.

Table 1. Difference between IK epistemologies and Western epistemologies

Dimension	Indigenous Epistemologies	Western Epistemologies
Nature of Knowledge	Knowledge is holistic, relational, cumulative, and grounded in lived experience; it integrates spiritual, ecological, social, and ancestral dimensions.	Knowledge is segmented, analytical, and often reductionist; it emphasises discrete categories, objectivity, and measurable variables.
Knowledge Source	Rooted in community memory, land-based experience, oral traditions, elders, and intergenerational transmission.	Rooted in empirical observation, experimentation, written texts, and formal institutional knowledge production.
Validation Criteria	Validity is based on usefulness to the community, ethical soundness, long-term sustainability, and alignment with cultural protocols.	Validity is based on logic, empiricism, replicability, peer review, and theoretical coherence.
Knowledge Transmission	Shared orally through storytelling, apprenticeship, ceremony, performance, and communal participation.	Shared through formal education, academic publications, lectures, and scientific reporting.
Relationship to Environment	Humans are deeply interconnected with land, animals, ancestors, and ecosystems; knowledge emerges from reciprocal relationships.	Humans are conceptually separate from nature; knowledge often positions humans as observers or controllers of the environment.
Ontology (nature of being)	Reality is relational and interdependent; the spiritual, material, and social worlds are inseparable.	Reality is often divided into material and immaterial domains; scientific inquiry tends to privilege material explanations.
Authority and Ownership	Knowledge is collectively held and governed; authority rests with elders and knowledge custodians following community protocols.	Knowledge is individually authored, copyrighted, published, and owned through intellectual property regimes.
Purpose of Knowledge	To sustain community well-being, cultural continuity, ecological balance, and collective identity.	To generate new theories, technologies, and applications; to advance scientific understanding and innovation.
Methods of Inquiry	Observation through participation, lived experience, ceremony, and ethical relationality; emphasis on long-term immersion.	Controlled experiments, data collection, formal analysis, and a systematic scientific method.
Ethical Foundations	Ethics centred on reciprocity, respect, responsibility, and relational accountability.	Ethics is centred on individual autonomy, consent, neutrality, and universal principles.

RESPECTING CULTURAL WORLDVIEWS AND ONTOLOGIES

Respect for cultural worldviews and ontologies is essential when engaging with indigenous communities and their knowledge systems, particularly within LIS, where Western epistemologies have long shaped practices of classification, archiving, and knowledge representation. Cultural worldviews reflect how different groups understand existence, identity, and the relationships that define their reality.

For indigenous peoples, these worldviews are deeply rooted in ancestral history, language, spirituality, and profound relationships with the land, all of which influence how knowledge is transmitted, embodied, and preserved. Chigwada, Ngulube and Dewah (2024) observe that IK often appears in forms such as poetry, folklore, dance, drama, and artefacts, formats not easily accommodated in text-centred LIS frameworks. Respecting these worldviews challenges Eurocentric assumptions and demands culturally responsive approaches to preservation, description, and access (Ruwhiu & Cone, 2010; Adams & Larrinaga, 2019).

At the core of respecting indigenous worldviews is acknowledging their ontological pluralism, the understanding that multiple realities coexist and are equally valid. This stands in contrast to the reductive tendencies of Western science, which privileges empirical evidence and objective methods while overlooking the metaphysical, relational, and spiritual narratives central to indigenous epistemologies (Williams & Shipley, 2023; Shipley & Williams, 2019). This recognition directly strengthens earlier arguments about the need for culturally grounded LIS practices. Balogun and Kalusopa's work on Indigenous knowledge systems in southern Africa illustrates that ethical preservation depends on honouring indigenous ontologies (Balogun & Kalusopa, 2021, 2022). Their research on South Africa's National Recorded System reveals that Western digital preservation models, which focus narrowly on metadata and storage, often fragment or misrepresent Indigenous knowledge by overlooking relational and spiritual meanings (Balogun, 2023; Balogun & Kalusopa, 2021, 2022). They argue that digitisation must follow Indigenous authority, cultural protocols, and place-based context to prevent epistemic dominance (Balogun & Kalusopa, 2021). Such ontological commitments enable LIS practitioners to preserve knowledge with cultural integrity and fidelity to indigenous ways of knowing.

Respectful engagement with indigenous worldviews also requires acknowledging the historical trauma and ongoing colonial impacts that shape indigenous communities' relationships with researchers, institutions, and knowledge systems. These communities have endured centuries of marginalisation, dispossession, and epistemic violence, making it essential for researchers and practitioners to understand the socio-political dynamics that continue to inform their interactions (Bansal et al., 2024). Meaningful recognition, therefore, requires listening to indigenous peoples and enabling them to co-create knowledge rather than positioning them as research subjects. As Huaman and Mataira (2019) emphasise, Indigenous communities possess unique socio-historical contexts that must guide participatory approaches grounded in autonomy, respect, and cultural authority.

Frameworks such as "Two-eyed seeing" offer a respectful pathway for integrating indigenous and Western knowledge systems (Broadhead & Howard, 2021). This approach values the strengths of both paradigms, fostering holistic understandings in areas such as health, education, environmental management, and increasingly,

LIS practice (Forbes et al., 2020). In LIS, Two-eyed seeing aligns with calls for indigenous-led metadata design, culturally governed digitisation, and classification systems that honour indigenous relationality and protocol (Belarde-Lewis et al., 2024). Embracing such frameworks enables practitioners to move beyond Eurocentric biases, recognise the legitimacy of indigenous epistemologies, and contribute to broader goals of decolonisation, ethical knowledge stewardship, and social justice.

IK system research designs

IK system research designs encompass four principles, relationality, respect, spirituality, and community-based knowledge, that are central to indigenous epistemologies and have direct implications for LIS (see figure 1). These elements clarify the way in which indigenous peoples understand their identity, their environment and the interdependence that links them to their neighbours and their country. Recognising the importance of these dimensions promotes a greater respect for and integration of indigenous perspectives in academic discourse and community practice.

Figure 1. A diagram of a research process (Designed by Authors)

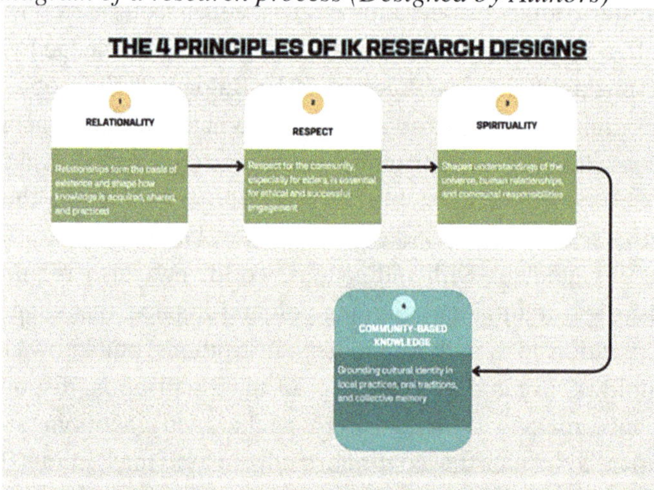

Relationality

Relationality is a central principle in indigenous ways of knowing, asserting that relationships form the basis of existence and shape how knowledge is acquired, shared, and practised. Indigenous epistemologies view knowledge as emerging from relationships among people, land, animals, plants, and spiritual entities (Chigwada & Ngulube, 2025b; Tarisayi, 2024). This interdependence affirms that humans belong to a broader ecological system where all beings hold intrinsic value and rights essential for harmonious coexistence (Gannaway, 2020).

In practice, relationality guides community-based approaches to education and research. Participatory action research grounded in relational frameworks positions community members as co-creators of knowledge and honours cultural practices and land-based insight (Hunt & Young, 2021). These approaches challenge hierarchical structures typical of Western scholarship. In LIS, relationality underpins IK organisation systems, participatory metadata creation, community-led collection development, and culturally governed archival stewardship (Belarde-Lewis et al., 2024; Littletree et al., 2020). It also aligns with indigenous data governance frameworks, including CARE and contextual integrity models, which emphasise relational accountability to people, land, ancestors, and future generations (Palmer et al., 2024).

Respect

When conducting research with indigenous communities, respect for the community, particularly for its elders, is essential for both ethical and successful engagement. Elders serve as cultural authorities, knowledge custodians, and key gatekeepers whose guidance ensures that research aligns with community values and protocols (Bourassa et al., 2020). Their involvement is vital for mobilising wider community participation and ensuring culturally grounded consent. In LIS, similar principles apply: elders play central roles in community-led archival appraisal, participatory metadata creation, and decisions about culturally appropriate access, reinforcing the need for respectful, relational, and protocol-driven research practice (Littletree et al., 2023).

Spirituality

Spirituality is a vital dimension of IK systems, shaping understandings of the universe, human relationships, and communal responsibilities. It is inseparable from land and nature, linking ecological stewardship with spiritual practice and daily decision-making (Ngaka, 2025). Recognising these challenges, Western paradigms separate the spiritual from the empirical. In LIS, scholars argue that indigenous spirituality should inform preservation, description, and access, as spiritual meaning is deeply embedded in oral traditions, artefacts, and community memory (Balogun, 2023; Eyre, 2023; Shiri et al., 2022). Integrating such perspectives enables culturally grounded, holistic knowledge stewardship and supports community wellbeing (Kovach, 2021).

Community-based knowledge

Community-based knowledge is central to indigenous resilience and self-determination, as it grounds cultural identity in local practices, oral traditions, and collective memory (Huaman & Mataira, 2019). The knowledge shared through storytelling, riddles, proverbs, songs, ceremonies, and community gatherings sustains continuity and fosters a shared sense of belonging (Mendenhall et al., 2024; Shiri et al., 2022). Recognising such knowledge means acknowledging indigenous peoples as experts within their own contexts, with insights shaped by generations of lived experience (Gannaway, 2020; Tarisayi, 2024). Within LIS, these insights inform community-led collection development, participatory archival appraisal, and culturally grounded metadata that reflect indigenous categories, languages, and relational worldviews (Balogun & Kalusopa, 2021; Belarde-Lewis et al., 2024; Chigwada & Ngulube, 2025a; Littletree & Krise, 2024). Collaborative engagement with community knowledge aligns research and LIS practice with indigenous values and self-determination (Huaman & Mataira, 2019). Such partnerships enhance research relevance, support ethical knowledge stewardship, and contribute to the decolonisation of information systems.

Ultimately, foregrounding relationality, spirituality, and community-based knowledge shifts LIS beyond individualistic Western frameworks toward holistic, just, and culturally revitalising approaches.

DECOLONIZING RESEARCH APPROACHES

Decolonising research in LIS necessitates a decisive shift away from Western-centric methods and extractive data practices toward indigenous-led approaches grounded in sovereignty, relationality, and community-defined priorities. Scholars consistently argue that decolonisation cannot be achieved by merely inserting indigenous content into western frameworks (Melo, 2025; Thambinathan & Kinsella, 2021); such tokenism preserves the epistemic hierarchies it claims to disrupt. Instead, decolonisation requires rethinking the structural practices that determine how information is accessed, classified, represented, and governed within LIS institutions (Littletree et al., 2023).

A core step in this shift is recognising the limitations of Western knowledge systems. Traditional LIS infrastructures such as classification schemes, metadata standards, and archival protocols are built on Eurocentric assumptions about universality, textual authority, and linear categorisation. These systems often misrepresent or erase indigenous worldviews that emphasise relationality, land-based meaning, and spiritual knowledge. Patin et al. (2021) argue that revising these colonial foundations is essential for enabling indigenous epistemologies to shape information practices. Mdhluli et al. (2021) demonstrate how digital library systems can be redesigned by integrating African indigenous ways of knowing, advocating for community participation and culturally relevant metadata standards that reflect local classifications, values, and oral traditions.

Indigenous leadership is therefore central to any decolonised LIS approach. Scholars emphasise that Indigenous communities must guide research design, governance, and dissemination to ensure alignment with cultural priorities and intergenerational responsibilities (Stevens-Uninsky et al., 2023). This aligns with the development of Indigenous Knowledge Organisation Systems (IKOS), which embed indigenous languages, kinship networks, and place-based ontologies directly into classification structures.

Equally crucial is the adoption of indigenous-led data governance frameworks. The CARE Principles provide a sovereignty-centred alternative to Western FAIR (Findable, Accessible, Interoperable, and Reusable) principles (Carroll et al., 2020). Tools such as traditional knowledge (TK) Labels operationalise these principles by enabling communities to assert cultural protocols over access, use, and circulation of materials.

Finally, decolonised LIS research requires relational accountability. Frameworks like contextual integrity (Palmer et al., 2024) ensure metadata reflects community values and cultural context. Participatory metadata creation, community-led archival appraisal, and culturally governed digitisation practices help prevent fragmentation or misrepresentation of Indigenous knowledge.

Together, these commitments form a coherent pathway toward ethical, sovereignty-centred, and culturally grounded LIS research and practice.

INDIGENOUS RESEARCH METHODOLOGIES

In order to promote indigenous scholarship, indigenous research methodologies emphasise participatory, relational and place-based practices that focus on indigenous knowledge, voices, management and priorities. The emphasis is shown in Figure 2. The following summary summarises the main claims and practical implications of these methods, drawing on the cited works to illustrate how these approaches work.

Figure 2. Indigenous research methodologies (Designed by Authors)

Participatory action research (PAR)

Participatory action research (PAR) is closely aligned with the principle that indigenous communities must lead research processes to ensure that projects reflect local knowledges, cultural priorities, and self-determination. This community-led orientation counters extractive or externally imposed agendas that have historically shaped research involving indigenous peoples (Stevens-Uninsky et al., 2023). The leadership imperative within PAR resonates strongly with indigenous standpoint theory and other community-driven frameworks that position indigenous voices at the centre of shaping research questions, governance structures, and dissemination pathways (Cox et al., 2021). They further observe that under PAR, community partners participate as co-researchers rather than research subjects, influencing data collection, interpretation, and the development of actionable outcomes that benefit their communities.

Across health, education, and social science fields, PAR is widely recognised as a key decolonising methodology. It moves researchers beyond tokenistic inclusion of indigenous elements toward reciprocal partnerships grounded in mutual accountability and shared authority (Jordan & Hall, 2023; Raymond, 2024). PAR's insistence on co-constructed processes affirms the epistemic authority of indigenous ways of knowing, enabling the generation of culturally resonant and contextually relevant evidence (Thambinathan & Kinsella, 2021). Indigenous-led governance throughout the research cycle aligns with calls for culturally responsive evaluation and indigenous evaluation frameworks that redefine what counts as valid evidence in applied settings (Jordan & Hall, 2023).

Within LIS, PAR offers a powerful mechanism for enacting indigenous leadership in the design of information systems and protocols. PAR-inspired collaboration supports co-creation of cataloguing schemas and IKOS that reflect indigenous languages, relational ontologies, and storytelling traditions (Begay & Klor, 2024; Duarte & Belarde-Lewis, 2015; Littletree et al., 2020). Community-defined metadata standards, such as those used in Indigenous digital heritage projects, ensure that cultural protocols and rights of access guide how materials are represented and shared (Chaulk et al., 2025). PAR also facilitates Indigenous-led decisions around digitisation priorities, archival appraisal, and the contextualisation of oral histories, areas where conventional Western LIS approaches often misrepresent or decontextualise Indigenous knowledge.

By embedding sustained partnership, shared governance, and co-authorship, PAR provides a practical and ethical pathway for integrating Indigenous leadership into LIS research ecosystems, advancing both decolonisation and culturally grounded knowledge stewardship.

Storytelling, oral histories, and narrative methods

Storytelling and oral histories are foundational to IK systems, providing culturally grounded methodologies that affirm indigenous sovereignty and epistemic authority. These narrative approaches honour relationality, memory, and place-based meaning while challenging western research traditions that privilege textual, quantitative, or elite-produced forms of evidence (Raymond, 2024; Thambinathan & Kinsella, 2021). Manyane (2024) argues for harmonising oral and written histories as mutually enriching epistemic approaches. Similarly, the African Oral Traditional Storytelling

Framework demonstrates how storytelling strengthens linguistic identity and cultural preservation in contemporary research (Osei-Tutu, 2023).

Narrative-based methodologies also function as ethical and decolonising tools. They foreground community voice, reciprocity, and self-determination, countering the historical silencing of marginalised groups, such as those affected by apartheid-era exclusions (Osei-Tutu, 2023). Gathogo (2021) illustrates how individuals narrate their past from their own perspective, validating oral testimony as an authentic and authoritative source. Wiegratz et al. (2022) similarly highlight how oral life histories recover the experiences of "ordinary people" omitted from official textual accounts.

Within LIS, storytelling shapes community archiving, oral heritage digitisation, and culturally appropriate metadata. Indigenous and African oral traditions inform descriptive practices that embed relational context, cultural status, and access protocols (Littletree et al., 2023; Palmer et al., 2024). Tools such as TK Labels and community-defined metadata schemas support ethical representation of stories (Montenegro, 2019). Efforts to convert oral narratives into written heritage in Limpopo, South Africa (Setumu & Ngoepe, 2016), underscore the urgency of documenting disappearing knowledge while ensuring cultural authority in interpretation. Together, these narrative methods strengthen epistemic sovereignty, reshape LIS practices, and preserve indigenous and African knowledge systems with integrity.

Land-based and experiential learning methods

Land-based methodologies and experiential learning are central to indigenous epistemologies, which emphasise place, relationality, and embodied forms of knowledge. These approaches position the land not merely as a backdrop but as an active teacher, sustaining cultural memory, identity, and governance. Research has shown that experiential, indigenous-led learning fosters cultural safety, responsiveness, and deeper awareness among health, education, and social service professionals when guided by community priorities rather than externally imposed frameworks (Macklin et al., 2021; Riediger et al., 2020). Such interventions create measurable shifts in practice, for example, in perinatal and clinical care, demonstrating that land-based experiences can meaningfully inform policy, service delivery, and research outcomes (Bruno et al., 2022).

Experiential, community-driven initiatives such as heritage gardens, language revitalisation programs, and stewardship projects illustrate how learning occurs through direct engagement with land, ceremony, and collective memory. These activities support Indigenous education, language preservation, and cultural continuity while generating knowledge rooted in community-defined priorities (Lys et al., 2022; Manning, 2023). They also align naturally with participatory and Indigenous-led evaluation approaches that position participants as co-designers of knowledge.

Within LIS, land-based and experiential methodologies translate into community mapping projects, field-based documentation, and language-keeper collaborations that produce place-embedded knowledge artefacts. These materials enrich community-led information infrastructures such as digital storytelling archives, ethnobiological collections, and land-based metadata schemas that honour Indigenous temporalities, spatialities, and cultural authority (Chigwada & Ngulube, 2025b; Hunter, 2021; Littletree et al., 2023; Osei-Tutu, 2023). By embedding land-based perspectives into cataloguing, metadata design, and digital preservation, LIS practice becomes more relational, culturally grounded, and aligned with indigenous knowledge sovereignty.

INTEGRATING PAR, STORYTELLING, AND LAND-BASED METHODS INTO A COHERENT INDIGENOUS RESEARCH POSTURE

A critically grounded Indigenous research posture demands more than the juxtaposition of indigenous methods with Western frameworks; it requires a structural reorientation toward indigenous governance, epistemologies, and relational accountability. Integrating Participatory Action Research (PAR), storytelling/oral histories, and land-based experiential methods forms a coherent methodological constellation through which Indigenous communities assert epistemic authority and steer all phases of the research process (see Figure 3). These approaches, when combined, reveal the inadequacy of token indigenous elements that remain subordinate to Western methodological assumptions. Instead, the synthesis compels a move toward indigenous-led, context-specific protocols and practices that reflect distinct histories, governance structures, and place-based worldviews (Cox et al., 2021; Thambinathan & Kinsella, 2021).

Figure 3. A diagram of the indigenous governance and epistemic authority (Designed by Authors)

A decolonized methodological posture is anchored in several critical commitments: indigenous epistemologies must serve as primary, not supplementary, knowledge sources; indigenous leadership must direct the formulation of research questions, methodological design, and dissemination; relationality, reciprocity, and ethical accountability must govern engagements; and researchers must actively resist the universalisation of western templates (Jordan & Hall, 2023; Raymond, 2024; Stevens-Uninsky et al., 2023). These commitments are not optional enhancements but structural imperatives for countering the epistemic violence inherent in extractive research traditions.

Operationalising this posture requires aligning PAR with indigenous governance systems, ensuring that research agendas, data interpretations, and action-oriented outcomes remain community-controlled. Storytelling and oral histories extend this work by privileging storied knowledge, relational memory, and community narrative sovereignty, reclaiming spaces where indigenous voices have historically been marginalised. land-based and experiential approaches further ground research in lived reality, anchoring knowledge generation in place, ceremony, and embodied practice (Lys et al., 2022; Macklin et al., 2021; Riediger et al., 2020).

Collectively, these intertwined methodologies produce knowledge that reflects indigenous priorities and resists the abstraction or decontextualisation characteristic of Western evidence paradigms. They demand trust-building, long-term relational engagement, and co-constructed knowledge artefacts governed by Indigenous sovereignty (Ndimande, 2012; Thambinathan & Kinsella, 2021).

Research ethics and sovereignty

Indigenous research ethics are grounded in the understanding that indigenous peoples are the rightful authors, custodians, and stewards of their own knowledge systems. Ethical practice, therefore, requires acknowledging and ensuring that research processes and outputs follow community governance structures, and repositioning researchers as partners or allies rather than primary knowledge producers (Hudson et al., 2023; Ignace et al., 2023). A robust indigenous ethics framework places indigenous leadership at the centre of all decisions, operationalizes Free, Prior, and Informed Consent (FPIC) as an ongoing governance mechanism rather than a one-time approval, and embeds the Ownership, Control, Access, and Possession (OCAP) principles throughout project design, data stewardship, and dissemination (Akpa-Inyang & Chima, 2021; Walter et al., 2020). This ethical orientation aligns closely with decolonising methodologies and Indigenous data sovereignty scholarship, which emphasise reciprocal benefit, community control over the production and circulation of knowledge, and continual ethical reflexivity across the entire research lifecycle (Hayward et al., 2021; Walter et al., 2020). Such frameworks ensure that research respects Indigenous laws, cultural protocols, and collective rights, supporting outcomes that uphold community wellbeing, self-determination, and long-term stewardship of knowledge.

Respect for indigenous sovereignty over knowledge

Indigenous sovereignty over knowledge asserts that communities retain authority over what is studied, how knowledge is produced, who may access it, and who benefits. This principle underpins the Ownership, Control, Access, and Possession (OCAP) framework and informs the governance of datasets, archives, and cultural materials (Walker et al., 2018; Walter et al., 2020). Sovereignty requires community-led governance, ongoing consent for data use and interpretation, and mechanisms for approving reuse (Williamson et al., 2022). In LIS, culturally governed metadata, community-led digitisation, and Indigenous stewardship of oral and material heritage uphold this authority (Balogun & Kalusopa, 2021; Chigwada, Ngulube & Dewah, 2024). Tools such as IKOS, TK Labels, and culturally specific archival protocols ensure benefits return to communities through sustainable information infrastructures (Marshall et al., 2020).

Free, prior, and informed consent (FPIC)

FPIC is a fundamental right and procedural mechanism for community-level authorisation that encompasses the initial and ongoing consent across all stages of research, from design through dissemination and secondary use of data. Institutional ethics review boards often lack formal FPIC procedures, and FPIC should be anchored in community-specific governance rather than only individual consent, ensuring collective rights and priorities are respected (Ignace et al., 2023; Walker et al., 2018; Walter et al., 2020).

FPIC practices entail iterative engagement, transparent information-sharing about risks and benefits, and negotiated terms of collaboration, data sharing, and eventual benefits. In practice, FPIC entails co-designing study aims, co-creating data governance agreements, and maintaining ongoing dialogue to renegotiate terms as projects evolve, particularly in longitudinal or cross-jurisdictional work (Mackey et al., 2022).

When FPIC is operationalised effectively, research maps onto indigenous governance timelines and decision-making processes, enabling mutual accountability between researchers and Indigenous communities and ensuring that research outputs align with community priorities and protocols (Walter et al., 2020).

Data ownership, control, access, and possession (OCAP) and Indigenous data governance

OCAP principles stipulate that communities own, control, access, and possess their own data, thereby enabling governance over data collection, storage, sharing, and interpretation. In Canada, OCAP has been foundational for First Nations' data governance, influencing research design, repository development, and data-sharing agreements (Konczi & Bill, 2024; Paul, 2023).

Beyond formal OCAP, Indigenous data governance encompasses community-defined metadata, data stewardship roles, and governance structures that empower communities to decide who may access data and for what purposes. This includes digital archives, biobanks, genomic datasets, environmental and health data, and repositories of language or cultural knowledge (Chigwada, Ngulube, & Dewah, 2024; Cunningham-Reimann et al., 2025).

Ethical practice requires concrete mechanisms, including formal data-sharing agreements, community data governance boards, consent terms that specify reuse and secondary analysis, clear provenance and benefit-sharing plans, and accountability measures to ensure data stewardship remains in indigenous hands (Carroll et al., 2020; Williamson et al., 2022).

Incorporating OCAP-like practices into information governance also reshapes cataloguing, access controls, and reuse policies to respect community rights, language sovereignty, and cultural protocols within information systems (Paul, 2023; Walter et al., 2020).

Ethical considerations rooted in indigenous protocols

Indigenous protocols emphasise relational accountability, reciprocity, respect for self-determination, and the primacy of community interests. Ethics align with culturally aligned evaluation frameworks that prioritise community-defined outcomes, governance, and co-interpretation of results (Etchart, 2022; Johnson et al., 2023; Shrinkhal, 2021).

Reflexivity, power-sharing, and relational methodologies are central: researchers must continually reflect on how their positionality, funding structures, and institutional norms shape research dynamics and actively dismantle colonising tendencies by centring indigenous voices and leadership (O'Connor et al., 2022).

Practical implications include designing research with Indigenous-led governance, ensuring co-authorship and co-dissemination with Indigenous partners, building capacity within communities to sustain information systems, and aligning outputs with Indigenous-defined measures of well-being and sovereignty rather than solely academic impact metrics (Marshall et al., 2020; Morisano et al., 2024; Tone-Pah-Hote & Redvers, 2022; Waddell-Henowitch et al., 2020).

Collaborative and community-based approaches

Collaborative and community-based approaches form the foundation of effective and ethical research with Indigenous communities, emphasising genuine partnerships, co-creation of knowledge, reciprocity, and trust. These approaches challenge extractive and top-down Western research paradigms, aligning instead with Indigenous worldviews that privilege relationality, consent, and collective governance. In indigenous scholarship, meaningful collaboration is described as a process that must be embedded in long-term engagement, respect for cultural protocols, and commitment to community priorities.

Building genuine partnerships with indigenous communities

Building genuine partnerships requires researchers to invest deeply in relationship-building, cultural immersion, and understanding local governance structures. Thorpe et al. (2021) argue that partnerships grounded in long-term engagement, not sporadic consultation, are central to effective community-based participatory research. This model redistributes decision-making power, enabling Indigenous communities to set priorities, guide research design, and shape the dissemination of findings. Balogun and Kalusopa (2021) demonstrate that indigenous knowledge projects in South Africa's National Recordal System were successful only when researchers established long-standing rapport with communities and respected cultural governance over knowledge. Chigwada, Ngulube, and Dewah (2024) similarly demonstrate that in Zimbabwe, indigenous communities value relational engagement, with elders and cultural custodians serving as the primary gatekeepers of oral and material heritage. These examples underscore the need for researchers to engage through locally meaningful avenues, such as village meetings, clan gatherings, and cultural ceremonies, rather than imposing external research timelines.

Co-creating knowledge and ensuring community benefits

Co-creation of knowledge is central to indigenous methodologies because it ensures that research outcomes directly benefit communities and reflect their epistemologies. Struik et al. (2022) emphasise that co-designing research from initial conception to dissemination promotes relevance and cultural integrity. Hutchinson et al. (2014) illustrate this through Métis knowledge-sharing sessions where researchers, elders, and youth co-developed knowledge frameworks. Datta and Kairy (2024) demonstrate how integrating traditional ecological knowledge into environmental planning fosters culturally appropriate and sustainable strategies. Within LIS, co-creation appears in community-led archival appraisal, participatory metadata creation, and co-designed IKOS, which embeds indigenous languages, kinship relations, and place-based categories into classification (Barité & Rauch, 2025). Community participation in building oral history archives, digitising heritage collections, or designing access protocols ensures that knowledge systems remain culturally grounded.

Beyond producing outputs, co-creation fosters community benefits, including capacity building, cultural revitalisation, documentation of endangered languages, and increased agency in decision-making regarding cultural resources (Maar et al., 2022). LIS projects illustrate how co-creation strengthens local control and enhances digital heritage preservation (Fullmer, 2021; Machin-Mastromatteo, 2023; Mdhluli et al., 2021).

Emphasising reciprocity and trust

Reciprocity and trust are fundamental to Indigenous collaborative research. Datta and Starlight (2024) and Arriola-Pacheco et al. (2025) emphasise that trust emerges through transparency, shared responsibilities, and continuous dialogue. Researchers must adopt an accountability stance in which benefits are negotiated upfront, and community feedback shapes research over time. Hudson et al. (2023) highlight the rising assertion of Indigenous rights to narrative control, demanding ethical frameworks that prioritise consent, governance, and community authority. In LIS, tools such as TK Labels, the CARE Principles and OCAP frameworks support trust by giving communities decision-making power over data, metadata, access, and reuse (Carroll et al., 2020).

Sekiwu et al. (2024) note that indigenous communities share knowledge only when trust is established through relational engagement. Similarly, African oral history projects emphasise returning knowledge to communities in accessible forms like storybooks, digital archives, and community exhibitions, ensuring reciprocal exchange rather than one-way extraction (Chipangura et al., 2025).

METHODOLOGICAL INNOVATIONS AND ADAPTATIONS

Methodological innovation in Indigenous research is essential for addressing the diverse cultural, social, and epistemological contexts of Indigenous communities. Because indigenous ways of knowing differ from Western scientific paradigms, research designs must be flexible, relational, and grounded in cultural relevance (David-Chavez, 2019). Combining indigenous and Western methods, when done ethically and under indigenous governance, creates research processes that uphold indigenous sovereignty while generating meaningful outcomes for communities (see Table 2). These braided approaches strengthen research by integrating community priorities, cultural protocols, and long-standing Indigenous epistemologies.

Table 2. Methodological spectrum in indigenous research

Indigenous Approaches	Hybrid / Braided Approaches	Western Approaches
Storywork / storytelling	Two-Eyed Seeing	Experimental designs
Oral histories	Participatory Action Research (PAR)	Randomised control trials (RCTs)
Talking circles	Community-Based Participatory Research (CBPR)	Surveys & statistical modelling
Land-based learning	Mixed methods beginning with Indigenous methods	Standardised measurement tools
Ceremonial knowledge	Co-designed evaluation frameworks	Positivist/empirical frameworks
Relational & place-based epistemologies	Digital storytelling, photovoice	Objectivist epistemology

continued on following page

Table 2. Continued

Indigenous Approaches	Hybrid / Braided Approaches	Western Approaches
Indigenous Knowledge Organisation Systems (IKOS)	Indigenous-led technology innovations	Generalizable, decontextual knowledge frameworks
TK Labels & cultural protocols	Blended methodologies guided by Indigenous governance	Researcher-driven interpretation

INTEGRATING INDIGENOUS AND WESTERN METHODOLOGIES

Hybrid research designs can leverage the strengths of both indigenous and Western approaches. Clapham et al. (2021) demonstrate how digital tools such as photovoice, participatory mapping, and digital storytelling can support Indigenous-led documentation of land-based knowledge and cultural narratives. These tools are especially powerful in participatory frameworks, where indigenous communities maintain interpretive authority.

Simonds and Christopher (2013) emphasise that PAR and community-based participatory research are particularly suited to indigenous contexts because they redistribute decision-making power, involve communities throughout the research cycle, and counter historical patterns of extractive research. In African IK systems, methodological traditions such as participatory rural appraisal, community ecological inventories, and oral tradition reconstruction have been instrumental in environmental and cultural research (Balogun & Kalusopa, 2021; Chigwada, Ngulube & Dewah, 2024). These designs validate knowledge held by elders, healers, and artisans; individuals recognised by their communities as authoritative knowledge keepers.

Flexible and context-specific research designs

Flexibility is fundamental to indigenous research because IK is relational, place-based, and responsive to social and ecological rhythms. David-Chavez (2019) proposes a decolonising environmental framework that blends Indigenous and scientific knowledge to support community priorities. Jamieson (2025) further argues that indigenous research designs must accommodate shifting community contexts, obligations, and timelines.

Ensuring cultural relevance and sensitivity

Cultural relevance and sensitivity underpin all Indigenous research methodologies. The CARE Principles guide ethical data governance by centering indigenous rights (Carroll et al., 2020). Cultural safety frameworks, which train researchers to recognise colonial histories, Indigenous sovereignty, and cultural protocols, are essential for ethical engagement (Trepanier et al., 2019).

In LIS contexts, culturally sensitive methodologies appear through IKOS, TK Labels, community-led archival appraisal, and place-based metadata schemas that reflect indigenous relationality (Littletree et al., 2023; Montenegro, 2019). These approaches safeguard cultural integrity and prevent misrepresentation or misuse of Indigenous knowledge.

Dissemination and knowledge translation

Effective dissemination and knowledge translation in Indigenous research are crucial for ensuring that research findings are meaningful, accessible, and beneficial to Indigenous communities. Central to IK translation is the commitment to share findings in culturally grounded ways that align with indigenous worldviews, languages, and storytelling traditions. Scholars such as Huria et al. (2019) emphasise that dissemination must be embedded within a broader relational strategy that reflects local social realities and maintains community engagement throughout the research process. This entails developing knowledge-sharing pathways that enable communities to utilise evidence in advocacy, policy dialogue, and community well-being.

Collaborative model-building is critical. Shiri et al. (2022) demonstrate that culturally tailored dissemination emerges from co-designed approaches, listening sessions, community forums, and iterative feedback loops that ensure findings resonate with community experiences. Creative and contextualised portrayals, including visual storytelling, digital story maps, and narrative-based reports, can enhance relevance. Gathogo (2021) argue that these forms respect indigenous storywork traditions and act as bridges across knowledge systems.

Accessibility remains a central challenge. Williams and Shipley (2023) demonstrate that resources often fail because dissemination relies on Western communication strategies that overlook indigenous contexts. Kennedy et al. (2024) highlight the importance of transparent, community-driven knowledge translation frameworks that incorporate local feedback as an ethical and methodological requirement. Similarly, Bennett (2023) notes that local indigenous pathways of knowledge circulation vary widely, requiring researchers to understand community-specific transmission systems to design effective knowledge translation strategies.

Brinckley et al. (2022) emphasise that simply generating resources is insufficient; tailored dissemination strategies are needed to ensure uptake. Failing to integrate Indigenous epistemologies and feedback can limit the acceptability and impact of research (Harfield et al., 2020). Thus, effective knowledge translation requires a relational, participatory, and culturally anchored approach.

Data gathering strategies in indigenous research reflect this same relational and culturally grounded orientation. Common methods include one-on-one interviews, group conversations, and talking circles, approaches rooted in Indigenous communal life and social spaces (Begay & Klor, 2024; Chigwada & Ngulube, 2025b; Hunter, 2021). Talking circles may occur around fireplaces, in village squares, at market days, during ceremonies, or at community meetings. Oral histories, storytelling, and life narratives are equally significant, enabling participants to share lived experience in ways that affirm cultural identity and collective memory.

Crucially, all data collection is organised through consultation with elders, who serve as cultural authorities and guide protocols for interviews and gatherings. Their guidance ensures that knowledge collection respects community norms, relational obligations, and culturally appropriate procedures.

Together, these approaches reveal that Indigenous research is effective only when dissemination and data collection honour Indigenous governance, relational ethics, community priorities, and culturally grounded modes of knowledge production.

SAMPLING METHODS

Sampling in Indigenous research is grounded in relationality, community trust, and culturally appropriate engagement. Unlike conventional sampling, which prioritises randomness, statistical representation, or standardised frames, indigenous sampling emphasises relationships, community legitimacy, and the ethical involvement of knowledge holders. Relational sampling, also called network or connection-based sampling, is one of the most effective strategies. Here, participants are recruited through social bonds, where an initial participant, often identified through an elder or cultural leader, invites trusted friends, relatives, or clan members to participate (Raifman et al., 2022). This method respects cultural expectations of familiarity and mitigates the discomfort that Indigenous peoples may feel when approached by unfamiliar researchers (Sekiwu et al., 2024).

Another widely used method is purposive sampling, which involves intentionally selecting knowledge keepers such as elders, healers, artisans, or oral historians whose expertise is essential for the study. For example, research on IK dissemination utilised purposive sampling of librarians with recognised community authority (Chigwada & Ngulube, 2024). Similarly, snowball sampling supports recruitment in dispersed or marginalised communities, where each participant recommends additional individuals with relevant experiences or cultural knowledge.

Indigenous methodologies also employ community-nominated sampling, where elders or councils collectively identify who should participate, ensuring cultural appropriateness and legitimacy (McCausland et al., 2025). In contexts involving communal knowledge, such as land stewardship or clan histories, talking circle sampling is used to engage intact social groups who customarily discuss matters together (Rieth, 2023).

Together, these sampling approaches honor Indigenous relational ethics, cultural authority, and community-defined protocols.

CHALLENGES IN IMPLEMENTING INDIGENOUS EPISTEMOLOGIES WITHIN INSTITUTIONAL LIS FRAMEWORKS

Implementing Indigenous epistemologies within institutional LIS frameworks presents significant challenges rooted in structural, epistemic, and cultural misalignments. Most LIS systems, such as classification schemes, cataloguing rules, and digital repositories, are built upon Western epistemologies that privilege universality, textuality, and hierarchical knowledge structures. These systems struggle to accommodate indigenous relational worldviews, place-based knowledge, and oral or embodied forms of transmission (Barité & Rauch, 2025; Littletree et al., 2020). As a result, IK is often misrepresented, decontextualised, or forced into inappropriate categories.

Institutional protocols also conflict with Indigenous governance principles. Standard LIS practices assume open access, data neutrality, and researcher-driven interpretation, which contradict cultural protocols requiring restricted access, community authority, and contextual decision-making (Carroll et al., 2020). Many institutions lack mechanisms for community control over metadata, digitisation priorities, or the withdrawal of sensitive materials.

Resource constraints pose further barriers. Many African IK communities lack digital infrastructure, long-term funding, or preservation tools suited to their contexts (Balogun & Kalusopa, 2021). Additionally, LIS professionals often lack training in indigenous methodologies, cultural safety, and ethical engagement, limiting their ability to implement culturally grounded practices.

Finally, institutional resistance to change, alongside colonial legacies embedded in LIS education, continues to impede the meaningful integration of indigenous epistemologies into mainstream practice.

FURTHER RESEARCH DIRECTIONS

Future research should prioritise African-generated models and theories that strengthen Indigenous research methodologies and guide sustainable development within African contexts. In LIS, this includes developing locally grounded knowledge-organisation systems, culturally relevant metadata standards, and community-driven archival frameworks. Greater collaboration between Indigenous communities, LIS scholars, and scientists, particularly in health, biodiversity, and environmental knowledge, can help document, preserve, and mobilise rich but under-recorded Indigenous knowledge systems, generating innovations responsive to African cultures, genotypes, and information needs.

ACKNOWLEDGEMENT

This publication acknowledges financial support from Carnegie New York Corporation which has funded my post doctoral studies and given me the opportunity to be mentored as a scholar in Library and information Science.

REFERENCES

Acharibasam, J. B., & McVittie, J. (2022). Connecting children to nature through the integration of indigenous ecological knowledge into early childhood environmental education. *Australian Journal of Environmental Education, 39*(3), 349–361. DOI: 10.1017/aee.2022.37

Akpa-Inyang, F., & Chima, S. C. (2021). South African traditional values and beliefs regarding informed consent and limitations of the principle of respect for autonomy in African communities: A cross-cultural qualitative study. *BMC Medical Ethics, 22*(1), 111. DOI: 10.1186/s12910-021-00678-4

Arriola-Pacheco, F., Ness, A., Sihuay-Torres, K., Garcia-Quintana, A., Proaño, D., & Lawrence, H. P. (2025). Community-based participatory research: Lessons and challenges. symposium special communication. *JDR Clinical and Translational Research, 10*(2), 96–99. DOI: 10.1177/23800844241266505

Balogun, T. (2023). Digitization of Indigenous knowledge systems in Africa: The case of South Africa's National Recorded System (NRS). *Records Management Journal, 33*(1), 88–102. DOI: 10.1108/RMJ-10-2022-0029

Balogun, T., & Kalusopa, T. (2021). A framework for digital preservation of Indigenous knowledge system (IKS) in repositories in South Africa. *Records Management Journal, 31*(2), 176–196. DOI: 10.1108/RMJ-12-2020-0042

Balogun, T., & Kalusopa, T. (2022). Web archiving of indigenous knowledge systems in South Africa. *Information Development, 38*(4), 658–671. DOI: 10.1177/02666669211005522

Bansal, S., Sarker, T., Yadav, A., Garg, I., Gupta, M., & Sarvaiya, H. (2024). Indigenous communities and sustainable development: A review and research agenda. *Global Business and Organizational Excellence, 43*(4), 65–87. https://doi.org/https://doi.org/10.1002/joe.22237. DOI: 10.1002/joe.22237

Barité, M., & Rauch, M. (2025). Native peoples and knowledge organization: Perspective from the Indigenous subject representation to promote Latin American approaches. *Knowledge Organization*, *51*(8), 626–641. DOI: 10.5771/0943-7444-2024-8-626

Bawack, R., Roderick, S., Badhrus, A., Dennehy, D., & Corbett, J. (2025). Indigenous knowledge and information technology for sustainable development. *Information Technology for Development*, *31*(2), 233–250. DOI: 10.1080/02681102.2025.2472495

Begay, V., & Klor, K. M. (2024). Provenance through storytelling: Application of Indigenous relationality toward arrangement and description. *Archival Science*, *24*(4), 611–635. DOI: 10.1007/s10502-024-09451-z

Belarde-Lewis, M., Littletree, S., Braine, I. S., Srader, K., Guerrero, N., & Palmer, C. L. (2024). Centering relationality and CARE for stewardship of indigenous research data. *Data Science Journal*, *23*(32), 1–16. https://doi.org/https://doi.org/10.5334/dsj-2024-032. DOI: 10.5334/dsj-2024-032

Bennett, K. (2023). Approaches to knowledge translation. In *The Routledge handbook of translation theory and concepts* (pp. 443–462). Routledge. DOI: 10.4324/9781003161448-27

Bourassa, C., Billan, J., Starblanket, D., Anderson, S., Legare, M., Hagel, M. C., Oakes, N., Jardine, M., Boehme, G., Dubois, E., Spencer, O., Hotomani, M., & McKenna, B. (2020). Ethical research engagement with Indigenous communities. *Journal of Rehabilitation and Assistive Technologies Engineering*, *7*, 2055668320922706. DOI: 10.1177/2055668320922706

Brinckley, M.-M., Bourke, S., Watkin Lui, F., & Lovett, R. (2022). Knowledge translation in Aboriginal and Torres Strait Islander research contexts in Australia: Scoping review protocol. *BMJ Open*, *12*(7), e060311. DOI: 10.1136/bmjopen-2021-060311

Broadhead, L.-A., & Howard, S. (2021). Confronting the contradictions between Western and Indigenous science: A critical perspective on two-eyed seeing. *Alternative*, *17*(1), 111–119. DOI: 10.1177/1177180121996326

Bruno, G., Bell, R. C., Parlee, B., Lightning, P., Bull, I., Cutknife, B., & Oster, R. T. (2022). Mâmawihitowin (bringing the camps together): Perinatal healthcare provider and staff participation in an indigenous-led experiential intervention for enhancing culturally informed care—A mixed methods study. *International Journal for Equity in Health*, *21*(1), 164. Advance online publication. DOI: 10.1186/s12939-022-01764-8

Carroll, S. R., Garba, I., Figueroa-Rodríguez, O. L., Holbrook, J., Lovett, R., Materechera, S. A., Parsons, M. A., Raseroka, K., Rodriguez-Lonebear, D., Rowe, R., Sara, R., Walker, J., Anderson, J., & Hudson, M. (2020). The CARE Principles for Indigenous data governance. *Data Science Journal*, *19*, 43. Advance online publication. DOI: 10.5334/dsj-2020-043

Chaulk, K. G., Ballard, M., Hill, S., Wolfrey, D., Campbell, M. L., Sutherland, M., Wawatie, S., & Auger, L. (2025). Bridging, braiding, and weaving Indigenous and Western science to understand and make predictions about weather and climate change. *Ecology and Evolution*, *15*(9), e72085. Advance online publication. DOI: 10.1002/ece3.72085

Chigwada, J., & Ngulube, P. (2024). Librarians' role in the preservation and dissemination of indigenous knowledge. *IFLA Journal*, *50*(2), 242–256. DOI: 10.1177/03400352231217270

Chigwada, J., & Ngulube, P. (2025a). Integrating CARE and FAIR Principles in the digital preservation of indigenous knowledge to achieve the Sustainable Development Goals. In Ngulube, P. (Ed.), *Harnessing Indigenous Epistemologies for Sustainable Progress* (pp. 197–230). IGI Global Scientific Publishing., https://doi.org/DOI, DOI: 10.4018/979-8-3693-8447-3.ch010

Chigwada, J., & Ngulube, P. (2025b). Stakeholders in the acquisition, preservation, and dissemination of indigenous knowledge projects. *Information Development*, *41*(4), 1281–1298. DOI: 10.1177/02666669231192851

Chigwada, J., Ngulube, P., & Dewah, P. (2024). Preserving indigenous knowledge in culture houses in the digital era in Masvingo Province, Zimbabwe. *African Journal of Library, Archives and Information Science*, 34(2), 205–216. Retrieved on 4 December from https://www.ajol.info/index.php/ajlais/article/view/281737

Chipangura, N., Chipangura, P. M., Mataga, J., & Chabata, F. M. (2025). Truth-telling as a form of decolonial care: Stories from a collaborative object biography research with the Ndau community of Eastern Zimbabwe. *Museums & Social Issues*, *19*(1), 26–44. DOI: 10.1080/15596893.2025.2463657

Choudhury, G., & Kalita, D. (2025). Collaborative stewardship: Sustainable approaches to indigenous knowledge in academic libraries. *Journal of Academic Librarianship*, *51*(2), 103007. https://doi.org/https://doi.org/10.1016/j.acalib.2025.103007. DOI: 10.1016/j.acalib.2025.103007

Clapham, K., Hasan, H., Fredericks, B., Bessarab, D., Kelly, P. J., Harwood, V., Senior, K., Longbottom, M., & Dale, E. (2021). Digital support for Indigenous research methodologies. *AJIS. Australasian Journal of Information Systems, 25.* Advance online publication. DOI: 10.3127/ajis.v25i0.2885

Cox, G., Firemoon, P., Anastario, M., Ricker, A., Thunder, R. E., Baldwin, J. A., & Rink, E. (2021). Indigenous Standpoint Theory as a theoretical framework for decolonizing social science health research with American Indian communities. *Alternative*, *17*(4), 460–468. DOI: 10.1177/11771801211042019

Cunningham-Reimann, S., Doshi, A., Stutz, S., Tjong, G. B., & Mashford-Pringle, A. (2025). Flipping data on its head: Differing conceptualisations of data and the implications for actioning Indigenous data sovereignty principles. *Global Public Health: An International Journal for Research, Policy and Practice*, *20*(1), 2450395. DOI: 10.1080/17441692.2025.2450395

Datta, R., & Kairy, B. (2024). Decolonizing climate change adaptations from indigenous perspectives: Learning reflections from Munda Indigenous communities, coastal areas in Bangladesh. *Sustainability (Basel)*, *16*(2), 769. DOI: 10.3390/su16020769

Datta, R., & Starlight, T. (2024). Building a meaningful bridge between Indigenous and Western worldviews: Through decolonial conversation. *International Journal of Qualitative Methods*, *23*, 16094069241235564. DOI: 10.1177/16094069241235564

David-Chavez, D. (2019). *A guiding model for decolonizing environmental science research and restoring relational accountability with Indigenous communities*. PhD Diss, Colorado State University, Colorado. https://doi.org/DOI: 10.31237/osf.io/ec9s5

DePuy, W., Weger, J., Foster, L. C., Bonanno, A., Kumar, S., Lear, K., Basilio, R., & German, L. (2021). Environmental governance: Broadening ontological spaces for a more livable world. *Environment and Planning. E, Nature and Space*, *5*(2), 947–975. DOI: 10.1177/25148486211018565

Duarte, M. E., & Belarde-Lewis, M. (2015). Imagining: Creating spaces for Indigenous ontologies. *Cataloging & Classification Quarterly*, *53*(5-6), 677–702. DOI: 10.1080/01639374.2015.1018396

Etchart, L. (2022). Indigenous peoples and international law in the Ecuadorian Amazon. *Laws*, *11*(4), 55. DOI: 10.3390/laws11040055

Eyre, M. (2023). The spirit of knowledge: Creating spiritual competency for critical information spaces. *Public Services Quarterly*, *19*(1), 69–76. DOI: 10.1080/15228959.2022.2160857

Forbes, A., Ritchie, S., Walker, J., & Young, N. (2020). Applications of two-eyed seeing in primary research focused on Indigenous health: A scoping review. *International Journal of Qualitative Methods*, *19*, 1609406920929110. DOI: 10.1177/1609406920929110

Fullmer, M. (2021). Are we there yet? Visualizing Indigenous culture in today's library. *IFLA Journal*, *47*(3), 313–320. DOI: 10.1177/0340035220987577

Gannaway, J. (2020). Knocking, Unsettling, ceding: A non-Indigenous teacher's journey towards decolonizing teaching practice in a "remote Indigenous community". [NJCIE]. *Nordic Journal of Comparative and International Education*, *4*(1), 102–117. DOI: 10.7577/njcie.3553

Gathogo, J. M. (2021). Memory and history: Oral techniques in the East African context. *Hervormde Teologiese Studies*, *77*(2). Advance online publication. https://www.ajol.info/index.php/hts/article/view/212210. DOI: 10.4102/hts.v77i2.6477

Harfield, S., Pearson, O., Morey, K., Kite, E., Canuto, K., Glover, K., Gomersall, J. S., Carter, D., Davy, C., Aromataris, E., & Braunack-Mayer, A. (2020). Assessing the quality of health research from an Indigenous perspective: The Aboriginal and Torres Strait Islander quality appraisal tool. *BMC Medical Research Methodology*, *20*(1), 79. Advance online publication. DOI: 10.1186/s12874-020-00959-3

Hatfield, H. R., & Ahn, S. J. (2025). Integrating community-based participatory research into immersive narratives. *Journal of Media Psychology*, *37*(2), 101–115. DOI: 10.1027/1864-1105/a000463

Hayward, A., Sjoblom, E., Sinclair, S., & Cidro, J. (2021). A new era of indigenous research: Community-based Indigenous research ethics protocols in Canada. *Journal of Empirical Research on Human Research Ethics; JERHRE*, *16*(4), 403–417. DOI: 10.1177/15562646211023705

Huaman, E. S., & Mataira, P. (2019). Beyond community engagement: Centering research through Indigenous epistemologies and peoplehood. *Alternative*, *15*(3), 281–286. DOI: 10.1177/1177180119871705

Hudson, M., Carroll, S. R., Anderson, J., Blackwater, D., Cordova-Marks, F. M., Cummins, J., David-Chavez, D., Fernandez, A., Garba, I., Hiraldo, D., Jäger, M. B., Jennings, L. L., Martinez, A., Sterling, R., Walker, J. D., & Rowe, R. K. (2023). Indigenous Peoples' Rights in Data: A contribution toward Indigenous research sovereignty [perspective]. *Frontiers in Research Metrics and Analytics*, *8*, 2023. DOI: 10.3389/frma.2023.1173805

Hunt, S. C., & Young, N. L. (2021). Blending Indigenous sharing circle and Western focus group methodologies for the study of Indigenous children's health: A systematic review. *International Journal of Qualitative Methods, 20*, 16094069211015112. DOI: 10.1177/16094069211015112

Hunter, A. (2021). The Shortcomings of Bibliographic description in service of Indigenous peoples in Canada. *Emerging Library & Information Perspectives, 4*(1), 88–116. DOI: 10.5206/elip.v4i1.13439

Huria, T., Palmer, S. C., Pitama, S., Beckert, L., Lacey, C., Ewen, S., & Smith, L. T. (2019). Consolidated criteria for strengthening reporting of health research involving Indigenous peoples: The CONSIDER Statement. *BMC Medical Research Methodology, 19*(1), 173. Advance online publication. DOI: 10.1186/s12874-019-0815-8

Hutchinson, P., Dingwall, C., Kurtz, D., Evans, M., Jones, G. R., & Corbett, J. (2014). Maintaining the iIntegrity of Indigenous knowledge: Sharing metis knowing through mixed methods. *International Journal of Critical Indigenous Studies, 7*(1), 1–14. DOI: 10.5204/ijcis.v7i1.118

Ignace, L., Burton, L., Mynott, S., Meehan, M., Olson, E., Steel, J., Ojeda, J., Harper, S., Ramirez, L., Baker, D., Sleigh, L., Frenkel, C., Rhodes, C., & Ban, N. C. (2023). Researchers' responsibility to uphold Indigenous rights. *Science, 381*(6654), 129–131. https://doi.org/doi:10.1126/science.adh4470. DOI: 10.1126/science.adh4470

Jamieson, V., & Chu, C. M. (2025). Disrupting informed consent: Refusal and cultural sovereignty in Indigenous knowledge practices. *Proceedings of the Association for Information Science and Technology, 62*(1), 1496–1498. DOI: 10.1002/pra2.1446

Johnson, L. R., Wilcox, A., Alexander, S. M., Bowles, E., Castleden, H., Henri, D., Herc, C., King, L., Provencher, J. F., & Orihel, D. M. (2023). Weaving Indigenous and Western ways of knowing in ecotoxicology and wildlife health: A review of Canadian studies. *Environmental Reviews, 31*(3), 452–470. DOI: 10.1139/er-2022-0087

Jordan, L. S., & Hall, J. N. (2023). Framing anticolonialism in evaluation: Bridging decolonizing methodologies and culturally responsive evaluation. *Journal of Multidisciplinary Evaluation, 19*(44), 102–116. Advance online publication. DOI: 10.56645/jmde.v19i44.769

Kennedy, M., Ninomiya, M. E. M., Ninomiya, M. M., Brascoupé, S., Smylie, J., Calma, T., Mohamed, J., Stewart, P., & Maddox, R. (2024). Knowledge translation in Indigenous health research: Voices from the field. *The Medical Journal of Australia, 221*(1), 61–67. DOI: 10.5694/mja2.52357

Konczi, A. E., & Bill, L. (2024). Advancing first nations principles of OCAP. In Garvey, G. (Ed.), *Indigenous and tribal peoples and cancer* (pp. 37–39). Springer Nature Switzerland., DOI: 10.1007/978-3-031-56806-0_8

Kovach, M. (2021). *Indigenous methodologies: Characteristics, conversations, and contexts*. University of Toronto press.

Littletree, S., and, M. B.-L., & Duarte, M. (2020). Centering relationality: A conceptual model to advance Indigenous knowledge organization practices. *Knowledge Organization*, *47*(5), 410–426. DOI: 10.5771/0943-7444-2020-5-410

Littletree, S., Andrews, N., & Loyer, J. (2023). Information as a relation: Defining Indigenous information literacy. *Journal of Information Literacy*, *17*(2), 4–23. https://doi.org/https://doi.org/10.11645/17.2.8. DOI: 10.11645/17.2.8

Littletree, S., & Krise, C. (2024). The Squaxin Island Museum Library and Research Center: A report from the field of tribal libraries and museums. *International Journal of Librarianship*, *9*(2), 51–59. DOI: 10.23974/ijol.2024.vol9.2.361

López-Quiñones, A., Martínez-López, M., Sandoval, C. D. M., Carroll-Miranda, J., Lindala, A., Chatman, M. C., Fleming, J. S., Shockley, E. T., Cadeau, D., & Flores-Reyes, E. (2023). Ancestral computing for sustainability: Centering Indigenous epistemologies in researching Computer Science Education. *TechTrends*, *67*(3), 435–445. DOI: 10.1007/s11528-022-00820-y

Lys, I., Quinnell, R., Weyand, P., Phillips, B., & Mochrie, K. (2022). Students as co-creators of an Aboriginal and Torres Strait heritage garden and interactive maps. *Ascilite Publications*, e22209. DOI: 10.14742/apubs.2022.209

Maar, M., Ominika, T., & Manitowabi, D. (2022). Community-led recovery from the opioid crisis through culturally-based programs and community-based data governance. *International Indigenous Policy Journal*, *13*(2). Advance online publication. DOI: 10.18584/iipj.2022.13.2.13792

Machin-Mastromatteo, J. D. (2023). Community-driven and social initiatives. *Information Development*, *39*(3), 393–401. DOI: 10.1177/02666669231197243

Mackey, T. K., Calac, A. J., Keshava, B. S. C., Yracheta, J., Tsosie, K. S., & Fox, K. (2022). Establishing a blockchain-enabled Indigenous data sovereignty framework for genomic data. *Cell*, *185*(15), 2626–2631. DOI: 10.1016/j.cell.2022.06.030

Macklin, C. D., Marchand, C., Mitchell, E., Price, R., Mitchell, V., & Bryant, L. (2021). Planting the seeds: Insights for researchers interested in working with Indigenous peoples. *International Journal of Indigenous Health*, *16*(1). Advance online publication. DOI: 10.32799/ijih.v16i1.33193

Manning, B. R. M. (2023). Relationships, respect, and reciprocity: Approaches to learning and teaching about indigenous cultural burning and landscape stewardship, 145-156. DOI: 10.4337/9781789905069.00019

Manyane, R. M. (2024). Centuries-long African oral traditions and history: Revisiting the debate. *Journal of African History, Archaeology and Tourism, 2*(1), 99-116. https://doi.org/doi:10.31920/2753-3204/2024/v2n1a5

Marshall, R., Desjardine, M., Levison, J., Anderson, K., & McBean, E. A. (2020). Moving towards effective first nations' source water protection: Barriers, opportunities, and a framework. *Water (Basel), 12*(11), 2957. DOI: 10.3390/w12112957

McCausland, R., Jamieson, S. K., Robinson, V., Spencer, W., MacGillivray, P., & Andersen, M. (2025). Elders' perspectives and priorities for ageing well in a remote Aboriginal community. *Ageing and Society*, *45*(1), 31–54. DOI: 10.1017/S0144686X23000156

McMullin, J. (2025). Decolonization. *Annals of Anthropological Practice*, *49*(1), e70002. Advance online publication. DOI: 10.1111/napa.70002

Mdhluli, T. D., Mokgoatšana, S., Kugara, S. L., & Vuma, L. (2021). Knowledge management: Preserving, managing and sharing indigenous knowledge through digital library. *Hervormde Teologiese Studies*, *77*(2). Advance online publication. https://www.ajol.info/index.php/hts/article/view/232746. DOI: 10.4102/hts.v77i2.6795

Melo, N. (2025). *Decolonizing research methods: Indigenous, Afrocentric, and participatory approaches: Challenging colonial epistemologies and centering community-led knowledge production*. Nouridin Melo.

Mendenhall, A., Hicks, C., Holder, M., Holmes, C., Jung, E., Ramirez, L., Fairman, M., Bulls, S. T., & Levy, M. (2024). Growing stronger together: Sharing a story of culturally responsive evaluation with indigenous families and communities. *Child Abuse & Neglect*, *148*, 106344. https://doi.org/https://doi.org/10.1016/j.chiabu.2023.106344. DOI: 10.1016/j.chiabu.2023.106344

Montenegro, M. (2019). Subverting the universality of metadata standards: The TK labels as a tool to promote Indigenous data sovereignty. *The Journal of Documentation*, *75*(4), 731–749. DOI: 10.1108/JD-08-2018-0124

Morisano, D., Robinson, M., Rush, B., & Linklater, R. (2024). Conducting research with Indigenous peoples in Canada: Ethical and policy considerations. *Frontiers in Psychology*, *14*, 1214121. Advance online publication. DOI: 10.3389/fpsyg.2023.1214121

Ndimande, B. S. (2012). Decolonizing research in postapartheid South Africa. *Qualitative Inquiry*, *18*(3), 215–226. DOI: 10.1177/1077800411431557

Ngaka, M. N. (2025). A deeper look into Indigenous people's spirituality epistemologies and creating new social and economic systems for sustainability. In Ngulube, P. (Ed.), *Harnessing Indigenous Epistemologies for Sustainable Progress* (pp. 1–24). IGI Global Scientific Publishing., https://doi.org/DOI, DOI: 10.4018/979-8-3693-8447-3.ch001

O'Connor, D., Scoville, D., Donald, R., Donald, V., Witney, J., Ellefson, O., Jewell, G., Jewell, L., & Silkie, V. (2022). A multiple evidence-based approach to métis community-based monitoring: A case study from the South Athabasca oil sands area, Alberta, Canada. *Ecology and Society*, *27*(4), art27. Advance online publication. DOI: 10.5751/ES-13678-270427

Osei-Tutu, A. A. Z. (2023). Developing African oral traditional storytelling as a framework for studying with African peoples. *Qualitative Research*, *23*(6), 1497–1514. DOI: 10.1177/14687941221082263

Palmer, C. L., Sebastian, K., Belarde-Lewis, M., Littletree, S., & Guerrero, N. (2024). Stewarding contextual integrity in data services for Indigenous scholarship. *Proceedings of the Association for Information Science and Technology*, *61*(1), 626–631. https://doi.org/https://doi.org/10.1002/pra2.1070. DOI: 10.1002/pra2.1070

Patin, B., Oliphant, T., Allard, D., Gray, L., Clarke, R. I., Tacheva, J., & Lar-Son, K. (2021). At the margins of epistemology: Amplifying Alternative ways of knowing in Library and Information Science. *Proceedings of the Association for Information Science and Technology*, *58*(1), 630–633. https://doi.org/https://doi.org/10.1002/pra2.515. DOI: 10.1002/pra2.515

Paul, M. (2023). Looking to the future: Indigenous data sovereignty and policy in Canada. *Pathfinder: A Canadian Journal for Information Science Students and Early Career Professionals*, *4*(1), 54-67. DOI: 10.29173/pathfinder71

Raifman, S., DeVost, M. A., Digitale, J. C., Chen, Y.-H., & Morris, M. D. (2022). Respondent-driven sampling: A sampling method for hard-to-reach populations and beyond. *Current Epidemiology Reports*, *9*(1), 38–47. DOI: 10.1007/s40471-022-00287-8

Raymond, C. (2024). Decolonizing the presentation of research findings: Amplifying epistemic authority through poetic re-storying. *Narrative Works*, *11*, 149–173. DOI: 10.7202/1108959ar

Riediger, N. D., Cyr, M., & Mignone, J. (2020). An evaluation of an experiential learning program in global and Indigenous health: The University of Manitoba's Queen Elizabeth II Diamond Jubilee Scholarship Program. *Inquiry : a Journal of Medical Care Organization, Provision and Financing*, *57*, 0046958020951002. Advance online publication. DOI: 10.1177/0046958020951002

Rieth, B. (2023). Empowered circles: Using talking circles in research. *New England College Journal of Applied Educational Research*, *4*, 1–23.

Sekiwu, D., Adyanga, F. A., Genza, G. M., Rugambwa, N. O., & Muwagga, A. M. (2024). Lifelong educational and decolonization programs for the Batwa Indigenous people of southwestern Uganda. *East African Journal of Interdisciplinary Studies*, *7*(1), 38–52. DOI: 10.37284/eajis.7.1.1838

Setumu, T., & Ngoepe, M. (2016). Converting oral narratives into written literature: Lessons learnt from heritage projects undertaken in the Makgabeng area, Limpopo, South Africa. *Mousaion*, *34*(4), 148–164. https://doi.org/doi:10.25159/0027-2639/940

Shiri, A., Howard, D., & Farnel, S. (2022). Indigenous digital storytelling: Digital Interfaces supporting cultural heritage preservation and access. *The International Information & Library Review*, *54*(2), 93–114. DOI: 10.1080/10572317.2021.1946748

Shrinkhal, R. (2021). "Indigenous sovereignty" and right to self-determination in international law: A critical appraisal. *Alternative*, *17*(1), 71–82. DOI: 10.1177/1177180121994681

Simonds, V., & Christopher, S. (2013). Adapting Western research methods to indigenous ways of knowing. *American Journal of Public Health*, *103*(12), 2185–2192. DOI: 10.2105/AJPH.2012.301157

Stevens-Uninsky, M., Barkhad, A., MacDonald, T., Perez, A., & Mbuagbaw, L. (2023). Decolonization in sexual and reproductive health research methods: Protocol for a scoping review. *JMIR Research Protocols*, *12*, e45771. DOI: 10.2196/45771

Struik, L., Werstuik, S.-T., Sundstrom, A., Dow-Fleisner, S., & Ben-David, S. (2022). Factors that influence the decision to vape among Indigenous youth. *BMC Public Health*, *22*(1), 641. Advance online publication. DOI: 10.1186/s12889-022-13095-y

Tarisayi, K. S. (2024). Integrating Indigenous knowledge in South African Geography education curricula for social justice and decolonization. *E-Journal of Humanities Arts and Social Sciences*, 1195-1206. DOI: 10.38159/ehass.20245711

Thambinathan, V., & Kinsella, E. A. (2021). Decolonizing methodologies in qualitative research: Creating spaces for transformative praxis. *International Journal of Qualitative Methods*, *20*, 16094069211014766. Advance online publication. DOI: 10.1177/16094069211014766

Thorpe, K., Christen, K., Booker, L., & Galassi, M. (2021). Designing archival information systems through partnerships with Indigenous communities. *AJIS. Australasian Journal of Information Systems*, *25*. Advance online publication. DOI: 10.3127/ajis.v25i0.2917

Tone-Pah-Hote, T., & Redvers, N. (2022). The commercialization of biospecimens from Indigenous peoples: A scoping review of benefit-sharing. *Frontiers in Medicine*, *9*, 978826. Advance online publication. DOI: 10.3389/fmed.2022.978826

Trepanier, C., Shiri, A., & Samek, T. (2019). An examination of IFLA and Data Science Association ethical codes. *IFLA Journal*, *45*(4), 289–301. DOI: 10.1177/0340035219849614

Waddell-Henowitch, C., Herron, R., Gobeil, J., Tacan, F., Jager, M. D., Allan, J. A., & Roger, K. (2020). Grounded in culture: Reflections on sitting outside the circle in community-based research with Indigenous men. *Qualitative Health Research*, *30*(14), 2343–2350. DOI: 10.1177/1049732320960050

Walker, J., Pyper, E., Jones, C., Khan, S., Chong, N. W., Legge, D., Schull, M. J., & Henry, D. (2018). Unlocking first nations health information through data linkage. *International Journal of Population Data Science*, *3*(1). Advance online publication. DOI: 10.23889/ijpds.v3i1.450

Walter, M., Lovett, R., Maher, B., Williamson, B., Prehn, J., Bodkin-Andrews, G., & Lee, V. (2020). Indigenous data sovereignty in the era of big data and open data. *The Australian Journal of Social Issues*, *56*(2), 143–156. DOI: 10.1002/ajs4.141

Wiegratz, J., Mujere, J., & Fontein, J. (2022). The case for oral histories of neoliberal Africa. *Third World Thematics: A TWQ Journal, 7*(4-6), 199-221. DOI: 10.1080/23802014.2023.2208561

Williams, D., & Shipley, G. P. (2023). Indigenous research methodologies: Challenges and opportunities for broader recognition and acceptance. *Open Journal of Social Sciences*, *11*(05), 467–500. DOI: 10.4236/jss.2023.115030

Williamson, B., Provost, S., & Price, C. (2022). Operationalising Indigenous data sovereignty in environmental research and governance. *Environment and Planning. F, Philosophy, Theory, Models, Methods and Practice*, *2*(1-2), 281–304. DOI: 10.1177/26349825221125496

ADDITIONAL READING

Chisita, C. T. (2024). The library amid an epistemic storm: Research descriptions on Indigenous knowledge services. In *Evaluating Indigenous African tradition for cultural reconstruction and mind decolonization* (pp. 26–44). IGI Global. DOI: 10.4018/978-1-6684-8827-0.ch003

Mhlongo, M. (2021). Addressing epistemicide through the integration of indigenous knowledge: Are South African public libraries ready? *Alternation* (Special Edition), 38c, 65 – 94 65. DOI: 10.29086/2519-5476/2021/sp38c06

Moahi, K. H. (2022). The research process and Indigenous epistemologies. In *Research anthology on racial equity, identity, and privilege* (pp. 1387–1407). IGI Global Scientific Publishing. DOI: 10.4018/978-1-6684-4507-5.ch074

Ngulube, P. (2025). Indigenous theories and theorisation in African scholarship in the quest for epistemic freedom. In Ngulube, P. (Ed.), *Harnessing Indigenous epistemologies for sustainable progress* (pp. 139–172). IGI Global., DOI: 10.4018/979-8-3693-8447-3.ch008

Seifi, L., & Ghasemipoor, M. (2025). Librarians as stewards of Indigenous knowledge: A meta-synthesis of roles, challenges, and strategies. [IJISM]. *International Journal of Information Science and Management, 23*(4), 189–207.

KEY TERMS AND DEFINITIONS

Indigenous Knowledge Research Designs: are culturally grounded methodological frameworks that structure how research with Indigenous communities is conducted, guided by Indigenous worldviews, relational principles, and community governance.

Indigenous Knowledge Sampling: refers to culturally grounded participant selection methods based on relationships, community authority, and the identification of recognised knowledge holders rather than statistical criteria.

Indigenous Epistemologies: are ways of knowing rooted in Indigenous worldviews, emphasizing relationality, land, spirituality, community, and holistic understandings of knowledge.

Chapter 10
Developing a Conceptual Framework for Digital Record Trustworthiness in the Public Sector of Zimbabwe

Munyika Sibanda
http://orcid.org/0000-0002-5253-7948
Great Zimbabwe University, Zimbabwe

ABSTRACT

This chapter outlines the development of a context-sensitive conceptual framework to enhance the trustworthiness of digital records in Zimbabwe's public sector. Centred on accuracy, authenticity, reliability, and usability, it employs a qualitative approach, combining literature review and Delphi method validation. Findings indicate that global frameworks like ISO 15489, MoReq, and InterPARES often mismatch Zimbabwe's resource-constrained context. To address this, the proposed framework incorporates both technical and human factors, such as metadata standards, version control, capacity building, and secure data recovery. Expert validation refined the framework, ensuring relevance and sustainability. The chapter concludes by highlighting the model's implications for theory development, practical application, and research in computing and informatics.

DOI: 10.4018/979-8-2600-0434-0.ch010

INTRODUCTION

Across the globe, governments are embracing digital systems to improve transparency, accountability, and service delivery, making the creation of authentic, accurate, reliable, and usable digital records a critical requirement (Baheer, Lamas & Sousa, 2020; Albar et al., 2017; INTERNATIONAL ORGANIZATION FOR STANDARDIZATION, 2016; InterPARES, 2007). In Zimbabwe, fragmented systems and capacity gaps have limited records' legal and evidential value (Hunie & Dewah, 2019; Sibanda & Schellnack-Kelly, 2024). To address this, a context-sensitive conceptual framework, informed by international models such as MoReq, NOARK, and VERS and developed using Holton and Lowe's (2007) theory-building method, provides both methodological and practical guidance for digital records governance in the Zimbabwean public sector (Cain, 2002; Hagen-Sataslaatten, 2014; Leviston, 1999).

BACKGROUND

The success of digital transformation in government largely depends on the effective management of digital records, which preserve institutional memory and provide a foundation for informed decision-making, reliable public services, and compliance with legal requirements (Albar et al., 2017; Baheer, Lamas & Sousa, 2020; Cain, 2002; Hagen-Sataslaatten, 2014; Leviston, 1999). Public sector records, whether paper-based or electronic, must be trustworthy, defined by their authenticity, accuracy, reliability, and usability (InterPARES, 2007). Managing digital records is therefore central to realizing the benefits of e-government and digital public administration. International standards and models, such as ISO 15489-1, MoReq, NOARK, and the Victorian Electronic Records Strategy (VERS), articulate requirements for creating, maintaining, and preserving records to ensure their evidential value (Boudrez, 2007; InterPARES, 2007). However, these frameworks were primarily designed for contexts with established institutional maturity, technological infrastructure, and policy coherence, raising concerns about their direct applicability in less developed environments like Zimbabwe (Ambira, Kemoni & Ngulube, 2019).

In Zimbabwe, digital records management faces systemic challenges. Most institutions operate with hybrid or fragmented systems that lack consistency, interoperability, and legal recognition (Chikomba, Rodrigues & Ngoepe, 2021; Sigauke, 2022). Policy disjointedness, poor metadata standards, outdated information and communication technology (ICT) infrastructure, limited skilled personnel, and insufficient financial investment hinder the effective implementation of electronic records management systems (Masuku, Bhebhe & Khumalo, 2017; Mutsagondo,

2017; Sithole, 2022). Consequently, records are often incomplete, inaccessible, or legally inadmissible, resulting in delays in service delivery, administrative inefficiencies, and undermined institutional integrity, particularly in sensitive domains such as land, health, and finance (Bhebhe, 2015; Hunie & Dewah, 2019; Magama & Nduna, 2020).

To address these challenges, context-specific frameworks that draw on global best practices while responding to local realities are essential. Studies from Kenya (Ambira, 2016), Namibia (Nengomasha & Chikomba, 2018), and Botswana (Mosweu & Ngoepe, 2020) illustrate the value of developing indigenous frameworks for reliable digital records. In response, this study proposes a conceptual framework to enhance the trustworthiness of digital records in Zimbabwe's public sector. Developed using Holton and Lowe's (2007) seven-step theory-building methodology, the framework integrates empirical insights, theoretical constructs, and expert validation, focusing on authenticity, accuracy, reliability, and usability. Conceptual frameworks provide structured networks of interrelated concepts that link abstract ideas to empirical realities (Jabareen, 2009; Ngulube, 2018; Ngulube & Mosha, 2023), selectively drawing from multiple theories, models, and literature to clarify constructs, map relationships, and guide research design, data collection, and analysis (Imenda, 2014; Ngulube, 2020b; Collis & Hussey, 2014; Mahlangu, 2020; Maxwell, 2013). As Ngulube (2020a) and Rocco and Plakhotnik (2009) emphasize, a well-developed conceptual framework ensures cohesion, prevents fragmentation, and strengthens scholarly rigor, making it a critical tool for improving records governance and public trust in Zimbabwe's digital era.

PROBLEM STATEMENT

Zimbabwe's public sector continues to struggle with ensuring trustworthy digital records due to fragmented systems, inconsistent practices, weak policies, and poorly coordinated institution-specific initiatives (Masuku et al., 2017; Chikomba et al., 2021; Mutsagondo, 2021). Consequently, many digital records are incomplete, inaccessible, or legally and administratively inadmissible (Sigauke, 2022; Sithole, 2022). While international frameworks such as MoReq, NOARK, and VERS provide guidance, their direct implementation is constrained by local infrastructural and contextual challenges (Magama & Nduna, 2020; Nengomasha & Chikomba, 2018). This underscores the need for a locally grounded theoretical framework that integrates global best practices with Zimbabwe's unique socio-technical and institutional realities. Using Holton and Lowe's (2007) theory-building approach, this study proposes a framework to design trustworthy digital records systems that enhance accountability and public service delivery.

Purpose Of The Study

This study aims to develop a context-specific framework for managing trustworthy digital records in Zimbabwe's public sector. Using Holton and Lowe's (2007) theory-building method, it adapts international standards to help institutions ensure authenticity, accuracy, reliability, and usability for enhanced governance. This chapter has three main objectives:

i. To demonstrate the process of developing a conceptual framework through an integrative literature review and thematic analysis.
ii. To illustrate how constructs are operationalized into dimensions that form the backbone of a trustworthiness framework.
iii. To discuss the implications of the proposed framework for theory development, practice, and research methodology in computing and informatics.

METHODOLOGY

The conceptual framework was developed through a structured and iterative process, combining literature review and expert validation. The process was guided by the principles of systematic theory-building to ensure that the framework was both research-grounded and practically applicable to the Zimbabwean public sector context.

Literature review

A literature review was conducted to synthesize empirical, theoretical, and methodological studies, combining international digital records management frameworks such as MoReq, NOARK and VERS with Zimbabwean contextual realities (Cain, 2002; Hagen-Sataslaatten, 2014; Leviston, 1999; Torraco, 2016; Kraus, Breier & Dasi-Rodriguez, 2020; Rocco & Plakhotnik, 2009). Systematic searches and thematic synthesis identified authenticity, accuracy, reliability, and usability as core constructs, alongside ICT skills, infrastructure, metadata, and regulatory factors. Iterative searches, including snowballing, continued until theoretical saturation (Saunders et al., 2018; Mahlangu, 2020). These insights provided the empirical and theoretical foundation for developing a context-sensitive trustworthiness framework for Zimbabwe's public sector, guiding subsequent framework development, thematic analysis, and expert validation

The Delphi Method

To ensure the validity, relevance, and practical applicability of the proposed framework, the Delphi Method was employed. This structured communication technique facilitates iterative collection and refinement of expert opinions (Landeta, 2006). A panel of ten purposively selected experts, each holding a master's degree in a relevant field and at least ten years of practical experience in digital records management, reviewed the framework. Experts assessed the framework using a three-part instrument:

i. Segment A – demographic and professional background data
ii. Segment B – evaluation of clarity, relevance, coherence, and applicability
iii. Segment C – qualitative feedback on strengths, weaknesses, and suggestions for refinement

The Delphi process allowed independent input, minimizing group influence. The first round yielded a strong consensus on the framework's relevance and comprehensiveness. Minor refinements were incorporated, including improved terminology, clarification of dimensions, and structural adjustments, with no major conceptual disagreements. The process confirmed expert consensus and data saturation, enhancing the framework's credibility, contextual fit, and practical utility.

CONCEPTUAL FRAMEWORK DEVELOPMENT PROCESS

The development of a conceptual framework requires a systematic and rigorous approach to avoid "crash landing" methods that compromise theoretical and practical relevance (Elangovan & Rajendran, 2015; Mahlangu, 2020; Mutsagondo, 2021; Ngulube, 2018, 2020a; Ngulube & Mosha, 2023). Conceptual frameworks should follow scientific principles, such as theory building. While Dubin's (1978) eight-step method provides a two-stage process which involves defining units, constructs, laws of interaction, and system boundaries in the first stage, then formulating propositions, identifying empirical indicators, developing hypotheses, and testing them in the second, this study adopted Holton and Lowe's (2007) seven-step method. This approach is more flexible and suited to qualitative research, making it appropriate for developing a context-specific framework for trustworthy digital records in Zimbabwe's public sector. Holton and Lowe's method involves understanding the phenomenon through literature and interviews, identifying and analysing constructs, developing an initial theory by defining units, interactions, boundaries, system states, and propositions, evaluating the theory with scholarly criteria, synthesizing feedback, and modifying the framework accordingly. This iterative process, oscillating between theoretical abstraction and empirical reality (Eriksson, Johannesson & Bergholtz, 2018), ensures both the rigour and contextual relevance of the resulting framework, guiding the construction of a model tailored to Zimbabwe's public sector needs. Below is the presentation of the stages that were followed to come up with the conceptual framework

Phenomenon Understanding

Understanding the phenomenon of digital records' trustworthiness, covering authenticity, reliability, accuracy, and usability, is central to developing a conceptual framework. In Zimbabwe, the absence of a unified records management model and outdated legal frameworks has led to records often being inadmissible in legal proceedings (Hunie & Dewah, 2019; Chikomba et al., 2021; Mutsagondo, 2021). In contrast, countries like the UK, Australia, Canada, and the USA have advanced Electronic Records Management Systems (ERMS) supported by standards such as VERS and MoReq (Cain, 2002; Hagen-Sataslaatten, 2014; Leviston, 1999). This comparative gap underscores the need for a context-sensitive framework tailored to Zimbabwe's public sector. A concept map (Figure 1) was developed to visually represent the dimensions of trustworthiness in this context.

Figure 1. Pictorial presentation of the concept (Author's compilation, 2025)

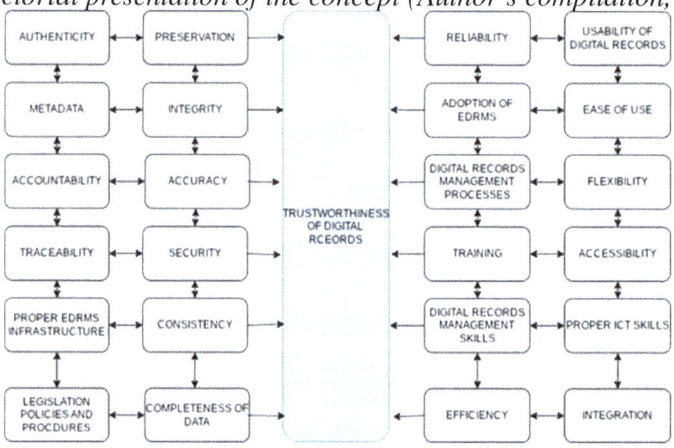

Figure 1. illustrates the key concepts underpinning the phenomenon of digital records' trustworthiness in Zimbabwe's public sector. As shown, the trustworthiness of digital records is shaped by multiple interrelated factors. These include the core constructs of authenticity, accuracy, reliability, and usability, which are supported

by essential enablers such as record security, long-term preservation strategies, regulatory and legislative frameworks, and the ICT competence of records personnel. Additionally, infrastructure readiness and the metadata-capturing capabilities of records management systems further influence trustworthiness. This conceptual map provides a holistic visual framework that highlights how these elements collectively contribute to or undermine the creation and maintenance of trustworthy digital records in a developing country context. Figure 1.1 represents the relationship and contribution of these factors to the trustworthiness of digital records.

Identification and Retrieval of Relevant Studies

The literature highlights several recurring themes regarding digital records management in public sector contexts. The core constructs of trustworthiness, authenticity, reliability, accuracy, and usability are widely recognized internationally as essential for ensuring that digital records serve as credible sources of information for accountability, decision-making, and legal purposes (INTERNATIONAL ORGANIZATION FOR STANDARDIZATION, 2016; InterPARES, 2007). However, the successful implementation of these principles in African contexts is often constrained by resource limitations, fragmented technological infrastructure, and organizational inefficiencies (Bwalya, 2009; Nengomasha & Chikomba, 2018). The absence of fully integrated electronic records management systems and inconsistent access to reliable ICT tools compromise both the integrity and accessibility of digital records.

The human dimension, including staff skills, competencies, and training, is frequently overlooked but is critical for maintaining the trustworthiness of records (Ambira, 2016; Mosweu & Ngoepe, 2020). Without adequate training in digital records management principles, personnel may fail to implement proper procedures for authentication, preservation, and access, resulting in gaps that undermine the reliability and usability of records. This underscores the necessity of integrating capacity-building initiatives, professional development, and continuous training programs into records management frameworks to ensure practical effectiveness and sustainability.

The literature also emphasizes the importance of context-specific adaptations. Frameworks developed in well-resourced environments, such as MoReq, NOARK, VERS, ISO 15489, and InterPARES, provide robust theoretical guidance, but their assumptions often do not align with the operational realities of Zimbabwe's public sector (Sithole, 2022; Sigauke, 2022). Local conditions, including limited financial resources, fragmented information systems, and uneven policy enforcement, necessitate frameworks that are flexible, adaptable, and sensitive to technological, human, and organizational constraints.

Integrating conceptual and theoretical insights, particularly those offered by Ngulube (2018; 2020a; 2020b; 2025) and Ngulube and Mosha (2023), further strengthens the development of context-sensitive frameworks. His work emphasizes that frameworks should be grounded in rigorous theory, informed by empirical evidence, and structured to accommodate both operational practicality and scholarly robustness. By combining international best practices with regional and local experiences, and explicitly addressing human, technological, and organizational challenges, Zimbabwe's public sector can develop practical, sustainable, and trustworthy digital records management frameworks that enhance service delivery, accountability, and legal compliance.

Construct Analysis

Constructs, defined as abstract representations of significant phenomena (Polites et al., 2012), were analyzed using thematic analysis to identify patterns and organize dimensions clearly (Maguire & Delahunt, 2017; Clarke & Braun, 2013). The process involved researching digital records trustworthiness, defining, and describing measurement dimensions, identifying key constructs, and mapping these constructs to their dimensions. Below is an explanation of the steps.

Step 1: Researching the Trustworthiness of Digital Records

To understand the factors affecting the trustworthiness of digital records in Zimbabwe's public sector, the researcher employed a triangulated approach, combining an extensive literature review, professional experience, and theoretical insights. Sources included peer-reviewed articles, theses, technical reports, and media publications, with a focus on identifying core constructs and dimensions for a conceptual framework. By synthesizing empirical and theoretical evidence, the study aimed to produce a framework grounded in both scholarship and local contextual realities.

Step 2: Identify, Define and Describe Measurement Dimensions

The researcher systematically identified and extracted 24 key dimensions (Table 1.1) from a comprehensive review of digital records management models and scholarly literature to develop the initial constructs of the conceptual framework. Each dimension was clearly defined to ensure conceptual clarity and enable comparative and thematic analysis. Organized in a table with concise descriptions, this process minimized overlap, ensured consistency, and laid the foundation for grouping related dimensions in later stages of framework development.

Table 1. Definitions of Dimensions Linked to the Trustworthiness of Digital Records

Dimension	Description
Reliability	Reliability is the authority and trustworthiness of records as proof and memory of the activity and their ability to stand for the facts they represent (Rogers, 2015; NARA, 2005).
Authenticity	This means that a record is what it purports to be and has been created by the organization with which it is identified (Duranti, 2015).
Accuracy	This refers to the degree to which the data from digital records is free from error (Boudrez, 2007).
Integrity	Digital records integrity denotes the accuracy, consistency, and completeness of data throughout its lifecycle (Boudrez, 2007).
Usability	Usability is the degree to which records can be easily and quickly located, retrieved, presented, and interpreted (Duranti, 2015).
Ease of use	Refers to the simplicity of accessing digital records (Ahmad et al., 2019).
Integration	This refers to the extent to which digital records from various departments and ministries can be shared using one access point (Waller *et al.,* 2014).
Metadata	This is information about the creator, unique identifiers, title of the records, description, date and time, keywords and index terms, format type and provisions for disposal, among other elements (MacNeil, 2000; McLeod & Gormly, 2017).
Flexibility	This is the ability of digital records management systems to adapt to newer technologies (Abu-Shanab et al., 2014).
Security	This is the protection of digital records from physical damage, external data breaches and internal theft or fraud. Security can be enhanced using passwords, Digital signatures, Data encryption, Access Control Lists (ACL), error detection mechanisms and robust backup systems (Magama, 2018).
Preservation	Preservation is the ongoing process of maintaining records in a manner that ensures their long-term accessibility, authenticity, and integrity. In digital contexts, it involves protecting electronic records from technological obsolescence, data loss, and alteration to ensure they remain trustworthy and usable over time (Magama, 2018).
Accessibility	This is the extent to which digital records are available to the public, considering the restrictions and IT capacities (Roberts & Hernandez, 2019).
Efficiency	Refers to cost-saving measures in terms of money, time, and effort when using digital records (Patsioura, 2014).
Correct	The quality of being free from error (Boudrez, 2007).
Traceability	The ability to identify and measure all the stages that led to a particular point in a process that consists of a chain of interrelated events (Duranti, 2015).
Truthfulness	The quality of being honest or not telling lies (McLeod & Gormly, 2017).
Records management processes	Refers to the systematic and professional practice of managing the records of an organization throughout its lifecycle. Management of records includes (registration, profiling, classification, indexing, preservation, security, appraisal, and disposal (Magama & Nduna 2020).
Timeliness	Refers to the degree to which digital records can be accessed without delay (Palvia and Sharma 2007).
Transparency	A process in which it is clear or traceable how records are managed and by whom (Duranti, 2015).

continued on following page

Table 1. Continued

Dimension	Description
Completeness	Refers to the adequateness or whether all the required information is included on the record (Zhou et al. 2019)
Proper infrastructure	Refers to adequate ICT equipment for digital records management (Mutsagondo 2017).
ICT Skills	These are the technological expertise required to manage digital records (Mutsagondo 2021).
EDRMS	This is an Electronic Document Records Management System used to manage digital records from creation to disposal (Magama & Nduna 2020).
Regulatory frameworks and legislation	Refers to guidelines, policies, and laws to guide the management of digital records (Franks, 2018).

Source: Authors compilation (2025)

Step 3. Identifying Constructs

After extracting and defining 24 dimensions from the literature and digital records management models, the researcher employed thematic analysis to examine the key meanings and concepts associated with each dimension (Maguire & Delahunt, 2017; Clarke & Braun, 2013). This process led to the development of a taxonomy that grouped the dimensions into four central constructs: authenticity, accuracy, reliability, and usability. Dimensions related to originality, genuineness, and integrity were classified under authenticity; those reflecting truthfulness, consistency, and completeness were assigned to accuracy; reliability captured evidential value, authority, and accountability; and usability encompassed accessibility, retrievability, and long-term use. These four constructs constitute the foundational pillars of the conceptual framework for enhancing the trustworthiness of digital records in Zimbabwe's public sector. Below are the identified constructs.

Authenticity of Digital Records

Authenticity refers to the quality of a digital record that confirms it is what it claims to be and that it can be reliably linked to its identified creator (Duranti, 2015). It implies that the record has not been altered in an unauthorized manner and that its creation and management processes uphold its integrity. Authenticity is a foundational component of trustworthiness, as it ensures that records can be relied upon as accurate representations of facts or actions. According to Duranti (1991), demonstrating the authenticity of a digital record involves verifying its continuing identity, a principle that enables the record to be distinguished from others. This includes establishing critical metadata such as the creator, date of creation, context, purpose, and version status, all of which contribute to determining whether the record is original, complete, and untampered.

Accuracy of Digital Records

Accuracy in digital records refers to their truthfulness, precision, and completeness, ensuring they faithfully represent the facts, events, or transactions they document (Duranti, 2010). It is shaped by the competence and integrity of the record's author, as well as by organizational and technical controls such as metadata standards, validation checks, and secure transmission processes (Duranti, 2015). These mechanisms reduce errors and safeguard the integrity of information throughout its lifecycle. Accurate records are indispensable for accountability, legal admissibility, decision-making, and institutional efficiency. Conversely, inaccuracies undermine trust, distort evidence, and weaken governance processes, making accuracy both a technical necessity and a cornerstone of credible digital recordkeeping.

Reliability of Digital Records

Reliability refers to the capacity of digital records to serve as authoritative evidence of activities and accurately reflect the facts they document (Rogers, 2015). It is achieved through systematic records management practices such as classification, indexing, registration, and appraisal, as well as the expertise of personnel involved (Duranti & MacNeil, 1997). Ensuring reliability involves safeguarding records against manipulation or falsification post-creation (Duranti & MacNeil, 1997). Key indicators include metadata that contextualizes the record, links to the creator or sender, and the use of digital signatures to affirm accountability (ICA, 2021).

Usability of Digital Records

Usability refers to how easily and efficiently digital records can be located, retrieved, and interpreted for both immediate and long-term purposes. An effective Electronic Document and Records Management System (EDRMS) should therefore not only enable quick search and access but also ensure records remain understandable and usable across different technological platforms and over time (Sigauke, 2022). Key features that enhance usability include security protocols, version control to track changes, intuitive classification schemes, metadata to provide context, and preservation strategies that protect records from technological obsolescence. When usability is maintained, records support transparency, accountability, and informed decision-making. However, if records are difficult to access, poorly organized, or incompatible with evolving systems, their trustworthiness is compromised, and the very objectives of digital adoption, efficiency, service delivery, and accountability are undermined.

Step 4. Mapping Constructs and Dimensions

The study mapped 24 dimensions extracted from the literature into four key constructs, authenticity, accuracy, reliability, and usability, to develop a conceptual model for digital records trustworthiness. This mapping was guided by thematic similarity, contextual alignment, and the ability of grouped dimensions to collectively articulate a construct. Table 1.2 presents the resulting construct-dimension alignment.

Table 2. The constructs and their related dimensions

Construct	Dimension
Authenticity	Metadata.
	Accountability.
	Traceability.
	Security of digital records (Use of passwords, Digital signatures, Data encryption, Access Control Lists (ACL) and Error detection mechanisms and robust backup systems).
	Long term preservation of digital records
	Transparency.
	Proper EDRMS infrastructure.
Accuracy	Correct
	Consistency.
	Completeness of data
	Training
	Truthfulness
Reliability	Adoption of EDRMS
	Digital records management processes (registration, profiling, classification, indexing, preservation, appraisal, and disposal).
	Training in digital records management skills.
Preservation	Proper storage, integrity checks, metadata management, and controlled access.
Regulatory compliance	Legislation, policies, and procedures.
Usability of records	Ease of use.
	Flexibility.
	Accessibility.
	Timeliness
	Efficiency.
	Integration.
	Proper ICT skills.

Source: Authors compilation (2025)

The conceptual model developed in this study is based on constructs and dimensions from the literature on digital records' trustworthiness. Guided by Lynham's (2002) principles of theory and model development, the model defines both its boundaries and the laws of interaction. The boundary is limited to the trustworthiness of digital records within Zimbabwe's public sector, recognizing that models simplify reality by focusing on specific contexts. The study also established the laws of interaction, which explain how the four constructs, authenticity, accuracy, reliability, and usability, interact with their dimensions to influence overall trustworthiness. As emphasized by Lynham (2002) and Holton and Lowe (2007), these interactions are critical for demonstrating how a conceptual model operates. In this framework, four laws of interaction were identified, each outlining the logical relationship between constructs and dimensions, providing a structured basis for strengthening trust in digital records management in the public sector. Below is an explanation of the laws of interaction.

Law of Interaction 1

The authenticity of digital records is strengthened through the preservation of their identity and integrity. This is achieved by capturing and maintaining comprehensive metadata, as well as ensuring accountability, traceability, and transparency throughout the records management lifecycle. These elements collectively support the verification of a record's origin, context, and unaltered status, which are essential indicators of authenticity. Figure 2 presents the law of interaction explaining how the construct of authenticity interacts with preservation, integrity, identity, metadata, traceability and transparency dimensions.

Figure 2. Law of interaction 1 (Author's compilation, 2025)

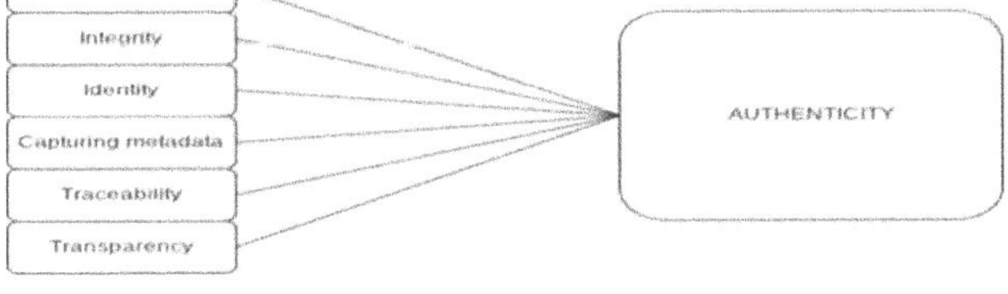

Law of Interaction 2

Accuracy refers to the truthfulness, exactness, precision, and completeness of a digital record (Duranti, 2010). It is primarily determined by the correctness of the record's content, the competence of its creator, and the effectiveness of controls governing its creation, recording, and transmission (Duranti, 2015). Ensuring accuracy involves implementing practices such as regular data backups to prevent loss or corruption and enforcing access controls to restrict editing rights to authorized personnel, thereby preserving the integrity of the record's content. Figure 3 presents the law of interaction explaining how the construct of accuracy interacts with correctness, truthfulness, consistency, completeness of digital data and security of records dimensions.

Figure 3. Law of interaction 2 (Author's compilation, 2025)

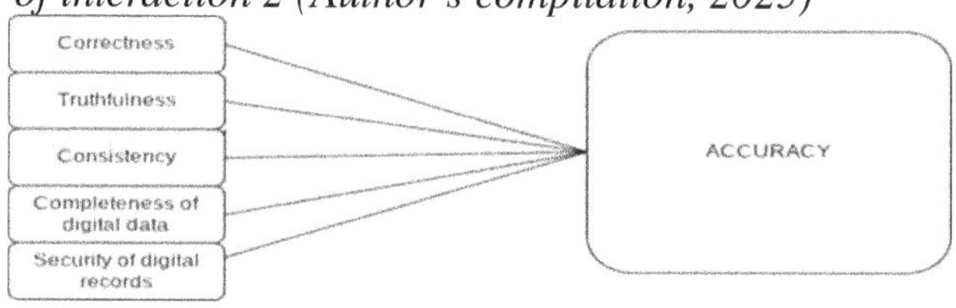

Law of Interaction 3

The reliability of digital records is reinforced through their completeness and the consistent application of formal records management procedures. This includes the adoption of Electronic Document and Records Management Systems (EDRMS) and the implementation of processes such as registration, profiling, classification, indexing, preservation, appraisal, and disposal. Additionally, the presence of skilled personnel in digital records management is essential to ensure that these procedures are effectively executed, thereby upholding the credibility and evidentiary value of the records. Figure 4 presents the law of interaction explaining how the construct of reliability interacts with completeness, adoption of EDRMS, records management procedures and records management skills dimensions.

Figure 4. Law of Interaction 3 (Author's compilation, 2025)

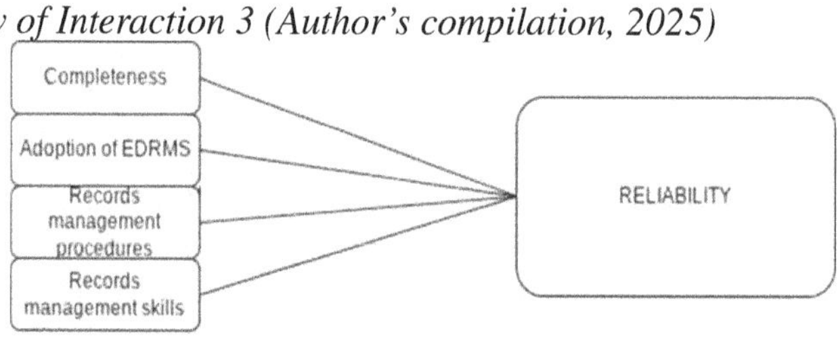

Law of Interaction 4

The usability of digital records is enhanced when the records management system is user-friendly, interactive, flexible, accessible, efficient, and capable of supporting timely retrieval and use. A system that facilitates intuitive navigation and rapid access to records not only improves operational efficiency but also ensures that records remain functional and meaningful over time. Figure 5 presents the law of interaction explaining how the construct of usability of digital records interacts with ease of use, efficiency, interactivity, flexibility and timeliness dimensions.

Figure 5. Law of interaction 4 (Author's compilation, 2025)

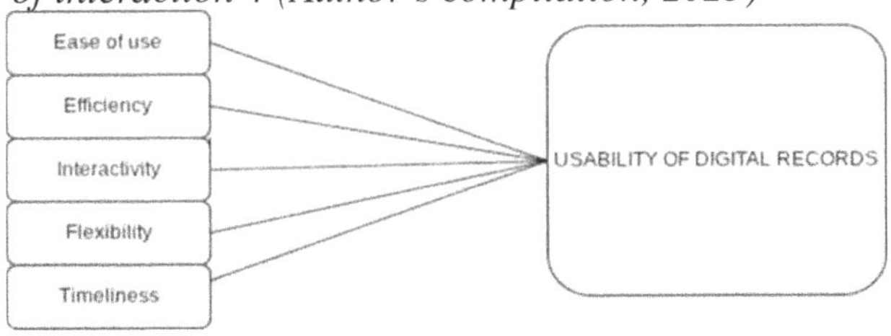

PROPOSED CONCEPTUAL MODEL FOR THE TRUSTWORTHINESS OF PUBLIC SECTOR DIGITAL RECORDS

Figure 6 presents the proposed conceptual model for the trustworthiness of digital records, structured according to the laws of interaction.

Figure 6. Conceptual model for the trustworthiness of digital records (Author's compilation, 2025)

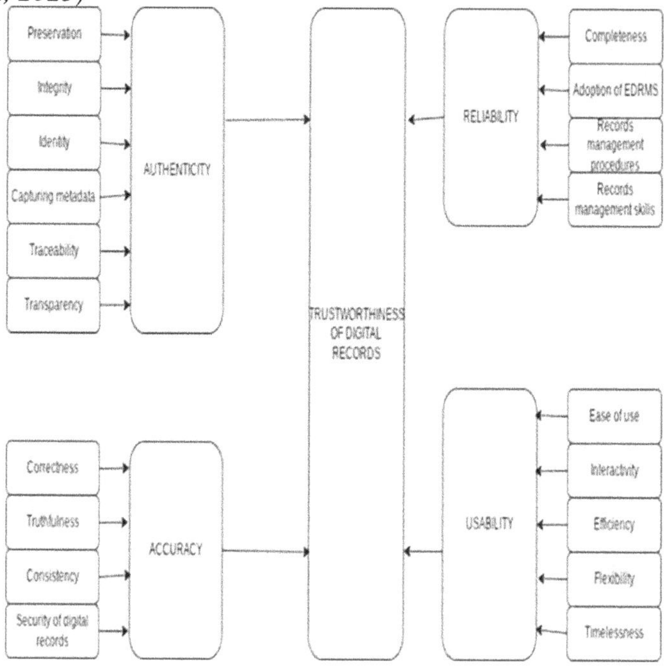

Figure 6 illustrates a conceptual framework for the trustworthiness of digital records in Zimbabwe's public sector, initially developed from a literature review. In this framework, the constructs of authenticity, accuracy, reliability, and usability function as independent variables. Each of these constructs exerts a direct influence on the overall trustworthiness of digital records within Zimbabwe's public sector. The model posits that the degree to which digital records can be considered trustworthy is contingent upon the collective strength and proper implementation of all four constructs. This interdependent relationship underscores the need for a holistic approach in developing and managing digital records systems to ensure they are credible, functional, and legally admissible. Since the proposed framework in Figure 6 lacked contextual grounding, the framework was validated using the Delphi method. Expert input enhanced its relevance and applicability, resulting in a refined, context-sensitive model presented after validation.

FRAMEWORK VALIDATION: THE DELPHI METHOD

Model validation is essential to ensure alignment with real-world conditions (Ling & Mahadevan, 2013). In this study, the initial framework derived from literature was refined through expert feedback using the Delphi method. Experts independently reviewed its strengths and weaknesses, and their input, verified against existing literature, guided modifications to improve its contextual relevance and applicability to digital records trustworthiness in Zimbabwe's public sector.

Responses on Accuracy

Experts agreed that the framework strengthens digital record accuracy by ensuring truthfulness, completeness, and protection from unauthorized changes. Key features such as digital signatures, encryption, backups, and error-detection align with international standards (ISO 15489-1, 2016; InterPARES, 2007) and scholarly work (Duranti, 2010, 2015). Its policies against post-creation alterations and integration of EDRMS functionalities were also praised, echoing findings from studies in African contexts (Bwalya, 2009; Nengomasha & Chikomba, 2018; Sigauke, 2022; Sibanda & Schellnack-Kelly, 2024).

Responses on Reliability

Experts confirmed that the proposed framework enhances the reliability of digital records by detailing procedures for creation, receipt, control, appraisal, and disposal, and by integrating skilled personnel with systems like BIS and ERMS. These features are supported by ISO 15489-1 (2016), which highlights the importance of structured records management processes, and by studies in African contexts emphasizing classification, indexing, metadata capture, and preservation as critical for safeguarding record integrity (Mosweu & Ngoepe, 2020; Sigauke, 2022; Sibanda & Schellnack-Kelly, 2024)

Response on the Authenticity of Records

Experts emphasized that authenticity is vital for digital records' trustworthiness and agreed that the proposed framework supports this effectively. They highlighted features such as date stamps, letterheads, digital signatures, and metadata, which link records to their creators and reduce risks of manipulation or forgery. The inclusion of Commissioners of Oaths was recommended to verify authenticity via authorized third parties, while One Time Password (OTP) notifications were suggested to prevent impersonation and ensure access by authorized personnel. These recommendations are consistent with ISO 15489-1 (2016) and InterPARES (2007), which emphasize the importance of technical and procedural controls in safeguarding record authenticity. Overall, the framework incorporates essential mechanisms supported by international standards and scholarly research to strengthen the authenticity of public sector digital records.

Response on Usability

All experts unanimously agreed that the proposed framework effectively enhances the usability of digital records, defined as the ease and speed of locating, retrieving, presenting, and interpreting records. Experts suggested several improvements to strengthen usability, including adopting standardized metadata schemas (e.g., Dublin Core, MODS), robust search functions (full-text, faceted search), user-friendly interfaces, multiple download options, browsing previews, and integrated help resources. Additional recommendations included ensuring accessibility for diverse users, incorporating feedback and analytics, integrating with other systems, implementing version control and collaborative tools, allowing customization, and maintaining continuous updates to keep the framework current. These recommendations are supported by ISO 15489-1 (2016), which underscores accessibility, retrievability, and usability as key components of trustworthy records, and by Duranti (2010), who highlights the importance of structured metadata and system integration for effective digital records management. The researcher incorporated these expert recommendations to refine the framework.

IDENTIFIED WEAKNESSES AND AREAS FOR IMPROVEMENT IN THE PROPOSED FRAMEWORK

Experts highlighted key weaknesses in the proposed framework, including the lack of version control, certification and validation mechanisms, non-repudiation features, and alignment with standards and local laws. They also noted gaps in transparency, defined roles, regular audits, training, and funding. These evidence-based recommendations, supported by ISO 15489-1 (2016), ISO 27001 (2013), Duranti (2010), and InterPARES (2007), emphasize the need to revise and strengthen the framework for Zimbabwe's public sector.

JUSTIFICATION FOR FRAMEWORK REDESIGN

The framework was redesigned based on expert feedback and cross-checked with literature and international standards to enhance its comprehensiveness. Key improvements included version control (ISO 15489-1; Duranti, 2010), certification and validation of records (InterPARES, 2007), secure data recovery and backups (ISO 27001, 2013), non-repudiation mechanisms like digital signatures and audit logs, and clearer transparency and role definitions (Sibanda & Schellnack-Kelly, 2024). Continuous monitoring, user training (Mosweu & Ngoepe, 2020; Sithole, 2022), and adequate funding (Masuku et al., 2017) were also incorporated. These evidence-based enhancements made the framework a robust, contextually relevant tool for managing trustworthy digital records.

FRAMEWORK REDESIGN: PUBLIC SECTOR DIGITAL TRUSTWORTHINESS FRAMEWORK

The redesigned framework, validated by experts, strengthens digital record trustworthiness in Zimbabwe's public sector by combining technical controls and institutional measures to ensure accuracy, reliability, authenticity, and usability (ISO 15489-1, 2016; InterPARES, 2007). Figure 7 below presents this Public Sector Digital Trustworthiness Framework.

Figure 7. Public Sector Digital Trustworthiness Framework (Author's compilation, 2025)

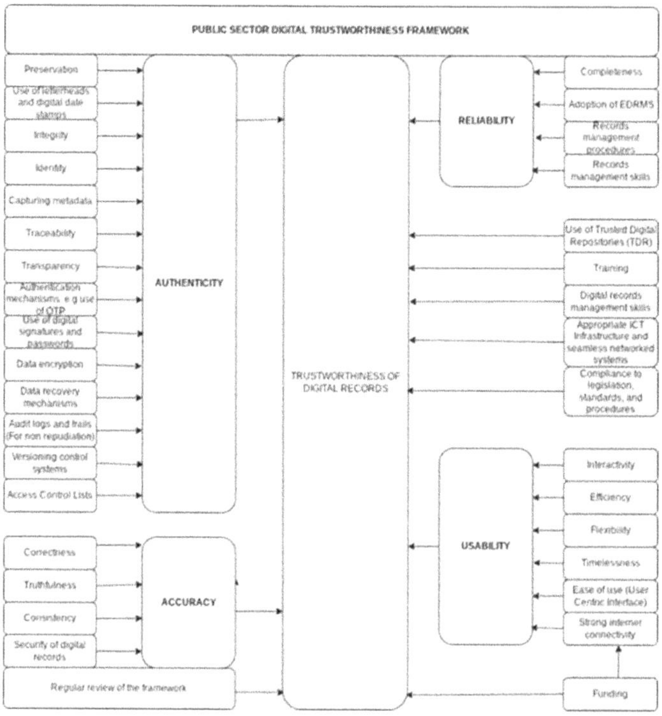

The redesigned framework represents a significant advancement over existing generic models by explicitly addressing the trustworthiness of digital records within Zimbabwe's public sector, a context characterized by constrained financial and technological resources, fragmented information systems, and a shortage of skilled personnel (Sigauke, 2022; Magama & Nduna, 2020). While established international frameworks, including VERS (Leviston, 1999), ISO 15489 (INTERNATIONAL ORGANIZATION FOR STANDARDIZATION, 2016), MoReq (Cain, 2002), and InterPARES (Duranti, 2015; InterPARES, 2007), provide comprehensive guidelines for records management, they assume well-resourced environments with integrated systems and personnel possessing advanced ICT and records management competencies. In Zimbabwe, public sector organizations often deploy Business Information Systems without adequate integration of records management, resulting in inconsistent record-keeping, diminished accountability, and vulnerabilities in the authenticity, reliability, and accessibility of digital records (Chikomba et al., 2021; Sigauke, 2022; Sithole, 2022).

A distinguishing feature of this framework is its deliberate emphasis on human capacity development, particularly the training and competency of records personnel, an aspect often underrepresented in prior models (Ambira, 2016; Mosweu & Ngoepe, 2020; Sithole, 2022). Recognizing the persistent ICT skills gap among public sector staff, the framework integrates targeted capacity-building initiatives, structured training programs, and continuous professional development strategies, designed to equip personnel with the knowledge and practical skills required to manage digital records effectively and uphold standards of trustworthiness. Complementing the human dimension, the framework embeds mechanisms for policy enforcement, workflow standardization, and procedural compliance, ensuring organizational practices align with local regulatory requirements and international best practices, even in resource-constrained environments (Sibanda & Schellnack-Kelly, 2024; Sigauke, 2022).

The framework also adopts a holistic approach to technological integration, providing actionable guidance for aligning digital records systems with operational processes, addressing system fragmentation, and ensuring that electronic records remain authentic, reliable, and usable throughout their lifecycle (INTERNATIONAL ORGANIZATION FOR STANDARDIZATION, 2016; InterPARES, 2007; Duranti, 2010). By concurrently addressing human and technological factors, the framework offers a context-sensitive, sustainable, and adaptable model capable of enhancing records governance, promoting accountability, and safeguarding the integrity of public sector digital records in Zimbabwe. Beyond compliance, it provides a strategic roadmap for fostering a culture of record reliability, transparency, and long-term preservation, thereby contributing to more effective public sector administration in challenging operational contexts (Baheer, Lamas & Sousa, 2020; Mahlangu, 2020).

IMPLICATIONS FOR THEORY DEVELOPMENT

The proposed framework contributes significantly to theory development in computing and informatics by extending existing records management and digital preservation theories to resource-constrained contexts with ICT skills gaps (Ambira, 2016; Mosweu & Ngoepe, 2020; Sibanda & Schellnack-Kelly, 2024; Sigauke, 2022). Unlike traditional frameworks, which assume stable infrastructure and skilled personnel (INTERNATIONAL ORGANIZATION FOR STANDARDIZATION, 2016; Leviston, 1999; Cain, 2002), this model integrates human, organizational, and technological dimensions, enhancing theoretical understandings of digital trustworthiness (Duranti, 2010; InterPARES, 2007). By emphasizing capacity building, policy enforcement, and socio-technical alignment, it encourages scholars to incorporate contextual realities into future models (Ngulube, 2020a; Sibanda & Schellnack-Kelly, 2024). Overall, this holistic perspective bridges the gap between normative standards and practical challenges in emerging economies, promoting more inclusive, adaptable, and context-sensitive conceptual frameworks.

IMPLICATIONS FOR PRACTICE

Practically, the framework provides Zimbabwean public sector organizations and similar contexts with an actionable guide for improving digital records management, integrating technical and human-centric measures (INTERNATIONAL ORGANIZATION FOR STANDARDIZATION, 2016; InterPARES, 2007; Magama & Nduna, 2020). Technical safeguards include metadata standards, version control, and security protocols, while human-focused strategies emphasize training, governance, and accountability (Ambira, 2016; Mosweu & Ngoepe, 2020; Sigauke, 2022). Continuous monitoring, auditing, and updates enhance adaptability and resilience, allowing organizations to respond effectively to evolving technological and regulatory requirements (Sibanda & Schellnack-Kelly, 2024). The framework supports transparency, efficiency, and good governance, thereby improving service delivery and compliance in resource-constrained public sector environments.

IMPLICATIONS FOR RESEARCH METHODOLOGY

From a research methodology perspective, the framework demonstrates a rigorous qualitative approach suitable for complex informatics research (Torraco, 2016; Kraus, Breier & Dasi-Rodriguez, 2020; Rocco & Plakhotnik, 2009). By triangulating literature review, conceptual analysis, and Delphi validation, the study combined multiple data sources to develop a context-sensitive model, enhancing validity and relevance in under-studied environments such as Zimbabwe's public sector (Ngulube, 2020b; Mahlangu, 2020; Sibanda & Schellnack-Kelly, 2024). Expert validation and contextual adaptation further highlight the importance of stakeholder engagement in framework design, providing a methodological precedent for iterative, participatory approaches in applied computing and informatics research (Holton & Lowe, 2007; Eriksson, Johannesson & Bergholtz, 2018).

FUTURE RESEARCH DIRECTIONS

The development of the Public Sector Digital Trustworthiness Framework opens several avenues for further research. Future studies could empirically test the framework's effectiveness across multiple public sector institutions in Zimbabwe to evaluate its impact on digital records accuracy, reliability, authenticity, and usability. Comparative studies could explore its adaptability in other resource-constrained contexts within Africa, assessing how socio-technical and organizational factors influence implementation. Researchers could also investigate the integration of emerging technologies, such as blockchain, artificial intelligence, and cloud-based solutions, to enhance record verification, security, and automated auditing. Additionally, longitudinal studies examining the role of continuous training, policy enforcement, and stakeholder engagement could provide insights into sustaining trustworthiness in dynamic digital environments. Finally, further theoretical work could refine the socio-technical dimensions of digital records management frameworks, bridging gaps between normative standards and practical challenges in emerging economies.

CONCLUSION

This chapter discussed the development and validation of a contextually relevant Public Sector Digital Trustworthiness Framework tailored for Zimbabwe's unique challenges. Through a rigorous process involving literature review and expert validation, the framework was refined to encompass the critical dimensions of accuracy, reliability, authenticity, and usability. Importantly, the redesign incorporated expert feedback to address gaps such as version control, certification, non-repudiation, transparency, continuous improvement, training, and funding, elements essential for enhancing the trustworthiness of digital records in resource-constrained environments. By bridging global standards with local realities, the framework offers both theoretical advancement and practical utility, providing a sustainable model to improve digital records management in Zimbabwe's public sector. The integration of technological, human, and organizational factors ensures that the framework is not only comprehensive but also adaptable to evolving needs. This foundation sets the stage for future research and practical application aimed at strengthening governance, accountability, and service delivery through trustworthy digital records.

REFERENCES

Abu-Shanab, E., & Khasawneh, R. (2014). E-government adoption: The challenge of digital divide based on Jordanians' perceptions. *Theoretical and Empirical Researches in Urban Management, 9*(4), 5–19.

Ahmad, K. M., Campbell, J., Pathak, R. D., Belwal, R., Singh, G., Naz, R., Smith, R. F., Al-Zoubi, K., Carter, L., Bélanger, F., Lim, A. L., Masrom, M., Din, S., Chen, Y. C., Dimitrova, D. V., Dodeen, W., Adolph, A., Bortier, S., & Munyoka, W. (2019). Satisfaction with e-participation: A model from the citizens' perspective, expectations, and effective ties to the place. *African Journal of Business Management, 7*(1), 157–166.

Albar, H. M., Dahlan, H. A., Ahmad, A., Yuhefizar, E., & Napitupulu, D. (2017). E-government service quality based on e-GovQual approach: Case study in West Sumatera Province. *International Journal on Advanced Science, Engineering and Information Technology, 7*(6), 2337–2342. DOI: 10.18517/ijaseit.7.6.4226

Ambira, C. M. (2016). A framework for management of electronic records in support of e-government in Kenya. PhD thesis, University of South Africa, Pretoria.

Ambira, C. M., Kemoni, H., & Ngulube, P. (2019). A framework for electronic records management in support of e-government in Kenya. *Records Management Journal, 29*(3), 305–319. DOI: 10.1108/RMJ-03-2018-0006

Baheer, B. A., Lamas, D., & Sousa, S. (2020). A systematic literature review on existing digital government architectures: State-of-the-art, challenges, and prospects. *Administrative Sciences, 10*(2), 1–28. DOI: 10.3390/admsci10020025

Bhebhe, S. (2015). Contemporary diplomatics of the civil and deceased estate case files found at the national archives of Zimbabwe. *Records Management Journal, 25*(1), 107–120. DOI: 10.1108/RMJ-03-2014-0019

Boudrez, F. (2007). *Digital signatures and electronic records.* Springer Science: Business Media.

Bwalya, K. J. (2009). Factors affecting adoption of e-government in Zambia. *The Electronic Journal on Information Systems in Developing Countries, 38*(1), 1–13. DOI: 10.1002/j.1681-4835.2009.tb00267.x

Cain, P. (2002). Model Requirements for the Management of Electronic Records (MoReq): A critical evaluation. *Records Management Journal, 12*(1), 14–18. DOI: 10.1108/09565690210427752

Chikomba, A., Rodrigues, A., & Ngoepe, M. (2021). Management of digital records in selected financial services parastatals in Zimbabwe. *Information Development, 37*(2), 245–257. DOI: 10.1177/0266666920906858

Clarke, V., & Braun, V. (2013). Teaching thematic analysis: Overcoming challenges and developing strategies for effective learning. *The Psychologist, 26*(2), 120–123.

Collis, J., & Hussey, R. (2014). *Business research: A practical guide for undergraduate and postgraduate students*. Palgrave. DOI: 10.1007/978-1-137-03748-0

Duranti, L. (1991). Diplomatics. New users for an old science (Part V). *Archivaria, 3*, 6–24.

Duranti, L. (2010). From digital diplomatics to digital records forensics. *Archivaria, 68*, 39–66.

Duranti, L. (2015). Overview of InterPARES Trust-Work to date. Third International Symposium of InterPARES Trust. Retrieved on 12 June 2025 from https://interparestrust.org/system/files/publications/files/Basis_for_Conclusions_ISA_220_Redrafted_andISQC_1_Redrafted_0.pdf

Duranti, L., & MacNeil, H. (1997). The protection of the integrity of electronic records: An overview of the UBC-MAS research project. *Archivaria, 42*, 46–67.

Elangovan, R., & Rajendran, R. (2015). Conceptual model: A framework for institutionalising the vigor in business research. *Proceedings of the 3rd National on Indian Business Management*, 1–35.

Eriksson, O., Johannesson, P., & Bergholtz, M. (2018). Institutional ontology for conceptual modelling. *Journal of Information Technology, 33*(2), 105–123. DOI: 10.1057/s41265-018-0053-2

Franks, P. C. (2018). *Records and information management*. American Library Association.

Hagen-Sataslaatten, O. (2014). The Norwegian Noark model requirements for EDRMS in the context of open government and access to governmental information. *Records Management Journal, 24*(3), 189–204. DOI: 10.1108/RMJ-09-2014-0041

Holton, E. F., & Lowe, J. S. (2007). Toward a general research process for using Dublin's theory building model. *Human Resource Development Review, 6*(3), 297–320. DOI: 10.1177/1534484307304219

Hunie, P., & Dewah, P. (2019). Admissibility of digital records as evidence in Bulawayo High Court in Zimbabwe. *Journal of the South African Society of African Archivists, 52*, 133–148.

Imenda, S. (2014). Is there a conceptual difference between theoretical and conceptual frameworks? *Journal of Social Sciences*, *38*(2), 185–195. DOI: 10.1080/09718923.2014.11893249

International Council on Archives. (2021). *Guidelines for digital recordkeeping*. ICA. Retrieved on 12 June 2025 from https://www.ica.org

International Organization for Standardization. (2013). *ISO/IEC 27001:2013 Information technology — Security techniques — Information security management systems — Requirements*. ISO. Retrieved on 12 June 2025 from https://www.iso.org/standard/54534.html

International Organization for Standardization. (2016). *ISO 15489-1:2016 Information and documentation — Records management — Part 1: Concepts and principles*. ISO. Retrieved on 12 June 2025 from https://www.iso.org/standard/62542.html

InterPARES Authenticity Taskforce. (2007). Requirements for assessing and maintaining the authenticity of electronic records. In *The long-term preservation of authentic electronic records: Findings of the InterPARES project*. Vancouver. Retrieved on 12 June 2025 from https://interpares.org/book/interpares_book_d_part1.pdf

Jabareen, Y. (2009). Building a conceptual framework: Philosophy, definitions, and procedure. *International Journal of Qualitative Methods*, *8*(4), 49–62. DOI: 10.1177/160940690900800406

Kraus, S., Breier, M., & Dasi-Rodriguez, S. (2020). The art of crafting a systematic literature review in entrepreneurship research. *The International Entrepreneurship and Management Journal*, *16*(3), 1023–1043. DOI: 10.1007/s11365-020-00635-4

Landeta, J. (2006). Current validity of the Delphi method in social sciences. *Technological Forecasting and Social Change*, *73*(5), 467–482. DOI: 10.1016/j.techfore.2005.09.002

Leviston, T. (1999). Victorian electronic records strategy final report. *Archives and Manuscripts*, *27*(2), 92–94.

Ling, Y., & Mahadevan, S. (2013). Quantitative model validation techniques: New insights. *Reliability Engineering & System Safety*, *111*, 217–231. DOI: 10.1016/j.ress.2012.11.011

Lynham, S. A. (2002). Theory building: Dubin's method. *Advances in Developing Human Resources*, *4*(3), 242–276. DOI: 10.1177/15222302004003003

MacNeil, H. (2000). *Trusting records. Legal, historical and diplomatic perspectives*. Kluwer.

Magama, B. (2018). Strategies for the preservation of digital records in Masvingo province of Zimbabwe. Master's Thesis, University of South Africa, Pretoria.

Magama, B., & Nduna, V. (2020). Journey to EDRMS solution in Zimbabwe public sector. In Keakopa, S. M., & Mosweu, T. L. (Eds.), *Cases on electronic record management in the ESARBICA Region* (pp. 142–159). IGI Global., DOI: 10.4018/978-1-7998-2527-2.ch007

Maguire, M., & Delahunt, B. (2017). Doing a thematic analysis: A practical, step-by-step guide for learning and teaching scholars. *AISHE-J, 9*, 3351. https://ojs.aishe.org/index.php/aishe-j/article/view/3354

Mahlangu, G. (2020). *A multi-dimensional model for assessing e-government service gaps in the context of a developing country: A critical realist perspective*. Cape Peninsula University of Technology.

Masuku, M., Bhebhe, S., & Khumalo, N. B. (2017). An information management framework for the support of e-government in Zimbabwe. *Tidsskriftet Arkiv, 8*(1), 1–14. DOI: 10.7577/ta.1960

Maxwell, J. A. (2013). Qualitative research design: An interactive approach. *Sage (Atlanta, Ga.)*.

Mcleod, J., & Gormly, B. (2017). Using the cloud for records storage: Issues of trust. *Archival Science, 17*(4), 349–370. DOI: 10.1007/s10502-017-9280-5

Mosweu, O., & Ngoepe, M. (2020). Skills and competencies for authenticating digital records to support auditing process in Botswana public sector. *African Journal of Library and Information Science, 29*(1), 17–28.

Mutsagondo, S. (2017). Electronic records management in public departments. Master's thesis, University of South Africa, Pretoria.

Mutsagondo, S. (2021). Use and management of the electronic mail in the central government of Zimbabwe. DPhil thesis, University of South Africa, Pretoria.

NARA (US National Archives and Records Administration). (2005), NARA Guidance on Managing Web Records. Retrieved on 10 October 2025 from www.archives.gov/recordsmanagement/policy/managing-webrecords-index.html

Nengomasha, C. T., & Chikomba, A. (2018). Status of EDRMS implementation in the public sector in Namibia and Zimbabwe. *Records Management Journal, 28*(3), 252–264. DOI: 10.1108/RMJ-08-2017-0023

Ngulube, P. (2018). Overcoming the difficulties associated with conceptual and theoretical frameworks in heritage studies. In Ngulube, P. (Ed.), *Handbook of research in heritage management and preservation* (pp. 1–25). IGI Global. DOI: 10.4018/978-1-5225-3137-1.ch001

Ngulube, P. (2020a). Theory and theorising in information science scholarship. In Ngulube, P. (Ed.), *Handbook of research on connecting research methods for information science research* (pp. 18–39). IGI Global., DOI: 10.4018/978-1-7998-1471-9.ch002

Ngulube, P. (2020b). The movement of mixed methods research and the role of information science professionals. In Ngulube, P. (Ed.), *Handbook of research on connecting research methods for information science research* (pp. 425–455). IGI Global., DOI: 10.4018/978-1-7998-1471-9.ch022

Ngulube, P. (2025). Indigenous theories and theorisation in African scholarship in the quest for epistemic freedom, P. Ngulube (Ed.), *Harnessing Indigenous epistemologies for sustainable progress* (pp. 139-172). IGI Global. DOI: 10.4018/979-8-3693-8447-3.ch008

Ngulube, P., & Mosha, N. F. (2023). Transparency in the use of theoretical frameworks for the advancement of knowledge in core library and information science journals: A systematic review. *African Journal of Library Archives and Information Science, 33*(2), 129–146. DOI: 10.4314/ajlais.v33i2.2

Palvia, S.C.J. & Sharma, S.S. (2007). E-Government and e-governance: definitions/domain framework and status around the world. *Foundations of E-Government*, 1–12.

Patsioura, F. (2014). Evaluating e-government. In Yannacopoulos, D., Manolitzas, P., Matsatsinis, N., & Grigoroudis, E. (Eds.), *Evaluating websites and web services: interdisciplinary perspectives on user satisfaction* (pp. 1–18). IGI Global. DOI: 10.4018/978-1-4666-5129-6.ch001

Polites, G. L., Roberts, N., & Thatcher, J. (2012). Conceptualizing models using multidimensional constructs: A review and guidelines for their use. *European Journal of Information Systems, 21*(1), 22–48. DOI: 10.1057/ejis.2011.10

Roberts, T., & Hernandez, K. (2019). Digital access is not binary: The 5 "A"s of technology access in the Philippines. *The Electronic Journal on Information Systems in Developing Countries, 85*(4), 1–14. DOI: 10.1002/isd2.12084

Rocco, S. T., & Plakhotnik, M. S. (2009). Literature reviews, conceptual frameworks, and theoretical frameworks: Terms, functions and distinctions. *Human Resource Development Review, 8*(1), 120–130. DOI: 10.1177/1534484309332617

Rogers, C. (2015). *Virtual authenticity: Authenticity of digital records from theory to practice (Electronic theses and dissertations).* University of British Columbia.

Saunders, M., Lewis, P., & Thornhill, A. (2018). *Research methods for business students* (8th ed.). Pearson.

Sibanda, M., & Schellnack-Kelly, I. (2024). Trustworthiness of digital records in the public sector of Zimbabwe. *ESARBICA Journal, 43*, 124–146.

Sigauke, O. (2022). Digital records management practices in the public sector of Manicaland province of Zimbabwe. *ESARBICA Journal, 41*, 172–187. DOI: 10.4314/esarjo.v41i.12

Sithole, M. W. (2022). Electronic records management model for improving e-government service delivery in Zimbabwe. Master's thesis, Midlands State University.

Torraco, R. J. (2016). Writing integrative literature reviews: Using the past and present to explore the future. *Human Resource Development Review, 15*(4), 404–428. DOI: 10.1177/1534484316671606

Waller, P., Irani, Z., Lee, H., & Weerakkody, V. (2014). Lessons on measuring e-government satisfaction: An experience from surveying government agencies in the UK. *International Journal of Electronic Government Research, 10*(3), 37–46. DOI: 10.4018/ijegr.2014070103

Zhou, R., Wang, X., Shi, Y., Zhang, R., Zhang, L., & Guo, H. (2019). Measuring e-service quality and its importance to customer satisfaction and loyalty: An empirical study in a telecom setting. *Electronic Commerce Research, 19*(3), 477–499. DOI: 10.1007/s10660-018-9301-3

ADDITIONAL READING

Abu Bakar, N., Mohd Zahibi, M. F. H., Marzuki, N. F. N., Hamidon, H., & Mohd Affandi, N. A. (2025). Theoretical framework for improving the records management of retirees in public sector. *Journal of Information and Knowledge Management (JIKM), 15*(SI), 118-130.

Darji, D. A., Pancholi, P. D., Patel, S. B., Patel, S. J., & Patel, J. N. (2025). Theoretical and conceptual frameworks for understanding artificial intelligence in records and information management. In *Artificial Intelligence in Records and Information Management* (pp. 181–212). IGI Global Scientific Publishing. DOI: 10.4018/979-8-3693-9795-4.ch007

Duranti, L., Rogers, C., & Thibodeau, K. (2022). Authenticity. *Archives and Records (Abingdon, England)*, *43*(2), 188–203. DOI: 10.1080/23257962.2022.2054406

Farooq, R. (2019). Developing a conceptual framework of knowledge management. *International Journal of Innovation Science*, *11*(1), 139–160. DOI: 10.1108/IJIS-07-2018-0068

Kamatula, G., & Kemoni, H. (2018). A framework for e-records in support of e-government implementation in the Tanzanian public service. *ESARBICA Journal: Journal of the Eastern and Southern Africa Regional Branch of the International Council on Archives*, *37*, 69–88.

Kemoni, H. N. (2008). Theoretical framework and literature review in graduate records management research. *African Journal of Library Archives and Information Science*, *18*(2), 103–117.

Poshai, L., & Shava, E. (2025). Digitalization of public sector institutions in zimbabwe: Towards transparent, accountable and sustainable public service delivery. *OIDA International Journal of Sustainable Development*, *18*(11), 383–394.

Semantha, F. H., Azam, S., Shanmugam, B., Yeo, K. C., & Beeravolu, A. R. (2021). A conceptual framework to ensure privacy in patient record management system. *IEEE Access : Practical Innovations, Open Solutions*, *9*, 165667–165689. DOI: 10.1109/ACCESS.2021.3134873

Tsvuura, G. (2022). Knowledge and skills for managing digital records at selected state universities in Zimbabwe. *Journal of the South African Society of Archivists*, *55*, 110–123. DOI: 10.4314/jsasa.v55i.8

Tsvuura, G., & Ngulube, P. (2021). A framework for the digitisation of records and archives at selected state universities in Zimbabwe. *South African Journal of Information Management*, *23*(1), 1–9. DOI: 10.4102/sajim.v23i1.1312

KEY TERMS AND DEFINITIONS

Authenticity: is the quality of a digital record that confirms it is what it claims to be, was created, or sent by the stated source, and has not been altered or tampered with after creation.

Conceptual Framework: is a structured, theoretical model that outlines the key variables, constructs, and relationships needed to explain or address a specific phenomenon, in this case, the factors that contribute to the trustworthiness of digital records in Zimbabwe's public sector.

Digital RECORDs: are information created, received, and maintained in electronic form as evidence of transactions, decisions, or activities. They may include emails, databases, scanned documents, and files generated through business information systems (BIS).

Reliability: is the quality of a digital record that ensures it can be trusted as a full and accurate representation of the activities or facts it documents, and that it has been maintained in a dependable system over time.

Trustworthiness: refers to the degree to which digital records are considered reliable, authentic, accurate, and usable over time, such that they can serve as dependable evidence for administrative, legal, and historical purposes.

Public Sector: comprises government departments, parastatals, and public institutions responsible for delivering services and implementing policies using public resources. In Zimbabwe, this includes ministries, local authorities, and state-owned enterprises.

Usability: refers to the ease and effectiveness with which digital records can be located, retrieved, read, interpreted, and used by authorized users for decision-making, accountability, or service delivery.

Chapter 11
Data Analysis and Effective Presentation Methods

Santosh Ramkrishna Durugkar
http://orcid.org/0000-0002-5079-2224
Independent Researcher, India

Atul B. Jondhale
http://orcid.org/0000-0001-6756-7026
Pravara Rural Engineering College, Loni, India

ABSTRACT

This chapter emphasizes on different methods analyzing the data and presents the results in effective way. These methods also enable the users to focus on the required variables and visualize the results in effective manner. Data collection (integration) is the primary task and data can be collected from multiple heterogeneous sources. Therefore, once the data is collected, it requires application of the 'data cleaning' step. In the initial phase, researchers must divide the given variables into 'dependent' and 'independent' variables. Once the 'data cleaning' task completes next is to work on 'qualitative', 'quantitative', 'descriptive', 'predictive' and 'prescriptive' analysis. There are many sectors where data analysis plays very important role like healthcare, space, education, manufacturing, finance, supply chain etc. Integrating data from multiple sources is very critical task and has to face few challenges like highly configured machines are required, skillful human resource and most important powerful algorithms.

DOI: 10.4018/979-8-2600-0434-0.ch011

1. INTRODUCTION

It is crucial to present integrated data (data retrieved from multiple sources) in a clear and meaningful manner. Researchers and academics are continually working to present the processed data, aiming to gain better insights. There are various benefits to analysing data and presenting it in an abstract, meaningful form, such as gaining better insights, making more informed decisions, and facilitating continuous improvement in various applications, including businesses, healthcare, and education. Due to the rapidly increasing number of internet users, and data-driven applications 'data size' is also increasing, which is hard for human beings to process in less time. Being human, it is quite a challenging task to process such large amounts of data within a short time with great accuracy. Therefore, effective algorithms and methods must be used for data analysis (Tang et al., 2023). and processed data must be represented in an effective manner. 'Visualization' (i.e., charts, figures, tables, etc.) plays a very important role in presenting the processed and required data for better understanding. There are many tools available to present the required data effectively, such as MS Excel, Power BI, Tableau, SPSS and other statistical data processing tools. These tools help represent data in an abstract form using pie charts, line charts, column charts, and bar charts. With the help of the 'Data Analysis Tool Pack' in MS Excel, one can execute various options, such as ANOVA, Covariance, Correlation, P-test, F-test, regression, sampling, and histograms.

'Exploration' is the key idea in data analysis (Abughazala & Muccini, 2023), as it enables the identification of correlations between two or more variables. Exploring the data and identifying correlations is beneficial for businesses to improve weaker areas, help retain employees, and maximise profits, among other benefits. Researchers are working on various topics, such as 'climate change', 'rainfall statistics', and 'population growth', where analysing data over a specific time period is crucial. With the help of historical data from the specific period (time series data), one can develop a 'time series-based model' (Xindong, 2024) for more accurate future predictions. However, recent machine learning (ML) and artificial intelligence (AI) techniques and methods are very helpful for researchers to process large volumes of data in less time with better accuracy.

There are many ML applications developed for various industries, including businesses, healthcare, manufacturing, space, communication technologies, education, and transportation. ML applications are very popular, and ML methods like regression (Masteali et al., 2025), clustering, decision trees, neural networks, and support vector machines (SVMs), supervised learning, reinforcement learning, etc., are widely used to analyse the data. This chapter emphasises the various tests and methods (Longpre et al., 2023) to present the data for better visualisation. One can use these methods and tests as needed. However, any specific test or method has its own 'strengths' and 'weaknesses'. Therefore, it is suggested to apply the tests and methods wisely for effective visualisation. In every organisation, processing the 'data' and effective presentation are very important for many reasons, such as to gain better clarification and understanding. Instead of 'numbers (data)', it is a good way to engage the audience with the help of better visualisation and presentation, and help stakeholders in better decision-making.

The chapter is organised into different sections, including the first section, which provides an introduction to the applications of data analysis, a list of data analysis tools, and data analytics approaches. The second section focuses on variables used in research methodology, i.e., dependent variables, independent variables, and controlled variables. A major and important part of the chapter is the third section, which elaborates on the different tests/methods that can be used to analyse the data, such as ANOVA, ANOVA with single and two-factor summaries, correlation, co-variance, t-test, histogram, descriptive statistics, sampling, moving averages, and F-test with input data and step-by-step executions. The fourth section introduces recent techniques to analyze and represent the data with classification, regression, clustering, and time series analysis. The fifth section presents the data using various charts, including 2D column charts, line charts, pie charts, 3D clustered bar charts, radar charts, and doughnut charts.

Applications of Data Analysis

- Managing the businesses
- Improving marketing strategies
- Enhancing healthcare facilities
- Managing risks in the finance sector
- Introducing new ways of supply chain management (SCM)
- Improvising government services
- Education sector

List of Data Analysis Tools

- KNIME
- Power BI
- Excel
- Python
- Tableau
- R programming
- Google analytics
- Rapid miner

- SAS
- QlikView
- Talend
- Tensorflow

Data Analytics Approaches

- Descriptive Analysis - As its name suggests, 'descriptive' - it identifies the trends and patterns of the given data. Trends and patterns within a specific time period can be observed through descriptive analysis.
- Diagnostic Analysis - Diagnostic approach studies the data processed after the descriptive analysis. To gain a better understanding and meaningful insights, one can employ a diagnostic analytic approach (Bose et al., 2023).
- Predictive Analysis - This approach can be used to predict future values. It can be used in forecasting applications (e.g. weather forecasting) and applies statistical analysis (if required). Customer retention, employee retention, sales forecasting, and other applications are a few areas where predictive analytics can be applied.
- Prescriptive Analysis - Upon studying all the solutions, prescriptive analytics (Del Giorgio Solfa, et al., 2023). can suggest optimal 'recommendations' to improve the business processes, manufacturing industry, etc. Prescriptive analysis can be used in decision-making models.

VARIABLES USED IN RESEARCH METHODOLOGY

Analyzing the data (big data in the recent ML and AI applications) (Bose, S. et al., 2023) consists of a few steps, like integrating data from multiple sources, cleaning the integrated data (KDD process), transforming the data (if required) and developing a decision-making model and gaining the proper insights. Data analysis helps analyzing the qualitative (Khoa, B.T. et al., 2023) as well as quantitative data. This section describes the different variables used in research methodology, specifically dependent, independent, and controlled variables. Authors (Akhai et al., 2016) have focused on an unhealthy work environment, which affects the worker's productivity and directly impacts the manufacturing output. Authors have also focused on the impact of an unhealthy work environment on health. This research work emphasised the presence of CO2 in the work environment, which hampers the workers. This research work considered several parameters, including occupiers in the work environment, the operation of the air conditioning system, and, most importantly, the observation of changing CO2 levels in relation to these parameters.

Dependent Variable

The name itself suggests 'dependent' variable directly depends on other variables in the given dataset (Rittmann et al., 2025). If the given dataset consists of a few variables like 'patient name', 'age', 'gender', 'high blood pressure', 'diabetes', and 'mortality rate' then one can treat 'mortality rate' as a 'dependent variable' that could depend on other variables like 'high blood pressure' and 'diabetes' etc. Researchers observe the impact of other variables (independent) on these dependent variables. Therefore, it helps identify the possible effects, association and correlation between these variables (independent and dependent). Dependent variables are also known as 'response' variables because they depend on changes in other variables. A few researchers also treat these variables as an 'outcome' variable.

Independent Variable

In contrast to the dependent variables, these independent variables (Lim et al., 2024) do not depend on other variables. One can observe the impact of these variables on other variables, such as the 'dependent variable' (as discussed in the previous example). Independent variables, such as 'age' and 'gender', do not depend on other variables. Research hypotheses must process these 'dependent and independent variables' carefully (Rivalda et al., 2024). One can also term these variables as 'explanatory' variables capable of explaining the impact on the dependent variables. These variables can also be known as 'predictor' variables, e.g. 'high blood pressure', 'diabetes' and 'age' can be treated as predictor variables to identify the root cause of death of the patient, or other comorbidities.

Controlled Variable

Constant variables can be treated as a 'controlled' variable. Researchers and academicians keep one or more variables as a 'constant' in their case studies. These variables could have a significant impact on the output, and if these variables (control) change periodically, then it can hamper the results. If someone distinguishes variables from the given dataset into 03 categories like 'dependent', 'independent' and 'controlled' variables and if the controlled variable changes, then the correlation result between the 'dependent and independent' variables could be invalidated. 'Temperature', 'humidity' and similar variables are examples of controlled variables. These variables can be held constant and treated as a 'controlled variable'. Controlled variables have impact on the outcome. Therefore, it is advised to carefully select the controlled variable from the given datasets.

TESTS FOR ANALYZING THE DATA

ANOVA

ANOVA is an analysis of variance (Domonkos & Joelsson, 2024) and calculates the difference between the given samples and variables. It helps determine the relationship between the outcome and the considered factors (variables). The following tables (1, 2 and 3) show the ANOVA (Okoye & Hosseini, 2024) using single-factor analysis, 02 factors with replication, respectively.

Figure 1. ANOVA (data analysis - MS-Excel)

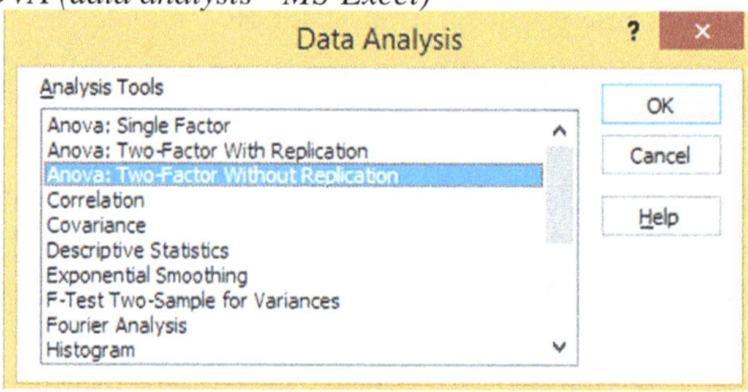

[Source: Microsoft Excel / Figure is based on the data of the authors]

Table 1. Dataset for ANOVA

Manufacturer 1	Manufacturer 2	Manufacturer 3
9.6	9.7	10
10	9.8	9
9.5	9.7	9.6
10	9.1	9.4
9.4	9.5	9.9
9.3	9	10.2
8	9.9	10.3
10	10	10.2

Single-factor ANOVA, also termed one-way ANOVA, determines the effect of the independent variable on the dependent variable. Researchers can identify the potential variations and impacts of independent variables on dependent variables.

The objective is to find the difference between the given elements in order to accept or reject the null hypotheses. If the test indicates a difference between the given elements, it means this approach is referred to as an alternative hypothesis.

Table 2. ANOVA: Single Factor summary

ANOVA: Single Factor						
SUMMARY						
Groups	**Count**	**Sum**	**Average**	**Variance**		
Manufacturer 1	8	75.8	9.475	0.436429		
Manufacturer 2	8	76.7	9.5875	0.132679		
Manufacturer 3	8	78.6	9.825	0.207857		
ANOVA						
Source of Variation	SS	df	MS	F	P-value	F crit
Between Groups	0.510833	2	0.255417	**0.98621**	**0.389611**	3.4668
Within Groups	5.43875	21	0.258988			
Total	5.949583	23				

Two factors ANOVA with replication is also known as *two-way* ANOVA as shown in the following table 3. Two-factor ANOVA helps determine if the given samples are similar in 'mean' or 'average' values. Two-way ANOVA determines the effect of the given variables from the given dataset on the results. It also helps find if the given variables interact with each other. It is a test describing the effect of 'predictor' variables on the outcome i.e., effects of independent variables on the dependent variable.

Table 3. ANOVA: Two-Factor with Replication summary

ANOVA: Two-Factor With Replication				
SUMMARY	Manufacturer 1	Manufacturer 2	Manufacturer 3	Total
C1				
Count	14	14	14	42
Sum	137.3	134.3	136.2	407.8
Average	**9.807143**	**9.592857**	**9.728571**	**9.709524**
Variance	0.439176	0.12533	0.539121	0.357956
Total				
Count	14	14	14	

continued on following page

Table 3. Continued

ANOVA: Two-Factor With Replication				
Sum	137.3	134.3	136.2	
Average	9.807143	9.592857	9.728571	
Variance	0.439176	0.12533	0.539121	

Correlation

As its name suggests, correlation finds the correlation between two variables. One can apply the CORREL function from the data analysis (Excel). Correlation between given variables could be categorised under 'positive' or 'negative' correlation. If the correlation is '+1', it means variable 1 increases & variable 2 increases; it also means if variable 1 decreases, variable 2 decreases.

In contrast to the positive correlation, there could be a perfect negative correlation between the given variables, i.e., if variable 1 increases, then variable 2 decreases; if variable 1 decreases, then variable 2 increases.

Table 4. Dataset for correlation

A	B	C
0	2	2
13	5	10
1	7	3
10	5	12
4	6	4

Figure 2. Correlation (data analysis - MS-Excel)

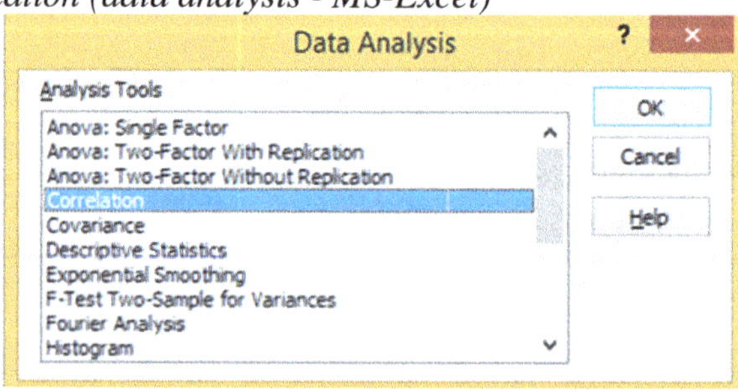

[Source: Microsoft Excel / Figure is based on the data of the authors]

Table 5. Output of correlation

	A	*B*	*C*
A	1		
B	0.141077	1	
C	**0.933709**	0.11893	1

Table 5 above describes the correlation between the given variables. One can conclude that variables A and C are positively correlated at 93.37% whereas variables A and B are not correlated. Similarly, variables B and C are not correlated, as shown in Figure 3.

Figure 3. Output of correlation

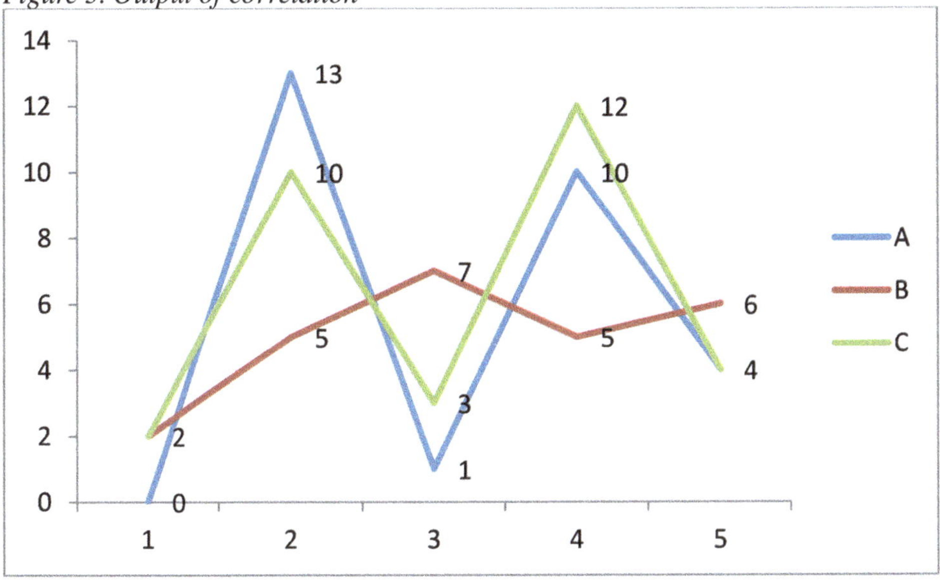

[Source: Microsoft Excel / Figure is based on the data of the authors]

Covariance

Covariance finds the relationship between given variables. It helps determine if there is a linear relationship between given variables. With the help of a covariance, researchers determine the direction of the relationship between the variables. Change in those variables can be measured with the help of covariance. There could be positive and negative covariances between the given variables.

Figure 4. Covariance (data analysis - MS-Excel)

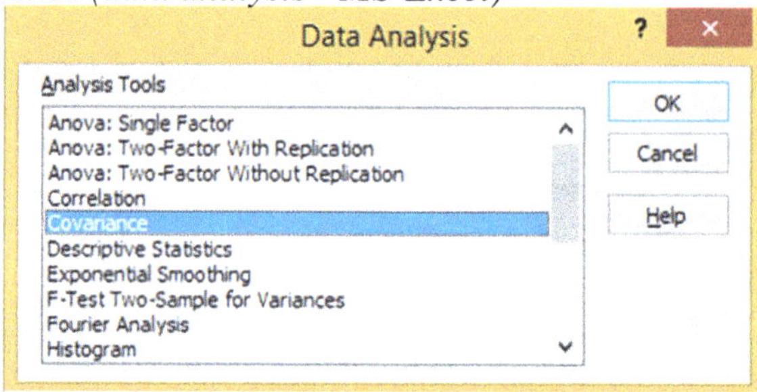

[Source: Microsoft Excel / Figure is based on the data of the authors]

Covariance helps calculate the possible relationship between the given elements to check whether these variables differ in the same direction or in the opposite direction, as shown in Tables 6 and 7. E.g., if the organization invest maximum in the marketing of the products which increase the sales, then it is to be treated as 'variables (marketing strategy and sales) vary directly i.e., positive covariance else if the variables vary in inverse direction, then it is termed as 'negative covariance'.

Table 6. Dataset for covariance

Student name	**Physics marks**	**Maths marks**	**Social science marks**
sam	87	93	87
ana	90	75	83
ravi	75	86	85
robert	80	91	93
philip	82	77	89
peter	95	91	85

Table 7. Covariance matrix

	Physics marks	*Maths marks*	*Social science marks*
Physics marks	43.80555556		
Maths marks	1.75	49.91666667	
Social science marks	-9.333333333	7.666666667	10.66666667

T-Test

The t-test (data analysis tool pack – Excel, as shown in Figure 5) is a statistical test used to determine the association between two variables. The t-test can be applied to test the null hypothesis and determine if the given data (variables) are equal. For the following data, one can calculate the 'paired two samples' with the given formula (Excel). Similarly, one can calculate the 'two-sample assuming equal variance' with the given formula & illustrated with the following tables 8, 9 and 10.

Table 8. Dataset for T-test

Unit test marks 1	Unit test marks 2
252	261
288	247
245	263
235	277
278	290
282	304
301	326
344	344
353	355
370	360

0.296970519
Paired two samples for a means =T.TEST (A2:A11, B2:B11, 2, 1)
0.695028
Two-sample assuming Equal variances =T.TEST (A2:A11, B2:B11, 2, 2)

Figure 5. T-test (data analysis- MS-Excel)

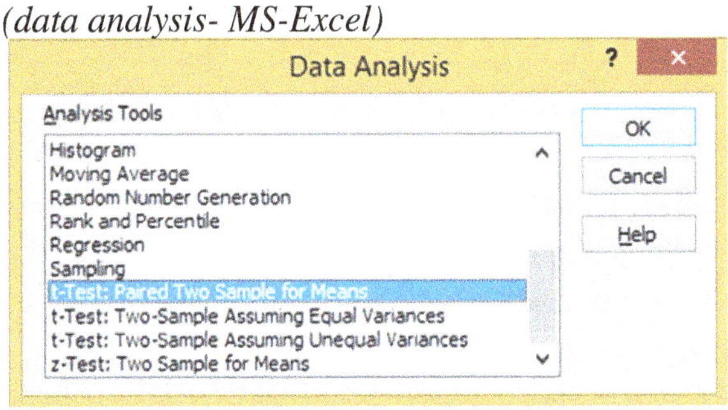

[Source: Microsoft Excel / Figure is based on the data of the authors]

Table 9. t-Test: Paired two sample for means

t-Test: Paired Two Sample for Means	Variable 1	Variable 2
Mean	**294.8**	**302.7**
Variance	2209.066667	1723.122222
Observations	10	10
Pearson Correlation	0.877226274	
Hypothesized Mean Difference	0	
df	9	
t Stat	**-1.107077834**	
P(T<=t) one-tail	0.14848526	
t Critical one-tail	1.833112933	
P(T<=t) two-tail	0.296970519	
t Critical two-tail	**2.262157163**	

Table 10. t-Test: Two-Sample Assuming Equal Variances

t-Test: Two-Sample Assuming Equal Variances	Variable 1	Variable 2
Mean	**294.8**	**302.7**
Variance	2209.066667	1723.122222
Observations	10	10
Pooled Variance	1966.094444	
Hypothesized Mean Difference	0	
df	18	
t Stat	**-0.398391355**	
P(T<=t) one-tail	0.347513854	
t Critical one-tail	1.734063607	
P(T<=t) two-tail	**0.695027708**	
t Critical two-tail	**2.10092204**	

'T-Test: Paired Two Samples for Means', and 'T-Test: Two-Sample Assuming Equal Variances' are given in the 'data analysis' (Excel) to find the inequality. One can find the equality or inequality between t-stat, t-critical two tail to reject or accept the proposed hypotheses. As shown in the above tables, the sample means for the given variable 1 and variable 2 are '294.8' and '302.7', which is not enough to describe the relation between the unit test marks.

Histogram

To illustrate the frequency of given numerical data, a histogram (data analysis tool pack – Excel, as shown in Figure 6) can be used, 2025) representation as shown in the following example, as shown in Table 11.

Table 11. Dataset for histogram

no. of employees	
21	
28	
39	**Bin range**
29	20
47	25
23	30
20	35
18	40
23	45
21	
24	
51	
34	
39	
30	
36	
20	
22	
21	
44	
43	
22	

Figure 6. Histogram (data analysis- MS-Excel)

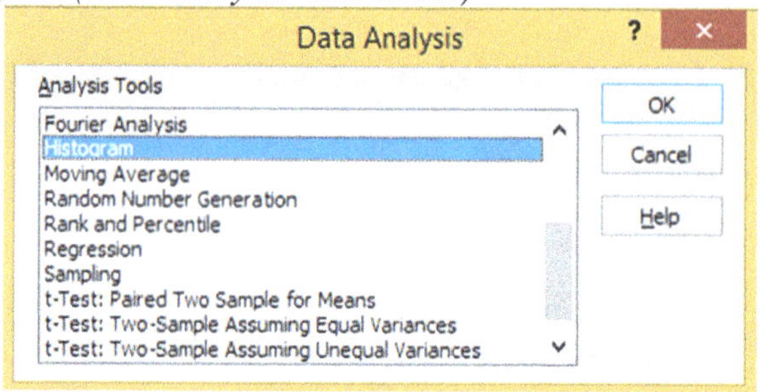

[Source: Microsoft Excel / Figure is based on the data of the authors]

Figure 7. Histogram

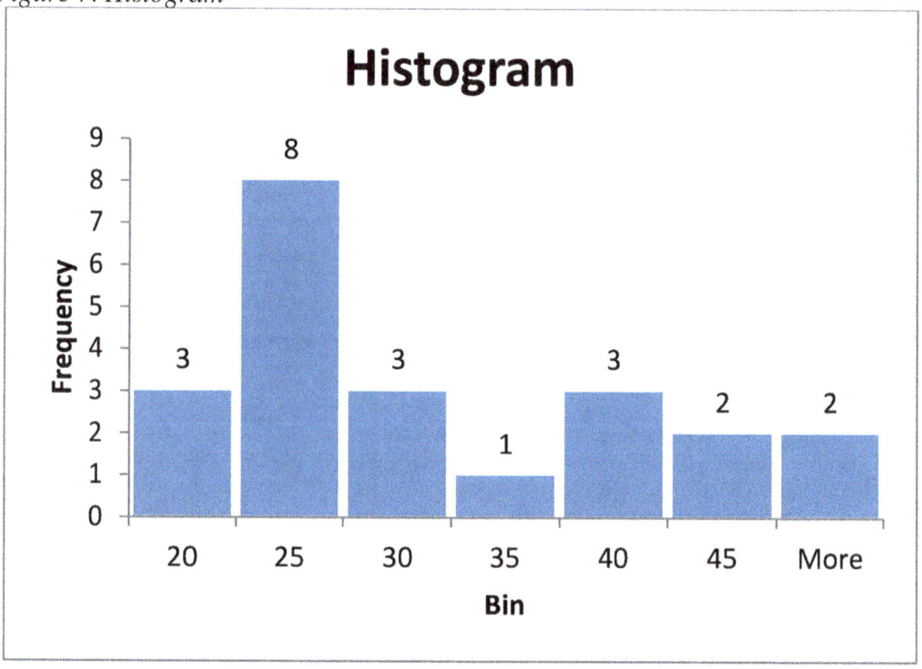

[Source: Microsoft Excel / Figure is based on the data of the authors]

For the given data, as shown in the previous table 11 - 'no. of employees', one can use histogram representation with the specified bin range. It is a rectangular bar representation of the boundaries. Identifying the frequency according to the specified bin range is an important task. The specified bin range is 20, 25, 30, 35, 40, and 45, and according to the same data, the frequency is calculated and represented as shown in Figure 7.

Descriptive Statistics

The descriptive statistics (Data Analysis ToolPack – Excel, as shown in Figure 8) help summarise the given statistical data with various options, as shown in the following tables: 12, 13, 14, and 15. The various options in the result include mean, standard error, median, mode, standard deviation, sample variance, skewness, minimum, maximum, range, sum and count.

Table 12. Dataset for descriptive statistics

Emp_ID	Name of Emp.	Designation	Department	Sub_Unit	Gender	Age	Hire Date	Annual Salary	Country
E001	Sam	Manger	IT	Research & Development	Female	56	4/8/2016	$140,500	United States
E002	Dinha	Tech Architect	IT	Manufacturing	Male	58	11/29/1997	$99,978	China
E003	Luna	Director Operations	Finance	Special Products	Female	51	10/26/2006	$180,098	China
E004	Philip	Systems Manager	IT	Manufacturing	Female	28	9/27/2019	$84,920	United States
E005	Austin	Sr. Analyst	Finance	Manufacturing	Male	56	11/20/1995	$95,410	United States
E007	Joshua	Accountant	Sales	Corporate	Male	58	1/24/2017	$51,000	China
E008	Barnard	Manager	IT	Corporate	Female	27	7/1/2020	$119,750	United States
E009	Martin	Data Analyst	Finance	Manufacturing	Male	26	5/16/2020	$41,350	United States
E010	Easton	Manager	Accounting	Manufacturing	Male	30	1/25/2019	$113,620	United States
E011	Michel	Sr. Analyst	Finance	Speciality Products	Female	35	6/13/2018	$77,210	United States
E012	John	Senior Manger	Human Resources	Manufacturing	Female	36	2/11/2009	$157,350	Brazil
E013	Rogers	Controls Engineer	Engineering	Speciality Products	Female	28	10/21/2021	$109,860	United States
E014	Elias Alvarado	Manager	Human Resources	Manufacturing	Male	58	3/14/1999	$105,100	United States
E015	Eli	Senior Manger	Finance	Research & Development	Female	52	6/10/2021	$146,800	China
E016	Smit	Sr. Analyst	Accounting	Special Products	Male	32	11/4/2017	$97,090	India
E017	Rivera	Vice President	Marketing	Research & Development	Female	42	3/13/2013	$249,300	United States
E018	Powell	Director Finance	Finance	Research & Development	Female	65	3/4/2002	$175,850	United States
E019	Silva	Senior Manger	Marketing	Special Products	Female	63	12/1/2003	$154,900	United States

continued on following page

Table 12. Continued

Emp_ID	Name of Emp.	Designation	Department	Sub_Unit	Gender	Age	Hire Date	Annual Salary	Country
E020	David	Director IT	IT	Corporate	Male	62	11/3/2013	$186,560	India
E021	Adam	Director Sales	Sales	Research & Development	Male	46	7/9/2002	$166,400	China
E022	Alvarado	Senior Manger	IT	Manufacturing	Male	57	1/9/2012	$146,250	Brazil
E023	Robert	Asst. Director	Sales	Manufacturing	Female	38	4/2/2021	$151,750	United States
E024	Logan	Asst. Director IT	IT	Research & Development	Male	60	5/24/2002	$160,790	Brazil
E025	Dixon	Analyst	Sales	Speciality Products	Male	38	9/5/2019	$51,000	United States

Table 13. Descriptive Statistics summary

Descriptive Statistics										
	Emp_ID	Name of Emp.	Designation	Department	Sub_ Unit	Gender	Age	Hire Date	Annual Salary	Country
Valid	24	24	24	24	24	24	24	24	24	24
Missing	0	0	0	0	0	0	0	0	0	0
Mean							45.917			
Std. Deviation							13.419			
Minimum							26.000			
Maximum							65.000			

Figure 8. Descriptive statistics (data analysis- MS-Excel)

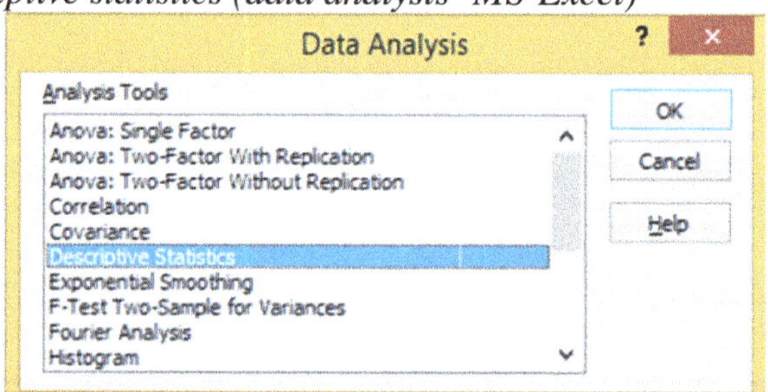

[Source: Microsoft Excel / Figure is based on the data of the authors]

Table 14. Descriptive statistics for single variable

Annual Salary	
Mean	127618.1667
Standard Error	10109.69709
Median	130125
Mode	51000
Standard Deviation	49527.19866
Sample Variance	2452943407
Kurtosis	0.194301162
Skewness	0.227844203
Range	207950
Minimum	41350
Maximum	249300
Sum	3062836
Count	24

Table 15. Descriptive statistics for single variable (2)

Age	
Mean	45.91666667
Standard Error	2.739219096
Median	48.5
Mode	58
Standard Deviation	13.41937816
Sample Variance	180.0797101
Kurtosis	-1.605144367
Skewness	-0.162486267
Range	39
Minimum	26
Maximum	65
Sum	1102
Count	24
Largest(1)	65
Smallest(1)	26
Confidence Level (95.0%)	5.66650643

Sampling

In many applications, it is necessary to select random data from the given dataset. Many tools provide a 'sampling' method (data analysis tool pack – Excel, as shown in Figures 9 and 10) to select data from the dataset. Sampling in the Excel tool offers 'random' and 'periodic' methods for selecting data from the given dataset, as illustrated in Tables 16 and 17.

Table 16. dataset for sampling

boys	girls
3	2
6	4
8	7
10	11
12	13
16	25
18	17
21	19
22	23

Figure 9. Sampling (data analysis MS-Excel) Random method

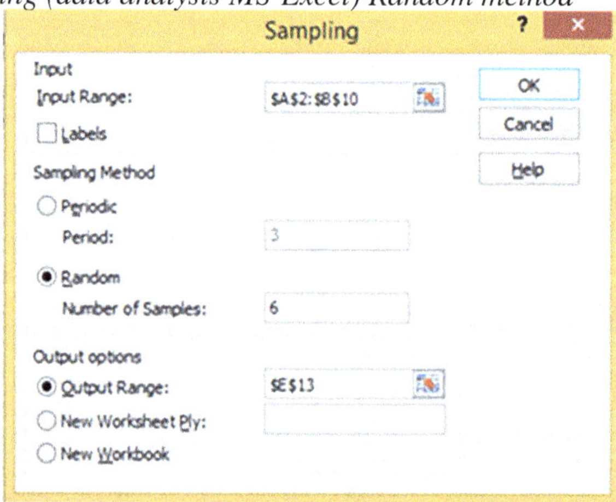

[Source: Microsoft Excel / Figure is based on the data of the authors]

Figure 10. Sampling (data analysis MS-Excel) Periodic method

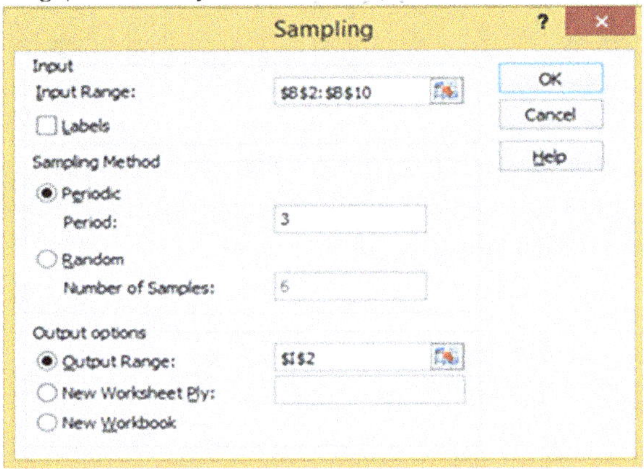

[Source: Microsoft Excel / Figure is based on the data of the authors]

Table 17. Output of sampling (Random and Periodic)

Random	Periodic
13	7
12	25
12	23
16	
13	
22	

Moving Averages

It is also known as *calculating the rolling averages* for the given data. One can apply the 'interval-i ' for the given data to observe the possible trends and patterns. Best examples of moving averages are time series analysis (Mao et al., 2024) and stock value prediction. In stock value prediction, it is very important to calculate the 'exponential moving average' for 05 days, 10 days, 20 days, 50 days, 100 days, etc. In the following example, the stock prices for the different months are shown. With the help of the 'moving average' from the 'data analysis (Excel) as shown in figures 11 and 12, it is possible to calculate the moving averages of the price of the stock as shown in table 18.

Table 18. Dataset for moving averages

interval (months)	**0**	**1**	**2**	**3**	**4**	**5**	**6**	**7**	**8**	**9**	**10**
price of the stock	180	177	190	167	188	192	210	224	198	230	235

This approach is very useful in stock value prediction as the stock price for different DEMA (day's exponential moving averages) can be calculated. Figure 13 illustrates the output of moving averages.

Figure 11. Moving average (data analysis MS-Excel)

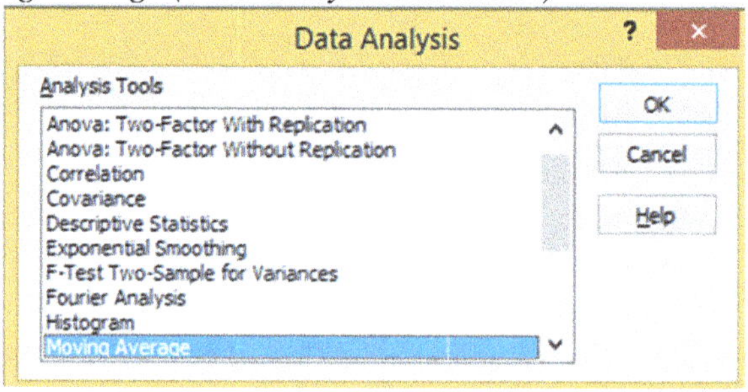

[Source: Microsoft Excel / Figure is based on the data of the authors]

Figure 12. Moving average (data analysis MS-Excel) (2)

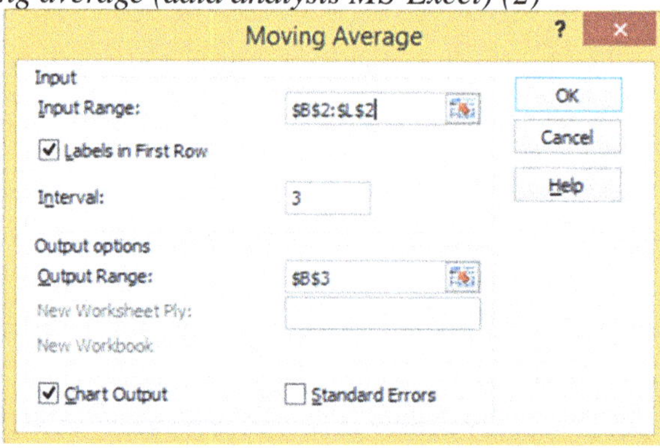

[Source: Microsoft Excel / Figure is based on the data of the authors]

Table 19. Output of moving averages with specified interval

interval (months)	**0**	**1**	**2**	**3**	**4**	**5**	**6**	**7**	**8**	**9**	**10**
price of the stock	180	177	190	167	188	192	210	224	198	230	235
	#N/A	**#N/A**	**178**	**181.6667**	**182.3333**	**196.6667**	**208.6667**	**210.6667**	**217.3333**	**221**	

Figure 13. Output of moving averages

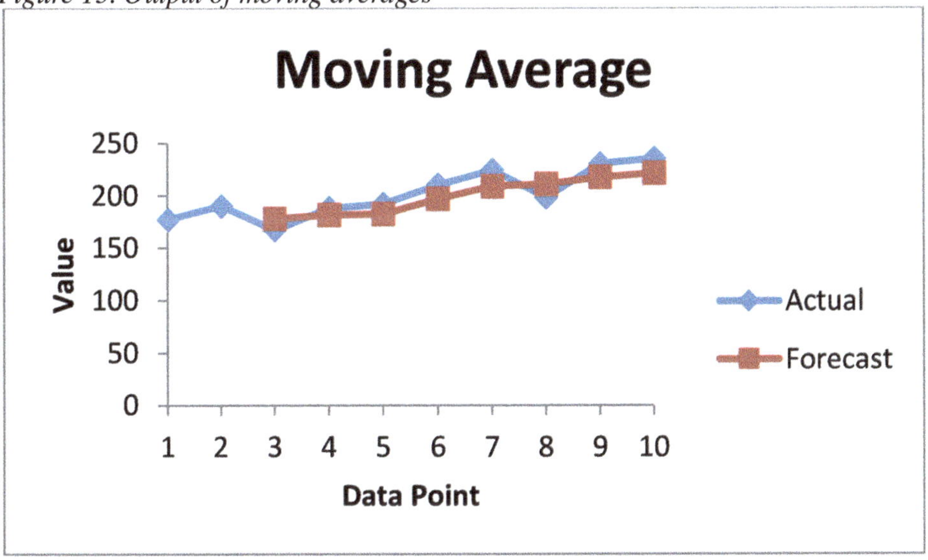

[Source: Microsoft Excel / Figure is based on the data of the authors]

F test

It (data analysis tool pack – Excel, as shown in Figures 14 and 15) is used to find the probability of the variance for the given elements, i.e. specified arrays 'unit test marks 1' and 'unit test marks 2' as shown in Table 20. The F-test helps determine whether they are different. In research and development, researchers present hypotheses for their research and categorise them into two types: null hypotheses and alternative hypotheses. In Excel, the 'F-test' is available in the 'data analysis' tool pack, where it is required to state the 'variable 1 range' and 'variable 2 range' with the alpha 0.05, i.e. considering 95% probability.

Table 20. Dataset for F-test

Unit test marks 1	Unit test marks 2
252	261
288	247
245	263
235	277
278	290
282	304
301	326
344	344
353	355
370	360

Figure 14. F-Test Two-Sample for Variances

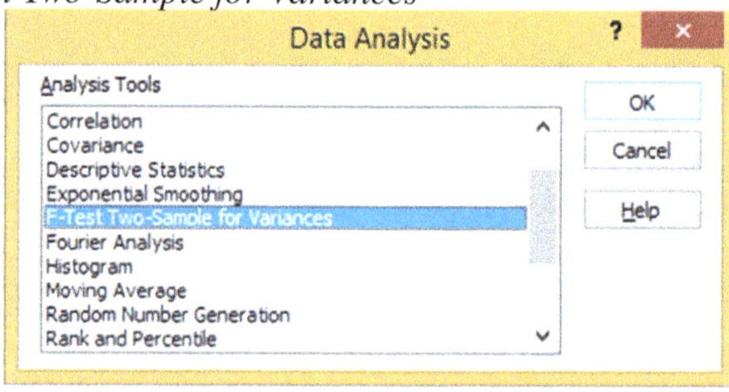

[Source: Microsoft Excel / Figure is based on the data of the authors]

Figure 15. F-Test Two-Sample for Variances (2)

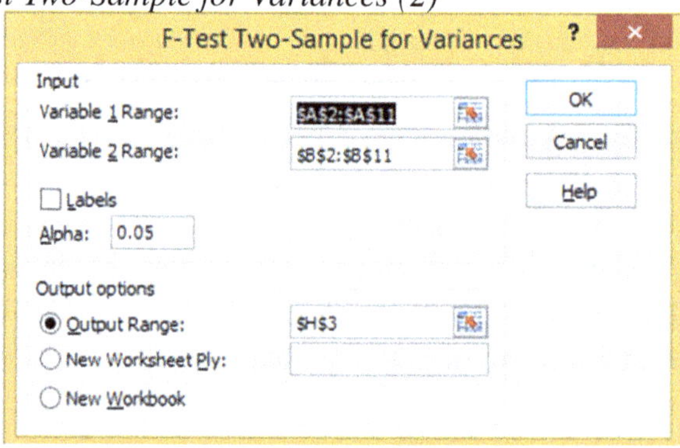

[Source: Microsoft Excel / Figure is based on the data of the authors]

Table 21. F-Test Two-Sample for Variances

F-Test Two-Sample for Variances	Variable 1	Variable 2
Mean	294.8	302.7
Variance	2209.066667	1723.122222
Observations	10	10
df	9	9
F	1.282013915	
P(F<=f) one-tail	0.35867267	
F Critical one-tail	3.178893104	

As discussed earlier, there is a NULL hypothesis to be accepted or rejected based on the results obtained upon execution of the F-test. As shown in the above table 21, F-test results, one can observe the Mean values for given variable 1 and variable 2 are 294.8 and 302.7, respectively, and the F value is 1.282013915, and the F critical one-tail value is 3.178893104, i.e., F<F critical one-tail, which helps in accepting the NULL hypotheses.

RECENT TECHNIQUES TO ANALYZE AND REPRESENT THE DATA

Classification

One can organise the given data into different categories (classes). Once the data is classified into the predetermined categories, it is very useful in further sorting and retrieving the required data within a minimum time and with less effort. If the data is in a structured format, it can be classified easily according to the labels. However, if the data is unstructured, then 'labels' must be designed prior to assigning the individual elements from the dataset to the specific class. If the labels are properly assigned and elements are correctly added to the specific class, then optimising the queries would be an easier task. The most important advantage of classification (Lokman et al., 2025) of the data is 'avoiding the redundancy'. Redundancy can be avoided easily, which also enhances the query optimisation process. However, there are different types of classification, such as geographical classification, qualitative (Bingham, 2023), and quantitative data classification.

Classification is a supervised machine learning method, and a few examples of the classification algorithm are as follows-

- Logistic regression

- Support vector machine
- K-nearest neighbours
- Naïve Bayes
- Decision tree classification
- Random forest classification

KNN – K Nearest Neighbour

KNN is a supervised learning approach (Elnanda & Gunawan, 2025) used in classification. KNN specifies 'k' nearest data points and other observations 'n'. The objective is to correlate the elements which are similar (closest) to the data point. Based on this criterion of closest similarity, one can classify the elements from the given dataset.

Assume the following are the data points from 02 datasets 'x' and 'y'; Classes are predefined as shown below, and upon execution of the 'KNN' classifier, one can get the following output as shown in Figure 16.

x = [5, 6, 11, 5, 4, 12, 15, 7, 11, 13]

y = [22, 20, 25, 18, 17, 26, 25, 23, 22, 22]

Classes = [0, 0, 1, 0, 0, 1, 1, 0, 1, 1]

Figure 16.

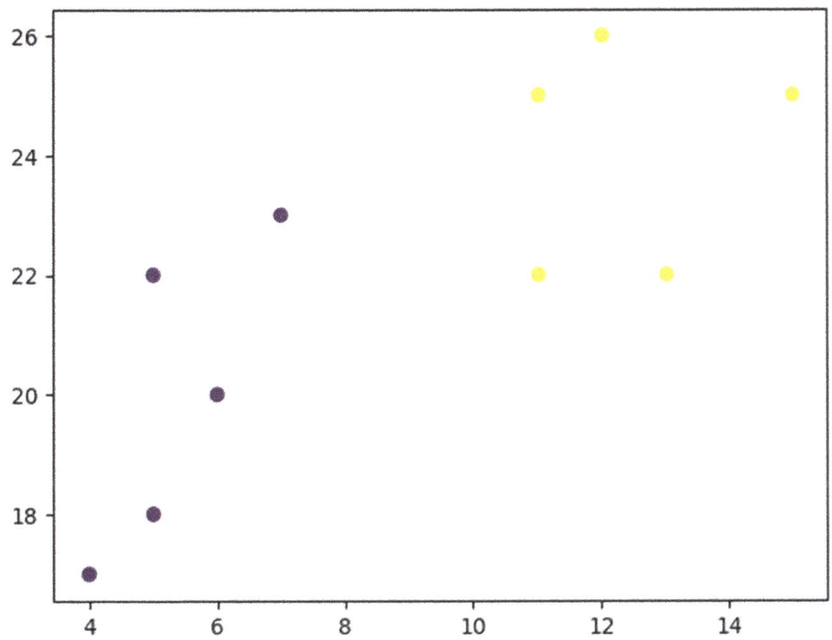

[Source: Figure is based on the data of the authors]

'X' and 'Y' are the input features, and 'classes' is the target class. KNN helps researchers predict the 'class' for the given data elements.

Regression

The dataset consists of dependent and independent variables. Regression (Hünermund & Louw, 2025) examines the relationship between these variables. Researchers will treat these dependent variables like 'target variables' and the independent variables as predictors. Machine learning employs various regression models to predict the possible relationship between these variables. Simple linear regression and multiple linear regression models are examples of 'regression' used in machine learning. As discussed in the earlier section, 'dependent variables' depend on other variables like independent variables. Independent variables have an impact on the results, and changes in these variables will, in turn, affect the dependent variables. Researchers must focus on 'outliers' when processing data, as they may have very low or high values that can impact the results. After processing the given dataset, one can decide whether the proposed model is 'over-fitted' or 'under-fitted'. Given dataset is divided into 'training' and 'testing' sets, and if the model processes the 'training' set but does not process the 'test' set, then it is treated as 'overfitting. If the proposed model does not process the 'training', then it is termed as 'under-fitting'. The dataset is used as shown in Table 22, and the regression output is presented in Figure 17. The dataset used for multiple linear regression (MLR) is shown in Table 23 with the specified equation for MLR.

Table 22. Dataset for regression

comp_sales	printer_sales
15	10
20	15
20	14
22	17
18	12
14	14
13	10
22	18
28	14
18	15
17	12
13	11

Figure 17. Output of linear regression

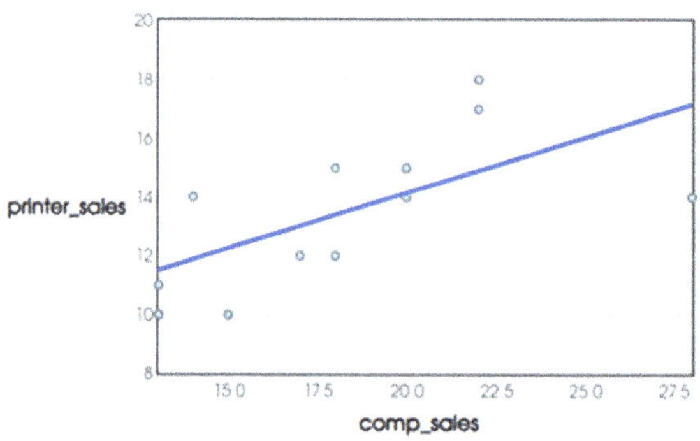

[Source: Figure is based on the data of the authors]

Table 23. Dataset for multiple linear regression

comp_sales	mobile_sales	printer_sales
15	25	10
20	15	15
20	30	14
22	32	17
18	30	12
14	16	14
13	17	10
22	18	18
28	18	14
18	17	15
17	16	12
13	15	11

$y=\beta_0+\beta_1X_1+\beta_2X_2+......+\beta qXq$ .... Equation (a), where **y** is the dependent variable and X1, X2, and X_q are the independent variables.

Clustering

Clustering (Oskouei et al., 2025) is nothing but the grouping of given data elements based on similarity. Based on either the 'threshold' value or the similarity criteria, elements can be grouped. Similarity patterns can be observed when the given elements are successfully clustered. Classification is a supervised learning method, whereas clustering is an unsupervised learning method that deals with unlabeled elements.

One can use any clustering algorithm to process the given dataset in order to form different clusters. There are many clustering algorithms available, like –

- K-means clustering
- DBSCAN
- Mean shift
- BIRCH

K-Means Clustering

'Clustering', based on the similarity criteria, groups the elements. K-means uses 'k' clusters (groups) and assigns elements to these clusters. Once the algorithm successfully groups the elements next task is to compute the 'centroid' or cluster representative for the cluster. Calculating the centroid or cluster representative is an iterative process. In the subsequent steps, the remaining elements will be added to the clusters. K-means clustering (Zhang et al., 2025) also employs the 'elbow' method to visualise the points (elements) where the decrease is linear, as shown in Figures 18 and 19.

Figure 18. Output of K means clustering

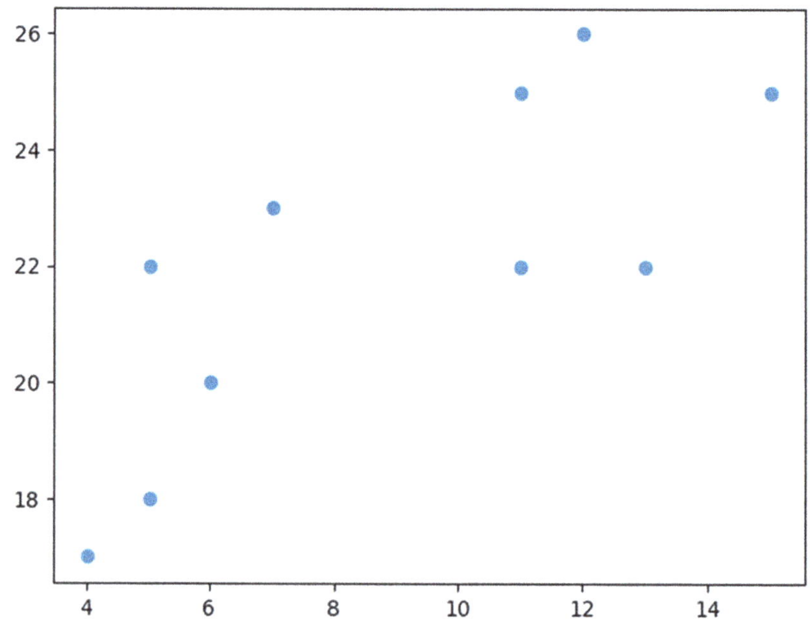

[Source: Figure is based on the data of the authors]

Figure 19. Output of K means clustering - elbow method

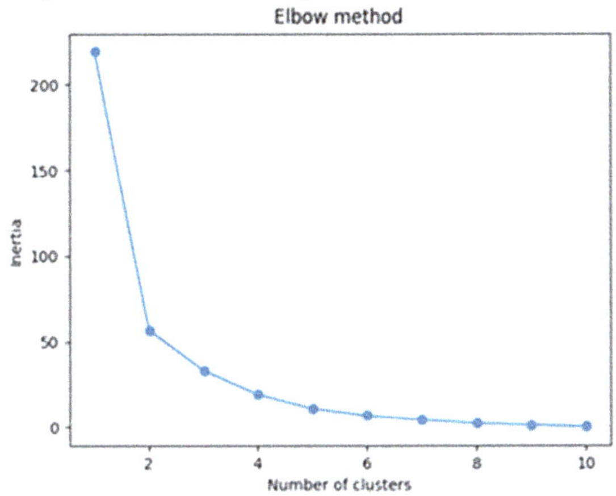

[Source: Figure is based on the data of the authors]

Time Series Analysis

With the help of time series analysis (as shown in Table 24), historical data, such as weather data, stock market data, and rainfall statistics, can be effectively represented. It consists of data points over a specific period and helps predict future values. Time series analysis (Tawakuli, A., et al., 2025) plays a crucial role in forecasting as it observes changes in data over a specific period. However, it is beneficial to have maximum data collected over a specific period to ensure the consistency and reliability of the data for accurate forecasting, as shown in Figure 20.

Table 24. dataset for time series analysis

Month	Passenger traffic
Jan	10000
Feb	9000
March	8500
April	8700
May	9200
June	9500
July	6500

Figure 20. Time series analysis

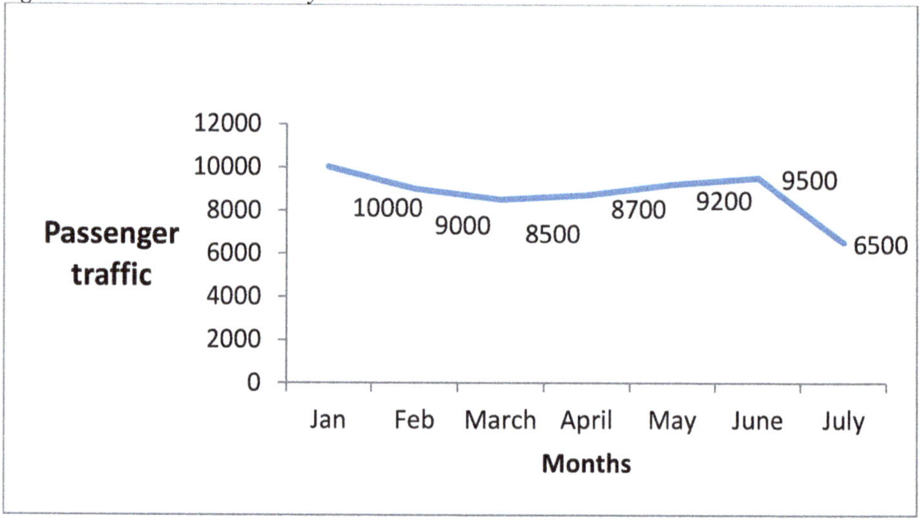

[Source: Figure is based on the data of the authors]

Charts

Charts are used to better understand the given data (Krishnan, A. R., 2025). One can draw a line chart, a bar chart, a pie chart, a radar chart, a clustered bar chart, and a doughnut chart to represent the given data. With the help of various charts, one can gain better insights into the given data and identify potential trends or patterns. The Table 25 dataset is used to create the various charts shown in this section.

Table 25. dataset for charts

Emp_ID	Annual Salary
E001	$140,500
E002	$99,978
E003	$180,098
E004	$84,920
E005	$95,410
E007	$51,000
E008	$119,750
E009	$41,350

continued on following page

Table 25. Continued

Emp_ID	Annual Salary
E010	$113,620
E011	$77,210
E012	$157,350
E013	$109,860
E014	$105,100
E015	$146,800
E016	$97,090
E017	$249,300
E018	$175,850
E019	$154,900
E020	$186,560
E021	$166,400
E022	$146,250
E023	$151,750
E024	$160,790
E025	$51,000

2D column chart

Figure 21. 2-D Column chart

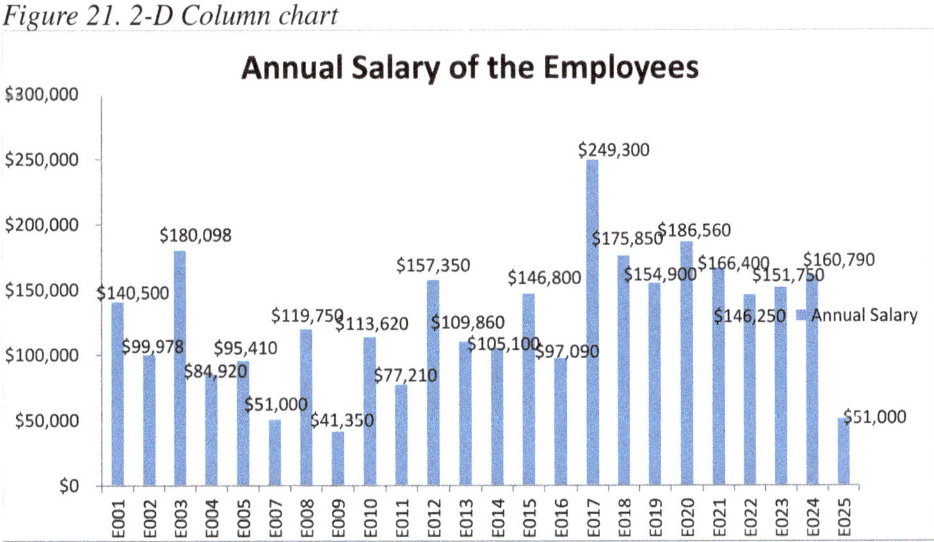

[Source: Microsoft Excel / Figure is based on the data of the authors]

Line chart

Figure 22. Line Chart

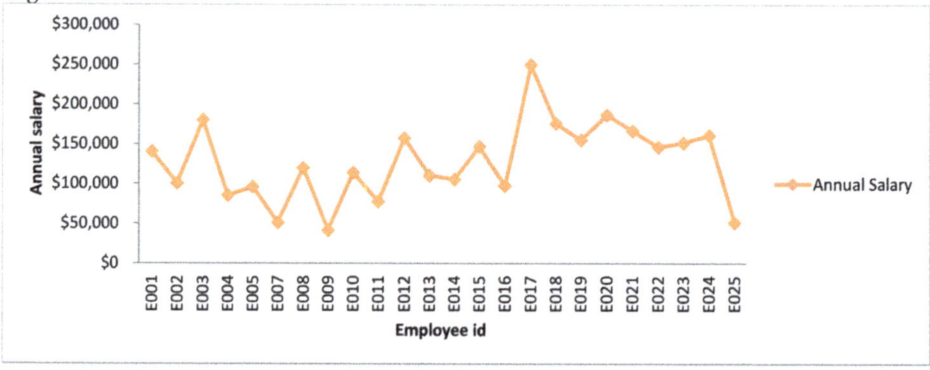

[Source: Microsoft Excel / Figure is based on the data of the authors]

Pie chart

Table 26. dataset for charts (2)

Country	Employees
Brazil	3
United States	14
India	2
China	5

Figure 23. Pie chart

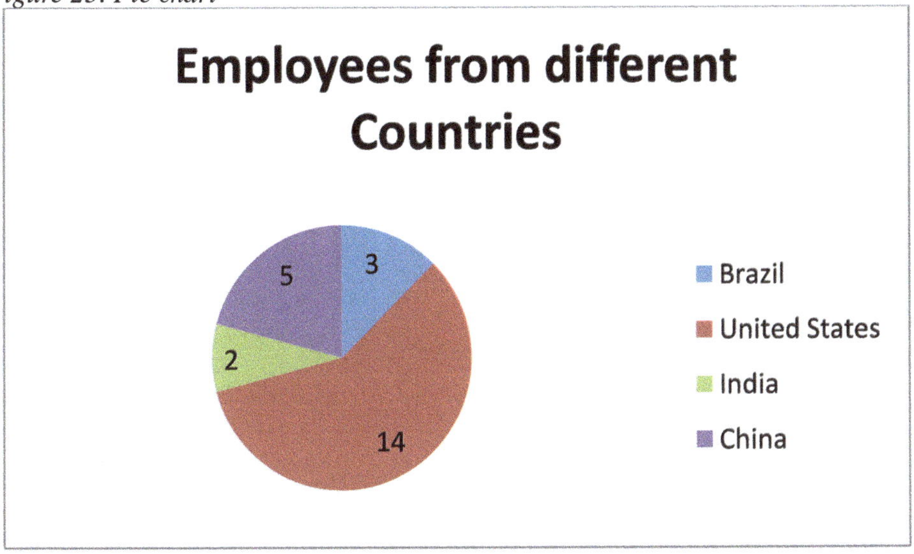

[Source: Microsoft Excel / Figure is based on the data of the authors]

3D clustered bar chart

Table 27. dataset for charts (3)

Month	mobile_sales	computer_sales	Printer_sales	total
january	25	15	10	50
february	15	20	5	40
march	30	20	5	55
april	32	22	14	68
may	30	18	10	58
june	16	14	20	50
july	17	13	23	53
august	18	22	18	58
sept	18	28	14	60
oct	17	18	15	50
nov	16	17	12	45
dec	15	13	18	46

Figure 24. 3D clustered bar chart

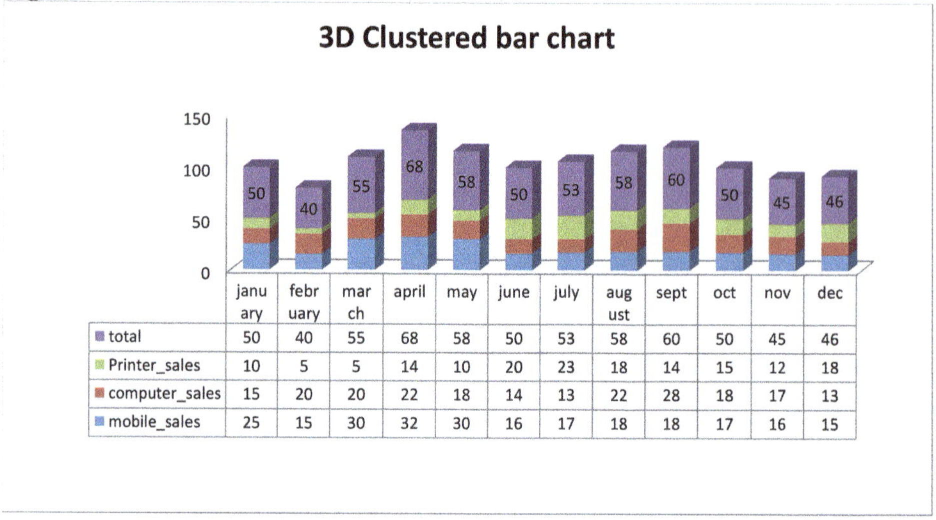

[Source: Microsoft Excel / Figure is based on the data of the authors]

Radar Chart

A radar chart (Ovchynnykova et al., 2025) can be prepared for the data shown in Table 28. It is used to present the multivariate data and can have 'n' dimensions. To achieve multidimensional views (representation), multiple variables from the given datasets can be utilised (Pérez-Benedito, 2025).

Table 28. dataset for charts (4)

Student name	Physics marks	Maths marks	Social science marks
sam	87	93	87
ana	90	75	83
ravi	75	86	85
robert	80	91	93
philip	82	77	89
peter	95	91	85

Figure 25. Radar chart

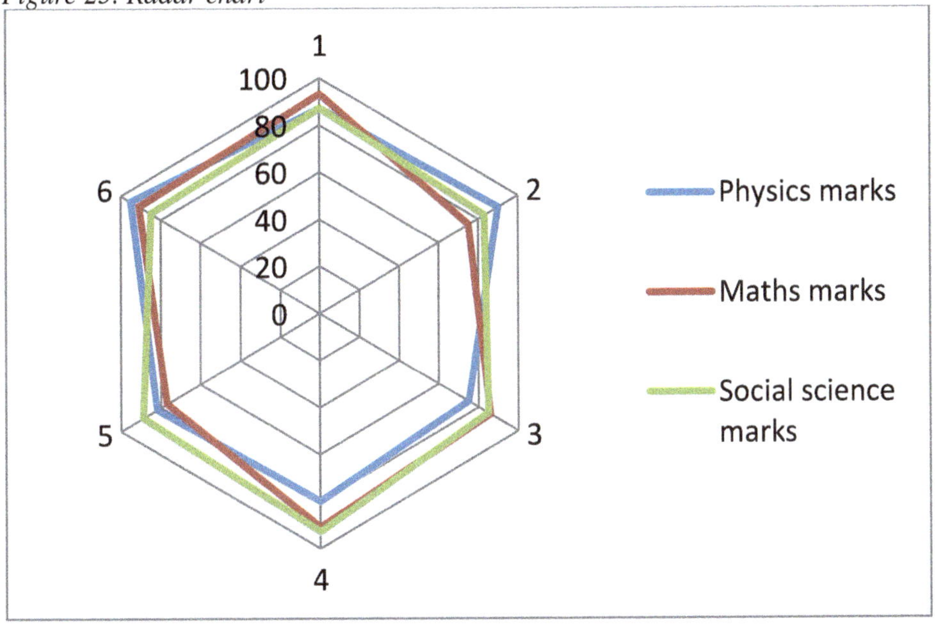

[Source: Microsoft Excel / Figure is based on the data of the authors]

Doughnut chart

Table 29. Dataset for doughnut chart

Student name	Physics marks
sam	87
ana	90
ravi	75
robert	80
philip	82
peter	95

Figure 26. doughnut chart

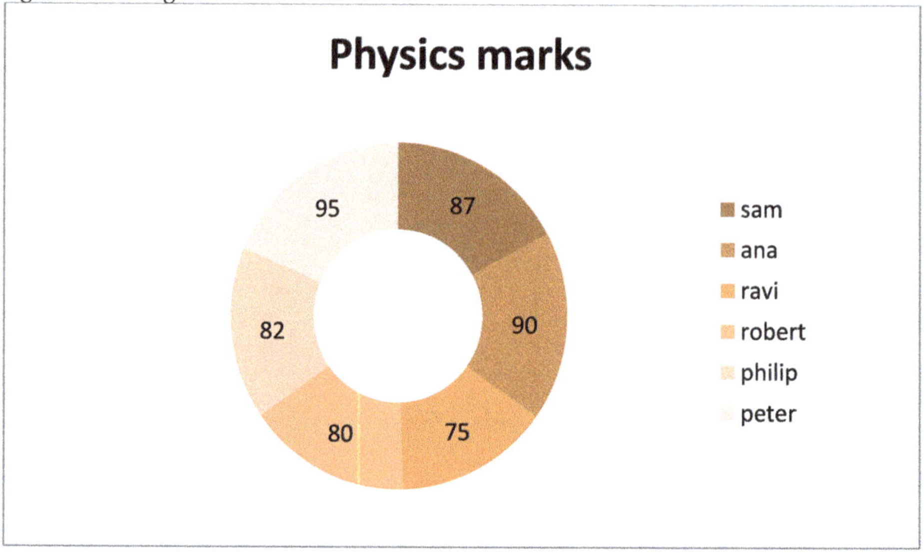

[Source: Microsoft Excel / Figure is based on the data of the authors]

CHALLENGES IN THE DATA VISUALIZATION AND ANALYSIS

However, researchers and academicians are facing a few challenges to visualise the retrieved data due to the following reasons –

- Data quality
- Dealing with vast and diverse data
- Data privacy
- Dealing with real-time data
- Integrating data from multiple sources
- Interpreting the test/method results
- Deciding the correct chart type to represent the data
- Failed to handle missing data
- Skillful persons to represent the data wisely
- As technology is changing rapidly – the decision to select the appropriate recent technology

FURTHER RESEARCH DIRECTIONS

- Developing best predictive models based on AI and ML methods.
- Integrating data from variety of heterogeneous sources.
- Analyzing data and metadata to retrieve meaningful data.
- Helping businesses to grow rapidly by taking automated decisions.
- Taking real-time decisions by processing data from variety of sources.
- Designing and developing a better decision-making model.

CONCLUSION

In this chapter, an attempt has been made to represent various tests and machine learning methods useful for data analysis. The chapter also introduces the different tools available for data analysis. With the help of different 'tests', one can process the given variables. For a better understanding and to gain important insights from the datasets, it is necessary to correctly identify the 'dependent', 'independent', and 'controlled' variables. Once a user identifies these variables, it is helpful to know the impact of independent variables on the dependent variables.

There are many data analysis tools (Radanliev & De Roure, 2023) available in the market, like Power BI, Tableau, and SPSS. Even MS-Excel provides a 'data analysis' tool pack consisting of many useful 'tests' for data analysis. Correlation, Covariance, t-test, f-test, ANOVA with various types, Descriptive statistics, moving averages, sampling, and histograms are available in MS Excel. It is also necessary to represent the 'results' with appropriate charts and figures for a better understanding of the end users. Line charts, bar charts, pie charts, and stacked charts can be used to represent the given data. Recent machine learning techniques, such as clustering, regression, and classification, are also very useful in representing the given dataset.

In the future, with the help of recent technologies like AI and ML, applications will be required to integrate data from multiple, heterogeneous sources, store it in a repository, and process it to represent it in a uniform format. Future research should focus on automating these steps. Providing real-time decisions could be the next challenge in future applications. In the future, businesses will require the best predictive models to execute daily activities smoothly and make better decisions.

REFERENCES

Abughazala, M., & Muccini, H. (2023, March). Modeling data analytics architecture for IOT applications using DAT. In *2023 IEEE 20th International Conference on Software Architecture Companion (ICSA-C)* (pp. 284-291). IEEE. DOI: 10.1109/ICSA-C57050.2023.00066

Akhai, S., Singh, V. P., & John, S. (2016). Human performance in industrial design centers with small unit air conditioning systems. *Journal of Advanced Research in Production Industrial Engineering*, *3*(2), 5–11. DOI: 10.13140/RG.2.2.21422.23361

Bingham, A. J. (2023). From data management to actionable findings: A five-phase process of qualitative data analysis. *International Journal of Qualitative Methods*, *22*, 16094069231183620. DOI: 10.1177/16094069231183620

Bose, S., Dey, S. K., & Bhattacharjee, S. (2023). Big data, data analytics and artificial intelligence in accounting: An overview. *Handbook of big data research methods (EE Elgar Online)*, 32-51. DOI: 10.4337/9781800888555.00007

Del Giorgio Solfa, F., & Simonato, F. R. (2023). Big data analytics in healthcare: Exploring the role of machine learning in predicting patient outcomes and improving healthcare delivery. *International Journal of Computations* [IJCIM]. *Information and Manufacturing*, *3*(1), 1–9. Advance online publication. DOI: 10.54489/ijcim.v3i1.235

Domonkos, P., & Joelsson, L. M. T. (2024). ANOVA (Benova) correction in relative homogenization: Why it is indispensable. *International Journal of Climatology*, *44*(12), 4515–4528. DOI: 10.1002/joc.8594

Elnanda, V. R., & Gunawan, P. H. (2025, February). Machine learning solutions for stunting classification in Karya Jaya: A case study using KNN and SVM. In *2025 International Conference on Advancement in Data Science, E-learning and Information System (ICADEIS)* (pp. 1-5). IEEE. DOI: 10.1109/ICADEIS65852.2025.10933024

He, X. (2024). Time Series Analysis. In Geographic Data Analysis Using R (pp. 105-134). Springer Nature Singapore. DOI: 10.1007/978-981-97-4022-2_6

Hünermund, P., & Louw, B. (2025). On the nuisance of control variables in causal regression analysis. *Organizational Research Methods*, *28*(1), 138–151. DOI: 10.1177/10944281231219274

Khoa, B. T., Hung, B. P., & Hejsalem-Brahmi, M. (2023). Qualitative research in social sciences: Data collection, data analysis and report writing. *International Journal of Public Sector Performance Management, 12*(1-2), 187–209. DOI: 10.1504/IJPSPM.2023.132247

Krishnan, A. R. (2025). Research trends in criteria importance through intercriteria correlation (CRITIC) method: a visual analysis of bibliographic data using the Tableau software. *Information Discovery and Delivery, Emerald publication, 53*(2), 233-247. DOI: 10.1108/IDD-02-2024-0030

Lim, J. S., Kim, Y. D., & Woo, J. C. (2024). Approximated uncertainty propagation of correlated independent variables using the ordinary least squares estimator. *Molecules. Molecules (Basel, Switzerland), 29*(6), 1248. DOI: 10.3390/molecules29061248

Lokman, A., Ismail, W. Z. W., & Aziz, N. A. A. (2025). A review of water quality forecasting and classification using machine learning models and statistical analysis. *Water. Water (Basel), 17*(15), 2243. DOI: 10.3390/w17152243

Longpre, S., Hou, L., Vu, T., Webson, A., Chung, H. W., Tay, Y., & Roberts, A. (2023, July). The flan collection: Designing data and methods for effective instruction tuning. In *International Conference on Machine Learning* (pp. 22631-22648). PMLR (Proceedings of Machine Learning Research). DOI: 10.48550/arXiv.2301.13688

Mao, S., Zhang, C., Song, Y., Wang, J., Zeng, X. J., Xu, Z., & Wen, Q. (2024). Time series analysis for education: Methods, applications, and future directions. *arXiv preprint arXiv:2408.13960*. https://doi.org//arXiv.2408.13960DOI: 10.48550

Masteali, S. H., Bayat, M., Bettinger, P., & Ghorbanpour, M. (2025). Uncertainty analysis of linear and non-linear regression models in the modeling of water quality in the Caspian Sea basin: Application of Monte-Carlo method. *Ecological Indicators, 170*, 112979. DOI: 10.1016/j.ecolind.2024.112979

Okoye, K., & Hosseini, S. (2024). Analysis of variance (ANOVA) in R: one-way and two-way ANOVA. In *R programming: statistical data analysis in research* (pp. 187-209). Springer Nature Singapore. DOI: 10.1007/978-981-97-3385-9_9

Oskouei, A. G., Samadi, N., Khezri, S., Moghaddam, A. N., Babaei, H., Hamini, K., & Arasteh, B. (2025). Feature-weighted fuzzy clustering methods: An experimental review. *Neurocomputing, 619*, 129176. DOI: 10.1016/j.neucom.2024.129176

Ovchynnykova, O., Svazas, M., & Navickas, V. (2025). Assessing economic profiles of coastal regions in the blue economy: A radar chart approach. *Challenges in Sustainability, 13*(2), 177–192. DOI: 10.56578/cis130203

Pérez-Benedito, M. A. (2025). Alternative measure of dispersion in the polar-radar chart methodology for activity management. *Archives of Business Research*, Services *for Science and Education, 13*(1), 07-20. https://doi.org/DOI: 10.14738/abr.131.18145

Radanliev, P., & De Roure, D. (2023). New and emerging forms of data and technologies: Literature and bibliometric review. *Multimedia Tools and Applications, 82*(2), 2887–2911. DOI: 10.1007/s11042-022-13451-5

Rittmann, O., Neunhoeffer, M., & Gschwend, T. (2025). How to improve the substantive interpretation of regression results when the dependent variable is logged. *Political Science Research and Methods, 13*(1), 203–211. DOI: 10.1017/psrm.2023.29

Rivalda, M. R., Ma'ruf, J. J., & Djalil, M. A. (2024). What drives repurchase intention? Mediating role of customer satisfaction and customer relationship management with marketing mix as an independent variable. *International Journal of Business Management and Economic Research, 7*(4), 117–131. DOI: 10.35409/IJBMER.2024.3595

Tang, W., He, H., & Tu, X. M. (2023). *Applied categorical and count data analysis*. Chapman and Hall/CRC., DOI: 10.1201/9781003109815

Tawakuli, A., Havers, B., Gulisano, V., Kaiser, D., & Engel, T. (2025). Survey: Time-series data preprocessing: A survey and an empirical analysis. *Journal of Engineering Research, 13*(2), 674–711. DOI: 10.1016/j.jer.2024.02.018

Zhang, Z., Chen, X., Wang, C., Wang, R., Song, W., & Nie, F. (2025). Structured multi-view k-means clustering. *Pattern Recognition, 160*, 111113. DOI: 10.1016/j.patcog.2024.111113

ADDITIONAL READING

Ezechi, O. N., Famoti, O., Ewim, C. P. M., Eloho, O., Muyiwa-Ajayi, T. P., Igwe, A. N., & Ibeh, A. I. (2025). Service quality improvement in the banking sector: A data analytics perspective. *International Journal of Advanced Multidisciplinary Research and Studies, 5*(1), 958–971. DOI: 10.62225/2583049X.2025.5.1.3749

Huynh, N., Uddin, M., & Chu, M. C. (2025). Data analytics for intermodal freight transportation applications. In *Data Analytics for Intelligent Transportation Systems* (pp. 263–287). Elsevier. DOI: 10.1016/B978-0-443-13878-2.00013-8

Rahim, M. J., Afroz, A., & Akinola, O. (2025). Predictive analytics in healthcare: Big data, better decisions. *International Journal of Scientific Research and Modern Technology*, *4*(1), 1–21.

Sariyer, G., Mangla, S. K., Kazancoglu, Y., Ocal Tasar, C., & Luthra, S. (2025). Data analytics for quality management in Industry 4.0 from a MSME perspective. *Annals of Operations Research*, *350*(2), 365–393. DOI: 10.1007/s10479-021-04215-9

KEY TERMS AND DEFINITIONS

Analysis of Variance (ANOVA): it is used in determining if there is significant difference between the means of three or more independent variables.

Artificial intelligence (AI): is a collection of methods used to create systems that can execute tasks like human intelligence, learning, and helps in better decision-making.

Covariance: is a statistical method used to check whether two variables change together and helps in deciding their linear relationship.

Correlation: is a statistical method to check the relationship of two variables. Correlation can be positive, negative, or can have no linear relationship. It can be measured with the correlation coefficient between -1 and +1.

Data cleaning: is a method to identify and correct the errors, and inaccuracies in the given dataset & helps improving dataset quality, accuracy, and reliability.

Descriptive analysis: it is a kind of summarization of the features of a dataset. With the help of descriptive analysis one can visualize the data and perform exploratory data analysis.

Exploratory analysis: is an approach analyzing dataset to summarize the main characteristics, features using different visualization techniques.

Knowledge Discovery in Databases: (KDD): extracting the meaningful data, patterns and features from unknown patterns and large datasets.

Machine learning: is a collection of algorithms learn from data, identifies patterns, and helps making correct predictions.

Predictive analytics: it is the use of data, algorithms, and machine learning methods to make predictions about outcomes.

Prescriptive analytics: is the advanced data analytics which uses data and algorithms to recommend the actions to achieve the predicted outcomes.

Sampling: is the method of selecting a subset from a large group or dataset to estimate the characteristics of the whole population.

Supervised learning: is a machine learning method which uses labeled data to train predictive models.

Support Vector Machine (SVM): is a supervised machine learning algorithm used for classification and regression tasks.

Unsupervised learning: uses algorithms and methods finding patterns and structures in unlabeled data. It is expected that it does not require human supervision or pre-defined categories.

Compilation of References

Abad, Z. A. N., & Momayezi, F. (2024). Optimizing modular hub location in air and road transportation systems. *Journal of Project Management*, *9*(3), 277–300. DOI: 10.5267/j.jpm.2024.3.001

Abughazala, M., & Muccini, H. (2023, March). Modeling data analytics architecture for IOT applications using DAT. In *2023 IEEE 20th International Conference on Software Architecture Companion (ICSA-C)* (pp. 284-291). IEEE. DOI: 10.1109/ICSA-C57050.2023.00066

Abu-Shanab, E., & Khasawneh, R. (2014). E-government adoption: The challenge of digital divide based on Jordanians' perceptions. *Theoretical and Empirical Researches in Urban Management*, *9*(4), 5–19.

Acharibasam, J. B., & McVittie, J. (2022). Connecting children to nature through the integration of indigenous ecological knowledge into early childhood environmental education. *Australian Journal of Environmental Education*, *39*(3), 349–361. DOI: 10.1017/aee.2022.37

Adekoya, O. A., & Chen, W. (2023). Implications of algorithmic management on careers and employment relationships in the gig economy: A developing-country perspective. *The Journal of Industrial Relations*, *65*(4), 587–604.

Aggarwal, C. C. (2018). Machine learning for text: An introduction. In *Machine learning for text* (pp. 1–16). Springer International Publishing. DOI: 10.1007/978-3-319-73531-3_1

Agrawal, R., Gupta, T., Gupta, S., Chauhan, S. S., Patel, P., & Hamdare, S. (2025). Fostering trust and interpretability: Integrating explainable AI (XAI) with machine learning for enhanced disease prediction and decision transparency. *Diagnostic Pathology*, *20*(1), 105. Advance online publication. DOI: 10.1186/s13000-025-01686-3

Ahmad, K. M., Campbell, J., Pathak, R. D., Belwal, R., Singh, G., Naz, R., Smith, R. F., Al-Zoubi, K., Carter, L., Bélanger, F., Lim, A. L., Masrom, M., Din, S., Chen, Y. C., Dimitrova, D. V., Dodeen, W., Adolph, A., Bortier, S., & Munyoka, W. (2019). Satisfaction with e-participation: A model from the citizens' perspective, expectations, and effective ties to the place. *African Journal of Business Management, 7*(1), 157–166.

Aje, O. F., & Adie, A. J. (2020). Response surface methodology and Taguchi method based applications – A Review. Zenodo *(CERN European Organization for Nuclear Research)*. European Organization for Nuclear Research. DOI: 10.5281/zenodo.4467686

Akhai, S., & Khang, A. (2024). Innovations in medical manufacturing: A review of 3D printing, robotics, and internet of things (IoT). *The Quantum Evolution*, 226-241. DOI: 10.1201/9781032642079-10

Akhai, S., Srivastava, P., Sharma, V., & Bhatia, A. (2021, August). Investigating weld strength of AA8011-6062 alloys joined via friction-stir welding using the RSM approach. In *Journal of Physics: Conference Series*, 1950(1), 012016. IOP Publishing. DOI: 10.1088/1742-6596/1950/1/012016

Akhai, S., Singh, V. P., & John, S. (2016). Human performance in industrial design centers with small unit air conditioning systems. *Journal of Advanced Research in Production Industrial Engineering, 3*(2), 5–11.

Akhai, S., & Wadhwa, A. S. (2024). Wireless networking solutions for climate control and comfort in modern air conditioning systems. In *Intelligent Networks* (pp. 165–181). CRC Press., DOI: 10.1201/9781003541363-11

Akil, B., Zhou, Y., & Röhm, U. (2017). On the usability of Hadoop MapReduce, Apache Spark & Apache flink for data science. *Proceedings - 2017 IEEE International Conference on Big Data, Big Data 2017, 2018-January*. DOI: 10.1109/BigData.2017.8257938

Akpa-Inyang, F., & Chima, S. C. (2021). South African traditional values and beliefs regarding informed consent and limitations of the principle of respect for autonomy in African communities: A cross-cultural qualitative study. *BMC Medical Ethics, 22*(1), 111. DOI: 10.1186/s12910-021-00678-4

Alahmadi, S., Mohammed, T., Albeshri, A., Katib, I., & Mehmood, R. (2020). Performance analysis of sparse matrix-vector multiplication (Spmv) on graphics processing units (gpus). *Electronics (Basel), 9*(10), 1675. Advance online publication. DOI: 10.3390/electronics9101675

Albar, H. M., Dahlan, H. A., Ahmad, A., Yuhefizar, E., & Napitupulu, D. (2017). E-government service quality based on e-GovQual approach: Case study in West Sumatera Province. *International Journal on Advanced Science, Engineering and Information Technology*, *7*(6), 2337–2342. DOI: 10.18517/ijaseit.7.6.4226

Alberts, I. L., Mercolli, L., Pyka, T., Prenosil, G., Shi, K., Rominger, A., & Afshar-Oromieh, A. (2023). Large language models (LLMs) and ChatGPT: What will the impact on nuclear medicine be? *European Journal of Nuclear Medicine and Molecular Imaging*, *50*(6), 1549–1552. DOI: 10.1007/s00259-023-06172-w

Aldeghi, M., Häse, F., Hickman, R. J., Tamblyn, I., & Aspuru-Guzik, A. (2021). Golem: An algorithm for robust experiment and process optimisation. *Chemical Science (Cambridge)*, *12*(44), 14792–14807. DOI: 10.1039/D1SC01545A

Aldulaijan, N., Marsden, J. A., Manson, J. A., & Clayton, A. D. (2023). Adaptive mixed variable bayesian self-optimisation of catalytic reactions. *Reaction Chemistry & Engineering*, *9*(2), 308–316. DOI: 10.1039/D3RE00476G

Al-Husban, Y., Al-Ghriybah, M., Gaeid, K. S., Takialddin, A. S., Handam, A., & Alkhazaleh, A. H. (2023). Optimisation of the residential solar energy consumption using the taguchi technique and box-behnken design: A case study for Jordan. *International Journal on Energy Conversion*, *11*(1), 25. Advance online publication. DOI: 10.15866/irecon.v11i1.22672

Ali, H., Muthudoss, P., Chauhan, C., Ilango, K., Kumar, D., Paudel, A., & Ramasamy, G. (2023). Machine learning-enabled NIR Spectroscopy. Part 3: Hyperparameter by Design (HyD) Based ANN-MLP Optimisation, Model Generalizability, and Model Transferability. *AAPS PharmSciTech*, *24*(8), 254. Advance online publication. DOI: 10.1208/s12249-023-02697-3

Al-Kfairy, M., Mustafa, D., Kshetri, N., Insiew, M., & Alfandi, O. (2024). Ethical challenges and solutions of generative AI: An interdisciplinary perspective. *Informatics (MDPI)*, *11*(3), 58. DOI: 10.3390/informatics11030058

Alpaydin, E. (2020). *Introduction to machine learning* (4th ed.). MIT Press.

Alqahtani, H., Kavakli-Thorne, M., & Kumar, G. (2021). Applications of Generative Adversarial Networks (GANs): An Updated Review. *Archives of Computational Methods in Engineering*, *28*(2), 525–552. Advance online publication. DOI: 10.1007/s11831-019-09388-y

Al-Sibai, N. (2023, September 13). Companies are paying bonkers salaries for people good at ChatGPT: Expert prompt engineers can name their price. *The Byte*. https://futurism.com/the-byte/companies-paying-high-salaries-ChatGPT

Amatriain, X. (2024). Prompt design and engineering: Introduction and advanced methods. *arXiv*. https://doi.org//arXiv.2401.14423DOI: 10.48550

Ambira, C. M. (2016). A framework for management of electronic records in support of e-government in Kenya. PhD thesis, University of South Africa, Pretoria.

Ambira, C. M., Kemoni, H., & Ngulube, P. (2019). A framework for electronic records management in support of e-government in Kenya. *Records Management Journal*, *29*(3), 305–319. DOI: 10.1108/RMJ-03-2018-0006

Amershi, S., Weld, D., Vorvoreanu, M., Fourney, A., Nushi, B., Collisson, P., & Horvitz, E. (2019, May). Guidelines for human-AI interaction. In *Proceedings of the 2019 chi conference on human factors in computing systems* (pp. 1-13).

Amirjalili, F., Neysani, M., & Nikbakht, A. (2024, March). Exploring the boundaries of authorship: A comparative analysis of AI-generated text and human academic writing in English literature. *Frontiers in Education*, *9*(1), 1347421. Advance online publication. DOI: 10.3389/feduc.2024.1347421

Andersen, J. P., Degn, L., Fishberg, R., Graversen, E. K., Horbach, S. P. J. M., Schmidt, E. K., Schneider, J. W., & Sørensen, M. P. (2025). Generative artificial intelligence (GenAI) in the research process: A survey of researchers' practices and perceptions. *Technology in Society*, *81*, 1–11. DOI: 10.1016/j.techsoc.2025.102813

Anifa, M., & Mahmoud, A. B. (2024). Sports analytics: Data-driven sports and decision intelligence. In *Sports analytics: Data-driven sports and decision intelligence* (pp. 1–17). Springer., DOI: 10.1007/978-3-031-63573-1_1

Anis, S., & French, J. A. (2023). Efficient, explicatory, and equitable: Why qualitative researchers should embrace AI, but cautiously. *Business & Society*, *62*(6), 1139–1144. DOI: 10.1177/00076503231163286

Anwar, M. A. (2022). Platforms of inequality: Gender dynamics of digital labour in Africa. *Gender and Development*, *30*(3), 747–764. DOI: 10.1080/13552074.2022.2121059

Aoki, N., Tatsumi, T., Naruse, G., & Maeda, K. (2024). Explainable AI for government: Does the type of explanation matter to the accuracy, fairness, and trustworthiness of an algorithmic decision as perceived by those who are affected? *Government Information Quarterly*, *41*(4), 101965. Advance online publication. DOI: 10.1016/j.giq.2024.101965

Ao, S.-I., Gelman, L., Karimi, H. R., & Tiboni, M. (2022). Advances in machine learning for sensing and condition monitoring. *Applied Sciences (Basel, Switzerland)*, *12*(23), 12392. DOI: 10.3390/app122312392

Aqib, M., Mehmood, R., Alzahrani, A., Katib, I., Albeshri, A., & Altowaijri, S. M. (2019). Smarter traffic prediction using big data, in-memory computing, deep learning and gpus. *Sensors (Basel), 19*(9), 2206. Advance online publication. DOI: 10.3390/s19092206 PMID: 31086055

Arora, P. (2024). Creative data justice: A decolonial and indigenous framework to assess creativity and artificial intelligence. *Information Communication and Society*. Advance online publication.

Arriola-Pacheco, F., Ness, A., Sihuay-Torres, K., Garcia-Quintana, A., Proaño, D., & Lawrence, H. P. (2025). Community-based participatory research: Lessons and challenges. symposium special communication. *JDR Clinical and Translational Research, 10*(2), 96–99. DOI: 10.1177/23800844241266505

Arteaga-Cabrera, E., Ramírez-Márquez, C., Sánchez-Ramírez, E., Segovia-Hernández, J. G., Osorio, O., & Gómez-Salazar, J. A. (2025). Advancing optimisation strategies in the food industry: From traditional approaches to multi-objective and technology-integrated solutions. *Applied Sciences (Basel, Switzerland), 15*(7), 3846. DOI: 10.3390/app15073846

Artiushenko, V., Lang, S., Lerez, C., Reggelin, T., & Hackert-Oschätzchen, M. (2024). Resource-efficient Edge AI solution for predictive maintenance. *Procedia Computer Science, 232*, 348–357. DOI: 10.1016/j.procs.2024.01.034

Arubayi, D. E. (2022). *Algorithmic management and everyday mobilities: A case study of ride-hailing drivers in Lagos, Nigeria* (Doctoral dissertation, University of Manchester).

Arundel, S. T., Li, W., & Wang, S. (2020). GeoNat v1.0: A dataset for natural feature mapping with artificial intelligence and supervised learning. *Transactions in GIS, 24*(3), 556–572. Advance online publication. DOI: 10.1111/tgis.12633

Aslam, M. S., & Nisar, S. (2024). Ethical Considerations for artificial intelligence tools in academic research and manuscript preparation: A web content analysis. In *Digital Transformation in Higher Education, Part B: Cases, Examples and Good Practices* (pp. 155–196). Emerald Publishing Limited. DOI: 10.1108/978-1-83608-424-220241007

Awais, M., Wang, X., Hussain, S., Aziz, F., & Mahmood, M. Q. (2025). Advancing precision agriculture through digital twins and smart farming technologies: A Review [Review of *Advancing Precision Agriculture Through Digital Twins and Smart Farming Technologies: A Review*]. *AgriEngineering, 7*(5), 137. Multidisciplinary Digital Publishing Institute. DOI: 10.3390/agriengineering7050137

Ayentimi, D., Amankwaa, A., & Burgess, J. (2025). The Emerging Gig Economy and Sustainable Development in Sub-Saharan Africa. *Societies (Basel, Switzerland), 15*(10), 274. DOI: 10.3390/soc15100274

Azeroual, O. (2025). Beyond Black Boxes: Adaptive XAI for Dynamic Data Pipelines. In *Proceedings of the 17th International Joint Conference on Knowledge Discovery, Knowledge Engineering and Knowledge Management* - (Volume 2), pp. 428-437. DOI: 10.5220/0013736100004000

Azevedo, É. S. F., & Mendonça, J. R. C. (2023). Algorithmic management on digital labour platforms: A systematic literature review. *International Journal of Human-Computer Studies, 178*, 103048.

Aziz, A., Mehboob, S. S., Tayyab, A., Khan, D., Hayyat, K., Ali, A., & Qureshi, Q. B. I. L. (2024). Enhancing sustainability in self-compacting concrete by optimizing blended supplementary cementitious materials. *Scientific Reports, 14*(1), 12326. Advance online publication. DOI: 10.1038/s41598-024-62499-w

Bachorz, R. A., Nowak, D., & Ratajewski, M. (2024). QSPRmodeler - An open source application for molecular predictive analytics. *Frontiers in Bioinformatics, 4*, 1441024. Advance online publication. DOI: 10.3389/fbinf.2024.1441024

Baheer, B. A., Lamas, D., & Sousa, S. (2020). A systematic literature review on existing digital government architectures: State-of-the-art, challenges, and prospects. *Administrative Sciences, 10*(2), 1–28. DOI: 10.3390/admsci10020025

Bailey, D. E., & Barley, S. R. (2020). Beyond design and use: How scholars should study intelligent technologies. *Information and Organization, 30*(2), 100286. Advance online publication. DOI: 10.1016/j.infoandorg.2019.100286

Baker, M. (2016). 1,500 scientists lift the lid on reproducibility. *Nature, 533*(7604), 452–454. DOI: 10.1038/533452a

Bakhaidar, R. B., Naveen, N. R., Basim, P., Murshid, S. S. A., Kurakula, M., Alamoudi, A. J., Bukhary, D. M., Jali, A. M., Majrashi, M., Alshehri, S., Alissa, M., & Ahmed, R. A. (2022). Response surface methodology (RSM) powered formulation development, optimisation and evaluation of thiolated based mucoadhesive nanocrystals for local delivery of simvastatin. *Polymers, 14*(23), 5184. DOI: 10.3390/polym14235184

Balogun, T. (2023). Digitization of Indigenous knowledge systems in Africa: The case of South Africa's National Recorded System (NRS). *Records Management Journal, 33*(1), 88–102. DOI: 10.1108/RMJ-10-2022-0029

Balogun, T., & Kalusopa, T. (2021). A framework for digital preservation of Indigenous knowledge system (IKS) in repositories in South Africa. *Records Management Journal*, *31*(2), 176–196. DOI: 10.1108/RMJ-12-2020-0042

Balogun, T., & Kalusopa, T. (2022). Web archiving of indigenous knowledge systems in South Africa. *Information Development*, *38*(4), 658–671. DOI: 10.1177/02666669211005522

Banashooshtari, H., Khamehchi, E., Rashidi, F., & Dargi, M. (2024). An experimental study of epoxy-based nanocomposite for chemical consolidation in a sandstone reservoir with high clay content. *Journal of Petroleum Exploration and Production Technology*, *14*(10), 2769–2784. DOI: 10.1007/s13202-024-01853-0

Bangi, M. S. F., & Kwon, J. S. (2019). Deep hybrid modeling of chemical process: Application to hydraulic fracturing. *Computers & Chemical Engineering*, *134*, 106696. DOI: 10.1016/j.compchemeng.2019.106696

Bano, M., Hoda, R., Zowghi, D., & Treude, C. (2023). Large language models for qualitative research in software engineering: Exploring opportunities and challenges. *Automated Software Engineering*, *31*(1), 8. Advance online publication. DOI: 10.1007/s10515-023-00407-8

Bansal, S., Sarker, T., Yadav, A., Garg, I., Gupta, M., & Sarvaiya, H. (2024). Indigenous communities and sustainable development: A review and research agenda. *Global Business and Organizational Excellence*, *43*(4), 65–87. https://doi.org/https://doi.org/10.1002/joe.22237. DOI: 10.1002/joe.22237

Barité, M., & Rauch, M. (2025). Native peoples and knowledge organization: Perspective from the Indigenous subject representation to promote Latin American approaches. *Knowledge Organization*, *51*(8), 626–641. DOI: 10.5771/0943-7444-2024-8-626

Batra, N., Kaur, J., & Batra, N. K. (2020). SmartCOdetect: An automated car window opening system on detection of carbon monoxide. *International Journal of Vehicle Structures & Systems*, *12*(4), 398–404. DOI: 10.4273/ijvss.12.4.09

Baum, S. D. (2021). Artificial Interdisciplinarity: Artificial Intelligence for Research on Complex Societal Problems. *Philosophy & Technology*, *34*(S1), 45–63. Advance online publication. DOI: 10.1007/s13347-020-00416-5

Bawack, R., Roderick, S., Badhrus, A., Dennehy, D., & Corbett, J. (2025). Indigenous knowledge and information technology for sustainable development. *Information Technology for Development*, *31*(2), 233–250. DOI: 10.1080/02681102.2025.2472495

Bayuo, J., Abukari, M. A., & Pelig-Ba, K. B. (2020). Optimisation using central composite design (CCD) of response surface methodology (RSM) for biosorption of hexavalent chromium from aqueous media. *Applied Water Science, 10*(6), 135. Advance online publication. DOI: 10.1007/s13201-020-01213-3

Beam, A. L., & Kohane, I. S. (2018). Big data and machine learning in health care. *Journal of the American Medical Association, 319*(13), 1317–1318. DOI: 10.1001/jama.2017.18391

Begay, V., & Klor, K. M. (2024). Provenance through storytelling: Application of Indigenous relationality toward arrangement and description. *Archival Science, 24*(4), 611–635. DOI: 10.1007/s10502-024-09451-z

Bekker, A. (2017). *Spark vs. Hadoop MapReduce: Which big data framework to choose*. ScienceSoft.

Belarde-Lewis, M., Littletree, S., Braine, I. S., Srader, K., Guerrero, N., & Palmer, C. L. (2024). Centering relationality and CARE for stewardship of indigenous research data. *Data Science Journal, 23*(32), 1–16. https://doi.org/https://doi.org/10.5334/dsj-2024-032. DOI: 10.5334/dsj-2024-032

Belim, M., Meireles, T., Gonçalves, G., & Pinto, R. (2024). Forecasting models analysis for predictive maintenance. *Frontiers in Manufacturing Technology, 4*, 1475078. Advance online publication. DOI: 10.3389/fmtec.2024.1475078

Bell, N. L., Kupper, M., & Cronin, L. (2021). Design of experiments for optimisation of polyoxometalate syntheses. *Chemistry of Materials, 33*(18), 7263–7271. DOI: 10.1021/acs.chemmater.1c01401

Beltagy, I., Lo, K., & Cohan, A. (2019). SciBERT: A pretrained language model for scientific text. *arXiv preprint arXiv:1903.10676*. DOI: 10.18653/v1/D19-1371

Benbya, H., Pachidi, S., & Jarvenpaa, S. L. (2021). Special issue editorial: Artificial intelligence in organisations: Implications for information systems research. *Journal of the Association for Information Systems, 22*(2), 281–303. DOI: 10.17705/1jais.00662

Benbya, H., Strich, F., & Tamm, T. (2023). Navigating the promises and perils of generative artificial intelligence for knowledge and creative work. *Journal of the Association for Information Systems, 25*(1), 23–36. DOI: 10.17705/1jais.00861

Bennett, K. (2023). Approaches to knowledge translation. In *The Routledge handbook of translation theory and concepts* (pp. 443–462). Routledge. DOI: 10.4324/9781003161448-27

Berente, N., Gu, B., & Santhanam, R. (2021). Managing artificial intelligence. *Management Information Systems Quarterly*, *45*(3), 1433–1450. DOI: 10.25300/MISQ/2021/16274

Bhattamisra, S. K., Banerjee, P., Gupta, P., Mayuren, J., Patra, S., & Candasamy, M. (2023). Artificial Intelligence in Pharmaceutical and Healthcare Research. In *Big Data and Cognitive Computing* (Vol. 7, Issue 1). DOI: 10.3390/bdcc7010010

Bhebhe, S. (2015). Contemporary diplomatics of the civil and deceased estate case files found at the national archives of Zimbabwe. *Records Management Journal*, *25*(1), 107–120. DOI: 10.1108/RMJ-03-2014-0019

Bheel, N., Shams, M. A., Sohu, S., Buller, A. S., Najeh, T., Ismail, F. I., & Benjeddou, O. (2024). A comprehensive study on the impact of human hair fiber and millet husk ash on concrete properties: Response surface modeling and optimisation. *Scientific Reports*, *14*(1), 13569. Advance online publication. DOI: 10.1038/s41598-024-63050-7

Bielska, B., Kalinowska, K., Męcfal, S., & Surmiak, A. (2024). Strengths and limitations of an online qualitative survey in times of social crisis: Example of the COVID-19 pandemic. *Qualitative Sociology Review*, *20*(4), 78–100. Advance online publication. DOI: 10.18778/1733-8077.20.4.04

Bingham, A. J. (2023). From data management to actionable findings: A five-phase process of qualitative data analysis. *International Journal of Qualitative Methods*, *22*, 16094069231183620. DOI: 10.1177/16094069231183620

Binns, R., Stein, J., Datta, S., Van Kleek, M., & Shadbolt, N. (2025, June). Not Even Nice Work If You Can Get It: A Longitudinal Study of Uber's Algorithmic Pay and Pricing. In *Proceedings of the 2025 ACM Conference on Fairness, Accountability, and Transparency* (pp. 1484-1497). DOI: 10.1145/3715275.3732099

Birks, D. F., Fernandez, W., Levina, N., & Nasirin, S. (2013). Grounded theory method in information systems research: Its nature, diversity and opportunities. *European Journal of Information Systems*, *22*(1), 1–8. DOI: 10.1057/ejis.2012.48

Birks, M., & Mills, J. (2015). *Grounded theory: A practical guide* (2nd ed.). Sage Publications.

Bokányi, E., Bokányi, E., Hannak, A., Hannak, A., & Hannak, A. (2020). Understanding Inequalities in Ride-Hailing Services Through Simulations. *Scientific Reports*, *10*(1), 6500. DOI: 10.1038/s41598-020-63171-9 PMID: 32300117

Bolanos, F., Salatino, A., Osborne, F., & Motta, E. (2024). Artificial intelligence for literature reviews: Opportunities and challenges. *Artificial Intelligence Review, 57*(10), 259. DOI: 10.1007/s10462-024-10902-3

Book Aid International. (2013). Talking to users: a review of our work in five countries. Retrieved on 10 October from https://www.bookaid.org/our-work/publications-2/talking-to-users-2/

Bornmann, L., & Mutz, R. (2015). Growth rates of modern science: A bibliometric analysis based on the number of publications and cited references. *Journal of the Association for Information Science and Technology, 66*(11), 2215–2222. DOI: 10.1002/asi.23329

Bose, S., Dey, S. K., & Bhattacharjee, S. (2023). Big data, data analytics and artificial intelligence in accounting: An overview. *Handbook of big data research methods (EE Elgar Online)*, 32-51. DOI: 10.4337/9781800888555.00007

Botha, N., Inglis, H. M., Coetzer, R., & Labuschagne, F. J. (2021). Statistical design of experiments: An introductory case study for polymer composites manufacturing applications. *MATEC Web of Conferences, 347*, 28. DOI: 10.1051/matecconf/202134700028

Boudrez, F. (2007). *Digital signatures and electronic records.* Springer Science: Business Media.

Boukredera, F. S., Hadjadj, A., Youcefi, M. R., & Ouadi, H. (2025). AI-driven optimisation of drilling performance through torque management using machine learning and differential evolution. *Processes (Basel, Switzerland), 13*(5), 1472. DOI: 10.3390/pr13051472

Bourassa, C., Billan, J., Starblanket, D., Anderson, S., Legare, M., Hagel, M. C., Oakes, N., Jardine, M., Boehme, G., Dubois, E., Spencer, O., Hotomani, M., & McKenna, B. (2020). Ethical research engagement with Indigenous communities. *Journal of Rehabilitation and Assistive Technologies Engineering, 7*, 2055668320922706. DOI: 10.1177/2055668320922706

Bozkurt, A. (2024). Tell me your prompts, and I will make them true: The alchemy of prompt engineering and generative AI. *Open Praxis, 16*(2), 111–118. DOI: 10.55982/openpraxis.16.2.661

Brailovskaya, V., Onkundi, F. K., & Atela, M. (2025). *Economic lives of digital platform gig workers: Case of motorbike passenger and delivery drivers in Kenya.* IDInsight.

Brändle, S., Hanussek, M., Blohm, M., & Kintz, M. (2022). Evaluation of Representation Models for Text Classification with AutoML Tools. *Lecture Notes in Networks and Systems, 359 LNNS*. DOI: 10.1007/978-3-030-89880-9_24

Braun, V., & Clarke, V. (2006). Using thematic analysis in psychology. *Qualitative Research in Psychology*, *3*(2), 77–101. DOI: 10.1191/1478088706qp063oa

Braun, V., & Clarke, V. (2013). *Successful qualitative research: A practical guide for beginners*. SAGE Publications.

Braun, V., & Clarke, V. (2021). Can I use TA? Should I use TA? Should I not use TA? Comparing reflexive thematic analysis and other pattern-based qualitative analytic approaches. *Counselling & Psychotherapy Research*, *21*(1), 37–47. DOI: 10.1002/capr.12360

Braun, V., Clarke, V., Boulton, E., Davey, L., & McEvoy, C. (2021). The online survey as a qualitative research tool. *International Journal of Social Research Methodology*, *24*(6), 641–654. DOI: 10.1080/13645579.2020.1805550

Brinckley, M.-M., Bourke, S., Watkin Lui, F., & Lovett, R. (2022). Knowledge translation in Aboriginal and Torres Strait Islander research contexts in Australia: Scoping review protocol. *BMJ Open*, *12*(7), e060311. DOI: 10.1136/bmjopen-2021-060311

Broadhead, L.-A., & Howard, S. (2021). Confronting the contradictions between Western and Indigenous science: A critical perspective on two-eyed seeing. *Alternative*, *17*(1), 111–119. DOI: 10.1177/1177180121996326

Bromark, K., Knutsson, O., & Spånberger Weitz, Y. (2023). Co-designing a dynamic tool to enhance participation for young people: A participatory design project with young service users and social workers. *Children and Youth Services Review*, *149*, 106856. DOI: 10.1016/j.childyouth.2023.106856

Bronfenbrenner, U. (2005). Ecological systems theory (1992). In Bronfenbrenner, U. (Ed.), *Making human beings human: Bioecological perspectives on human development* (pp. 106–173). Sage Publications Ltd.

Brooke, S., & Rao, A. H. (2024). Designing for justice in freelancing: Testing platform interventions to minimise discrimination in online labour markets. *Big Data & Society*, *11*(1), 20539517241232631. Advance online publication. DOI: 10.1177/20539517241232631

Brownlee, A. E. I., Epitropakis, M. G., Mulder, J., Paelinck, M., & Burke, E. (2022). A systematic approach to parameter optimisation and its application to flight schedule simulation software. *Journal of Heuristics*, *28*(4), 509–538. DOI: 10.1007/s10732-022-09501-8

Bruno, G., Bell, R. C., Parlee, B., Lightning, P., Bull, I., Cutknife, B., & Oster, R. T. (2022). Mâmawihitowin (bringing the camps together): Perinatal healthcare provider and staff participation in an indigenous-led experiential intervention for enhancing culturally informed care—A mixed methods study. *International Journal for Equity in Health, 21*(1), 164. Advance online publication. DOI: 10.1186/s12939-022-01764-8

Bryant, A., & Charmaz, K. (2007). *The SAGE handbook of grounded theory*. Sage Publications., DOI: 10.4135/9781848607941

Bryman, A., & Bell, E. (2011). *Business research methods*. Oxford University Press.

Bryman, A., & Burgess, R. G. (1994). Reflections on qualitative data analysis. In Bryman, A., & Burgess, R. G. (Eds.), *Analysing qualitative data* (pp. 216–226). Routledge., DOI: 10.4324/9780203413081_chapter_11

Bryson, J. J., & Theodorou, A. (2019). How society can maintain human-centric artificial intelligence. In Toivonen-Noro, M., Saari, E., Melkas, H., & Hasu, M. (Eds.), *Human-centered digitalization and services* (pp. 305–323). Springer., DOI: 10.1007/978-981-13-7725-9_16

Bubeck, S., Chandrasekaran, V., Eldan, R., Gehrke, J., Horvitz, E., Kamar, E., . . . Zhang, Y. (2023). Sparks of artificial general intelligence: Early experiments with GPT-4. *arXiv preprint arXiv:2303.12712.*

Bubeník, P., Rakyta, M., Buzalka, M., Biňasová, V., & Kovaríková, Z. (2025). Optimisation of business processes using artificial intelligence. *Electronics (Basel), 14*(11), 2105. DOI: 10.3390/electronics14112105

Buczak, A. L., & Guven, E. (2015). A survey of data mining and machine learning methods for cyber security intrusion detection. *IEEE Communications Surveys and Tutorials, 18*(2), 1153–1176. DOI: 10.1109/COMST.2015.2494502

Burrell, J. (2016). How the machine "thinks": Understanding opacity in machine learning algorithms. *Big Data & Society, 3*(1), 1–12. DOI: 10.1177/2053951715622512

Butler, K. T., Davies, D. W., Cartwright, H., Isayev, O., & Walsh, A. (2018). Machine learning for molecular and materials science. *Nature, 559*(7715), 547–555. DOI: 10.1038/s41586-018-0337-2

Buwule, R. S., & Ponelis, S. R. (2017). Perspectives on university library automation and national development in uganda. *IFLA Journal, 43*(3), 256–265. DOI: 10.1177/0340035217710539

Bwalya, K. J. (2009). Factors affecting adoption of e-government in Zambia. *The Electronic Journal on Information Systems in Developing Countries*, *38*(1), 1–13. DOI: 10.1002/j.1681-4835.2009.tb00267.x

Cahill, M., Cole, A., Joo, S., LeFebvre, L., & Garcia, A. (2024). Public library virtual storytimes: The whys, why nots, and measures of success. *Journal of Library Administration*, *64*(2), 224–242. DOI: 10.1080/01930826.2024.2305074

Cai, H., Lin, J., Lin, Y., Liu, Z., Tang, H., Wang, H., Zhu, L., & Han, S. (2022). Enable Deep Learning on Mobile Devices: Methods, Systems, and Applications. *ACM Transactions on Design Automation of Electronic Systems*, *27*(3), 1–50. Advance online publication. DOI: 10.1145/3486618

Cain, P. (2002). Model Requirements for the Management of Electronic Records (MoReq): A critical evaluation. *Records Management Journal*, *12*(1), 14–18. DOI: 10.1108/09565690210427752

Calacci, D., & Pentland, A. (2022). Bargaining with the black-box: Designing and deploying worker-centric tools to audit algorithmic management. *Proceedings of the ACM on Human-Computer Interaction*, *6*(CSCW2), 1-24. DOI: 10.1145/3570601

Caldwell, K. (2023). A New Librarian's take on ECRR: Best practices for evidence-based early literacy storytimes. *Children & Libraries*, *21*(1), 23–25. DOI: 10.5860/cal.21.1.23

Capurro, R. (2006). Towards an ontological foundation of information ethics. *Ethics and Information Technology*, *8*(4), 175–186. DOI: 10.1007/s10676-006-9108-0

Carabajal, M., Teglia, C. M., Cerutti, S., Culzoni, M. J., & Goicoechea, H. C. (2019). Applications of liquid-phase microextraction procedures to complex samples assisted by response surface methodology for optimisation. *Microchemical Journal*, *152*, 104436. DOI: 10.1016/j.microc.2019.104436

Carlin, B. P., & Louis, T. A. (2008). *Bayesian methods for data analysis*. CRC press. DOI: 10.1201/b14884

Carobene, A., Padoan, A., Cabitza, F., Banfi, G., & Plebani, M. (2024). Rising adoption of artificial intelligence in scientific publishing: Evaluating the role, risks, and ethical implications in paper drafting and review process. [CCLM]. *Clinical Chemistry and Laboratory Medicine*, *62*(5), 835–843. DOI: 10.1515/cclm-2023-1136 PMID: 38019961

Carroll, S. R., Garba, I., Figueroa-Rodríguez, O. L., Holbrook, J., Lovett, R., Materechera, S. A., Parsons, M. A., Raseroka, K., Rodriguez-Lonebear, D., Rowe, R., Sara, R., Walker, J., Anderson, J., & Hudson, M. (2020). The CARE Principles for Indigenous data governance. *Data Science Journal*, *19*, 43. Advance online publication. DOI: 10.5334/dsj-2020-043

Carson, D., & Lee, H. (2025). AI-enabled coaching platforms: Integrating strategic analytics in sports training. *Journal of Applied Cloud Computing*, *7*(3), 233–250. DOI: 10.1111/jcal.13082

Carter, B., & Ford, K. (2013). Researching children's health experiences: The place for participatory, child-centered, arts-based approaches. *Research in Nursing & Health*, *36*(1), 95–107. DOI: 10.1002/nur.21517

Carter, L., & Nguyen, T. (2024). Integrating digital innovations and leadership to revolutionize the sports industry. In *Innovation and leadership in sports management* (Vol. 1, pp. 22–40). Springer., DOI: 10.1007/978-3-031-63573-1_3

Cartolano, A., Cuzzocrea, A., & Pilato, G. (2024). Analyzing and assessing explainable AI models for smart agriculture environments. *Multimedia Tools and Applications*, *83*(12), 37225–37246. Advance online publication. DOI: 10.1007/s11042-023-17978-z

Castéran, F., Ibáñez, R., Argerich, C., Delage, K., Chinesta, F., & Cassagnau, P. (2020). Application of machine learning tools for the improvement of reactive extrusion simulation. *Macromolecular Materials and Engineering*, *305*(12), 2000375. Advance online publication. DOI: 10.1002/mame.202000375

Chan, C. K. Y., & Hu, W. (2023). Students' voices on generative AI: Perceptions, benefits, and challenges in higher education. *International Journal of Educational Technology in Higher Education*, *20*(1), 43. Advance online publication. DOI: 10.1186/s41239-023-00411-8

Chance, R., Okello-Obura, C., & Luyombya, D. (2025). User's Satisfaction with School Library Services: A Case of Selected Government-Aided Secondary Schools in Uganda. *East African Journal of Education and Social Sciences*, *6*(2), 59–70. DOI: 10.46606/eajess2025v06i02.0436

Chandra, A., Yadav, A., Singh, S., & Arora, P. K.Pawan Kumar Arora. (2023). Optimisation of machining parameters for CNC milling of fibre reinforced polymers. *Evergreen*, *10*(2), 765–773. DOI: 10.5109/6792826

Chang, T., DeJonckheere, M., Vydiswaran, V. G. V., Li, J., Buis, L. R., & Guetterman, T. C. (2021). Accelerating mixed methods research with natural language processing of big text data. *Journal of Mixed Methods Research, 15*(3), 398–412. DOI: 10.1177/15586898211021196

Chaulk, K. G., Ballard, M., Hill, S., Wolfrey, D., Campbell, M. L., Sutherland, M., Wawatie, S., & Auger, L. (2025). Bridging, braiding, and weaving Indigenous and Western science to understand and make predictions about weather and climate change. *Ecology and Evolution, 15*(9), e72085. Advance online publication. DOI: 10.1002/ece3.72085

Cheligeer, C., Yang, L., Nandi, T., Doktorchik, C., Quan, H., Zeng, Y., & Singh, S. (2023). Natural language processing (NLP) aided qualitative methods in health research. *Journal of Integrated Design & Process Science, 27*(1), 41–58. DOI: 10.3233/JID-220013

Chelladurai, S. J. S., Murugan, K., Ray, A. P., Upadhyaya, M., Narasimharaj, V., & Gnanasekaran, S. (2020). Optimisation of process parameters using response surface methodology: Review of *Optimisation of process parameters using response surface methodology: A review. Materials Today Proceedings, 37*, 1301. Elsevier BV. DOI: 10.1016/j.matpr.2020.06.466

Chen, M., Tworek, J., Jun, H., Yuan, Q., Pinto, H. P. D. O., Kaplan, J., . . . Zaremba, W. (2021). Evaluating large language models trained on code. *arXiv preprint arXiv:2107.03374.*

Chen, D., & Huang, Y. (2025). Integrating reinforcement learning and large language models for crop production process management optimisation and control through a new knowledge-based deep learning paradigm. *Computers and Electronics in Agriculture, 232*, 110028. DOI: 10.1016/j.compag.2025.110028

Chen, K. Y., Lee, H. C., Lin, T. C., Lee, C. Y., & Ho, Z. P. (2023). Deep Learning Algorithms with LIME and Similarity Distance Analysis on COVID-19 Chest X-ray Dataset. *International Journal of Environmental Research and Public Health, 20*(5), 4330. Advance online publication. DOI: 10.3390/ijerph20054330 PMID: 36901338

Chen, Q. (2023). Response surface methodology in the optimisation of natural product extraction and separation techniques. *Academic Journal of Science and Technology, 5*(3), 149–151. DOI: 10.54097/ajst.v5i3.7902

Chen, W., & Hirschheim, R. (2004). A paradigmatic and methodological examination of information systems research from 1991 to 2001. *Information Systems Journal, 14*(3), 197–235. DOI: 10.1111/j.1365-2575.2004.00173.x

Cheong, B. C. (2024). Transparency and accountability in AI systems: Safeguarding wellbeing in the age of algorithmic decision-making. *Frontiers in Human Dynamics*, *6*, 1421273. DOI: 10.3389/fhumd.2024.1421273

Chetwynd, E. (2024). Ethical use of artificial intelligence for scientific writing: Current trends. *Journal of Human Lactation*, *40*(2), 211–215. DOI: 10.1177/08903344241235160 PMID: 38482810

Chhajer, H., & Roy, R. (2024). Rationalised experiment design for parameter estimation with sensitivity clustering. *Scientific Reports*, *14*(1), 25864. Advance online publication. DOI: 10.1038/s41598-024-75539-2

Chibanda, R., Tsibolane, P., & Nkohla-Ramunenyiwa, T. (2022). Gendered inequality on digital labour platforms in the Global South: Towards a freedom-based inclusion. In Zheng, Y. (Eds.), *ICT4D 2022* (pp. 1–14). Springer. DOI: 10.1007/978-3-031-19429-0_4

Chigwada, J., Ngulube, P., & Dewah, P. (2024). Preserving indigenous knowledge in culture houses in the digital era in Masvingo Province, Zimbabwe. *African Journal of Library, Archives and Information Science*, 34(2), 205–216. Retrieved on 4 December from https://www.ajol.info/index.php/ajlais/article/view/281737

Chigwada, J., & Ngulube, P. (2024). Librarians' role in the preservation and dissemination of indigenous knowledge. *IFLA Journal*, *50*(2), 242–256. DOI: 10.1177/03400352231217270

Chigwada, J., & Ngulube, P. (2025a). Integrating CARE and FAIR Principles in the digital preservation of indigenous knowledge to achieve the Sustainable Development Goals. In Ngulube, P. (Ed.), *Harnessing Indigenous Epistemologies for Sustainable Progress* (pp. 197–230). IGI Global Scientific Publishing., https://doi.org/DOI, DOI: 10.4018/979-8-3693-8447-3.ch010

Chigwada, J., & Ngulube, P. (2025b). Stakeholders in the acquisition, preservation, and dissemination of indigenous knowledge projects. *Information Development*, *41*(4), 1281–1298. DOI: 10.1177/02666669231192851

Chikomba, A., Rodrigues, A., & Ngoepe, M. (2021). Management of digital records in selected financial services parastatals in Zimbabwe. *Information Development*, *37*(2), 245–257. DOI: 10.1177/0266666920906858

Chi, O., Denton, G., & Gursoy, D. (2020). Artificially intelligent device use in service delivery: A systematic review, synthesis, and research agenda. *Journal of Hospitality Marketing & Management*, *29*(7), 757–786. DOI: 10.1080/19368623.2020.1721394

Chipangura, N., Chipangura, P. M., Mataga, J., & Chabata, F. M. (2025). Truth-telling as a form of decolonial care: Stories from a collaborative object biography research with the Ndau community of Eastern Zimbabwe. *Museums & Social Issues*, *19*(1), 26–44. DOI: 10.1080/15596893.2025.2463657

Chisom, O. N., Biu, P. W., Umoh, A. A., Obaedo, B. O., Adegbite, A. O., & Abatan, A. (2024). Reviewing the role of AI in environmental monitoring and conservation: A data-driven revolution for our planet. *World Journal of Advanced Research and Reviews*, *21*(1), 161–171. Advance online publication. DOI: 10.30574/wjarr.2024.21.1.2720

Chiu, T. K. F. (2024). Future research recommendations for transforming higher education with generative AI. *Computers and Education: Artificial Intelligence*, *6*, 100197. Advance online publication. DOI: 10.1016/j.caeai.2023.100197

Choi, H. S. (2023). Siamese Neural Network for User Authentication in Field-Programmable Gate Arrays (FPGAs) for Wearable Applications. *Electronics (Basel)*, *12*(19), 4030. Advance online publication. DOI: 10.3390/electronics12194030

Choudhury, G., & Kalita, D. (2025). Collaborative stewardship: Sustainable approaches to indigenous knowledge in academic libraries. *Journal of Academic Librarianship*, *51*(2), 103007. https://doi.org/https://doi.org/10.1016/j.acalib.2025.103007. DOI: 10.1016/j.acalib.2025.103007

Choustoulakis, E. (2024, November). AI-driven analysis of game tactics and player performance. In *Proceedings of the 17th International Conference of Education, Research and Innovation (ICERI 2024)* (pp. 3350–3359). IATED. https://doi.org/ DOI: 10.21125/iceri.2024.0871

Christensen, P., & James, A. (2020). *Research with children: Perspectives and Practices* (Vol. 3, Issue 5).

Christensen, M., Yunker, L. P. E., Adedeji, F., Häse, F., Roch, L. M., Gensch, T., Gomes, G. dos P., Zepel, T., Sigman, M. S., Aspuru-Guzik, A., & Hein, J. E. (2021). Data-science driven autonomous process optimisation. *Communications Chemistry*, *4*(1), 112. Advance online publication. DOI: 10.1038/s42004-021-00550-x

Clapham, K., Hasan, H., Fredericks, B., Bessarab, D., Kelly, P. J., Harwood, V., Senior, K., Longbottom, M., & Dale, E. (2021). Digital support for Indigenous research methodologies. *AJIS. Australasian Journal of Information Systems*, *25*. Advance online publication. DOI: 10.3127/ajis.v25i0.2885

Clark, A., & Moss, P. (2011). *Listening to young children: The mosaic approach.* Retrieved on 10 October from https://books.google.com/books?hl=en&lr=&id=K1eHBAAAQBAJ&oi=fnd&pg=PR3&dq=Clark,+A.,+%26+Moss,+P.+(2011).+Listening+to+young+children:+The+mosaic+approach.+National+Children%E2%80%99s+Bureau.&ots=BDNBxTYwhF&sig=RTvPaqPUKPYjvccTrF7F-MCaL6E

Clarke, A., & Jackson, P. (2025). Data-driven message optimization in dynamic sports media: An AI approach. *Journal of Media Analytics, 12*(1), 55–72. DOI: 10.1080/14413523.2024.2372122

Clarke, V., & Braun, V. (2013). Teaching thematic analysis: Overcoming challenges and developing strategies for effective learning. *The Psychologist, 26*(2), 120–123.

Coffee, A. J., & Atkinson, P. (1996). *Making sense of qualitative data: Complementary research strategies (and social thought)* (1st ed.). Sage Publications.

Cohen, L., Manion, L., & Morrison, K. (2007). *Research methods in education* (6th ed.). Routledge Falmer., DOI: 10.4324/9780203029053

Collins, C., Dennehy, D., Conboy, K., & Mikalef, P. (2021). Artificial intelligence in information systems research: A systematic literature review and research agenda. *International Journal of Information Management, 60*, 102383. Advance online publication. DOI: 10.1016/j.ijinfomgt.2021.102383

Collis, J., & Hussey, R. (2014). *Business research: A practical guide for undergraduate and postgraduate students.* Palgrave. DOI: 10.1007/978-1-137-03748-0

Collomosse, J., & Parsons, A. (2024). To authenticity, and beyond! Building safe and fair generative AI upon the three pillars of provenance. *IEEE Computer Graphics and Applications, 44*(3), 82–90. DOI: 10.1109/MCG.2024.3380168 PMID: 38905025

Conboy, K., Fitzgerald, G., & Mathiassen, L. (2012). Qualitative methods research in information systems: Motivations, themes, and contributions. *European Journal of Information Systems, 21*(2), 113–118. DOI: 10.1057/ejis.2011.57

Cooke, N. J., & Shope, S. M. (2017). *Designing a Synthetic Task Environment.* Routledge., DOI: 10.4324/9781315243771-21

Corbin, J., & Strauss, A. (2008). *Basics of qualitative research* (3rd ed.). Sage.

Correani, A., De Massis, A., Frattini, F., Petruzzelli, A., & Natalicchio, A. (2020). Implementing a digital strategy: Learning from the experience of three digital transformation projects. *California Management Review, 62*(4), 37–56. DOI: 10.1177/0008125620934864

Cossich, V. R. A., Carlgren, D., Holash, R. J., & Katz, L. (2023). Technological breakthroughs in sport: Current practice and future potential of artificial intelligence, virtual reality, augmented reality, and modern data visualization in performance analysis. *Applied Sciences (Basel, Switzerland), 13*(23), 12965. Advance online publication. DOI: 10.3390/app132312965

Cotton, D. R., Cotton, P. A., & Shipway, J. R. (2023). Chatting and cheating: Ensuring academic integrity in the era of ChatGPT. *Innovations in Education and Teaching International*, •••, 1–12. DOI: 10.1080/14703297.2023.2190148

Cox, G., Firemoon, P., Anastario, M., Ricker, A., Thunder, R. E., Baldwin, J. A., & Rink, E. (2021). Indigenous Standpoint Theory as a theoretical framework for decolonizing social science health research with American Indian communities. *Alternative, 17*(4), 460–468. DOI: 10.1177/11771801211042019

Crespo, M. L. (2022). Fair, transparent and accountable algorithmic decision-making: The premise, the proposed solutions, and the open challenges. *iSCHANNEL, 17*(1), 28–41.

Creswell, J. W., & Creswell, J. D. (2017). *Research design: Qualitative, quantitative, and mixed methods approaches.* Sage Publications.

Crusoe, J., Magnusson, J., & Eklund, J. (2024). Digital transformation decoupling: The impact of wilful ignorance on public sector digital transformation. *Government Information Quarterly, 41*(3), 101958. Advance online publication. DOI: 10.1016/j.giq.2024.101958

Cuesta, I. I., Díaz, A., Rojo, M. A., Peral, L. B., Martínez, J., & Alegre, J. M. (2022). Parameter optimisation in selective laser melting on C300 steel. *Applied Sciences (Basel, Switzerland), 12*(19), 9786. DOI: 10.3390/app12199786

Cunningham-Reimann, S., Doshi, A., Stutz, S., Tjong, G. B., & Mashford-Pringle, A. (2025). Flipping data on its head: Differing conceptualisations of data and the implications for actioning Indigenous data sovereignty principles. *Global Public Health: An International Journal for Research, Policy and Practice, 20*(1), 2450395. DOI: 10.1080/17441692.2025.2450395

Czyrski, A., & Jarzębski, H. (2020). Response Surface Methodology as a Useful Tool for Evaluation of the Recovery of the Fluoroquinolones from Plasma—The Study on Applicability of Box-Behnken Design, Central Composite Design and Doehlert Design. *Processes (Basel, Switzerland), 8*(4), 473. DOI: 10.3390/pr8040473

Danesh, M., & Danesh, S. (2023). Optimal design of adaptive neuro-fuzzy inference system using PSO and ant colony optimisation for estimation of uncertain observed values. *Soft Computing*, *28*(1), 135–152. DOI: 10.1007/s00500-023-09194-6

Dang, M. P., Dao, T.-P., Chau, N. L., & Le, H. G. (2019). Effective hybrid algorithm of Taguchi Method, FEM, RSM, and Teaching Learning-Based optimisation for multiobjective optimisation design of a compliant rotary positioning stage for nanoindentation tester. *Mathematical Problems in Engineering*, *2019*(1). DOI: 10.1155/2019/4191924

Darabkh, K. A., & Al-Akhras, M. (2025). Evolutionary cost analysis and computational intelligence for energy efficiency in internet of things-enabled smart cities: Multi-sensor data fusion and resilience to link and device failures. *Smart Cities*, *8*(2), 64. DOI: 10.3390/smartcities8020064

Datta, R., & Kairy, B. (2024). Decolonizing climate change adaptations from indigenous perspectives: Learning reflections from Munda Indigenous communities, coastal areas in Bangladesh. *Sustainability (Basel)*, *16*(2), 769. DOI: 10.3390/su16020769

Datta, R., & Starlight, T. (2024). Building a meaningful bridge between Indigenous and Western worldviews: Through decolonial conversation. *International Journal of Qualitative Methods*, *23*, 16094069241235564. DOI: 10.1177/16094069241235564

Davenport, T. H., & Miller, S. (2022). *Working with AI: Real stories of human-machine collaboration*. MIT Press. DOI: 10.7551/mitpress/14453.001.0001

David-Chavez, D. (2019). *A guiding model for decolonizing environmental science research and restoring relational accountability with Indigenous communities*. PhD Diss, Colorado State University, Colorado. https://doi.org/DOI: 10.31237/osf.io/ec9s5

Davis, F. D. (1989). Perceived usefulness, perceived ease of use, and user acceptance of information technology. *Management Information Systems Quarterly*, *13*(3), 319–340. DOI: 10.2307/249008

de Brito, T. G., Nunes, T. C., de Paiva, E. J., de Lacerda, J. C., Brito, R. F., & Brito, R. P. (2024). Optimisation of iron ore flotation parameters using response surface methodology. *Research Square (Research Square)*. DOI: 10.21203/rs.3.rs-5183292/v1

de Ruijter, A., Cats, O., & van Lint, H. (2024). Ridesourcing platforms thrive on socio-economic inequality. *Scientific Reports*, *14*(1), 7371. DOI: 10.1038/s41598-024-57540-x PMID: 38548793

De Souza, S. P. (2024). Decolonial data law and governance. *Technology and Regulation*, *6*, 1–20. DOI: 10.71265/rvwfyt51

Debener, J., Heinke, V., & Kriebel, J. (2023). Detecting insurance fraud using supervised and unsupervised machine learning. *The Journal of Risk and Insurance, 90*(3), 743–768. Advance online publication. DOI: 10.1111/jori.12427

Del Giorgio Solfa, F., & Simonato, F. R. (2023). Big data analytics in healthcare: Exploring the role of machine learning in predicting patient outcomes and improving healthcare delivery. *International Journal of Computations* [IJCIM]. *Information and Manufacturing, 3*(1), 1–9. Advance online publication. DOI: 10.54489/ijcim.v3i1.235

Deng,L. Yu,D. (2014) Deep *Learning: Methods and Applications.* Now Foundations and Trends.

Dengel, A., Gehrlein, R., Fernes, D., Görlich, S., Maurer, J., Pham, H. H., Großmann, G., & Eisermann, N. D. (2023). Qualitative research methods for large language models: Conducting semi-structured interviews with ChatGPT and BARD on computer science education. *Informatics (MDPI), 10*(4), 78. Advance online publication. DOI: 10.3390/informatics10040078

Dennis, A. R., Lakhiwal, A., & Sachdeva, A. (2023). AI agents as team members: Effects on satisfaction, conflict, trustworthiness, and willingness to work with. *Journal of Management Information Systems, 40*(2), 307–337. DOI: 10.1080/07421222.2023.2196773

Denzin, N. K. (1989). Interpretive biography. *Sage (Atlanta, Ga.).*

DePuy, W., Weger, J., Foster, L. C., Bonanno, A., Kumar, S., Lear, K., Basilio, R., & German, L. (2021). Environmental governance: Broadening ontological spaces for a more livable world. *Environment and Planning. E, Nature and Space, 5*(2), 947–975. DOI: 10.1177/25148486211018565

Dergaa, I., Chamari, K., Zmijewski, P., & Saad, H. B. (2023). From human writing to artificial intelligence generated text: Examining the prospects and potential threats of ChatGPT in academic writing. *Biology of Sport, 40*(2), 615–622. DOI: 10.5114/biolsport.2023.125623 PMID: 37077800

Devlin, J., Chang, M. W., Lee, K., & Toutanova, K. (2019). Bert: Pre-training of deep bidirectional transformers for language understanding. In *Proceedings of the 2019 conference of the North American chapter of the association for computational linguistics: human language technologies, volume 1 (long and short papers)* (pp. 4171-4186).

Dhiman, S., Seema, S., & Akhai, S. (2025). Performance evaluation of OPC 43 concrete incorporating micro silica and recycled ceramic tile waste as partial cement and coarse aggregate replacements. *Asian Journal of Civil Engineering*, 1-19. DOI: 10.1007/s42107-025-01504-7

Dikshit, I., Bhushan, G., Singh, R. P., & Batra, N. K. (2022). Statistical modeling and optimisation of density in a novel carbon/jute fiber reinforced hybrid composite. *Materials Today: Proceedings, 63*, 259-263. https://doi.org/ rights and contentDOI: 10.1016/j.matpr.2022.03.052Get

Dinçer, S. (2024). The use and ethical implications of artificial intelligence in scientific research and academic writing. *Educational Research & Implementation, 1*(2), 139–144. DOI: 10.14527/edure.2024.10

Dockett, S., & Perry, B. (2011). Researching with young children: Seeking assent. *Child Indicators Research, 4*(2), 231–247. DOI: 10.1007/s12187-010-9084-0

Domonkos, P., & Joelsson, L. M. T. (2024). ANOVA (Benova) correction in relative homogenization: Why it is indispensable. *International Journal of Climatology, 44*(12), 4515–4528. DOI: 10.1002/joc.8594

Dörfler, F., He, Z., Belgioioso, G., Bolognani, S., Lygeros, J., & Muehlebach, M. (2024). Toward a Systems Theory of Algorithms. *IEEE Control Systems Letters, 8*, 1198–1210. DOI: 10.1109/LCSYS.2024.3406943

dos Santos, J. F. P., & Williamson, P. J. (2024). Beyond connectivity: Artificial intelligence and the internationalisation of digital firms. *Information and Organization, 34*(4), 100538. Advance online publication. DOI: 10.1016/j.infoandorg.2024.100538

Doshi-Velez, F., & Kim, B. (2017). Towards a rigorous science of interpretable machine learning. *arXiv preprint arXiv:1702.08608.*

Doumpos, M., Zopounidis, C., Gounopoulos, D., Platanakis, E., & Zhang, W. (2023). Operational research and artificial intelligence methods in banking. In *European Journal of Operational Research* (Vol. 306, Issue 1). DOI: 10.1016/j.ejor.2022.04.027

Draghi, B., Wang, Z., Myles, P., & Tucker, A. (2024). Identifying and handling data bias within primary healthcare data using synthetic data generators. *Heliyon, 10*(2), e24164. Advance online publication. DOI: 10.1016/j.heliyon.2024.e24164 PMID: 38288010

Drakopoulos, S. X., Gholamipour-Shirazi, A., Macdonald, P. M., Parini, R., Reynolds, C. D., Burnett, D. L., Pye, B., O'Regan, K., Wang, G., Whitehead, T. M., Conduit, G. J., Cazacu, A., & Kendrick, E. (2021). Formulation and manufacturing optimisation of lithium-ion graphite-based electrodes via machine learning. *Cell Reports. Physical Science*, *2*(12), 100683. DOI: 10.1016/j.xcrp.2021.100683

Driessnack, M., & Furukawa, R. (2012). Arts-based data collection techniques used in child research. *Journal for Specialists in Pediatric Nursing*, *17*(1), 3–9. DOI: 10.1111/j.1744-6155.2011.00304.x

Drori, I., & Te'eni, D. (2024). Human-in-the-loop AI reviewing: Feasibility, opportunities, and risks. *Journal of the Association for Information Systems*, *25*(1), 98–109. Advance online publication. DOI: 10.17705/1jais.00867

Druga, M., Müller, S., & Zhai, Q. (2024). AI in digital sports coaching – a systematic review. *International Journal of Sports Coaching Science*, *1*(1), 1–23. DOI: 10.1080/23750472.2024.2449016

Duarte, M. E., & Belarde-Lewis, M. (2015). Imagining: Creating spaces for Indigenous ontologies. *Cataloging & Classification Quarterly*, *53*(5-6), 677–702. DOI: 10.1080/01639374.2015.1018396

Dudek, G. (2025). Computational intelligence and machine learning: Advances in models and applications. *Electronics (Basel)*, *14*(8), 1530. DOI: 10.3390/electronics14081530

Du, J., Mamo, Y. Z., Floyd, C., Karthikeyan, N., & James, J. D. (2024). Machine learning in sport social media research: Practical uses and opportunities. *International Journal of Sport Communication*, *1*(1), 97–106. DOI: 10.1123/ijsc.2023-0151

Duranti, L. (2015). Overview of InterPARES Trust-Work to date. Third International Symposium of InterPARES Trust. Retrieved on 12 June 2025 from https://interparestrust.org/system/files/publications/files/Basis_for_Conclusions_ISA_220_Redrafted_andISQC_1_Redrafted_0.pdf

Duranti, L. (1991). Diplomatics. New users for an old science (Part V). *Archivaria*, *3*, 6–24.

Duranti, L. (2010). From digital diplomatics to digital records forensics. *Archivaria*, *68*, 39–66.

Duranti, L., & MacNeil, H. (1997). The protection of the integrity of electronic records: An overview of the UBC-MAS research project. *Archivaria*, *42*, 46–67.

Dwivedi, R., Nerur, S., & Balijepally, V. (2023). Exploring artificial intelligence and big data scholarship in information systems: A citation, bibliographic coupling, and co-word analysis. *International Journal of Information Management Data Insights*, *3*(2), 100185. Advance online publication. DOI: 10.1016/j.jjimei.2023.100185

Dwivedi, Y. K., Kshetri, N., Hughes, L., Slade, E. L., Jeyaraj, A., Kar, A. K., Baabdullah, A. M., Koohang, A., Raghavan, V., Ahuja, M., Albanna, H., Albashrawi, M. A., Al-Busaidi, A. S., Balakrishnan, J., Barlette, Y., Basu, S., Bose, I., Brooks, L., Buhalis, D., & Wright, R. (2023). Opinion Paper: “So what if ChatGPT wrote it?” Multidisciplinary perspectives on opportunities, challenges and implications of generative conversational AI for research, practice and policy. *International Journal of Information Management*, *71*, 102642. Advance online publication. DOI: 10.1016/j.ijinfomgt.2023.102642

Dwork, C., Hardt, M., Pitassi, T., Reingold, O., & Zemel, R. (2012). Fairness through awareness. *Proceedings of the 3rd Innovations in Theoretical Computer Science Conference (ITCS ’12)*, 214–226. DOI: 10.1145/2090236.2090255

Eager, B., & Brunton, R. (2023). Prompting higher education towards AI-augmented teaching and learning Practice. *Journal of University Teaching & Learning Practice*, *20*(5), 1–19. DOI: 10.53761/1.20.5.02

Eckert, T., Klein, F. C., Frieler, P., Thunich, O., & Abetz, V. (2021). Experimental design in polymer chemistry—A guide towards true optimisation of a raft polymerization using design of experiments (DoE). *Polymers*, *13*(18), 3147. DOI: 10.3390/polym13183147

Editorial, N. (2023). Tools such as ChatGPT threaten transparent science; here are our ground rules for their use. *Nature*, *613*(7945), 612. DOI: 10.1038/d41586-023-00191-1 PMID: 36694020

Eisenhardt, K. M. (1989). Building theories from case study research. *Academy of Management Review*, *14*(4), 532–550. DOI: 10.2307/258557

Eke, D. O. (2023). ChatGPT and the rise of generative AI: Threat to academic integrity? *Journal of Responsible Technology*, *13*, 100060. DOI: 10.1016/j.jrt.2023.100060

Elahi, M., Afolaranmi, S. O., Lastra, J. L. M., & García, J. A. P. (2023). A comprehensive literature review of the applications of AI techniques through the lifecycle of industrial equipment. *Discover Artificial Intelligence*, *3*(1), 43. Advance online publication. DOI: 10.1007/s44163-023-00089-x

Elangovan, R., & Rajendran, R. (2015). Conceptual model: A framework for institutionalising the vigor in business research. *Proceedings of the 3rd National on Indian Business Management*, 1–35.

Elias, A., & Gomez, G. (2025). Digital platforms as institutional actors in the global south–Evidence from India's informal sector. *Digital Geography and Society*, *9*, 100146. DOI: 10.1016/j.diggeo.2025.100146

Elliott, V. (2018). Thinking about the coding process in qualitative data analysis. *The Qualitative Report*, *23*(11), 2850–2861. DOI: 10.46743/2160-3715/2018.3560

Elnanda, V. R., & Gunawan, P. H. (2025, February). Machine learning solutions for stunting classification in Karya Jaya: A case study using KNN and SVM. In *2025 International Conference on Advancement in Data Science, E-learning and Information System (ICADEIS)* (pp. 1-5). IEEE. DOI: 10.1109/ICADEIS65852.2025.10933024

Elsayed, M. E. A., Abdo, S., Attia, A. A. A., Attia, E.-A., & ElRahman, M. A. (2022). Parametric optimisation for the design of gravity energy storage system using Taguchi method. *Scientific Reports*, *12*(1), 19648. Advance online publication. DOI: 10.1038/s41598-022-20514-y

Elshaer, R. N., El-Aty, A. A., Sayed, E. M., Barakat, A. F., & Sobh, A. S. (2024). Optimisation of machining parameters for turning operation of heat-treated Ti-6Al-3Mo-2Nb-2Sn-2Zr-1.5Cr alloy by Taguchi method. *Scientific Reports*, *14*(1), 16494. Advance online publication. DOI: 10.1038/s41598-024-65786-8

Eriksson, O., Johannesson, P., & Bergholtz, M. (2018). Institutional ontology for conceptual modelling. *Journal of Information Technology*, *33*(2), 105–123. DOI: 10.1057/s41265-018-0053-2

Ernst, M. I.Michèle I. Ernst. (2023). The Crucial Role of School Libraries in Influencing Children's Literacy and Learning. *Journal of Childhood Literacy and Societal Issues*, *2*(1), 32–39. DOI: 10.71085/joclsi.02.01.26

Esteva, A., Robicquet, A., Ramsundar, B., Kuleshov, V., DePristo, M., Chou, K., Cui, C., Corrado, G., Thrun, S., & Dean, J. (2019). A guide to deep learning in healthcare. *Nature Medicine*, *25*(1), 24–29. DOI: 10.1038/s41591-018-0316-z

Etchart, L. (2022). Indigenous peoples and international law in the Ecuadorian Amazon. *Laws*, *11*(4), 55. DOI: 10.3390/laws11040055

Eticas Foundation. (2023). *Adversarial algorithmic auditing guide*.

Eticas Foundation. (2024). *Adversarial audit of ride-hailing platforms*. Taxi Project.

Eurofound. (2025). *Platform work: Algorithmic management and working conditions in Europe*. Publications Office of the European Union.

Eyre, M. (2023). The spirit of knowledge: Creating spiritual competency for critical information spaces. *Public Services Quarterly*, *19*(1), 69–76. DOI: 10.1080/15228959.2022.2160857

Faldet, A. C., & Nes, K. (2024). Valuing vulnerable children's voices in educational research. *International Journal of Inclusive Education*, *28*(5), 657–672. DOI: 10.1080/13603116.2021.1956602

Fallahi, A., Tehrani, Y., & Mokhtari, H. (2024). Optimisation of a coordinated sustainable multi-vendor multi-livestock multi-rancher supply chain for growing products. *Scientia Iranica*, *0*(0), 0. Advance online publication. DOI: 10.24200/sci.2024.62208.7714

Faraj, S., Pachidi, S., & Sayegh, K. (2018). Working and organizing in the age of the learning algorithm. *Information and Organization*, *28*(1), 62–70. DOI: 10.1016/j.infoandorg.2018.02.005

Farhi, F., Jeljeli, R., Aburezeq, I., Dweikat, F. F., Al-shami, S. A., & Slamene, R. (2023). Analyzing the students' views, concerns, and perceived ethics about ChatGPT usage. *Computers and Education: Artificial Intelligence*, *5*, 100180. DOI: 10.1016/j.caeai.2023.100180

Faridmehr, I., Nehdi, M. L., Huseien, G. F., Baghban, M. H., Sam, A. R. M., & Algaifi, H. A. (2021). Experimental and informational modeling study of sustainable self-compacting geopolymer concrete. *Sustainability (Basel)*, *13*(13), 7444. DOI: 10.3390/su13137444

Farrukh Moin, M., Behl, A., & Zhang, Z. (2025, August). AI in the Organizational Nexus: Building Trust, Cementing Commitment, and Evolving Psychological Contracts. *Information Systems Frontiers*, *27*(4), 1413–1424. Advance online publication. DOI: 10.1007/s10796-024-10561-3

Feng, F., Ghorbani, H., & Radwan, A. E. (2024). Predicting groundwater level using traditional and deep machine learning algorithms. *Frontiers in Environmental Science*, *12*, 1291327. Advance online publication. DOI: 10.3389/fenvs.2024.1291327

Fernández-García, R., Lalatsa, A., Statts, L., Bolás-Fernández, F., Ballesteros, M. P., & Serrano, D. R. (2020). Transferosomes as nanocarriers for drugs across the skin: Quality by design from lab to industrial scale. *International Journal of Pharmaceutics*, *573*, 118817. DOI: 10.1016/j.ijpharm.2019.118817

Ferreira, L., Pilastri, A., Martins, C. M., Pires, P. M., & Cortez, P. (2021). A Comparison of AutoML Tools for Machine Learning, Deep Learning and XGBoost. *Proceedings of the International Joint Conference on Neural Networks, 2021-July*. DOI: 10.1109/IJCNN52387.2021.9534091

Fischer, V., Kucia-Tran, R., Lewis, W. J., & Velayudhan, A. (2019). Hybrid optimisation of preparative chromatography for a ternary monoclonal antibody mixture. *Biotechnology Progress*, *35*(5), e2849. Advance online publication. DOI: 10.1002/btpr.2849

Fister, I., & Perc, M. (2016). Toward the discovery of citation cartels in citation networks. *Frontiers in Physics (Lausanne)*, *4*, 240569. DOI: 10.3389/fphy.2016.00049

Floridi, L., & Cowls, J. (2022). A unified framework of five principles for AI in society. *Machine learning and the city: Applications in Architecture and Urban Design*, 535-545.

Floridi, L. (2013). *The ethics of information*. Oxford University Press. DOI: 10.1093/acprof:oso/9780199641321.001.0001

Floridi, L. (2020). AI and its new winter: From myths to realities. *Philosophy & Technology*, *33*(1), 1–3. DOI: 10.1007/s13347-020-00396-6

Floridi, L., & Chiriatti, M. (2020). GPT-3: Its nature, scope, limits, and consequences. *Minds and Machines*, *30*(4), 681–694. DOI: 10.1007/s11023-020-09548-1

Fodale, F., & Bates, J. (2011). *What is the impact of the school library on pupils' personal development? A case study of a secondary school in Northern Ireland.* University of Ulster. https://www.researchgate.net/publication/254758802

Forbes, A., Ritchie, S., Walker, J., & Young, N. (2020). Applications of two-eyed seeing in primary research focused on Indigenous health: A scoping review. *International Journal of Qualitative Methods*, *19*, 1609406920929110. DOI: 10.1177/1609406920929110

Fouché, R. (2023). Sport and AI: Charting the connection between historical evolution and emerging technologies. *Sport History Review*, *54*(2), 112–126. DOI: 10.1080/09523367.2023.2228118

Fózer, D., Nimmegeers, P., Tóth, A. J., Varbanov, P. S., Klemeš, J. J., Mizsey, P., Hauschild, M. Z., & Owsianiak, M. (2023). Hybrid prediction-driven high-throughput sustainability screening for advancing waste-to-dimethyl ether valorization. *Environmental Science & Technology*, *57*(36), 13449–13462. DOI: 10.1021/acs.est.3c01892

Franks, P. C. (2018). *Records and information management.* American Library Association.

Freire, K., Pope, R., Jeffrey, K., Andrews, K., Nott, M., & Bowman, T. (2022). Engaging with children and adolescents: A systematic review of participatory methods and approaches in research informing the development of health resources and interventions. *Adolescent Research Review*, *7*(3), 335–354. DOI: 10.1007/s40894-022-00181-w

French, A. M., & Shim, J. P. (2024). From Artificial intelligence to augmented intelligence: A shift in perspective, application, and conceptualization of AI. *Information Systems Frontiers: A Journal of Research and Innovation*, 1–22. https://doi.org/DOI: 10.1007/s10796-024-10562-2

Frey, C. B., & Osborne, M. A. (2017). The future of employment: How susceptible are jobs to computerisation? *Technological Forecasting and Social Change*, *114*, 254–280. DOI: 10.1016/j.techfore.2016.08.019

Friederich, J., Francis, D. P., Lazarova-Molnar, S., & Mohamed, N. (2021). A framework for data-driven digital twins of smart manufacturing systems. *Computers in Industry*, *136*, 103586. DOI: 10.1016/j.compind.2021.103586

Frith, H. (2000). Focusing on sex: Using focus groups in sex research. *Sexualities*, *3*(3), 275–297. DOI: 10.1177/136346000003003001

Fügener, A., Grahl, J., Gupta, A., & Ketter, W. (2021). Will humans-in-the-loop become borgs? Merits and pitfalls of working with AI. *Management Information Systems Quarterly*, *45*(3), 1527–1556. DOI: 10.25300/MISQ/2021/16553

Fullmer, M. (2021). Are we there yet? Visualizing Indigenous culture in today's library. *IFLA Journal*, *47*(3), 313–320. DOI: 10.1177/0340035220987577

Furr, N., Ozcan, P., & Eisenhardt, K. M. (2022). What is digital transformation? Core tensions facing established companies on the global stage. *Global Strategy Journal*, *97*(4), 1–24. DOI: 10.1002/gsj.1442

Ganjavi, C., Eppler, M. B., Pekcan, A., Biedermann, B., Abreu, A., Collins, G. S., Gill, I. S., & Cacciamani, G. E. (2024). Publishers' and journals' instructions to authors on use of generative artificial intelligence in academic and scientific publishing: Bibliometric analysis. *BMJ (Clinical Research Ed.)*, *384*, e077192. Advance online publication. DOI: 10.1136/bmj-2023-077192 PMID: 38296328

Gannaway, J. (2020). Knocking, Unsettling, ceding: A non-Indigenous teacher's journey towards decolonizing teaching practice in a "remote Indigenous community". [NJCIE]. *Nordic Journal of Comparative and International Education*, *4*(1), 102–117. DOI: 10.7577/njcie.3553

Gathogo, J. M. (2021). Memory and history: Oral techniques in the East African context. *Hervormde Teologiese Studies*, *77*(2). Advance online publication. https://www.ajol.info/index.php/hts/article/view/212210. DOI: 10.4102/hts.v77i2.6477

Gerasymenko, O. (2024). Exploring children's picture book design: analysis of composition and color application. In Conferinţa tehnico-ştiinţifică a studenţilor, masteranzilor şi doctoranzilor (Vol. 4, pp. 2151-2154).

Gerling, C., Meier, P., & Köhler, C. (2022). AI meets digital: a critical review on artificial intelligence in digital entrepreneurship. *European Conference on Information Systems.*

Ghader, S. (2025, October). AI-powered geotechnics: Enhancing rock mass classification for safer engineering practices. *Rock Mechanics and Rock Engineering*, *58*(10), 11319–11349. Advance online publication. DOI: 10.1007/s00603-024-04189-7

Ghadoui, M. E., Mouchtachi, A., & Majdoul, R. (2023). A hybrid optimisation approach for intelligent manufacturing in plastic injection molding by using artificial neural network and genetic algorithm. *Scientific Reports*, *13*(1), 21817. Advance online publication. DOI: 10.1038/s41598-023-48679-0

Ghaebi, H., & Soleymani, E. (2024). Investigating the effect of parameters in the thermodynamic analysis of the solid oxide fuel cell cycle using response surface methodology. *Research Square (Research Square)*. DOI: 10.21203/rs.3.rs-5053080/v1

Ghaheh, F. S., Razbin, M., Tehrani, M., Vayghan, L. Z. A., & Sadrjahani, M. (2024). Modeling and optimisation of dyeing process of polyamide 6 and woolen fabrics with plum-tree leaves using artificial intelligence. *Scientific Reports*, *14*(1), 15067. Advance online publication. DOI: 10.1038/s41598-024-64761-7

Ghazizadeh, M., Lee, J. D., & Boyle, L. N. (2012). Extending the technology acceptance model to assess automation. *Cognition Technology and Work*, *14*(39), 49. DOI: 10.1007/s10111-011-0194-3

Gillespie, T. (2010). The politics of 'platforms'. *New Media & Society*, *12*(3), 347–364. DOI: 10.1177/1461444809342738

Gill, T. G., & Gill, T. R. (2020). What is research rigour? Lessons for a transdiscipline. *Informing Science*, *23*, 47–77. DOI: 10.28945/4528

Gil, Y., Greaves, M., Hendler, J., & Hirsh, H. (2014). Amplify scientific discovery with artificial intelligence. *Science*, *346*(6206), 171–172. DOI: 10.1126/science.1259439

Glaser, B. G., & Strauss, A. L. (1967). *The discovery of grounded theory*. strategies for qualitative research. Aldine

Glikson, E., & Woolley, A. W. (2020). Human trust in artificial intelligence: Review of empirical research. *Academy of Management Annals,* (14:2), 627–660. https://doi.org/DOI: 10.5465/annals.2018.0057

Goodarzian, F., Kumar, V., & Ghasemi, P. (2022). Investigating a citrus fruit supply chain network considering CO2 emissions using meta-heuristic algorithms. *Annals of Operations Research*. Advance online publication. DOI: 10.1007/s10479-022-05005-7

Goodfellow, I., Bengio, Y., & Courville, A. (2016). *Deep learning*. MIT Press.

Goodfellow, I., Bengio, Y., Courville, A., & Bengio, Y. (2016). Deep Learning: Vol. 1. *No. 2*. MIT press.

Goranov, A. I., Sleighter, R. L., Yordanov, D. A., & Hatcher, P. G. (2023). TEnvR: MATLAB-based toolbox for environmental research. *Analytical Methods*, *15*(40), 5390–5400. DOI: 10.1039/D3AY00750B

Gregor, S. (2024). Responsible artificial intelligence and journal publishing. *Journal of the Association for Information Systems*, *25*(1), 48–60. DOI: 10.17705/1jais.00863

Grilli, L., & Pedota, M. (2024). Creativity and artificial intelligence: A multilevel perspective. *Creativity and Innovation Management*, *33*(2), 234–247. Advance online publication. DOI: 10.1111/caim.12580

Guan, S., Zhang, C., Wang, Y., & Liu, W. (2024). Hadoop-based secure storage solution for big data in cloud computing environment. *Digital Communications and Networks*, *10*(1), 227–236. Advance online publication. DOI: 10.1016/j.dcan.2023.01.014

Gulumbe, B. H., Audu, S. M., & Hashim, A. M. (2025). Balancing AI and academic integrity: What are the positions of academic publishers and universities? *AI & Society*, *40*(3), 1775–1784. DOI: 10.1007/s00146-024-01946-8

Gupta, R., Nair, K., Mishra, M., Ibrahim, B., & Bhardwaj, S. (2024). Adoption and impacts of generative artificial intelligence: Theoretical underpinnings and research agenda. *International Journal of Information Management Data Insights*, *4*(1), 100232. Advance online publication. DOI: 10.1016/j.jjimei.2024.100232

Gupta, S., & Rao, P. (2024). High-performance computing strategies for AI in sports analytics. *Concurrency and Computation*. Advance online publication. DOI: 10.1002/cpe.70068

Gustilo, L., Ong, E., & Lapinid, M. R. (2024). Algorithmically-driven writing and academic integrity: Exploring educators' practices, perceptions, and policies in AI era. *International Journal for Educational Integrity*, *20*(1), 3. DOI: 10.1007/s40979-024-00153-8

Hacker, P. (2021). Teaching fairness to artificial intelligence: Existing and novel strategies against algorithmic discrimination under EU law. *Common Market Law Review*, *58*(4), 1143–1186.

Haenlein, M., & Kaplan, A. (2019). A brief history of artificial intelligence: On the past, present, and future of artificial intelligence. *California Management Review*, *61*(4), 5–14. DOI: 10.1177/0008125619864925

Hafid, M., Istiyanto, J., & Nasruddin, N. (2023). Multiobjective optimisation of dimension and position of elliptical crush initiator on crashworthiness performance of square tube using response surface methodology. *Frontiers of Mechanical Engineering*, *9*, 1273447. Advance online publication. DOI: 10.3389/fmech.2023.1273447

Hagen-Sataslaatten, O. (2014). The Norwegian Noark model requirements for EDRMS in the context of open government and access to governmental information. *Records Management Journal*, *24*(3), 189–204. DOI: 10.1108/RMJ-09-2014-0041

Hakim, B. A. E., Abdel-Goad, M., Awad, M. E., & Shoaib, A. M. (2025). AI enhanced model predictive control for optimizing LPG recovery through integrated computational modeling design of experiments and multivariate regression. *Scientific Reports*, *15*(1), 29249. Advance online publication. DOI: 10.1038/s41598-025-13899-z

Hammersley, M. (2012). *Methodological paradigms in educational research*. British Educational Research Association.

Hanelt, A., Bohnsack, R., Marz, D., & Antunes Marante, C. (2021). A systematic review of the literature on digital transformation: Insights and implications for strategy and organizational change. *Journal of Management Studies*, *58*(5), 1159–1197. DOI: 10.1111/joms.12639

Hansen, H. F., Lillesund, E., Mikalef, P., & Altwaijry, N. (2024). Understanding artificial intelligence diffusion through an ai capability maturity model. *Information Systems Frontiers: A Journal of Research and Innovation*, 26(6), 2147–2163. https://doi.org/DOI: 10.1007/s10796-024-10528-4

Hansen, A. M., Kraemmergaard, P., & Mathiassen, L. (2011). Rapid adaptation in digital transformation: A participatory process for engaging IS and business leaders. *MIS Quarterly Executive*, *10*(4), 175–185.

Harati, K. (2024). ChatGPT and AI-powered writing tools: Unveiling risks and ethical challenges in scientific writing. *Journal of Reviews in Medical Sciences, 4*(1), 1-6. https://.doi.org.DOI: 10.22034/jrms.2024.493520.1031

Hardian, R., Liang, Z., Zhang, X., & Székely, G. (2020). Artificial intelligence: The silver bullet for sustainable materials development. *Green Chemistry*, *22*(21), 7521–7528. DOI: 10.1039/D0GC02956D

Harfield, S., Pearson, O., Morey, K., Kite, E., Canuto, K., Glover, K., Gomersall, J. S., Carter, D., Davy, C., Aromataris, E., & Braunack-Mayer, A. (2020). Assessing the quality of health research from an Indigenous perspective: The Aboriginal and Torres Strait Islander quality appraisal tool. *BMC Medical Research Methodology*, *20*(1), 79. Advance online publication. DOI: 10.1186/s12874-020-00959-3

Hariharasakthisudhan, P., Rajendran, S., Khalkar, V., Devarajan, G., & Logesh, K. (2024). Data-driven healthcare: The role of computational methods in medical innovation. *Computer Modeling in Engineering & Sciences*, *1*. Advance online publication. DOI: 10.32604/cmes.2024.056605

Harudin, N., Jamaludin, K. R., Ramlie, F., Muhtazaruddin, M. N., Razali, C. M. C., & Muhamad, W. Z. A. W. (2020). Binary particle swarm optimisation for variables selection optimisation in Taguchi's T-Method. *MATEMATIKA*, *69*, 69–84. Advance online publication. DOI: 10.11113/matematika.v36.n1.1181

Hatfield, H. R., & Ahn, S. J. (2025). Integrating community-based participatory research into immersive narratives. *Journal of Media Psychology*, *37*(2), 101–115. DOI: 10.1027/1864-1105/a000463

Haynes, J. (2014). SEEN and HEARD: Challenging the perceptions of children in libraries in sub-Saharan Africa.

Hayward, A., Sjoblom, E., Sinclair, S., & Cidro, J. (2021). A new era of indigenous research: Community-based Indigenous research ethics protocols in Canada. *Journal of Empirical Research on Human Research Ethics; JERHRE*, *16*(4), 403–417. DOI: 10.1177/15562646211023705

He, X. (2024). Time Series Analysis. In Geographic Data Analysis Using R (pp. 105-134). Springer Nature Singapore. DOI: 10.1007/978-981-97-4022-2_6

Heddar, M. I. E., Mehdi, B., Matougui, N., Tahan, A., & Jahazi, M. (2023). Adaptable multi-objective optimisation framework: Application to metal additive manufacturing. *Research Square (Research Square)*. DOI: 10.21203/rs.3.rs-3751025/v1

Heeks, R. (2022). Digital inequality beyond the digital divide: Conceptualizing adverse digital incorporation in the global South. *Information Technology for Development*, *28*(4), 688–704. DOI: 10.1080/02681102.2022.2068492

Heirendt, L., Arreckx, S., Pfau, T., Mendoza, S. N., Richelle, A., Heinken, A., Haraldsdóttir, H. S., Wachowiak, J., Keating, S., Vlasov, V., Magnúsdóttir, S., Ng, C. Y., Preciat, G., Žagare, A., Chan, S. H. J., Aurich, M. K., Clancy, C. M., Modamio, J., Sauls, J. T., & Fleming, R. M. T. (2019). Creation and analysis of biochemical constraint-based models using the COBRA Toolbox v.3.0. *Nature Protocols*, *14*(3), 639–702. DOI: 10.1038/s41596-018-0098-2

Hejazi, T., & Moradpour, M. (2019). Developing a flexible methodology for modeling and solving multiple response optimisation problems. *Stochastics and Quality Control*, *34*(2), 103–113. DOI: 10.1515/eqc-2018-0024

Henning, S., & Hasselbring, W. (2024). Benchmarking scalability of stream processing frameworks deployed as microservices in the cloud. *Journal of Systems and Software*, *208*, 111879. Advance online publication. DOI: 10.1016/j.jss.2023.111879

Herman, N., & Abeza, G. (2024). The disruptive power of AI in sport and entertainment. In *the Routledge Handbook of Sport Communication* (pp. 350–362). Routledge. https://doi.org/DOI: 10.4324/9781003430278-35

Hermann, E., & Hermann, G. (2022). Artificial intelligence in research and development for sustainability: The centrality of explicability and research data management. *AI and Ethics*, *2*(1), 29–33. Advance online publication. DOI: 10.1007/s43681-021-00114-8

Hinings, B., Gegenhuber, T., & Greenwood, R. (2018). Digital innovation and transformation: An institutional perspective. *Information and Organization*, *28*(1), 52–61. DOI: 10.1016/j.infoandorg.2018.02.004

Hirosaki, M., & Patel, R. (2025). Artificial intelligence approach in biomechanics of gait and sport: A systematic literature review. *Journal of Biomedical Physics & Engineering*, *13*(5), 383–402. DOI: 10.31661/jbpe.v0i0

Hisam, M. W., Dar, A. A., Elrasheed, M. O., Khan, M. S., Gera, R., & Azad, I. (2024). The versatility of the Taguchi method: Optimizing experiments across diverse disciplines. *Journal of Statistical Theory and Applications : JSTA*, *23*(4), 365–389. Advance online publication. DOI: 10.1007/s44199-024-00093-9

Hitch, D. (2024). Artificial intelligence augmented qualitative analysis: The way of the future? *Qualitative Health Research, 34*(7), 595–606. DOI: 10.1177/10497323231217392

Hoff, K. A., & Bashir, M. (2015). Trust in automation: Integrating empirical evidence on factors that influence trust. *Human Factors, 57*(3), 407–434. DOI: 10.1177/0018720814547570

Hoffman, J., & Kumar, R. (2024). Game theory and AI in team sports decision-making. In *Advances in leadership and digital innovations in sports* (Vol. 3, pp. 45–62). Springer., DOI: 10.1007/978-3-031-76047-1_5

Holton, E. F., & Lowe, J. S. (2007). Toward a general research process for using Dublin's theory building model. *Human Resource Development Review, 6*(3), 297–320. DOI: 10.1177/1534484307304219

Hope, T., Downey, D., Weld, D. S., Etzioni, O., & Horvitz, E. (2023). A computational inflection for scientific discovery. *Communications of the ACM, 66*(8), 62–73. DOI: 10.1145/3576896

Hosseini, M., Resnik, D. B., & Holmes, K. (2023). The ethics of disclosing the use of artificial intelligence tools in writing scholarly manuscripts. *Research Ethics, 19*(4), 449–465. DOI: 10.1177/17470161231180449 PMID: 39749232

Hsieh, J., Zhang, A., Rasetarinera, M., Chou, E., Ngo, D., Lightman, K., Lee, M. K., & Zhu, H. (2024). Supporting gig worker needs and advancing policy through worker-centered data-sharing. *Proceedings of the ACM on Measurement and Analysis of Computing Systems, 8*(4), 111.

Huaman, E. S., & Mataira, P. (2019). Beyond community engagement: Centering research through Indigenous epistemologies and peoplehood. *Alternative, 15*(3), 281–286. DOI: 10.1177/1177180119871705

Huang, L. J., & Shieh, J. C. (2021). Elementary school students' perceptions of school library and expectations of library space. In *IASL Annual Conference Proceedings*. DOI: 10.29173/iasl7977

Huang, S., Yang, K., Qi, S., & Wang, R. (2024). When large language model meets optimisation. *Swarm and Evolutionary Computation, 90*, 101663. DOI: 10.1016/j.swevo.2024.101663

Huang, Y.-M., & Rocha, T. (2023). *Innovative technologies in learning* (Vol. 14785). Springer Nature Switzerland.

Hudson, M., Carroll, S. R., Anderson, J., Blackwater, D., Cordova-Marks, F. M., Cummins, J., David-Chavez, D., Fernandez, A., Garba, I., Hiraldo, D., Jäger, M. B., Jennings, L. L., Martinez, A., Sterling, R., Walker, J. D., & Rowe, R. K. (2023). Indigenous Peoples' Rights in Data: A contribution toward Indigenous research sovereignty [perspective]. *Frontiers in Research Metrics and Analytics*, *8*, 2023. DOI: 10.3389/frma.2023.1173805

Hünermund, P., & Louw, B. (2025). On the nuisance of control variables in causal regression analysis. *Organizational Research Methods*, *28*(1), 138–151. DOI: 10.1177/10944281231219274

Hunie, P., & Dewah, P. (2019). Admissibility of digital records as evidence in Bulawayo High Court in Zimbabwe. *Journal of the South African Society of African Archivists*, *52*, 133–148.

Hunter, A. (2021). The Shortcomings of Bibliographic description in service of Indigenous peoples in Canada. *Emerging Library & Information Perspectives*, *4*(1), 88–116. DOI: 10.5206/elip.v4i1.13439

Hunt, S. C., & Young, N. L. (2021). Blending Indigenous sharing circle and Western focus group methodologies for the study of Indigenous children's health: A systematic review. *International Journal of Qualitative Methods*, *20*, 16094069211015112. DOI: 10.1177/16094069211015112

Huria, T., Palmer, S. C., Pitama, S., Beckert, L., Lacey, C., Ewen, S., & Smith, L. T. (2019). Consolidated criteria for strengthening reporting of health research involving Indigenous peoples: The CONSIDER Statement. *BMC Medical Research Methodology*, *19*(1), 173. Advance online publication. DOI: 10.1186/s12874-019-0815-8

Hutchinson, P., Dingwall, C., Kurtz, D., Evans, M., Jones, G. R., & Corbett, J. (2014). Maintaining the iIntegrity of Indigenous knowledge: Sharing metis knowing through mixed methods. *International Journal of Critical Indigenous Studies*, *7*(1), 1–14. DOI: 10.5204/ijcis.v7i1.118

Iansiti, M., & Lakhani, K. R. (2020). Competing in the age of AI. *Harvard Business Review*, •••, 12–19.

IFLA School Libraries Section Standing Committee. (2015). IFLA school Library guidelines, 2nd Edition, International Federation of Library Associations and Institutions(IFLA).

Ignace, L., Burton, L., Mynott, S., Meehan, M., Olson, E., Steel, J., Ojeda, J., Harper, S., Ramirez, L., Baker, D., Sleigh, L., Frenkel, C., Rhodes, C., & Ban, N. C. (2023). Researchers' responsibility to uphold Indigenous rights. *Science, 381*(6654), 129–131. https://doi.org/doi:10.1126/science.adh4470. DOI: 10.1126/science.adh4470

Ilk, N., Shang, G., & Goes, P. B. (2020). Improving customer routing in contact centers: An automated triage design based on text analytics. *Journal of Operations Management, 66*(5), 553–577. DOI: 10.1002/joom.1084

Iman, M., Arabnia, H. R., & Rasheed, K. (2023). A Review of Deep Transfer Learning and Recent Advancements. In *Technologies* (Vol. 11, Issue 2). DOI: 10.3390/technologies11020040

Imenda, S. (2014). Is there a conceptual difference between theoretical and conceptual frameworks? *Journal of Social Sciences, 38*(2), 185–195. DOI: 10.1080/09718923.2014.11893249

International Council on Archives. (2021). *Guidelines for digital recordkeeping*. ICA. Retrieved on 12 June 2025 from https://www.ica.org

International Labour Organization. (2021). *The role of algorithmic management in the gig economy*. ILO Working Paper Series.

International Labour Organization. (2024a). *Digital labour platforms in Kenya*. ILO Prospects series.

International Labour Organization. (2024b). *Understanding and improving women's work on digital labour platforms – Stakeholder consultation workshop in India*.

International Organization for Standardization. (2013). *ISO/IEC 27001:2013 Information technology — Security techniques — Information security management systems — Requirements*. ISO. Retrieved on 12 June 2025 from https://www.iso.org/standard/54534.html

International Organization for Standardization. (2016). *ISO 15489-1:2016 Information and documentation — Records management — Part 1: Concepts and principles*. ISO. Retrieved on 12 June 2025 from https://www.iso.org/standard/62542.html

International Trade Union Confederation. (2025). *Time to deliver: Platform workers and migrant labour*.

InterPARES Authenticity Taskforce. (2007). Requirements for assessing and maintaining the authenticity of electronic records. In *The long-term preservation of authentic electronic records: Findings of the InterPARES project*. Vancouver. Retrieved on 12 June 2025 from https://interpares.org/book/interpares_book_d_part1.pdf

Isbah, M. F. (2022). Algorithmic Exploitation: Understanding Labor Process and Control among RideHailing Platform Workers. *Jurnal Sosioteknologi, 21*(2). Advance online publication. DOI: 10.5614/sostek.itbj.2022.21.2.5

Ishfaq, K., Sana, M., & Ashraf, W. M. (2023). Artificial intelligence–built analysis framework for the manufacturing sector: Performance optimisation of wire electric discharge machining system. *International Journal of Advanced Manufacturing Technology, 128*(11-12), 5025–5039. DOI: 10.1007/s00170-023-12191-6

Ishola, N. B., & McKenna, T. F. L. (2021). Influence of process parameters on the gas phase polymerization of ethylene: RSM or ANN statistical methods? *Macromolecular Theory and Simulations, 30*(6), 2100059. Advance online publication. DOI: 10.1002/mats.202100059

Istanboulian, L., Rose, L., Yunusova, Y., & Dale, C. (2023). Adapting co-design methodology to a virtual environment: Co-designing a communication intervention for adult patients in critical care. *Research Involvement and Engagement, 9*(1), 103. DOI: 10.1186/s40900-023-00514-6 PMID: 37957776

Jabagi, N., Croteau, A. M., Audebrand, L. K., & Marsan, J. (2025). Do algorithms play fair? Analysing the perceived fairness of HR-decisions made by algorithms and their impacts on gig-workers. *International Journal of Human Resource Management, 36*(2), 235–274. DOI: 10.1080/09585192.2024.2441448

Jabagi, N., Croteau, A., Audebrand, L. K., & Marsan, J. (2024). The effects of algorithmic transparency on gig workers. *Information Systems Journal, 34*(2), 345–367.

Jabareen, Y. (2009). Building a conceptual framework: Philosophy, definitions, and procedure. *International Journal of Qualitative Methods, 8*(4), 49–62. DOI: 10.1177/160940690900800406

Jaddi, H., Hmaidi, A. E., Ousmana, H., Berrada, M., Aouragh, M. H., Iallamen, Z., Kasse, Z., Ouali, A. E., Boufala, M., Ragragui, H., & Saouita, J. (2024). Predicting Nitrate levels in the saïss water table: A comparative study of machine learning methods. *BIO Web of Conferences, 115*, 3001. DOI: 10.1051/bioconf/202411503001

Jamieson, V., & Chu, C. M. (2025). Disrupting informed consent: Refusal and cultural sovereignty in Indigenous knowledge practices. *Proceedings of the Association for Information Science and Technology, 62*(1), 1496–1498. DOI: 10.1002/pra2.1446

Janković, A., Chaudhary, G., & Goia, F. (2021). Designing the design of experiments (DOE) – An investigation on the influence of different factorial designs on the characterization of complex systems. *Energy and Building, 250*, 111298. DOI: 10.1016/j.enbuild.2021.111298

Jardim, P. S. J., Rose, C. J., Ames, H. M., Echavez, J. F. M., Van de Velde, S., & Muller, A. E. (2022). Automating risk of bias assessment in systematic reviews: A real-time mixed methods comparison of human researchers to a machine learning system. *BMC Medical Research Methodology*, *22*(1), 167. DOI: 10.1186/s12874-022-01649-y

Jarvenpaa, S., & Klein, S. (2024). New frontiers in information systems: Theorising human-AI collaboration. *Journal of the Association for Information Systems*, *25*(1), 110–121. DOI: 10.17705/1jais.00868

Jeyaraj, A., & Zadeh, A. H. (2020). Evolution of information systems research: Insights from topic modelling. *Information & Management*, *57*(4), 103207. Advance online publication. DOI: 10.1016/j.im.2019.103207

Ji, S., Pan, S., Cambria, E., Marttinen, P., & Yu, P. S. (2021). A survey on knowledge graphs: Representation, acquisition, and applications. *IEEE Transactions on Neural Networks and Learning Systems*, *33*(2), 494–514. DOI: 10.1109/TNNLS.2021.3070843

Johnson, L. R., Wilcox, A., Alexander, S. M., Bowles, E., Castleden, H., Henri, D., Herc, C., King, L., Provencher, J. F., & Orihel, D. M. (2023). Weaving Indigenous and Western ways of knowing in ecotoxicology and wildlife health: A review of Canadian studies. *Environmental Reviews*, *31*(3), 452–470. DOI: 10.1139/er-2022-0087

Johnson, R. T., & Brown, M. P. (2024). Why do leaders matter? A revisit of player-turned-coaches from the perspective of elite basketball. *Applied Economics*, *56*(34), 3759–3775. DOI: 10.1080/00036846.2023.2266606

Jones, M. (2019). Design and optimisation of oleochemical processes. *Research Portal Denmark*, 150. Retrieved on 10 October 2025 from https://local.forskningsportal.dk/local/dki-cgi/ws/cris-link?src=dtu&id=dtu-f63e262f-6567-467d-bcdb-c97fbe7bde66&ti=Design%20and%20Optimisation%20of%20Oleochemical%20Processes

Jones, S., & Sen, K. (2022). Labour market effects of digital matching platforms: Experimental evidence from Sub-Saharan Africa (IZA Discussion Paper No. 15409). Institute of Labor Economics.

Jones, L. M., Mitchell, K. J., & Beseler, C. L. (2024). The impact of youth digital citizenship education: Insights from a cluster randomized controlled trial outcome evaluation of the be internet awesome (BIA) curriculum. *Contemporary School Psychology*, *28*(4), 509–523. DOI: 10.1007/s40688-023-00465-5

Jones, S., & Sen, K. (2025). The gendered impact of digital jobs platforms: Experimental evidence from Mozambique. *The World Bank Economic Review*, *39*(1), 143–172. DOI: 10.1093/wber/lhae019

Jong, W., Huang, Y.-M., Lin, Y.-Z., Chen, S., & Chen, Y. (2020). Integrating Taguchi method and artificial neural network to explore machine learning of computer aided engineering. *Zhongguo Gongcheng Xuekan*, *43*(4), 346–356. DOI: 10.1080/02533839.2019.1708804

Jordan, L. S., & Hall, J. N. (2023). Framing anticolonialism in evaluation: Bridging decolonizing methodologies and culturally responsive evaluation. *Journal of Multidisciplinary Evaluation*, *19*(44), 102–116. Advance online publication. DOI: 10.56645/jmde.v19i44.769

Jordan, M. I., & Mitchell, T. M. (2015). Machine learning: Trends, perspectives, and prospects. *Science*, *349*(6245), 255–260. DOI: 10.1126/science.aaa8415

Joseph, F. J. J., Nonsiri, S., & Monsakul, A. (2021). Keras and TensorFlow: A Hands-On Experience. In *EAI/Springer Innovations in Communication and Computing*. DOI: 10.1007/978-3-030-66519-7_4

Joshi, D. J., Kale, I., Gandewar, S., Korate, O., Patwari, D., & Patil, S. (2021). Reinforcement Learning: A Survey. *Advances in Intelligent Systems and Computing, 1311 AISC*. DOI: 10.1007/978-981-33-4859-2_29

Jumper, J., Evans, R., Pritzel, A., Green, T., Figurnov, M., Ronneberger, O., Tunyasuvunakool, K., Bates, R., Žídek, A., Potapenko, A., Bridgland, A., Meyer, C., Kohl, S. A. A., Ballard, A. J., Cowie, A., Romera-Paredes, B., Nikolov, S., Jain, R., Adler, J., & Hassabis, D. (2021). Highly accurate protein structure prediction with AlphaFold. *Nature*, *596*(7873), 583–589. DOI: 10.1038/s41586-021-03819-2

Juul-Nyholm, H. K. (2024). *Multi-objective design indicators: A quantitative paradigm for early engineering design of deeply integrated products* [Doctoral dissertation, Technical University of Denmark]. Technical University of Denmark. Retrieved on 10 October 2025 from https://local.forskningsportal.dk/local/dki-cgi/ws/cris-link?src=dtu&id=dtu-b7c12d31-0ef3-47a2-803d-a407053c22ac&ti=Multi-Objective%20Design%20Indicators%20%3A%20A%20Quantitative%20Paradigm%20for%20Early%20Engineering%20Design%20of%20Deeply%20Integrated%20Products

Kadolkar, I., Ali, S., & Vieira, D. (2024). Algorithmic management in the gig economy: A systematic review. *New Technology, Work and Employment*, *39*(1), 102–121.

Kalach, F. E., Yousif, I., Wuest, T., Sheth, A., & Harik, R. (2024). Cognitive manufacturing: Definition and current trends. *Journal of Intelligent Manufacturing, 36*(6), 3695–3715. DOI: 10.1007/s10845-024-02429-9

Kalota, F. (2024). A Primer on Generative Artificial Intelligence. In *Education Sciences* (Vol. 14, Issue 2). DOI: 10.3390/educsci14020172

Kanbach, D. K., Heiduk, L., Blueher, G., Schreiter, M., & Lahmann, A. (2023). The GenAI is out of the bottle: Generative artificial intelligence from a business model innovation perspective. *Review of Managerial Science*, •••, 1–32. DOI: 10.1007/s11846-023-00696-z

Kapoor, S. (2021). AI-driven decision support systems in sports project management: Enhancing strategic planning. *International Journal of Artificial Intelligence, Data Science, and Machine Learning*, *2*(3), 1–11. DOI: 10.63282/3050-9262.IJAIDSML-V2I3P101

Karazi, S., Moradi, M., & Benyounis, K. Y. (2019). Statistical and numerical approaches for modeling and optimizing laser micromachining process-review. In *Elsevier eBooks*. Elsevier BV., DOI: 10.1016/B978-0-12-803581-8.11650-9

Karsgaard, C. (2024). Democracy and planetary politics: Achille Mbembe and futures of digital citizenship education for life. *Critical Studies in Education*, *65*(5), 460–475. DOI: 10.1080/17508487.2024.2307573

Kasani, P. H., Cho, K. H., Jang, J. W., & Yun, C. H. (2024). Influence of artificial intelligence and chatbots on research integrity and publication ethics. *Science Editing*, *11*(1), 12–25. DOI: 10.6087/kcse.323

Kaspar, C., Ravoo, B. J., van der Wiel, W. G., Wegner, S. V., & Pernice, W. H. (2021). The rise of intelligent matter. *Nature*, *594*(7863), 345–355. DOI: 10.1038/s41586-021-03453-y

Kavak, A., & Yıldız, R. (2025). Library Image in Children's Imagination: A Qualitative Analysis Based on Drawings in Türkiye. *Public Library Quarterly*, •••, 1–22. DOI: 10.1080/01616846.2025.2473164

Kawalya, J. (2021). The national library of Uganda: challenges faced in performing its institutional practices. *Library Philosophy and Practice (e-journal), 5073.*

Kaziz, S., Mariem, I. B., Echouchene, F., Belkhiria, M., & Belmabrouk, H. (2022). Taguchi optimisation of integrated flow microfluidic biosensor for COVID-19 detection. *The European Physical Journal Plus*, *137*(11), 1235. Advance online publication. DOI: 10.1140/epjp/s13360-022-03457-1

Keim, D., Qu, H., & Ma, K. L. (2013). Big-data visualization. *IEEE Computer Graphics and Applications*, *33*(4), 20–21. DOI: 10.1109/MCG.2013.54

Kellogg, K. C., Valentine, M. A., & Christin, A. (2020). Algorithms at work: The new contested terrain of control. *The Academy of Management Annals*, *14*(1), 366–410. DOI: 10.5465/annals.2018.0174

Kelly, R. (2023). Over-reliance on ChatGPT could harm worker performance. ITPro. Retrieved on 29 October 2024 from https://www.itpro.com/technology/artificial-intelligence/over-reliance-on-chatgpt-could-harm-worker-performance

Kemp, A. (2024). Competitive advantage through artificial intelligence: Toward a theory of situated AI. *Academy of Management Review*, *49*(3), 618–635. DOI: 10.5465/amr.2020.0205

Kennedy, M., Ninomiya, M. E. M., Ninomiya, M. M., Brascoupé, S., Smylie, J., Calma, T., Mohamed, J., Stewart, P., & Maddox, R. (2024). Knowledge translation in Indigenous health research: Voices from the field. *The Medical Journal of Australia*, *221*(1), 61–67. DOI: 10.5694/mja2.52357

Kepdieu, J. M., Djangang, C. N., Njimou, J. R., Tchanang, G., Ekani, C. J., Maicaneanu, S. A., & Tizaoui, C. (2023). Mathematical modelling using full factorial design in response surface methodology applied in the adsorptive removal of dye basic blue 9 from synthetic aqueous solutions onto oryza sativa husk-derived nanosilica doped smectic clay. *Research Square (Research Square)*. DOI: 10.21203/rs.3.rs-3776135/v1

Khalifa, M., & Albadawy, M. (2024). Using artificial intelligence in academic writing and research: An essential productivity tool. *Computer Methods and Programs in Biomedicine Update*, *100145*, 100145. Advance online publication. DOI: 10.1016/j.cmpbup.2024.100145

Khan, M., Chuenchart, W., Surendra, K. C., & Khanal, S. K. (2023). Applications of artificial intelligence in anaerobic co-digestion: Recent advances and prospects. *Bioresource Technology*, *370*, 128501. DOI: 10.1016/j.biortech.2022.128501

Khoa, B. T., Hung, B. P., & Hejsalem-Brahmi, M. (2023). Qualitative research in social sciences: Data collection, data analysis and report writing. *International Journal of Public Sector Performance Management*, *12*(1-2), 187–209. DOI: 10.1504/IJPSPM.2023.132247

Kietzmann, J., & Park, A. (2024). Written by ChatGPT: AI, large language models, conversational chatbots, and their place in society and business. *Business Horizons*, *67*(5), 453–459. DOI: 10.1016/j.bushor.2024.06.002

King, R. D., Rowland, J., Oliver, S. G., Young, M., Aubrey, W., Byrne, E., & Whelan, K. E. (2009). The automation of science. *Science, 324*(5923), 85–89. DOI: 10.1126/science.1165620

Kiran, B. R., Sobh, I., Talpaert, V., Mannion, P., Sallab, A. A. A., Yogamani, S., & Perez, P. (2022). Deep Reinforcement Learning for Autonomous Driving: A Survey. *IEEE Transactions on Intelligent Transportation Systems, 23*(6), 4909–4926. Advance online publication. DOI: 10.1109/TITS.2021.3054625

Klein, H. K., & Myers, M. D. (1999). A set of principles for conducting and evaluating interpretive field studies in information systems. *Management Information Systems Quarterly, 23*(1), 67–93. DOI: 10.2307/249410

Kleinrichert, D. (2024). Empathy: An ethical consideration of AI & others in the workplace. *AI & Society, 39*(6), 2743–2757. Advance online publication. DOI: 10.1007/s00146-023-01831-w

Kodumuru, R., Sarkar, S., Parepally, V., & Chandarana, J.Publishing Institute. (2025). Artificial intelligence and internet of things integration in pharmaceutical manufacturing: A smart synergy. *Pharmaceutics, 17*(3), 290. DOI: 10.3390/pharmaceutics17030290

Kok, C. L., Ho, C. K., Lee, T. K., Loo, Z. Y., Koh, Y. Y., & Chai, J. P. (2024). A novel and low-cost cloud-enabled iot integration for sustainable remote intravenous therapy management. *Electronics (Basel), 13*(10), 1801. DOI: 10.3390/electronics13101801

Koli, T. A., Arajpure, V. G., & Patil, V. (2024). "Optimizing lathe process capability using non-metallic gears: A comparative analysis of machining performance." *Research Square (Research Square)*. DOI: 10.21203/rs.3.rs-5288850/v1

Koller, D., & San Juan, V. (2015). Play-based interview methods for exploring young children's perspectives on inclusion. *International Journal of Qualitative Studies in Education : QSE, 28*(5), 610–631. DOI: 10.1080/09518398.2014.916434

Konczi, A. E., & Bill, L. (2024). Advancing first nations principles of OCAP. In Garvey, G. (Ed.), *Indigenous and tribal peoples and cancer* (pp. 37–39). Springer Nature Switzerland., DOI: 10.1007/978-3-031-56806-0_8

Kovach, M. (2021). *Indigenous methodologies: Characteristics, conversations, and contexts*. University of Toronto press.

Kraus, S., Breier, M., & Dasi-Rodriguez, S. (2020). The art of crafting a systematic literature review in entrepreneurship research. *The International Entrepreneurship and Management Journal, 16*(3), 1023–1043. DOI: 10.1007/s11365-020-00635-4

Krauss, J. P., Kailer, N., & Weinmann, C.-B. (2021). Digital transformation: An overview of the current state of the art of research. *SAGE Open*, *11*(3), 1–15. DOI: 10.1177/21582440211047576

Krishnan, A. R. (2025). Research trends in criteria importance through intercriteria correlation (CRITIC) method: a visual analysis of bibliographic data using the Tableau software. *Information Discovery and Delivery, Emerald publication*, *53*(2), 233-247. DOI: 10.1108/IDD-02-2024-0030

Kronberger, G., Bachinger, F., & Affenzeller, M. (2020). Smart manufacturing and continuous improvement and adaptation of predictive models. *Procedia Manufacturing*, *42*, 528–531. DOI: 10.1016/j.promfg.2020.02.037

Kuhlthau, C. (1988a). Perceptions of the information search process in libraries: A study of changes from high school through college. *Information Processing & Management*, *24*(4), 419–427. DOI: 10.1016/0306-4573(88)90045-3

Kuhlthau, C. (1988b). Longitudinal case studies of the information search process of users in libraries. *Library & Information Science Research*, *10*(3), 257–304.

Kuhlthau, C. (1996). The concept of a Zone of Intervention for identifying the role of intermediaries in the information search process. *Proceedings of the American Society for Information Science Annual Meeting*, 33, 367-376.

Kumar, A. (2024). The role of simulation and modeling in artificial intelligence: A review. *International Journal of Modeling, Simulation and Scientific Computing*, *15*(03), 2430002. DOI: 10.1142/S1793962324300024

Kumar, G., Goyal, K. K., Batra, N. K., & Mehdi, H. (2025). Optimizing reconfigurable manufacturing system configuration selection with multi-objective grey wolf optimisation. [IJIDeM]. *International Journal on Interactive Design and Manufacturing*, *19*(8), 5567–5582. DOI: 10.1007/s12008-024-02150-0

Kumar, G., Goyal, K. K., Batra, N. K., & Rani, D. (2022). Single part reconfigurable flow line design using fuzzy best worst method. *Opsearch*, *59*(2), 603–631. DOI: 10.1007/s12597-021-00550-4

Kumar, G., Tomer, A., Kumar, M., Goel, P., & Singh, O. P. (2021). Optimisation of turning process variables for surface integrity using taguchi technique. *IOP Conference Series. Materials Science and Engineering*, *1149*(1), 12029. DOI: 10.1088/1757-899X/1149/1/012029

Kumari, P., & Lavanya, M. (2024). Optimization strategies for corrosion management in industries with artificial neural network and response surface technology: A comprehensive review. *Journal of Bio- and Tribo-Corrosion, 10*(3), 59. DOI: 10.1007/s40735-024-00863-z

Kumar, P., & Misra, J. P. (2020). Process modeling and optimisation using ANN and RSM during dry turning of titanium alloy used in automotive industry. *Proceedings of the Institution of Mechanical Engineers. Part D, Journal of Automobile Engineering, 235*(7), 2040–2050. DOI: 10.1177/0954407020969255

Kumar, R., & Reji, M. (2023). Response surface methodology (RSM): An overview to analyze multivariate data. *Indian Journal of Microbiology Research, 9*(4), 241–248. DOI: 10.18231/j.ijmr.2022.042

Kuo, C., & Shieh, C. (2020). Biocatalytic process optimisation. *Catalysts, 10*(11), 1303. DOI: 10.3390/catal10111303

Kusan, M., & Arin, A. (2025). Ethical and security issues: The impact of artificial intelligence in sports management. *Acta Scientiae et Intellectus, 10*(2), 78. Advance online publication. DOI: 10.53469/asi.2025.10.2.78

Kushwah, A., Sharma, P. K., Kushwaha, H. L., Sharma, B. B., Shrivastava, A. K., Nag, R. H., Chowdhury, M., Carpenter, G., & Yadav, R. (2024). Hybrid taguchi-grey relational analysis approach for optimizing cutter operational parameters in selective cauliflower harvesting. *Scientific Reports, 14*(1), 29709. Advance online publication. DOI: 10.1038/s41598-024-81323-z

Kwanya, T., & Wakunuma, K. (2023). Regulation of digital platforms for a socially-just gig economy in Kenya. *HIIG Impact Publication Series*. DOI: 10.5281/zenodo.7588795

Lai, N. S., Tew, Y. S., Zhong, X., Yin, J., Li, J., Yan, B., & Wang, X. (2023). Artificial Intelligence (AI) workflow for catalyst design and optimisation. *Industrial & Engineering Chemistry Research, 62*(43), 17835–17848. DOI: 10.1021/acs.iecr.3c02520

Lake, L., & Jamieson, L. (2016). Using a child rights approach to strengthen prevention of violence against children. *South African Medical Journal, 106*(12), 1168–1172. DOI: 10.7196/SAMJ.2017.v106i12.12128

Landeta, J. (2006). Current validity of the Delphi method in social sciences. *Technological Forecasting and Social Change, 73*(5), 467–482. DOI: 10.1016/j.techfore.2005.09.002

Langley, C., Cirstea, B. I., Cuzzolin, F., & Sahakian, B. J. (2022). Theory of Mind and Preference Learning at the Interface of Cognitive Science, Neuroscience, and AI: A Review. In *Frontiers in Artificial Intelligence* (Vol. 5). DOI: 10.3389/frai.2022.778852

LeCun, Y., Bengio, Y., & Hinton, G. (2015). Deep learning. *Nature, 521*(7553), 436–444. DOI: 10.1038/nature14539

Lee, S. H., Ye, I., Lee, C., Kim, J., Choi, G., Nam, S. C., & Park, I. (2025). accelerating battery material optimization through iterative machine learning. *arXiv preprint arXiv:2505.18162*. DOI: 10.48550/ARXIV.2505.18162

Lee, T., Singh, V. P., & Cho, K. H. (2021). *Tensorflow and Keras Programming for Deep Learning*. DOI: 10.1007/978-3-030-64777-3_11

Lee, C., Lin, K.-X., & Chou, C.-W. (2024). Surrogate-based optimisation framework for enhancing smt process quality and productivity in electronics manufacturing services. *Research Square (Research Square)*. DOI: 10.21203/rs.3.rs-4770565/v1

Lee, D. S. (2025). *Governing global gig platforms in the age of AI*. Oxford University Press.

Lee, S., Kim, J., & Park, H. (2024). AI-assisted scientific discovery: From data to hypotheses. *Nature Machine Intelligence*, *6*(1), 45–55.

Lei, B., Kirk, T., Bhattacharya, A., Pati, D., Qian, X., Arróyave, R., & Mallick, B. K. (2021). Bayesian optimisation with adaptive surrogate models for automated experimental design. *npj Computational Materials*, *7*(1), 194. Advance online publication. DOI: 10.1038/s41524-021-00662-x

Lennon, R. P., Fraleigh, R., Van Scoy, L. J., Keshaviah, A., Hu, X. C., Snyder, B. L., Miller, E. L., Calo, W. A., Zgierska, A. E., & Griffin, C. (2021). Developing and testing an automated qualitative assistant (AQUA) to support qualitative analysis. *Family Medicine and Community Health*, *9*(Suppl 1), e001287. Advance online publication. DOI: 10.1136/fmch-2021-001287

Lepri, B., Oliver, N., Letouzé, E., Mehrabi, N., Oliver, N., & Pentland, A. (2018). Fair, accountable, and transparent algorithms in the public sector. *Proceedings of the 1st ACM SIGCAS Conference on Computing and Sustainable Societies (COMPASS '18)*. https://doi.org/DOI: 10.1145/3209811.3209812

Leslie, D., Katell, M., Aitken, M., Singh, J., Briggs, M., Powell, R., . . . Mazumder, A. (2022). Advancing data justice research and practice: An integrated literature review. *arXiv preprint arXiv:2204.03090*.

Lettieri, N., Malandrino, D., Guarino, A., & Zaccagnino, R. (2022). The Eye of the Rider. Visualization and data-driven heuristics for the critical analysis of gig economy. *International Conference on Information Visualisation*, 373–378. DOI: 10.1109/IV56949.2022.00068

Leviston, T. (1999). Victorian electronic records strategy final report. *Archives and Manuscripts*, *27*(2), 92–94.

Levy, R. V., & Wasmuth, H. (2025). Developmentally Appropriate Practice and its Impact on the Field of Early Child Education and Care in the United States1. *International Perspectives on the History of Early Childhood Education: Institution–Family–Profession*, 109.

Li, A., & Huang, W. (2023). A comprehensive survey of artificial intelligence and cloud computing applications in the sports industry. *Wireless Networks*, *30*(8), 6973–6984. DOI: 10.1007/s11276-023-03567-3

Li, H., Wang, S., & Tang, R. (2019). Robust optimal design of zero/low energy buildings considering uncertainties and the impacts of objective functions. *Applied Energy*, *254*, 113683. DOI: 10.1016/j.apenergy.2019.113683

Lim, J. S., Kim, Y. D., & Woo, J. C. (2024). Approximated uncertainty propagation of correlated independent variables using the ordinary least squares estimator. *Molecules. Molecules (Basel, Switzerland)*, *29*(6), 1248. DOI: 10.3390/molecules29061248

Lincoln, Y. S., & Guba, E. G. (2000). Paradigmatic controversies, contradictions, and emerging confluences. In Denzin, N. K., & Lincoln, Y. S. (Eds.), *The handbook of qualitative research* (2nd ed., pp. 1065–1122). Sage Publications.

Ling, Y., & Mahadevan, S. (2013). Quantitative model validation techniques: New insights. *Reliability Engineering & System Safety*, *111*, 217–231. DOI: 10.1016/j.ress.2012.11.011

Lipski, J., & Zaleski, K. (2019). Implementation of artificial intelligence in optimisation of technological processes. *MATEC Web of Conferences*, *252*, 3008. DOI: 10.1051/matecconf/201925203008

Littletree, S., and, M. B.-L., & Duarte, M. (2020). Centering relationality: A conceptual model to advance Indigenous knowledge organization practices. *Knowledge Organization*, *47*(5), 410–426. DOI: 10.5771/0943-7444-2020-5-410

Littletree, S., Andrews, N., & Loyer, J. (2023). Information as a relation: Defining Indigenous information literacy. *Journal of Information Literacy*, *17*(2), 4–23. https://doi.org/https://doi.org/10.11645/17.2.8. DOI: 10.11645/17.2.8

Littletree, S., & Krise, C. (2024). The Squaxin Island Museum Library and Research Center: A report from the field of tribal libraries and museums. *International Journal of Librarianship*, *9*(2), 51–59. DOI: 10.23974/ijol.2024.vol9.2.361

Liu, N., Shapira, P., & Yue, X. (2021). Tracking developments in artificial intelligence research: Constructing and applying a new search strategy. *Scientometrics*, *126*(4), 3153–3192. Advance online publication. DOI: 10.1007/s11192-021-03868-4 PMID: 34720254

Li, Z. (2025). Innovation and transformation of sports event management driven by artificial intelligence: Athlete performance, referee decision support, security, and fan interaction. *Journal of Social Science Humanities and Literature*, *8*(2), 3. Advance online publication. DOI: 10.53469/jsshl.2025.08(03).03

Li, Z., Fei, F., & Zhang, G. (2022). Edge-to-Cloud IIoT for condition monitoring in manufacturing systems with ubiquitous smart sensors. *Sensors (Basel)*, *22*(15), 5901. DOI: 10.3390/s22155901

Lohith, N., Bhaskar, H. B., & Yogendrakumar, S. (2024). Optimisation of condensation heat transfer coefficients using Taguchi method: Influence of temperature, surface type, and coolant flow rate. *Research Square (Research Square)*. DOI: 10.21203/rs.3.rs-5421143/v1

Lokman, A., Ismail, W. Z. W., & Aziz, N. A. A. (2025). A review of water quality forecasting and classification using machine learning models and statistical analysis. *Water. Water (Basel)*, *17*(15), 2243. DOI: 10.3390/w17152243

Lo, L. S. (2023). The CLEAR path: A framework for enhancing information literacy through prompt engineering. *Journal of Academic Librarianship*, *49*(4), 102720. Advance online publication. DOI: 10.1016/j.acalib.2023.102720

Longkumer, I., & Hussain Mazumder, D. (2024). Improving cancer prediction using feature selection in spark environment. *Concurrency and Computation*, *36*(2), e7903. Advance online publication. DOI: 10.1002/cpe.7903

Longpre, S., Hou, L., Vu, T., Webson, A., Chung, H. W., Tay, Y., & Roberts, A. (2023, July). The flan collection: Designing data and methods for effective instruction tuning. In *International Conference on Machine Learning* (pp. 22631-22648). PMLR (Proceedings of Machine Learning Research). DOI: 10.48550/arXiv.2301.13688

López-Chila, R., Llerena-Izquierdo, J., Sumba-Nacipucha, N., & Cueva-Estrada, J. (2024). Artificial Intelligence in Higher Education: An Analysis of Existing Bibliometrics. In *Education Sciences* (Vol. 14, Issue 1). DOI: 10.3390/educsci14010047

López, G., Yate, C., Ramos, F. A., Cala, M. P., Restrepo, S., & Baena, S. (2019). Production of polyunsaturated fatty acids and lipids from autotrophic, mixotrophic and heterotrophic cultivation of galdieria sp. strain USBA-GBX-832. *Scientific Reports*, *9*(1), 10791. Advance online publication. DOI: 10.1038/s41598-019-46645-3

López-Quiñones, A., Martínez-López, M., Sandoval, C. D. M., Carroll-Miranda, J., Lindala, A., Chatman, M. C., Fleming, J. S., Shockley, E. T., Cadeau, D., & Flores-Reyes, E. (2023). Ancestral computing for sustainability: Centering Indigenous epistemologies in researching Computer Science Education. *TechTrends*, *67*(3), 435–445. DOI: 10.1007/s11528-022-00820-y

Lou, B., & Wu, L. (2021). AI on drugs: Can artificial intelligence accelerate drug development? Evidence from a large-scale examination of Bio-Pharma firms. *Management Information Systems Quarterly*, *45*(3), 1451–1482. DOI: 10.25300/MISQ/2021/16565

Loveridge, J., Wood, B. E., Davis-Rae, E., & McRae, H. (2024). Ethical challenges in participatory research with children and youth. *Qualitative Research*, *24*(2), 391–411. DOI: 10.1177/14687941221149594

Lozzi, D., Di Pompeo, I., Marcaccio, M., Alemanno, M., Krüger, M., Curcio, G., & Migliore, S. (2025). AI-powered analysis of eye tracker data in basketball games. *Sensors (Basel)*, *25*(11), 3572. DOI: 10.3390/s25113572

Lund, B. D., Wang, T., Mannuru, N. R., Nie, B., Shimray, S., & Wang, Z. (2023). ChatGPT and a new academic reality: Artificial intelligence-written research papers and the ethics of the large language models in scholarly publishing. *Journal of the Association for Information Science and Technology*, *74*(5), 570–581. DOI: 10.1002/asi.24750

Lundberg, S. (2019). *SHAP (SHapley Additive exPlanations)*. Github Repository.

Lundberg, S. M., & Lee, S. I. (2017). A unified approach to interpreting model predictions. *Advances in Neural Information Processing Systems*, •••, 30.

Lynham, S. A. (2002). Theory building: Dubin's method. *Advances in Developing Human Resources*, *4*(3), 242–276. DOI: 10.1177/15222302004003003

Lys, I., Quinnell, R., Weyand, P., Phillips, B., & Mochrie, K. (2022). Students as co-creators of an Aboriginal and Torres Strait heritage garden and interactive maps. *Ascilite Publications*, e22209. DOI: 10.14742/apubs.2022.209

Ma, C., Zhao, M., & Zhao, Y. (2023). An overview of Hadoop applications in transportation big data. In *Journal of Traffic and Transportation Engineering (English Edition)* (Vol. 10, Issue 5). DOI: 10.1016/j.jtte.2023.05.003

Maar, M., Ominika, T., & Manitowabi, D. (2022). Community-led recovery from the opioid crisis through culturally-based programs and community-based data governance. *International Indigenous Policy Journal, 13*(2). Advance online publication. DOI: 10.18584/iipj.2022.13.2.13792

Machin-Mastromatteo, J. D. (2023). Community-driven and social initiatives. *Information Development, 39*(3), 393–401. DOI: 10.1177/02666669231197243

MacKenzie, A. (2025). Credible, competent contributors: Children and young people as postdigital citizen social scientists. *Postdigital Science and Education, 7*(1), 224–245. DOI: 10.1007/s42438-024-00517-w

Mackey, T. K., Calac, A. J., Keshava, B. S. C., Yracheta, J., Tsosie, K. S., & Fox, K. (2022). Establishing a blockchain-enabled Indigenous data sovereignty framework for genomic data. *Cell, 185*(15), 2626–2631. DOI: 10.1016/j.cell.2022.06.030

Macklin, C. D., Marchand, C., Mitchell, E., Price, R., Mitchell, V., & Bryant, L. (2021). Planting the seeds: Insights for researchers interested in working with Indigenous peoples. *International Journal of Indigenous Health, 16*(1). Advance online publication. DOI: 10.32799/ijih.v16i1.33193

MacNeil, H. (2000). *Trusting records. Legal, historical and diplomatic perspectives.* Kluwer.

Macpherson, T., Churchland, A., Sejnowski, T., DiCarlo, J., Kamitani, Y., Takahashi, H., & Hikida, T. (2021). Natural and Artificial Intelligence: A brief introduction to the interplay between AI and neuroscience research. *Neural Networks, 144*, 603–613. Advance online publication. DOI: 10.1016/j.neunet.2021.09.018 PMID: 34649035

Madanchian, M., & Taherdoost, H. (2024). AI-Powered Innovations in High-Tech Research and Development: From Theory to Practice. *Computers, Materials & Continua, 81*(2), 2133–2159. DOI: 10.32604/cmc.2024.057094

Magama, B. (2018). Strategies for the preservation of digital records in Masvingo province of Zimbabwe. Master's Thesis, University of South Africa, Pretoria.

Magama, B., & Nduna, V. (2020). Journey to EDRMS solution in Zimbabwe public sector. In Keakopa, S. M., & Mosweu, T. L. (Eds.), *Cases on electronic record management in the ESARBICA Region* (pp. 142–159). IGI Global., DOI: 10.4018/978-1-7998-2527-2.ch007

Magara, E., & Batambuze, C. (2009). Reading promotion programmes in primary schools: A study of school library management practices in Pallisa District in Uganda. *Mousaion, 27*(si-1), 108–127.

Magara, E., & Bukirwa Nyumba, J. (2004). Towards a school library development policy for Uganda. *Library Review*, *53*(6), 313–322. DOI: 10.1108/00242530410544402

Maguire, M., & Delahunt, B. (2017). Doing a thematic analysis: A practical, step-by-step guide for learning and teaching scholars. *AISHE-J, 9*, 3351. https://ojs.aishe.org/index.php/aishe-j/article/view/3354

Ma, H. M., Huang, W., & Dennis, A. R. (2024). Unintended consequences of disclosing recommendations by artificial intelligence versus humans on true and fake news believability and engagement. *Journal of Management Information Systems*, *41*(3), 616–644. DOI: 10.1080/07421222.2024.2376381

Mahapatra, A. P. K., Saraswat, R., Botre, M., Paul, B., & Prasad, N. (2020). Application of response surface methodology (RSM) in statistical optimisation and pharmaceutical characterization of a patient compliance effervescent tablet formulation of an antiepileptic drug levetiracetam. *Future Journal of Pharmaceutical Sciences*, *6*(1), 82. Advance online publication. DOI: 10.1186/s43094-020-00096-0

Mahlangu, G. (2020). *A multi-dimensional model for assessing e-government service gaps in the context of a developing country: A critical realist perspective*. Cape Peninsula University of Technology.

Mahmood, A. M., Siponen, M., Straub, D. W., Rao, H. R., & Raghu, T. S. (2010). Moving toward black hat research in information systems security: An editorial introduction to the special issue. *Management Information Systems Quarterly*, *34*(3), 431–433. DOI: 10.2307/25750685

Mairpady, A., Mourad, A. I., & Mozumder, M. S. (2021). Statistical and machine learning-driven optimisation of mechanical properties in designing durable HDPE Nanobiocomposites. *Polymers*, *13*(18), 3100. DOI: 10.3390/polym13183100

Mala, S., Deep, U., Mala, E., & Akhai, S. (2025). Digital health technologies for preventive cardiology. In *Impact of Digital Solutions for Improved Healthcare Delivery* (pp. 379–400). IGI Global., DOI: 10.4018/979-8-3693-5237-3.ch014

Manning, B. R. M. (2023). Relationships, respect, and reciprocity: Approaches to learning and teaching about indigenous cultural burning and landscape stewardship, 145-156. DOI: 10.4337/9781789905069.00019

Manning, M. S., Hassanein, M., Partridge, M. A., Jawa, V., Mora, J., Ryman, J., Barker, B., Braithwaite, C., Carleton, K., Hay, L. S., Hottenstein, C., Kubiak, R., & Devanarayan, V. (2022). Comparison of titer and signal to noise (S/N) for determination of anti-drug antibody magnitude using clinical data from an industry consortium. *The AAPS Journal, 24*(4), 81. Advance online publication. DOI: 10.1208/s12248-022-00728-8

Manoharan, G., Ashtikar, S. P., & Nivedha, M. (2024). Integrating artificial intelligence in library management: an emerging trend. In Senthilkumar, K. (Ed.), *AI-assisted library reconstruction* (pp. 144–157). IGI Global. DOI: 10.4018/979-8-3693-2782-1.ch008

Manyane, R. M. (2024). Centuries-long African oral traditions and history: Revisiting the debate. *Journal of African History, Archaeology and Tourism, 2*(1), 99-116. https://doi.org/doi:10.31920/2753-3204/2024/v2n1a5

Mariani, M., & Dwivedi, Y. K. (2024). Generative artificial intelligence in innovation management: A preview of future research developments. *Journal of Business Research, 175*, 114542. Advance online publication. DOI: 10.1016/j.jbusres.2024.114542

Markham, A., & Buchanan, E. (2012). Ethical decision-making and internet research (Version 2.0): Recommendations from the AoIR Ethics Working Committee.

Marshall, I. J., & Wallace, B. C. (2019). Toward systematic review automation: A practical guide to using machine learning tools in research synthesis. *Systematic Reviews, 8*(1), 163. DOI: 10.1186/s13643-019-1074-9

Marshall, R., Desjardine, M., Levison, J., Anderson, K., & McBean, E. A. (2020). Moving towards effective first nations' source water protection: Barriers, opportunities, and a framework. *Water (Basel), 12*(11), 2957. DOI: 10.3390/w12112957

Marsland, M., Wilson, I., Abeyasekera, S., & Kleih, U. (2014). *Socioeconomic methodologies for natural resources research: Best practice guidelines combining quantitative (formal) and qualitative (informal) survey methods*. National Resources Institute.

Martínez, V., & Shadman, F. (2024). Mathematical Modeling for Optimizing the structural properties of catalysts used for series multiple reactions. *Zeitschrift für Angewandte Mathematik und Physik, 12*(11), 3923–3940. DOI: 10.4236/jamp.2024.1211239

Mashaly, A. A., Mahdy, M. G., & Elemam, W. E. (2023). Optimal design and characteristics of sustainable eco-friendly ultra-high-performance concrete. *Innovative Infrastructure Solutions*, *8*(12), 326. Advance online publication. DOI: 10.1007/s41062-023-01277-5

Mason, R. O., Collins, C. P., & Cox, E. L. (1986). Four ethical issues of the information age. *Management Information Systems Quarterly*, *10*(1), 5–12. DOI: 10.2307/248873

Masteali, S. H., Bayat, M., Bettinger, P., & Ghorbanpour, M. (2025). Uncertainty analysis of linear and non-linear regression models in the modeling of water quality in the Caspian Sea basin: Application of Monte-Carlo method. *Ecological Indicators*, *170*, 112979. DOI: 10.1016/j.ecolind.2024.112979

Masuku, M., Bhebhe, S., & Khumalo, N. B. (2017). An information management framework for the support of e-government in Zimbabwe. *Tidsskriftet Arkiv*, *8*(1), 1–14. DOI: 10.7577/ta.1960

Mateus, N., Abade, E., Coutinho, D., Gómez, M.-Á., Peñas, C. L., & Sampaio, J. (2025). Empowering the sports scientist with artificial intelligence in training, performance, and health management. *Sensors (Basel)*, *25*(1), 139. DOI: 10.3390/s25010139

Mavi, R. K., Shekarabi, S. A. H., Mavi, N. K., Arisian, S., & Moghdani, R. (2023). Multi-objective optimisation of sustainable closed-loop supply chain networks in the tire industry. *Engineering Applications of Artificial Intelligence*, *126*, 107116. DOI: 10.1016/j.engappai.2023.107116

Maxwell, J. A. (2013). Qualitative research design: An interactive approach. *Sage (Atlanta, Ga.)*.

Mbare, B. (2024). Algorithmic management, wellbeing and platform work. *Labour & Industry*, *34*(2), 145–166.

McCann, D., & Cruz-Santiago, A. (2022). Labour/data justice: A new framework for labour/regulatory datafication. *Journal of Law and Society*, *49*(4), 555–578. DOI: 10.1111/jols.12392

McCarty, W. (2010). Attending from and to the machine. Inaugural lecture, King's College London, 2 February.

McCausland, R., Jamieson, S. K., Robinson, V., Spencer, W., MacGillivray, P., & Andersen, M. (2025). Elders' perspectives and priorities for ageing well in a remote Aboriginal community. *Ageing and Society*, *45*(1), 31–54. DOI: 10.1017/S0144686X23000156

Mcleod, J., & Gormly, B. (2017). Using the cloud for records storage: Issues of trust. *Archival Science, 17*(4), 349–370. DOI: 10.1007/s10502-017-9280-5

McMullin, J. (2025). Decolonization. *Annals of Anthropological Practice, 49*(1), e70002. Advance online publication. DOI: 10.1111/napa.70002

McTaggart, V., & McLaughlin, C. (2024). Lessons learned from working from home: Exploring factors considered critical to supporting successful working from home during the pandemic. *Management and Labour Studies, 50*(1), 35–60. DOI: 10.1177/0258042X241283209

Mdhluli, T. D., Mokgoatšana, S., Kugara, S. L., & Vuma, L. (2021). Knowledge management: Preserving, managing and sharing indigenous knowledge through digital library. *Hervormde Teologiese Studies, 77*(2). Advance online publication. https://www.ajol.info/index.php/hts/article/view/232746. DOI: 10.4102/hts.v77i2.6795

Mehta, H., & Passi, K. (2022). Social Media Hate Speech Detection Using Explainable Artificial Intelligence (XAI). *Algorithms, 15*(8), 291. Advance online publication. DOI: 10.3390/a15080291

Melaku, N. S., & Bogale, T. M. (2023). Parameters optimisation in plasma arc cutting of AISI 1020 mild steel plate using hybrid genetic algorithm and artificial neural network. *International Journal for Simulation and Multidisciplinary Design Optimisation, 14*, 20. DOI: 10.1051/smdo/2023012

Melo, N. (2025). *Decolonizing research methods: Indigenous, Afrocentric, and participatory approaches: Challenging colonial epistemologies and centering community-led knowledge production*. Nouridin Melo.

Mendenhall, A., Hicks, C., Holder, M., Holmes, C., Jung, E., Ramirez, L., Fairman, M., Bulls, S. T., & Levy, M. (2024). Growing stronger together: Sharing a story of culturally responsive evaluation with indigenous families and communities. *Child Abuse & Neglect, 148*, 106344. https://doi.org/https://doi.org/10.1016/j.chiabu.2023.106344. DOI: 10.1016/j.chiabu.2023.106344

Mensah, R. A., Kirton, S. B., Cook, M. T., Styliari, I. D., Hutter, V., & Chau, D. (2019). Optimising poly(lactic-co-glycolic acid) microparticle fabrication using a Taguchi orthogonal array design-of-experiment approach. *PLoS One, 14*(9), e0222858. Advance online publication. DOI: 10.1371/journal.pone.0222858

Mergel, I., Edelmann, N., & Haug, N. (2019). Defining digital transformation: Results from expert interviews. *Government Information Quarterly, 36*(4), 101385. DOI: 10.1016/j.giq.2019.06.002

Mihalovits, M., & Kemény, S. (2022). Testing the fit of regression models estimated with extremely small samples: Application in pharmaceutical stability studies. *Periodica Polytechnica. Chemical Engineering*, *66*(2), 325–339. DOI: 10.3311/PPch.18793

Miikkulainen, R., Brundage, M., Epstein, J., Foster, T., Hodjat, B., Iscoe, N., Jiang, J., Legrand, D., Nazari, S., Qiu, X., Scharff, M., Schoolland, C., Severn, R., & Shagrin, A. (2020). Ascend by evolv: Artificial intelligence-based massively multivariate conversion rate optimisation. *AI Magazine*, *41*(1), 44–60. DOI: 10.1609/aimag.v41i1.5256

Miles, M. B., & Huberman, A. M. (1994). *Qualitative data analysis: an expanded sourcebook* (2nd ed.). Sage.

Miles, M. B., Huberman, A. M., & Saldaña, T. (2020). *Qualitative data analysis: A methods sourcebook* (4th ed.). Sage.

Mirbabaie, M., Brendel, A. B., & Hofeditz, L. (2022). Ethics and AI in information systems research. *Communications of the Association for Information Systems*, *50*(1), 38. DOI: 10.17705/1CAIS.05034

Mirbabaie, M., Stieglitz, S., & Marx, J. (2025). Transparency fallacy: Perceived fairness in algorithmic management. *Computers in Human Behavior*, *155*, 107347.

Modi, M., & Gopal, A. (2020). Modelling, Investigation of process responses, surface assessment and parametric optimisation in powder mixed electrical discharge diamond grinding of TI6AL4V utilizing grey-based Taguchi approach. *Transactions of FAMENA*, *44*(3), 93–112. DOI: 10.21278/TOF.44308

Moffatt, B., & Hall, A. (2024). Is AI my co-author? The ethics of using artificial intelligence in scientific publishing. *Accountability in Research*, •••, 1–17. DOI: 10.1080/08989621.2024.2386285 PMID: 39109816

Mogaji, E., & Jain, V. (2024). How generative AI is (will) change consumer behaviour: Postulating the potential impact and implications for research, practice, and policy. *Journal of Consumer Behaviour*, *23*(5), 2379–2389. DOI: 10.1002/cb.2345

Mohammadi, S., Pashaie, S., Ghaffarisadr, S. I., & Rayner, M. (2024). Opportunities and challenges of leveraging artificial intelligence (AI) in sport marketing. In Tran, M. T. (Ed.), *Advancing the marketing technology (MarTech) revolution* (pp. 233–272). IGI Global., DOI: 10.4018/979-8-3693-4361-6.ch009

Mohammed, M., Alqahtani, N. K., & Ali, S. (2024). Development, RSM-based modeling, and process optimisation of an ultrasonic coating system for extending the storage life of fresh fruits. *Frontiers in Sustainable Food Systems, 8*, 1403164. Advance online publication. DOI: 10.3389/fsufs.2024.1403164

Mohanadevi, M., & Dhanabalan, K. (2024). Enhanced Chromium removal from water using ball milled nano-biochar derived from Eichhornia crassipes (Water hyacinth). *Research Square (Research Square)*. DOI: 10.21203/rs.3.rs-5312229/v1

Monod, E., Mayer, A.-S., Straub, D., Joyce, E., & Qi, J. (2024). From worker empowerment to managerial control: The devolution of AI tools' intended positive implementation to their negative consequences. *Information and Organization, 34*(1), 100498. Advance online publication. DOI: 10.1016/j.infoandorg.2023.100498

Montenegro, M. (2019). Subverting the universality of metadata standards: The TK labels as a tool to promote Indigenous data sovereignty. *The Journal of Documentation, 75*(4), 731–749. DOI: 10.1108/JD-08-2018-0124

Morales-Álvarez, A., Valdés-Cuervo, A. A., & Parra-Pérez, L. G. (2025). Supportive parenting and adolescents digital citizenship behaviors: The mediating role of self-regulation. *Cyberpsychology (Brno), 19*(1). Advance online publication. DOI: 10.5817/CP2025-1-2

Moravec, H. (2008). Rise of the robots. [Special Edition]. *Scientific American, 18*(1), 12–19. DOI: 10.1038/scientificamerican0208-12sp

Morgan, D. L. (2023). Exploring the use of artificial intelligence for qualitative data analysis: The case of ChatGPT. *International Journal of Qualitative Methods, 22*, 1–10. DOI: 10.1177/16094069231211248

Morisano, D., Robinson, M., Rush, B., & Linklater, R. (2024). Conducting research with Indigenous peoples in Canada: Ethical and policy considerations. *Frontiers in Psychology, 14*, 1214121. Advance online publication. DOI: 10.3389/fpsyg.2023.1214121

Mosweu, O., & Ngoepe, M. (2020). Skills and competencies for authenticating digital records to support auditing process in Botswana public sector. *African Journal of Library and Information Science, 29*(1), 17–28.

Moussa, Z., Ghoniem, A. A., Elsayed, A., Alotaibi, A. S., Alenzi, A. M., Hamed, S. E., Elattar, K. M., & Saber, W. I. A. (2022). Innovative binary sorption of Cobalt(II) and methylene blue by Sargassum latifolium using Taguchi and hybrid artificial neural network paradigms. *Scientific Reports, 12*(1), 18291. Advance online publication. DOI: 10.1038/s41598-022-22662-7

Msakni, M. K., Risan, A., & Schütz, P. (2023). Using machine learning prediction models for quality control: A case study from the automotive industry. *Computational Management Science, 20*(1), 14. Advance online publication. DOI: 10.1007/s10287-023-00448-0

Mumuni, A., & Mumuni, F. (2024). Automated data processing and feature engineering for deep learning and big data applications: A survey. *Journal of Information and Intelligence.* DOI: 10.1016/j.jiixd.2024.01.002

Munanday, A. P., Sazali, N., Harun, W. S. W., Kadirgama, K., & Jamaludin, A. S. (2023). Analysis of Convolutional Neural Networks for Facial Expression Recognition on GPU, TPU and CPU. *Journal of Advanced Research in Applied Sciences and Engineering Technology, 31*(3), 50–67. Advance online publication. DOI: 10.37934/araset.31.3.5067

Musat, C. L., Mereuta, C., Nechita, A., Tutunaru, D., Voipan, A. E., Voipan, D., Mereuta, E., Gurau, T. V., Gurău, G., & Nechita, L. C. (2024). Diagnostic applications of AI in sports: A comprehensive review of injury risk prediction methods. *Diagnostics (Basel), 14*(22), 2516. DOI: 10.3390/diagnostics14222516

Mutsagondo, S. (2017). Electronic records management in public departments. Master's thesis, University of South Africa, Pretoria.

Mutsagondo, S. (2021). Use and management of the electronic mail in the central government of Zimbabwe. DPhil thesis, University of South Africa, Pretoria.

Myers, D., Mohawesh, R., Chellaboina, V. I., Sathvik, A. L., Venkatesh, P., Ho, Y. H., Henshaw, H., Alhawawreh, M., Berdik, D., & Jararweh, Y. (2024). Foundation and large language models: fundamentals, challenges, opportunities, and social impacts. In *Cluster Computing* (Vol. 27, Issue 1). DOI: 10.1007/s10586-023-04203-7

Myers, M. (2020). *Qualitative research in business and management* (3rd ed.). Sage Publications. DOI: 10.4135/9781036234539

Nakaziba, S., & Ngulube, P. (2024). Harnessing digital power for relevance: Status of digital transformation in selected university libraries in Uganda. *Collection and Curation, 43*(2), 33–44. DOI: 10.1108/CC-11-2023-0034

Nalbant, K. G., & Aydın, S. (2022). Literature review on the relationship between artificial intelligence technologies with digital sports marketing and sports management. *Indonesian Journal of Sport Management, 2*(2), 135–143. DOI: 10.31949/ijsm.v2i2.2876

NARA (US National Archives and Records Administration). (2005), NARA Guidance on Managing Web Records. Retrieved on 10 October 2025 from www.archives.gov/recordsmanagement/policy/managing-webrecords-index.html

NBA. (2023). *AI and Data Analytics in the NBA: Policy and Practice*. National Basketball Association.

Ndimande, B. S. (2012). Decolonizing research in postapartheid South Africa. *Qualitative Inquiry, 18*(3), 215–226. DOI: 10.1177/1077800411431557

Ndunda, E., Ouma, S., & Mwangi, J. (2024). *Digital labour platforms in Kenya: Working conditions and wellbeing*. ILO.

Nelson, A. S., Santamaría, P. V., & Javens, J. S. (2024, November). Students' Perceptions of Generative AI Use in Academic Writing. In *Conference Proceedings. Innovation in Language Learning 2024*.

Nengomasha, C. T., & Chikomba, A. (2018). Status of EDRMS implementation in the public sector in Namibia and Zimbabwe. *Records Management Journal, 28*(3), 252–264. DOI: 10.1108/RMJ-08-2017-0023

Neumann, O., Guirguis, K., & Steiner, R. (2024). Exploring artificial intelligence adoption in public organizations: A comparative case study. *Public Management Review, 26*(1), 114–141. Advance online publication. DOI: 10.1080/14719037.2022.2048685

Neupane, S., Mitra, S., Fernandez, I. A., Saha, S., Mittal, S., Chen, J., Pillai, N., & Rahimi, S. (2024). Security Considerations in AI-Robotics: A Survey of Current Methods, Challenges, and Opportunities. *IEEE Access : Practical Innovations, Open Solutions, 12*, 22072–22097. Advance online publication. DOI: 10.1109/ACCESS.2024.3363657

Neyazi, T. A., Nadaf, A. H., Tan, K. E., & Schroeder, R. (2024). Does trust in government moderate the perception towards deepfakes? Comparative perspectives from Asia on the risks of AI and misinformation for democracy. *Government Information Quarterly, 41*(4), 101980. Advance online publication. DOI: 10.1016/j.giq.2024.101980

Nga, H. T. T., & Thuy, A. N. T. (2024). Recommender System with Apache Spark. *Lecture Notes in Networks and Systems, 786*, 487–497. Advance online publication. DOI: 10.1007/978-981-99-6547-2_37

Ngulube, P. (2025). Indigenous theories and theorisation in African scholarship in the quest for epistemic freedom, P. Ngulube (Ed.), *Harnessing Indigenous epistemologies for sustainable progress* (pp. 139-172). IGI Global. DOI: 10.4018/979-8-3693-8447-3.ch008

Ngulube, P. (2018). Overcoming the difficulties associated with conceptual and theoretical frameworks in heritage studies. In Ngulube, P. (Ed.), *Handbook of research in heritage management and preservation* (pp. 1–25). IGI Global. DOI: 10.4018/978-1-5225-3137-1.ch001

Ngulube, P. (2020a). Theory and theorising in information science scholarship. In Ngulube, P. (Ed.), *Handbook of research on connecting research methods for information science research* (pp. 18–39). IGI Global., DOI: 10.4018/978-1-7998-1471-9.ch002

Ngulube, P. (2023). Improving the quality of reporting findings using computer data analysis applications in educational research in context. *Heliyon*, *9*(9), e19683. DOI: 10.1016/j.heliyon.2023.e19683

Ngulube, P., & Mosha, N. F. (2023). Transparency in the use of theoretical frameworks for the advancement of knowledge in core library and information science journals: A systematic review. *African Journal of Library Archives and Information Science*, *33*(2), 129–146. DOI: 10.4314/ajlais.v33i2.2

Ngulube, P., & Mosha, N. F. V. (2025). Integrating artificial intelligence-based technologies 'safely' in academic libraries: An overview through a scoping review. *Technical Services Quarterly*, *42*(1), 46–67. DOI: 10.1080/07317131.2024.2432093

Nguyen, H. T., Jeong, H.-Y., Kim, K.-D., Xuan-Mung, N., Dao, N., Tran, H.-N., Hoang, V.-K., Anh, N. N., & Vu, M. T. (2024). A new hyperparameter tuning framework for regression tasks in deep neural network: Combined-sampling algorithm to search the optimized hyperparameters. *Mathematics*, *12*(24), 3892. DOI: 10.3390/math12243892

Nguyen, T., & Morales, L. (2025). Accessibility to sports performance for greater inclusiveness through artificial intelligence: A systematic review. *Journal of Sports Engineering & Technology*, *239*(1), 1–15. DOI: 10.1080/10255842.2025.2457118

Niazi, S. K. (2023). The coming of age of AI/ML in drug discovery, development, clinical testing, and manufacturing: The FDA perspectives. *Drug Design, Development and Therapy*, *2691*, 2691–2725. Advance online publication. DOI: 10.2147/DDDT.S424991

Nikzad, M. H., Heidari-Rarani, M., & Mirkhalaf, M. (2024). A novel Taguchi-based approach for optimizing neural network architectures: Application to elastic short fiber composites. *Composites Science and Technology*, *259*, 110951. DOI: 10.1016/j.compscitech.2024.110951

Nisa, C. (2025, November). Child-friendly library management as a strategy to strengthen reading literacy in Islamic Elementary Education. In *Proceedings of the International Conference On Education, Society and Humanity* (Vol. 3, No. 1, pp. 1006-1018).

Nitin, & Gupta, S. B. (2024). Artificial Intelligence in Smart Agriculture: Applications and Challenges. In *Current Applied Science and Technology* (Vol. 24, Issue 2). DOI: 10.55003/cast.2023.254427

Nouri-Moghaddam, B., Ghazanfari, M., & Fathian, M. (2020). A novel filter-wrapper hybrid gene selection approach for microarray data based on multi-objective forest optimisation algorithm. *Decision Science Letters, 271*, 271–290. Advance online publication. DOI: 10.5267/j.dsl.2020.5.006

Nourmohammadzadeh, A., Sarhani, M., & Voß, S. (2023). A matheuristic approach for the family traveling salesman problem. *Journal of Heuristics, 29*(4-6), 435–460. DOI: 10.1007/s10732-023-09516-9

O'Connor, D., Scoville, D., Donald, R., Donald, V., Witney, J., Ellefson, O., Jewell, G., Jewell, L., & Silkie, V. (2022). A multiple evidence-based approach to métis community-based monitoring: A case study from the South Athabasca oil sands area, Alberta, Canada. *Ecology and Society*, *27*(4), art27. Advance online publication. DOI: 10.5751/ES-13678-270427

O'Dea, X. (2024). Generative AI: Is it a paradigm shift for higher education? *Studies in Higher Education*, *49*(5), 811–816. DOI: 10.1080/03075079.2024.2332944

Ofori-Boateng, R., Aceves-Martins, M., Wiratunga, N., & Moreno-Garcia, C. F. (2024). Towards the automation of systematic reviews using natural language processing, machine learning, and deep learning: A comprehensive review. *Artificial Intelligence Review*, *57*(8), 200. DOI: 10.1007/s10462-024-10844-w

Ojha, V., Abraham, A., & Snášel, V. (2019). Heuristic design of fuzzy inference systems: A review of three decades of research. *Engineering Applications of Artificial Intelligence*, *85*, 845–864. DOI: 10.1016/j.engappai.2019.08.010

Okello-Obura, C., & Kigongo-Bukenya, I. (2011). Library and information science education and training in Uganda: Trends, challenges, and the way forward. *Education Research International*, *2011*, 1–9. DOI: 10.1155/2011/705372

Okoye, K., & Hosseini, S. (2024). Analysis of variance (ANOVA) in R: one-way and two-way ANOVA. In *R programming: statistical data analysis in research* (pp. 187-209). Springer Nature Singapore. DOI: 10.1007/978-981-97-3385-9_9

Onokwai, A. O., Olusanya, O. O., Onifade, M. K., Anyaegbuna, B. E., Onoriode, A., & Agholor, D. (2024). Comparative analysis of response surface methodology and adaptive neuro-fuzzy inference system for predictive fault detection and optimisation in beverage industry. *Frontiers of Mechanical Engineering*, *10*, 1428717. Advance online publication. DOI: 10.3389/fmech.2024.1428717

Onyelowe, K. C., Moghal, A. A. B., Ebid, A. M., Rehman, A. U., Hanandeh, S., & Priyan, V. V. (2024). Estimating the strength of soil stabilized with cement and lime at optimal compaction using ensemble-based multiple machine learning. *Scientific Reports*, *14*(1), 15308. Advance online publication. DOI: 10.1038/s41598-024-66295-4

Ooi, K. B., Tan, G. W. H., Al-Emran, M., Al-Sharafi, M. A., Capatina, A., Chakraborty, A., Dwivedi, Y. K., Huang, T. L., Kar, A. K., Lee, V. H., Loh, X. M., Micu, A., Mikalef, P., Mogaji, E., Pandey, N., Raman, R., Rana, N. P., Sarker, P., Sharma, A., & Wong, L. W. (2023). The Potential of Generative Artificial Intelligence Across Disciplines: Perspectives and Future Directions. *Journal of Computer Information Systems*. Advance online publication. DOI: 10.1080/08874417.2023.2261010

Oosthuizen, R. (2022). The fourth industrial revolution - Smart technology, artificial intelligence, robotics and algorithms: Industrial psychologists in future workplaces. *Frontiers in Artificial Intelligence*, *5*, 913168. DOI: 10.3389/frai.2022.913168

Oppioli, M., Sousa, M. J., Sousa, M., & de Nuccio, E. (2023). The role of artificial intelligence for management decision: A structured literature review. *Management Decision*. Advance online publication. DOI: 10.1108/MD-08-2023-1331

Orlikowski, W. J., & Baroudi, J. J. (1991). Studying information technology in organisations: Research approaches and assumptions. *Information Systems Research*, *2*(1), 1–28. DOI: 10.1287/isre.2.1.1

Osei-Tutu, A. A. Z. (2023). Developing African oral traditional storytelling as a framework for studying with African peoples. *Qualitative Research*, *23*(6), 1497–1514. DOI: 10.1177/14687941221082263

Oskouei, A. G., Samadi, N., Khezri, S., Moghaddam, A. N., Babaei, H., Hamini, K., & Arasteh, B. (2025). Feature-weighted fuzzy clustering methods: An experimental review. *Neurocomputing*, *619*, 129176. DOI: 10.1016/j.neucom.2024.129176

Ovchynnykova, O., Svazas, M., & Navickas, V. (2025). Assessing economic profiles of coastal regions in the blue economy: A radar chart approach. *Challenges in Sustainability*, *13*(2), 177–192. DOI: 10.56578/cis130203

Öztürk, B., Kaymak, Ş., & Küçük, Ö. (2022). Taguchi and RSM based optimiation of energy consumption on internal gear pumps. *International Journal of 3D Printing Technologies and Digital Industry, 6*(1), 164-175. DOI: 10.46519/ij3dptdi.1001245

Pahija, E., Hwangbo, S., Saulnier-Bellemare, T., & Patience, G. S. (2024). Experimental methods in chemical engineering: Monte Carlo. *Canadian Journal of Chemical Engineering, 102*(10), 3308–3321. DOI: 10.1002/cjce.25374

Paladino, L. M., Hughes, A., Perera, A., Topsakal, O., & Akinci, T. C. (2023). Evaluating the Performance of Automated Machine Learning (AutoML) Tools for Heart Disease Diagnosis and Prediction. *AI, 4*(4), 1036–1058. Advance online publication. DOI: 10.3390/ai4040053

Palmer, C. L., Sebastian, K., Belarde-Lewis, M., Littletree, S., & Guerrero, N. (2024). Stewarding contextual integrity in data services for Indigenous scholarship. *Proceedings of the Association for Information Science and Technology, 61*(1), 626–631. https://doi.org/https://doi.org/10.1002/pra2.1070. DOI: 10.1002/pra2.1070

Pal, S. (2023). A Paradigm Shift in Research: Exploring the intersection of artificial intelligence and research methodology. *International Journal of Innovative Research in Engineering & Multidisciplinary Physical Sciences, 11*(3), 230125. Advance online publication. DOI: 10.37082/IJIRMPS.v11.i3.230125

Palvia, S.C.J. & Sharma, S.S. (2007). E-Government and e-governance: definitions/domain framework and status around the world. *Foundations of E-Government*, 1–12.

Pandi, A., Diehl, C., Kharrazi, A. Y., Scholz, S. A., Bobkova, E., Faure, L., Nattermann, M., Adam, D., Chapin, N., Foroughijabbari, Y., Moritz, C., Paczia, N., Cortina, N. S., Faulon, J., & Erb, T. J. (2022). A versatile active learning workflow for optimisation of genetic and metabolic networks. *Nature Communications, 13*(1), 3876. Advance online publication. DOI: 10.1038/s41467-022-31245-z

Pang, B., Nijkamp, E., & Wu, Y. N. (2020). Deep Learning With TensorFlow: A Review. In *Journal of Educational and Behavioral Statistics* (Vol. 45, Issue 2). DOI: 10.3102/1076998619872761

Panigrahi, S., Nanda, A., & Swarnkar, T. (2021). A Survey on Transfer Learning. *Smart Innovation. Systems and Technologies, 194*, 781–789. Advance online publication. DOI: 10.1007/978-981-15-5971-6_83

Parashar, M., Deblanc-Knowles, T., Gianchandani, E., Parker, L. E., & Lin, H. Y. (2023). Strengthening and Democratizing Artificial Intelligence Research and Development. *Computer, 56*(11), 85–90. Advance online publication. DOI: 10.1109/MC.2023.3284568

Park, J., & Choo, S. (2024). Generative AI prompt engineering for educators: Practical strategies. *Journal of Special Education Technology*, •••, 1–7. DOI: 10.1177/01626434241298954

Pasha, N., Mahdiraji, H. A., Hajiagha, S. H. R., Garza-Reyes, J. A., & Joshi, R. (2023). A multi-objective flexible manufacturing system design optimisation using a hybrid response surface methodology. *Operations Management Research : Advancing Practice Through Research*, *17*(1), 135–151. DOI: 10.1007/s12063-023-00412-w

Pasquale, F. (2015). *The Black Box Society: The Secret Algorithms That Control Money and Information*. Harvard University Press. DOI: 10.4159/harvard.9780674736061

Pasupuleti, V., Thuraka, B., Kodete, C. S., & Malisetty, S. (2024). Enhancing supply chain agility and sustainability through machine learning: Optimisation techniques for logistics and inventory management. *Logistics (Basel)*, *8*(3), 73. DOI: 10.3390/logistics8030073

Patin, B., Oliphant, T., Allard, D., Gray, L., Clarke, R. I., Tacheva, J., & Lar-Son, K. (2021). At the margins of epistemology: Amplifying Alternative ways of knowing in Library and Information Science. *Proceedings of the Association for Information Science and Technology*, *58*(1), 630–633. https://doi.org/https://doi.org/10.1002/pra2.515. DOI: 10.1002/pra2.515

Paton-Ash, M., & Wilmot, D. (2015). Issues and challenges facing school libraries in selected primary schools in Gauteng Province, South Africa. *South African Journal of Education*, *35*(1), 1–10. DOI: 10.15700/201503062342

Patsioura, F. (2014). Evaluating e-government. In Yannacopoulos, D., Manolitzas, P., Matsatsinis, N., & Grigoroudis, E. (Eds.), *Evaluating websites and web services: interdisciplinary perspectives on user satisfaction* (pp. 1–18). IGI Global. DOI: 10.4018/978-1-4666-5129-6.ch001

Paul, M. (2023). Looking to the future: Indigenous data sovereignty and policy in Canada. *Pathfinder: A Canadian Journal for Information Science Students and Early Career Professionals, 4*(1), 54-67. DOI: 10.29173/pathfinder71

Paulus, T., Lester, J. N., & Davis, C. (2025). The construction of the role of AI in qualitative data analysis in the social sciences. *AI & Society*. Advance online publication. DOI: 10.1007/s00146-025-02488-3

Paw, J. K. S., Kiong, T. S., Kamarulzaman, M. K., Adam, A., Hisham, S., Kadirgama, K., & Benedict, F. (2023). Advancing renewable fuel integration: A comprehensive response surface methodology approach for internal combustion engine performance and emissions optimization. *Heliyon, 9*(11). Advance online publication. DOI: 10.1016/j.heliyon.2023.e22238

Peng, R. D. (2011). Reproducible research in computational science. *Science, 334*(6060), 1226–1227. DOI: 10.1126/science.1213847

Peng, Y., Khaled, U., Al-Rashed, A. A. A. A., van der Meer, R. L. J., Goodarzi, M., & Sarafraz, M. M. (2020). Potential application of response surface methodology (RSM) for the prediction and optimisation of thermal conductivity of aqueous CuO (II) nanofluid: A statistical approach and experimental validation. *Physica A, 554*, 124353. DOI: 10.1016/j.physa.2020.124353

Pérez-Benedito, M. A. (2025). Alternative measure of dispersion in the polar-radar chart methodology for activity management. *Archives of Business Research*, Services *for Science and Education, 13*(1), 07-20. https://doi.org/DOI: 10.14738/abr.131.18145

Perkins, M., & Roe, J. (2024). The use of generative AI in qualitative analysis: Inductive thematic analysis with ChatGPT. *Journal of Applied Learning & Teaching, 7*(1). Advance online publication. DOI: 10.37074/jalt.2024.7.1.22

Pham, T.-D., Nguyen, N., Tran, C.-M., & Vuong, M. P. (2020). Constructing response surface designs with orthogonal quadratic effects using cyclic generators. *Chemometrics and Intelligent Laboratory Systems, 198*, 103918. DOI: 10.1016/j.chemolab.2019.103918

Pithan, L., Starostin, V., Mareček, D., Petersdorf, L., Völter, C., Munteanu, V., Jankowski, M., Konovalov, O., Gerlach, A., Hinderhofer, A., Murphy, B. M., Kowarik, S., & Schreiber, F. (2023). Closing the loop: Autonomous experiments enabled by machine-learning-based online data analysis in synchrotron beamline environments. *Journal of Synchrotron Radiation, 30*(6), 1064–1075. DOI: 10.1107/S160057752300749X

Podlubne, A., & Gohringer, D. (2023). A Survey on Adaptive Computing in Robotics: Modelling, Methods and Applications. *IEEE Access : Practical Innovations, Open Solutions, 11*, 53830–53849. Advance online publication. DOI: 10.1109/ACCESS.2023.3281190

Polites, G. L., Roberts, N., & Thatcher, J. (2012). Conceptualizing models using multidimensional constructs: A review and guidelines for their use. *European Journal of Information Systems, 21*(1), 22–48. DOI: 10.1057/ejis.2011.10

Polo, S., Rubio, E. M., Ayllón, J., & de Agustina, B. (2025). Emerging advances in sustainable manufacturing. *Processes (Basel, Switzerland), 13*(5), 1549. DOI: 10.3390/pr13051549

Popović, B. (2020). Planning, analyzing and optimizing experiments. *Journal of Engineering Management and Competitiveness, 10*(1), 15–30. DOI: 10.5937/jemc2001015P

Pourdarbani, R., Sabzi, S., Nadimi, M., & Paliwal, J. (2023). Interpretation of Hyperspectral Images Using Integrated Gradients to Detect Bruising in Lemons. *Horticulturae, 9*(7), 750. Advance online publication. DOI: 10.3390/horticulturae9070750

Prakash, N., Atiq, A., Shahid, M., Rani, J., & Dikshit, S. (2024). Merging Minds and Machines: The Role of Advancing AI in Robotics. *EAI Endorsed Transactions on Internet of Things, 10*. Advance online publication. DOI: 10.4108/eetiot.4658

Prasertpong, P., Jaroenkhasemmeesuk, C., Regalbuto, J. R., Lipp, J., & Tippayawong, N. (2019). Optimisation of process variables for esterification of bio-oil model compounds by a heteropolyacid catalyst. *Energy Reports, 6*, 1–9. DOI: 10.1016/j.egyr.2019.11.026

Prem, E. (2024). Principles of digital humanism: A critical post-humanist view. *Journal of Responsible Technology, 17*, 1–10. DOI: 10.1016/j.jrt.2024.100075

Punch, K. (2014). *Introduction to social research : quantitative and qualitative approaches* (3rd ed.). Sage Publications.

Radanliev, P. (2024). Artificial intelligence: Reflecting on the past and looking towards the next paradigm shift. *Journal of Experimental & Theoretical Artificial Intelligence*. Advance online publication. DOI: 10.1080/0952813X.2024.2323042

Radanliev, P., & De Roure, D. (2023). New and emerging forms of data and technologies: Literature and bibliometric review. *Multimedia Tools and Applications, 82*(2), 2887–2911. DOI: 10.1007/s11042-022-13451-5

Radford, A., Wu, J., Child, R., Luan, D., Amodei, D., & Sutskever, I. (2019). Language models are unsupervised multitask learners. *OpenAI blog, 1*(8), 9.

Raheem, B. R., Anjum, F., & Ghafar, Z. N. (2023). Exploring the profound impact of artificial intelligence applications (Quillbot, Grammarly and ChatGPT) on English academic writing: A systematic review. [IJIR]. *International Journal of Integrative Research, 1*(10), 599–622. DOI: 10.59890/ijir.v1i10.366

Rahman, M., Taib, N.-A. A. B., Matin, M. M., Rahman, M. M., Bakri, M. K. B., Alexanrovich, T. P., Vladimirovich, S. V., Sanaullah, K., Tazeddinova, D., & Khan, A. (2022). Optimisation of tensile strength and Young's modulus of CNT–CF/Epoxy composites using response surface methodology (RSM). *Materials (Basel)*, *15*(19), 6746. DOI: 10.3390/ma15196746

Rahm, L. (2021). Education, automation and AI: A genealogy of alternative futures. *Learning, Media and Technology*, *48*(1), 6–24. DOI: 10.1080/17439884.2021.1977948

Rahwan, I., Cebrian, M., Obradovich, N., Bongard, J., Bonnefon, J.-F., Breazeal, C., Crandall, J. W., Christakis, N. A., Couzin, I. D., Jackson, M. O., Jennings, N. R., Kamar, E., Kloumann, I. M., Larochelle, H., Lazer, D., McElreath, R., Mislove, A., Parkes, D. C., Pentland, A., & Wellman, M. (2019). **Machine behaviour.** Nature, 568(7753), 477–486. *Nature*, •••, 477–486. DOI: 10.1038/s41586-019-1138-y

Raifman, S., DeVost, M. A., Digitale, J. C., Chen, Y.-H., & Morris, M. D. (2022). Respondent-driven sampling: A sampling method for hard-to-reach populations and beyond. *Current Epidemiology Reports*, *9*(1), 38–47. DOI: 10.1007/s40471-022-00287-8

Rajpurohit, L., & Dobhada, S. (2024). Scientific writing and its ethical considerations using AI tools. In *Oral Oncology Reports* (Vol. 9). DOI: 10.1016/j.oor.2024.100196

Ramdurai, B., & Adhithya, P. (2023). The Impact, Advancements and Applications of Generative AI. *International Journal on Computer Science and Engineering*, *10*(6), 1–8. Advance online publication. DOI: 10.14445/23488387/IJCSE-V10I6P101

Ranjan, M. Kr., Barot, K., Khairnar, V., Rawal, V., Pimpalgaonkar, A., Saxena, S., & Sattar, A. Md. (2023). Python: Empowering data science applications and research. *Journal of Operating Systems Development & Trends*. DOI: 10.37591/joosdt.v10i1.576

Rao, B. A., Tak, M., Rao, R. N., & Bathe, R. (2023). Developing laser-assisted machining process for nickel based superalloy IN625 using experimental and statistical analysis. *Lasers in Manufacturing and Materials Processing*, *10*(4), 681–701. DOI: 10.1007/s40516-023-00230-9

Raposo, F., & Barceló, D. (2021). Assessment of goodness-of-fit for the main analytical calibration models: Guidelines and case studies. *Trends in Analytical Chemistry*, *143*, 116373. DOI: 10.1016/j.trac.2021.116373

Raymond, C. (2024). Decolonizing the presentation of research findings: Amplifying epistemic authority through poetic re-storying. *Narrative Works*, *11*, 149–173. DOI: 10.7202/1108959ar

Raza, Y., Raza, H., Ahmed, A., Quazi, M. M., Jamshaid, M., Anwar, M. T., Bashir, M. N., Younas, T., Jafry, A. T., & Soudagar, M. E. M. (2025). Integration of response surface methodology (RSM), machine learning (ML), and artificial intelligence (AI) for enhancing properties of polymeric nanocomposites-A review. *Polymer Composites, 46*(15), 13591–13627. Advance online publication. DOI: 10.1002/pc.30011

Rech, R. L.Junior, Malde, S., Cazzaniga, C., Kastriotou, M., Letiche, M., Frost, C., & Rech, P. (2022). High Energy and Thermal Neutron Sensitivity of Google Tensor Processing Units. *IEEE Transactions on Nuclear Science, 69*(3), 567–575. Advance online publication. DOI: 10.1109/TNS.2022.3142092

Redmiles, E. (2022). *Perspectives on Digital Humanism*. Springer.

Reis, L., Maier, C., Mattke, J., Creutzenberg, M., & Weitzel, T. (2020). Addressing user resistance would have prevented a healthcare AI project failure. *MIS Quarterly Executive, 19*(4), 279–236. DOI: 10.17705/2msqe.00038

Resnik, D. B., & Hosseini, M. (2025). The ethics of using artificial intelligence in scientific research: New guidance needed for a new tool. *AI and Ethics, 5*(2), 1499–1521. DOI: 10.1007/s43681-024-00493-8 PMID: 40337745

Retkowsky, J., Hafermalz, E., & Huysman, M. (2024). Managing a ChatGPT-empowered workforce: Understanding its affordances and side effects. *Business Horizons, 67*(5), 511–523. DOI: 10.1016/j.bushor.2024.04.009

Ribeiro, M. T., Singh, S., & Guestrin, C. (2016). " Why should I trust you?" Explaining the predictions of any classifier. In *Proceedings of the 22nd ACM SIGKDD International Conference on Knowledge Discovery and Data Mining* (pp. 1135-1144). DOI: 10.18653/v1/N16-3020

Richardson, H., & Robinson, B. (2007). The mysterious case of the missing paradigm: A review of critical information systems research, 1991–2001. *Information Systems Journal, 17*(3), 251–270. DOI: 10.1111/j.1365-2575.2007.00230.x

Riediger, N. D., Cyr, M., & Mignone, J. (2020). An evaluation of an experiential learning program in global and Indigenous health: The University of Manitoba's Queen Elizabeth II Diamond Jubilee Scholarship Program. *Inquiry : a Journal of Medical Care Organization, Provision and Financing, 57*, 0046958020951002. Advance online publication. DOI: 10.1177/0046958020951002

Rieth, B. (2023). Empowered circles: Using talking circles in research. *New England College Journal of Applied Educational Research, 4*, 1–23.

Ritterbusch, A. E., Boothby, N., Mugumya, F., Wanican, J., Bangirana, C., Nyende, N., Ampumuza, D., Apota, J., Mbabazi, C., Nabukenya, C., Kayongo, A., Ssembatya, F., & Meyer, S. R. (2020). Pushing the limits of child participation in research: Reflections from a youth-driven participatory action research (YPAR) initiative in Uganda. *International Journal of Qualitative Methods, 19*, 1609406920958962. DOI: 10.1177/1609406920958962

Rittmann, O., Neunhoeffer, M., & Gschwend, T. (2025). How to improve the substantive interpretation of regression results when the dependent variable is logged. *Political Science Research and Methods, 13*(1), 203–211. DOI: 10.1017/psrm.2023.29

Rivalda, M. R., Ma'ruf, J. J., & Djalil, M. A. (2024). What drives repurchase intention? Mediating role of customer satisfaction and customer relationship management with marketing mix as an independent variable. *International Journal of Business Management and Economic Research, 7*(4), 117–131. DOI: 10.35409/IJBMER.2024.3595

Roberts, T., & Hernandez, K. (2019). Digital access is not binary: The 5 "A"s of technology access in the Philippines. *The Electronic Journal on Information Systems in Developing Countries, 85*(4), 1–14. DOI: 10.1002/isd2.12084

Rocco, S. T., & Plakhotnik, M. S. (2009). Literature reviews, conceptual frameworks, and theoretical frameworks: Terms, functions and distinctions. *Human Resource Development Review, 8*(1), 120–130. DOI: 10.1177/1534484309332617

Rogers, A. W., Lane, A., Mendoza, Ć., Watson, S., Kowalski, A., Martin, P. A., & Zhang, D. (2024). Integrating knowledge-guided symbolic regression and model-based design of experiments to automate process flow diagram development. *Chemical Engineering Science, 300*, 120580. DOI: 10.1016/j.ces.2024.120580

Rogers, C. (2015). *Virtual authenticity: Authenticity of digital records from theory to practice (Electronic theses and dissertations)*. University of British Columbia.

Romelin, C. Zahedi, & Nusantara, B. C. (2024, September). Comparative analysis of response surface methodology (rsm) and taguchi method: optimisation hydraulic ram pump performance. In *Operations Research Forum* (Vol. 5, No. 4, p. 85). Cham: Springer International Publishing. DOI: 10.1007/s43069-024-00359-z

Rosa, A. C., Hammad, A. W. A., Boer, D., & Haddad, A. (2023). Use of operational research techniques for concrete mix design: A systematic review [Review of *Use of operational research techniques for concrete mix design: A systematic review*]. *Heliyon, 9*(4). Elsevier BV. DOI: 10.1016/j.heliyon.2023.e15362

Rosenblat, A. (2018). *Uberland: How algorithms are rewriting the rules of work*. University of California Press. DOI: 10.1525/9780520970632

Roy, N., Beauthier, C., Carletti, T., & Mayer, A. (2021). Setup of fuzzy hybrid particle swarms. *Proceedings of the Genetic and Evolutionary Computation Conference Companion*, 207. DOI: 10.1145/3449726.3459418

Rubinger, L., Gazendam, A., Ekhtiari, S., & Bhandari, M. (2023). Machine learning and artificial intelligence in research and healthcare. *Injury, 54*, S69–S73. Advance online publication. DOI: 10.1016/j.injury.2022.01.046 PMID: 35135685

Rudolph, M., Kurz, S., & Rakitsch, B. (2024). Hybrid modeling design patterns. *Journal of Mathematics in Industry, 14*(1), 3. Advance online publication. DOI: 10.1186/s13362-024-00141-0

Rudrapati, R., Bandyopadhyay, A., Pal, P. K., & Rathod, L. (2020). Analysis, modeling and optimisation of surface roughness in cylindrical traverse cut grinding using factorial design, RSM and simulated annealing algorithm. *IOP Conference Series. Materials Science and Engineering, 814*(1), 12016. DOI: 10.1088/1757-899X/814/1/012016

Ruiz-Real, J. L., Uribe-Toril, J., Torres, J. A., & Pablo, J. D. E. (2021). Artificial intelligence in business and economics research: Trends and future. *Journal of Business Economics and Management, 22*(1), 98–117. Advance online publication. DOI: 10.3846/jbem.2020.13641

Russell, S., & Norvig, P. (2010). *Artificial intelligence: A modern approach* (3rd ed.). Prentice Hall Press.

Saad, A., Su, B., & Bjørnson, F. O. (2023). A web-based platform for efficient and robust simulation of aquaculture systems using integrated intelligent agents. *Procedia Computer Science, 225*, 4560–4569. DOI: 10.1016/j.procs.2023.10.454

Sadeq, A. M., Homod, R. Z., Hasan, H. A., Alhasnawi, B. N., Hussein, A. K., Jahangiri, A., & Abbas, S. (2025). Advancements in combustion technologies: A review of innovations, methodologies, and practical applications. *Energy Conversion and Management, X*, 100964. DOI: 10.1016/j.ecmx.2025.100964

Sadollah, A., Nasir, M., & Geem, Z. W. (2020). Sustainability and optimisation: From conceptual fundamentals to applications. *Sustainability (Basel), 12*(5), 2027. DOI: 10.3390/su12052027

Säfken, B., & Rügamer, D. (2024). Editorial special issue: Bridging the gap between AI and Statistics. *AStA. Advances in Statistical Analysis, 108*(2), 225–229. DOI: 10.1007/s10182-024-00503-4

Sai, D. A., Nataraj, K. S., & Lakshmana, R. A. (2023). Response surface methodology: A statistical tool for the optimisation of responses. *Global Journal of Addiction & Rehabilitation Medicine*, *7*(1). Advance online publication. DOI: 10.19080/GJARM.2023.07.555705

Sajdak, M., Majewski, A. J., Sobek, S., Gałko, G., & Ouadi, M. (2025). Design of experiments method into upgrading pyrolytic oil for sustainable aviation fuel additives by hydrotreating and hydrocracking. *Waste Management (New York, N.Y.)*, *194*, 258–269. DOI: 10.1016/j.wasman.2025.01.012

Saldaña, J. (2021). The coding manual for qualitative researchers. *Sage (Atlanta, Ga.)*.

Salgotra, R., Sharma, P., Saravanakumar, R., & Gandomi, A. H. (2023). A contemporary systematic review on meta-heuristic optimisation algorithms with their MATLAB and Python code reference. *Archives of Computational Methods in Engineering*, *31*(3), 1749–1822. DOI: 10.1007/s11831-023-10030-1

Sallam, M. (2023). The utility of ChatGPT as an example of large language models in healthcare education, research and practice: Systematic review on the future perspectives and potential limitations. MedRxiv, 2023-02.

Sandberg, J., & Alvesson, M. (2021). Meanings of theory: Clarifying theory through typification. *Journal of Management Studies*, *58*(2), 487–516. DOI: 10.1111/joms.12587

Sansbury, S. (2025). The Role of School Librarians in Fostering School-Wide Positive Reading Culture. *School Libraries Worldwide*, *30*(2), 17–39. DOI: 10.29173/slw8811

Sarker, S., Susarla, A., Gopal, R., & Thatcher, J. B. (2023). Democratizing knowledge creation through human-AI collaboration in academic peer review. *Journal of the Association for Information Systems*, *25*(1), 158–171. DOI: 10.17705/1jais.00872

Sartori, L., & Theodorou, A. (2022). A sociotechnical perspective for the future of AI: Narratives, inequalities, and human control. *Ethics and Information Technology*, *24*(1), 4. Advance online publication. DOI: 10.1007/s10676-022-09624-3

Saunders, M. N. K., Lewis, P., & Thornhill, A. (2019). *Research methods for business students* (8th ed.). Pearson.

Schlagwein, D., & Willcocks, L. (2023). 'ChatGPT et al.': The ethics of using (generative) artificial intelligence in research and science. *Journal of Information Technology*, *38*(3), 232–238. DOI: 10.1177/02683962231200411

Schmidhuber, J. (2015). Deep learning in neural networks: An overview. *Neural Networks*, *61*, 85–117. DOI: 10.1016/j.neunet.2014.09.003

Schulze, L., Cai, Z., Trenz, M., & Tan, C.-W. (2022). Algorithmic unfairness on digital labor platforms: How algorithmic management practices disadvantage workers. In *Proceedings of the International Conference on Information Systems (ICIS 2022)*. AIS.

Schwaninger, M., & Ott, S. (2023). What is variety engineering and why do we need it? *Systems Research and Behavioral Science*, *41*(2), 235–246. DOI: 10.1002/sres.2964

Schwegler, M., Müller, C., & Reiterer, A. (2023). Integrated Gradients for Feature Assessment in Point Cloud-Based Data Sets. *Algorithms*, *16*(7), 316. Advance online publication. DOI: 10.3390/a16070316

Schwen, L. O., Schacherer, D., Geißler, C., & Homeyer, A. (2022). Evaluating generic AutoML tools for computational pathology. *Informatics in Medicine Unlocked*, *29*, 100853. Advance online publication. DOI: 10.1016/j.imu.2022.100853

Sekiwu, D., Adyanga, F. A., Genza, G. M., Rugambwa, N. O., & Muwagga, A. M. (2024). Lifelong educational and decolonization programs for the Batwa Indigenous people of southwestern Uganda. *East African Journal of Interdisciplinary Studies*, *7*(1), 38–52. DOI: 10.37284/eajis.7.1.1838

Setumu, T., & Ngoepe, M. (2016). Converting oral narratives into written literature: Lessons learnt from heritage projects undertaken in the Makgabeng area, Limpopo, South Africa. *Mousaion*, *34*(4), 148–164. https://doi.org/doi:10.25159/0027-2639/940

Sharkey, N., & Sharkey, A. (2020). The crying shame of robot nannies: an ethical appraisal. In *Machine ethics and Robot Ethics* (pp. 155–184). Routledge. DOI: 10.4324/9781003074991-16

Sharma, A., Singh, Y., Tyagi, A., Singh, N. K., & Singla, A. (2020). Application of Taguchi and response surface methodology approach to a sustainable model developed for a compression-ignition engine using polanga biodiesel/diesel blends. *Proceedings of the Institution of Mechanical Engineers. Part I, Journal of Systems and Control Engineering*, *235*(6), 854–868. DOI: 10.1177/0959651820965301

Shaw, C., Yuan, L., Brennan, D., & Martin, S. (2023). *Generative AI in higher education: Fall 2023 update (Time for Class study)*. Tyton Partners. https://tytonpartners.com/app/uploads/2023/10/GenAI-IN-HIGHER-EDUCATION-FALL-2023-UPDATE-TIME-FOR-CLASS-STUDY.pdf

Sheikhmohammadi, A., Alamgholiloo, H., Golaki, M., Khakzad, P., Asgari, E., & Rahimlu, F. (2024). Cefixime removal via WO3/Co-ZIF nanocomposite using machine learning methods. *Scientific Reports*, *14*(1), 13840. Advance online publication. DOI: 10.1038/s41598-024-64790-2

Shie, A., Lo, K.-H., Lin, W., Juan, C., & Jou, Y. (2019). An integrated model using the Taguchi method and artificial neural network to improve artificial kidney solidification parameters. *Biomedical Engineering Online*, *18*(1), 78. Advance online publication. DOI: 10.1186/s12938-019-0696-4

Shim, J. P., French, A., & Upreti, B. (2020). Speech analytics for actionable insights: Current status, recommendations, and guidance. *Foundations and Trends in Information Systems*, *4*(3), 213–274. DOI: 10.1561/2900000020

Shin, D., & Park, Y. J. (2019). Role of fairness, accountability, and transparency in algorithmic affordance. *Computers in Human Behavior*, *98*, 277–284. DOI: 10.1016/j.chb.2019.04.019

Shiri, A., Howard, D., & Farnel, S. (2022). Indigenous digital storytelling: Digital Interfaces supporting cultural heritage preservation and access. *The International Information & Library Review*, *54*(2), 93–114. DOI: 10.1080/10572317.2021.1946748

Shoaib, M., Shamer, A., & Iqbal, S. (2025). A Systematic Review of Academic Library Spaces as Facilitators of Student Engagement in Higher Education Learning. *Knowledge (Beverly Hills, Calif.)*, *4*(1), 123–134. DOI: 10.63062/tk/2k25a.41045

Shojaei, S., Shojaei, S., Band, S. S., Farizhandi, A. A. K., Ghoroqi, M., & Mosavi, A. (2021). Application of Taguchi method and response surface methodology into the removal of malachite green and auramine-O by NaX nanozeolites. *Scientific Reports*, *11*(1), 16054. Advance online publication. DOI: 10.1038/s41598-021-95649-5

Sholikha, W. L., & Wahyuni, A. (2023). Child Friendly Library Management in Fostering Students' Interest in Reading. *Academia Open*, *8*(1), 10-21070.Sanders, K., & Farago, F. (2018). Developmentally Appropriate Practice in the Twenty-First Century. In *Springer International Handbooks of Education: Vol. Part F1626* (pp. 1379–1400). Springer Nature. https://doi.org/DOI: 10.1007/978-94-024-0927-7_71

Shrinkhal, R. (2021). "Indigenous sovereignty" and right to self-determination in international law: A critical appraisal. *Alternative*, *17*(1), 71–82. DOI: 10.1177/1177180121994681

Sibanda, M., & Schellnack-Kelly, I. (2024). Trustworthiness of digital records in the public sector of Zimbabwe. *ESARBICA Journal*, *43*, 124–146.

Sigauke, O. (2022). Digital records management practices in the public sector of Manicaland province of Zimbabwe. *ESARBICA Journal, 41*, 172–187. DOI: 10.4314/esarjo.v41i.12

Silva, B. M. F., Todt, E., Nascimento, T. P., Curvelo, C., & Gonçalves, L. M. G. (2024). Robotics Meets AI and Vision in South America (Topical Collection). *Journal of Intelligent & Robotic Systems, 110*(1), 11. Advance online publication. DOI: 10.1007/s10846-023-02039-2

Silver, D., Huang, A., Maddison, C. J., Guez, A., Sifre, L., Van Den Driessche, G., Schrittwieser, J., Antonoglou, I., Panneershelvam, V., Lanctot, M., Dieleman, S., Grewe, D., Nham, J., Kalchbrenner, N., Sutskever, I., Lillicrap, T., Leach, M., Kavukcuoglu, K., Graepel, T., & Hassabis, D. (2016). Mastering the game of Go with deep neural networks and tree search. *Nature, 529*(7587), 484–489. DOI: 10.1038/nature16961

Silverman, D. (2020). *Interpreting qualitative data* (6th ed.). Sage Publications.

Simonds, V., & Christopher, S. (2013). Adapting Western research methods to indigenous ways of knowing. *American Journal of Public Health, 103*(12), 2185–2192. DOI: 10.2105/AJPH.2012.301157

Simone, M. D., Veneziano, S., Porcaro, A., & Guida, D. (2025). Design of experiments approach for structural optimisation of urban air mobility vehicles. *Actuators, 14*(4), 176. DOI: 10.3390/act14040176

Simons, H. (2009). *Case study research in practice.* Sage Publications. DOI: 10.4135/9781446268322

Sims, K. R., Maceren, J. P., Strand, A. I., He, B., Overby, C., & Benoit, D. S. W. (2020). Rigor and reproducibility in polymer nanoparticle synthesis and characterization. *RSC Advances, 10*(5), 2513–2518. DOI: 10.1039/C9RA10091A

Singhal, A. (2024). Toward fairness, accountability, transparency, and ethics in AI systems. *Journal of the American Medical Informatics Association : JAMIA, 31*(5), 1012–1025.

Singh, M., Singari, R. M., & Bholey, M. (2024). The Influence of Color on Visual Psychology and Cognitive Behavior: A study in paediatrics hospital environment. *Educational Administration: Theory and Practice, 30*(5), 13164–13177.

Singh, R., Kim, J. Y., Glassy, E. F., Dash, R. C., Brodsky, V., Seheult, J., de Baca, M. E., Gu, Q., Hoekstra, S., & Pritt, B. S. (2025). Introduction to generative artificial intelligence: Contextualizing the future. *Archives of Pathology & Laboratory Medicine, 149*(2), 112–122. DOI: 10.5858/arpa.2024-0221-RA PMID: 39631430

Singh, S., & Akhai, S. (2024, July). Electronics in precision control for mechanical engineering: A brief overview. In. AIP Conference Proceedings: Vol. 3121. *No. 1* (p. 020007). AIP Publishing LLC., DOI: 10.1063/5.0221464

Sipper, M. (2022). Combining deep learning with good old-fashioned machine learning. *SN Computer Science, 4*(1), 85. Advance online publication. DOI: 10.1007/s42979-022-01505-2

Sithole, M. W. (2022). Electronic records management model for improving e-government service delivery in Zimbabwe. Master's thesis, Midlands State University.

Smith, J. A., & Lee, H. K. (2025). Big data and artificial intelligence in sports analytics. In *Big data and artificial intelligence in sports analytics* (pp. 95–112). Springer., DOI: 10.1007/978-3-662-68241-8_8-1

Smyth, C., Dennehy, D., Fosso Wamba, S., Scott, M., & Harfouche, A. (2024). Artificial intelligence and prescriptive analytics for supply chain resilience: A systematic literature review and research agenda. *International Journal of Production Research, 62*(23), 8537–8561. DOI: 10.1080/00207543.2024.2341415

Someh, I., Wixom, B. H., Beath, C. M., & Zutavern, A. (2022). Building an artificial intelligence explanation capability. *MIS Quarterly Executive, 21*(2), 143–163. DOI: 10.17705/2msqe.00063

Sommer, R., & Paxson, V. (2010). Outside the closed world: On using machine learning for network intrusion detection. In *2010, IEEE Symposium on Security and Privacy* (pp. 305-316). IEEE.

Song, Z., Yang, X., Xu, Z., & King, I. (2023). Graph-Based Semi-Supervised Learning: A Comprehensive Review. *IEEE Transactions on Neural Networks and Learning Systems, 34*(11), 8174–8194. Advance online publication. DOI: 10.1109/TNNLS.2022.3155478 PMID: 35302941

Soori, M., Arezoo, B., & Dastres, R. (2023). Virtual manufacturing in Industry 4.0: A review [Review of Virtual manufacturing in Industry 4.0: A review]. *Data Science and Management, 7*(1), 47–63. DOI: 10.1016/j.dsm.2023.10.006

Stahl, B. C. (2014). Interpretive accounts and fairy tales: A critical polemic against the empiricist bias in interpretive IS research. *European Journal of Information Systems, 23*(1), 1–11. DOI: 10.1057/ejis.2012.58

Stegherr, H., Heider, M., & Hähner, J. (2020). Classifying metaheuristics: Towards a unified multi-level classification system. *Natural Computing, 21*(2), 155–171. DOI: 10.1007/s11047-020-09824-0

Stevens-Uninsky, M., Barkhad, A., MacDonald, T., Perez, A., & Mbuagbaw, L. (2023). Decolonization in sexual and reproductive health research methods: Protocol for a scoping review. *JMIR Research Protocols, 12*, e45771. DOI: 10.2196/45771

Stix, C. (2021). Actionable principles for artificial intelligence policy: Three pathways. *Science and Engineering Ethics, 27*(1), 15. DOI: 10.1007/s11948-020-00277-3 PMID: 33608756

Stokes, J. M., Yang, K., Swanson, K., Jin, W., Cubillos-Ruiz, A., Donghia, N. M., MacNair, C. R., French, S., Carfrae, L. A., Bloom-Ackermann, Z., Tran, V. M., Chiappino-Pepe, A., Badran, A. H., Andrews, I. W., Chory, E. J., Church, G. M., Brown, E. D., Jaakkola, T. S., Barzilay, R., & Collins, J. J. (2020). A deep learning approach to antibiotic discovery. *Cell, 181*(2), 475–483. DOI: 10.1016/j.cell.2020.04.001

Strauss, A. L., & Corbin, J. (1998). *Basics of qualitative research: techniques and procedures for developing grounded theory* (2nd ed.). Sage Publications.

Strobl, P., Schermer, E., Groetsch, D., Pointner-Gabriel, L., Voelkel, K., Pflaum, H., & Stahl, K. (2022). Identification and validation of linear friction models using ANOVA and stepwise regression. *Lubricants (Basel, Switzerland), 10*(11), 286. DOI: 10.3390/lubricants10110286

Struik, L., Werstuik, S.-T., Sundstrom, A., Dow-Fleisner, S., & Ben-David, S. (2022). Factors that influence the decision to vape among Indigenous youth. *BMC Public Health, 22*(1), 641. Advance online publication. DOI: 10.1186/s12889-022-13095-y

Subaveerapandiyan, A., Kalbande, D., & Ahmad, N. (2025). Perceptions of effectiveness and ethical use of AI tools in academic writing: A study among PhD scholars in India. *Information Development, 41*(3), 728–746. DOI: 10.1177/02666669251314840

Sufyan, M., Shokat, Z., & Ashfaq, U. A. (2023). Artificial intelligence in cancer diagnosis and therapy: Current status and future perspective. In *Computers in Biology and Medicine* (Vol. 165). DOI: 10.1016/j.compbiomed.2023.107356

Sun, B., Yang, C., Wang, Y., Gui, W., Craig, I. K., & Olivier, L. E. (2019). A comprehensive hybrid first principles/machine learning modeling framework for complex industrial processes. *Journal of Process Control, 86*, 30–43. DOI: 10.1016/j.jprocont.2019.11.012

Sung, J.-H., Ju, S.-H., Cho, S.-W., Joo, H. W., Seo, K.-M., & Kang, B. G. (2025). Modeling and exploratory analysis of discrete event simulations for optimizing overhead hoist transport systems and logistics in semiconductor manufacturing. *Mathematics, 13*(7), 1167. DOI: 10.3390/math13071167

Susarla, A., Gopal, R., Thatcher, J. B., & Sarker, S. (2023). The Janus effect of generative AI: Charting the path for responsible conduct of scholarly activities in information systems. *Information Systems Research, 34*(2), 399–408. DOI: 10.1287/isre.2023.ed.v34.n2

Swanborn, P. G. (1996). A common base for quality control criteria in quantitative and qualitative research. *Quality and Quantity. Quality & Quantity, 30*(1), 19–35. DOI: 10.1007/BF00139833

Sweeney, S. (2023). Who wrote this? Essay mills and assessment – Considerations regarding contract cheating and AI in higher education. *International Journal of Management Education, 21*(2), 100818. Advance online publication. DOI: 10.1016/j.ijme.2023.100818

Talati, D. V. (2021). Python: The alchemist behind AI's intelligent evolution. *International Journal of Science and Research Archive, 3*(1), 235. DOI: 10.30574/ijsra.2021.3.1.0169

Tammaro, A. M. (2011). Library and Information Science (LIS) education: A conceptual framework towards "Europeisation". *Journal of the Bangladesh Association of Young Researchers, 1*(1), 1–13. DOI: 10.3329/jbayr.v1i1.6950

Tănase, M., Portoacă, A. I., Dini ă, A., Brănoiu, G., Zamfir, F., S rbu, E.-E., & Călin, C. (2024). Optimizing mechanical properties of recycled 3D-printed pla parts for sustainable packaging solutions using experimental analysis and machine learning. *Polymers, 16*(23), 3268. DOI: 10.3390/polym16233268

Tangjitsitcharoen, S., Tangjitsitcharoen, S., & Faccia, A. (2025). Leveraging intelligent machines for sustainable and intelligent manufacturing systems. *Journal of Manufacturing and Materials Processing, 9*(5), 153. DOI: 10.3390/jmmp9050153

Tang, W., He, H., & Tu, X. M. (2023). *Applied categorical and count data analysis.* Chapman and Hall/CRC., DOI: 10.1201/9781003109815

Tarisayi, K. S. (2024). Integrating Indigenous knowledge in South African Geography education curricula for social justice and decolonization. *E-Journal of Humanities Arts and Social Sciences*, 1195-1206. DOI: 10.38159/ehass.20245711

Tawakuli, A., Havers, B., Gulisano, V., Kaiser, D., & Engel, T. (2025). Survey: Time-series data preprocessing: A survey and an empirical analysis. *Journal of Engineering Research, 13*(2), 674–711. DOI: 10.1016/j.jer.2024.02.018

Taylor, J., & Smith, R. (2025). Ethical implications of AI in sports analytics. *Journal of Sport and Social Issues, 49*(2), 123–145. DOI: 10.1177/14727978251337959

Taylor, L. (2017). What is data justice? *Big Data & Society*, *4*(2), 1–14.

Taylor, L., & Nguyen, P. (2024). Tools and open datasets to support, sustain, and secure critical digital infrastructure in sports. In *Proceedings in Sports Technology* (pp. 45–60). Springer., DOI: 10.1007/978-3-031-67256-9_3

Teh, S. Y., Chua, K. B., Hong, B. H., Ling, A. J. W., Andiappan, V., Foo, D. C. Y., Hassim, M. H., & Ng, D. K. S. (2019). A hybrid multi-objective optimisation framework for preliminary process design based on health, safety and environmental impact. *Processes (Basel, Switzerland)*, *7*(4), 200. DOI: 10.3390/pr7040200

Teng, S. Y., How, B. S., Leong, W. D., Teoh, J. H., Cheah, A. C. S., Motavasel, Z., & Lam, H. L. (2019). Principal component analysis-aided statistical process optimisation (PASPO) for process improvement in industrial refineries. *Journal of Cleaner Production*, *225*, 359–375. DOI: 10.1016/j.jclepro.2019.03.272

Thambinathan, V., & Kinsella, E. A. (2021). Decolonizing methodologies in qualitative research: Creating spaces for transformative praxis. *International Journal of Qualitative Methods*, *20*, 16094069211014766. Advance online publication. DOI: 10.1177/16094069211014766

Theodorou, A., & Dignum, V. (2020). Towards ethical and socio-legal governance in AI. *Nature Machine Intelligence*, *2*(1), 10–12. DOI: 10.1038/s42256-019-0136-y

Thorpe, K., Christen, K., Booker, L., & Galassi, M. (2021). Designing archival information systems through partnerships with Indigenous communities. *AJIS. Australasian Journal of Information Systems*, *25*. Advance online publication. DOI: 10.3127/ajis.v25i0.2917

Tipu, R. K., Rathi, P., Pandya, K. S., & Panchal, V. R. (2025). Optimizing sustainable blended concrete mixes using deep learning and multi-objective optimisation. *Scientific Reports*, *15*(1), 16356. Advance online publication. DOI: 10.1038/s41598-025-00943-1

Tone-Pah-Hote, T., & Redvers, N. (2022). The commercialization of biospecimens from Indigenous peoples: A scoping review of benefit-sharing. *Frontiers in Medicine*, *9*, 978826. Advance online publication. DOI: 10.3389/fmed.2022.978826

Topol, E. (2019). *Deep medicine: how artificial intelligence can make healthcare human again.*

Torraco, R. J. (2016). Writing integrative literature reviews: Using the past and present to explore the future. *Human Resource Development Review*, *15*(4), 404–428. DOI: 10.1177/1534484316671606

Treloar, N. J., Braniff, N., Ingalls, B., & Barnes, C. (2022). Deep reinforcement learning for optimal experimental design in biology. *PLoS Computational Biology*, *18*(11), e1010695. Advance online publication. DOI: 10.1371/journal.pcbi.1010695

Trepanier, C., Shiri, A., & Samek, T. (2019). An examination of IFLA and Data Science Association ethical codes. *IFLA Journal*, *45*(4), 289–301. DOI: 10.1177/0340035219849614

Triller, J., Lopez, M. L., Nossek, M., & Frenzel, M. (2023). Multidisciplinary optimisation of automotive mega-castings merging classical structural optimisation with response-surface-based optimisation enhanced by machine learning. *Scientific Reports*, *13*(1), 21678. Advance online publication. DOI: 10.1038/s41598-023-47937-5

Trimberger, S. M., & Moore, J. J. (2014). FPGA security: Motivations, features, and applications. *Proceedings of the IEEE*, *102*(8), 1248–1265. Advance online publication. DOI: 10.1109/JPROC.2014.2331672

Trist, E. L., & Bamforth, K. W. (1951). Social and psychological consequences of the longwall method of coal-getting. *Human Relations*, *4*(1), 3–38. DOI: 10.1177/001872675100400101

Truby, J. (2020). Governing artificial intelligence to benefit the UN sustainable development goals. *Sustainable Development (Bradford)*, *28*(4), 946–959. DOI: 10.1002/sd.2048

Tsafnat, G., Glasziou, P., Choong, M. K., Dunn, A., Galgani, F., & Coiera, E. (2014). Systematic review automation technologies. *Systematic Reviews*, *3*(1), 74. DOI: 10.1186/2046-4053-3-74

Tsai, S. (2023). Dual-orthogonal arrays for order-of-addition two-level factorial experiments. *Technometrics*, *65*(3), 388–395. DOI: 10.1080/00401706.2023.2173303

Tsoulos, I. G., Charilogis, V., Kyrou, G., Stavrou, V. N., & Tzallas, A. T. (2025). OPTIMUS: A multidimensional global optimisation package. *Journal of Open Source Software*, *10*(108), 7584. DOI: 10.21105/joss.07584

Tucker, V. M., Weedman, J., Bruce, C. S., & Edwards, S. L. (2014). Learning portals: Analyzing threshold concept theory for LIS education. *Journal of Education for Library and Information Science*, •••, 150–165.

Tufekci, Z. (2015). Algorithmic harms beyond Facebook and Google: Emergent challenges of computational agency. *Journal of the Association for Information Science and Technology*, *66*(11), 2195–2205. DOI: 10.1002/asi.23499

Turel, O., & Kalhan, S. (2023). Prejudiced against the machine? Implicit associations and the transience of algorithm aversion. *Management Information Systems Quarterly*, *47*(4), 1369–1394. DOI: 10.25300/MISQ/2022/17961

Turuba, R., Katan, C., Marchand, K., Brasset, C., Ewert, A., Tallon, C., & Barbic, S. (2024). Weaving community-based participatory research and co-design to improve opioid use treatments and services for youth, caregivers, and service providers. *PLoS One*, *19*(4), e0297532. DOI: 10.1371/journal.pone.0297532 PMID: 38635804

Ullah, F., Dhingra, S., Xia, X., & Babar, M. A. (2024). Evaluation of distributed data processing frameworks in hybrid clouds. *Journal of Network and Computer Applications*, *224*, 103837. Advance online publication. DOI: 10.1016/j.jnca.2024.103837

UNHCR Innovation Service. (2024). *Emerging practices to improve access to and working conditions on digital labour platforms for refugees and host communities.*

United Nations. (1989). *Convention on the Rights of the Child.* Geneva: Office of High Commissioner for Human Rights (OHCHR). Retrieved on 10 October from https://www.ohchr .org/EN/Profe ssion allnt erest /Pages/CRC.aspx

Usama, M., Qadir, J., Raza, A., Arif, H., Yau, K. L. A., Elkhatib, Y., Hussain, A., & Al-Fuqaha, A. (2019). Unsupervised Machine Learning for Networking: Techniques, Applications and Research Challenges. *IEEE Access : Practical Innovations, Open Solutions*, *7*, 65579–65615. Advance online publication. DOI: 10.1109/ACCESS.2019.2916648

Ütük Bayılmış, O., & Orhan, S. (2025). Decoding digital labor: A topic modeling analysis of platform work experiences. *Systems*, *13*(9), 819. DOI: 10.3390/systems13090819

Vaithianathan, M. (2025, January 19). The future of heterogeneous computing: Integrating CPUs, GPUs, and FPGAs for high-performance applications. International. *Journal of Emerging Trends in Computer Science and Information Technology*, *6*(1), 12–23. DOI: 10.63282/3050-9246.IJETCSIT-V6I1P102

Van Dis, E. A., Bollen, J., Zuidema, W., Van Rooij, R., & Bockting, C. L. (2023). ChatGPT: Five priorities for research. *Nature*, *614*(7947), 224–226. DOI: 10.1038/d41586-023-00288-7

van Engelen, J. E., & Hoos, H. H. (2020). A survey on semi-supervised learning. *Machine Learning*, *109*(2), 373–440. Advance online publication. DOI: 10.1007/s10994-019-05855-6

Van Norren, D. E. (2023). The ethics of artificial intelligence, UNESCO and the African Ubuntu perspective. *Journal of Information. Communication and Ethics in Society*, *21*(1), 112–128. DOI: 10.1108/JICES-04-2022-0037

Varian, H. R. (2014). Big data: New tricks for econometrics. *The Journal of Economic Perspectives*, *28*(2), 3–28. DOI: 10.1257/jep.28.2.3

Vasilev, I., Slater, D., Spacagna, G., Roelants, P., & Zocca, V. (2019). *Python Deep Learning: Exploring deep learning techniques and neural network architectures with Pytorch, Keras, and TensorFlow*. Packt Publishing Ltd.

Vaswani, A., Shazeer, N., Parmar, N., Uszkoreit, J., Jones, L., Gomez, A. N., & Polosukhin, I. (2017). Attention is all you need. *Advances in Neural Information Processing Systems*, •••, 30.

Veen, A., Barratt, T., & Goods, C. (2024). Platform-mediated work and the politics of algorithmic control. *Work, Employment and Society*, *38*(2), 432–450.

Venkatesh, B., Kiran, A., Khan, M., Rahmani, M. K. I., Upadhyay, L., Babu, J. C., & Narayana, T. L. (2024). Performance optimisation for an optimal operating condition for a shell and heat exchanger using a multi-objective genetic algorithm approach. *PLoS One*, *19*(6), e0304097. Advance online publication. DOI: 10.1371/journal.pone.0304097

Venkatesh, V., & Davis, F. D. (2000). A theoretical extension of the technology acceptance model: Four longitudinal field studies. *Management Science*, *46*(2), 186–204. DOI: 10.1287/mnsc.46.2.186.11926

Verhoef, P., Broekhuizen, T., Bart, Y., Bhattacharya, A., Qi Dong, J., Fabian, N., & Haenlein, M. (2021). Digital transformation: A multidisciplinary reflection and research agenda. *Journal of Business Research*, *122*, 889–901. DOI: 10.1016/j.jbusres.2019.09.022

Verma, P., & Singh, A. (2024). Game tactics and strategy optimization through AI and machine learning. In *Artificial intelligence and machine learning in sport research* (pp. 150–172). IGI Global., DOI: 10.4018/978-1-978-369-385-1.ch006

Victoria, A., Hine, P. J., Ward, K., & Ries, M. E. (2023). Design of experiments in the optimisation of all-cellulose composites. *Cellulose (London, England)*, *30*(17), 11013–11039. DOI: 10.1007/s10570-023-05535-8

Waardenburg, L., & Huysman, M. (2022). From coexistence to co-creation: Blurring boundaries in the age of AI. *Information and Organization*, *32*(4), 100432. Advance online publication. DOI: 10.1016/j.infoandorg.2022.100432

Waddell-Henowitch, C., Herron, R., Gobeil, J., Tacan, F., Jager, M. D., Allan, J. A., & Roger, K. (2020). Grounded in culture: Reflections on sitting outside the circle in community-based research with Indigenous men. *Qualitative Health Research, 30*(14), 2343–2350. DOI: 10.1177/1049732320960050

Wade, M., & Shan, J. (2020). COVID-19 has accelerated digital transformation, but it may have made it harder, not easier. *MIS Quarterly Executive, 19*(3), 199–211. DOI: 10.17705/2msqe.00034

Wadhwa, S. A., Mahapara, A., Akhai, S., Kumar, D., & Kumar, P. (2024). Integrating Taguchi optimisation for multi-criteria decision making in engineering applications. In *Recent Theories and Applications for Multi-Criteria Decision-Making* (Vol. 1, No. 1, pp. 125-150). IGI Global. DOI: 10.4018/979-8-3693-6502-1.ch005

Walker, J., Pyper, E., Jones, C., Khan, S., Chong, N. W., Legge, D., Schull, M. J., & Henry, D. (2018). Unlocking first nations health information through data linkage. *International Journal of Population Data Science, 3*(1). Advance online publication. DOI: 10.23889/ijpds.v3i1.450

Waller, P., Irani, Z., Lee, H., & Weerakkody, V. (2014). Lessons on measuring e-government satisfaction: An experience from surveying government agencies in the UK. *International Journal of Electronic Government Research, 10*(3), 37–46. DOI: 10.4018/ijegr.2014070103

Walsham, G. (1995). The Emergence of Interpretivism in IS Research. *Information Systems Research, 6*(4), 376–394. https://www.jstor.org/stable/23010981. DOI: 10.1287/isre.6.4.376

Walter, M., Lovett, R., Maher, B., Williamson, B., Prehn, J., Bodkin-Andrews, G., & Lee, V. (2020). Indigenous data sovereignty in the era of big data and open data. *The Australian Journal of Social Issues, 56*(2), 143–156. DOI: 10.1002/ajs4.141

Walter, Y. (2024). Embracing the future of artificial intelligence in the classroom: The relevance of AI literacy, prompt engineering, and critical thinking in modern education. *International Journal of Educational Technology in Higher Education, 21*(1), 15. DOI: 10.1186/s41239-024-00448-3

Wang, C., Zhou, H., Chang, K., Li, B., Mu, Y., Xiao, T., Liu, T., & Zhu, J. (2024). Hybrid alignment training for large language models. *arXiv (Cornell University)*. / arxiv.2406.15178DOI: 10.18653/v1/2024.findings-acl.676

Wang, K., Yan, L. Z., Li, W. Z., Jiang, C., Wang, N. N., Zheng, Q., Dong, N. G., & Shi, J. W. (2022). Comparison of Four Machine Learning Techniques for Prediction of Intensive Care Unit Length of Stay in Heart Transplantation Patients. *Frontiers in Cardiovascular Medicine*, *9*, 863642. Advance online publication. DOI: 10.3389/fcvm.2022.863642 PMID: 35800164

Wang, S., & Liang, Z. (2024). What does the public think about artificial intelligence? An investigation of technological frames in different technological contexts. *Government Information Quarterly*, *41*(2), 101939. Advance online publication. DOI: 10.1016/j.giq.2024.101939

Water, T. (2024). Ethical issues in participatory research with children and young people. In *Being Participatory: Researching with children and young people: co-constructing knowledge using creative, digital and innovative techniques* (pp. 39–64). Springer International Publishing. DOI: 10.1007/978-3-031-47787-4_3

Weerts, H., Pfisterer, F., Feurer, M., Eggensperger, K., Bergman, E., Awad, N., Vanschoren, J., Pechenizkiy, M., Bischl, B., & Hutter, F. (2024). Can Fairness be Automated? Guidelines and Opportunities for Fairness-aware AutoML. *Journal of Artificial Intelligence Research*, *79*, 639–677. Advance online publication. DOI: 10.1613/jair.1.14747

Wegener, K., Weikert, S., Mayr, J., Maier, M., Akbari, V. O. A., & Postel, M. (2021). Operator integrated – concept for manufacturing intelligence. *Journal of Machine Engineering*, *21*(4), 5–28. DOI: 10.36897/jme/144359

Weidele, D. K. I., Afzal, S., Valente, A. N., Makuch, C., Cornec, O., Vu, L., & Haehn, D. (2023, March). AutoDOViz: Human-centered automation for decision optimization. In *Proceedings of the 28th international conference on intelligent user interfaces* (pp. 664-680). DOI: 10.1145/3581641.3584094

Wei, Y., Peng, B., Xie, R., Chen, Y., Yu, Q., Wen, P., Bauer, S., Tung, P., & Raabe, D. (2024). Deep active learning for complex systems. *Research Square (Research Square)*. DOI: 10.21203/rs.3.rs-5434645/v1

Weldon, M. N., Thomas, G., & Skidmore, L. (2024). Establishing a future-proof framework for AI regulation: Balancing ethics, transparency, and innovation. *Transactions: The Tennessee Journal of Business Law*, *25*(2), 2.

Wesely, S., Hofer, E., Curth, R., Paryani, S., Mills, N., Ueberschär, O., & Westermayr, J. (2025). Artificial intelligence for objective assessment of acrobatic movements: Machine learning identification of tumbling elements in cheer sports. *Sensors (Basel)*, *25*(7), 2260. Advance online publication. DOI: 10.3390/s25072260

Wessel, M., Adam, M., Benlian, A., Majchrzak, A., Thies, F. (2023). Generative AI and its transformative value for digital platforms. Call for Papers to the Special Issue—Journal *of Management Information Systems*. Vol.np pp 1-6, https://doi.org/ DOI: 10.1080/07421222.2025.2487315

Wichapa, N., Pawaree, N., Nasawat, P., Chourwong, P., Sriburum, A., & Khanthirat, W. (2024). Process of solving multi-response optimisation problems using a novel data envelopment analysis variant-taguchi method. *International Journal of Technology, 15*(6), 2038. DOI: 10.14716/ijtech.v15i6.7134

Wiegratz, J., Mujere, J., & Fontein, J. (2022). The case for oral histories of neoliberal Africa. *Third World Thematics: A TWQ Journal, 7*(4-6), 199-221. DOI: 10.1080/23802014.2023.2208561

Williams, D., & Shipley, G. P. (2023). Indigenous research methodologies: Challenges and opportunities for broader recognition and acceptance. *Open Journal of Social Sciences, 11*(05), 467–500. DOI: 10.4236/jss.2023.115030

Williams, K., Berman, G., & Michalska, S. (2023). Investigating hybridity in artificial intelligence research. *Big Data & Society, 10*(2), 20539517231180577. Advance online publication. DOI: 10.1177/20539517231180577

Williamson, B., Provost, S., & Price, C. (2022). Operationalising Indigenous data sovereignty in environmental research and governance. *Environment and Planning. F, Philosophy, Theory, Models, Methods and Practice*, 2(1-2), 281–304. DOI: 10.1177/26349825221125496

Williams, R. T. (2024). Paradigm shifts: Exploring AI's influence on qualitative inquiry and analysis. *Frontiers in Research Metrics and Analytics, 9*, 1331589. Advance online publication. DOI: 10.3389/frma.2024.1331589

Willis, J., Hughes, H., & Bland, D. (2019). Students reimagining school libraries as spaces of learning and wellbeing. In Hughes, H., Franz, J., & Willis, J. (Eds.), *School spaces for student well-being and learning: Insights from research and practice* (pp. 121–137). Springer. DOI: 10.1007/978-981-13-6092-3_7

Winograd, T. (2006). Shifting viewpoints: Artificial intelligence and human–computer interaction. *Artificial Intelligence, 170*(18), 1256–1258. DOI: 10.1016/j.artint.2006.10.011

Wohlin, C., Runeson, P., Höst, M., Ohlsson, M. C., Regnell, B., & Wesslén, A. (2012). *Experimentation in software engineering* (Vol. 236). Springer. DOI: 10.1007/978-3-642-29044-2

Wood, A. J., Lehdonvirta, V., & Graham, M. (2024). Workers of the gig economy unite? Algorithmic management and collective action. *Socio-economic Review, 22*(1), 109–132.

World Economic Forum. (2025). *Pathways for decent work in Kenya's digital economy.*

World Journal, A. I. (2024, January 10). *Challenge between professors and university students: Using AI for theses.* Retrieved on 12 October 2025 from https://aiworldjournal.com/challenge-between-professors-and-university-students-using-ai-for-theses/#:~:text=In%202023%2C%20a%20graduate%20student,the%20field%20of%20social%20sciences

Wright, L. A., & Patel, M. (2025). Enhancing decision-making in team sports through AI-driven computing solutions. *International Journal of Sports Computing, 12*(1), 45–67. DOI: 10.1002/cpe.8372

Xu, T., & Baghaei, S. (2025). Reshaping the future of sports with artificial intelligence: Challenges and opportunities in performance enhancement, fan engagement, and strategic decision-making. *Engineering Applications of Artificial Intelligence, 142*, 109912. Advance online publication. DOI: 10.1016/j.engappai.2024.109912

Xu, X., Li, D., Bi, J., & Moeckel, M. (2024). AutoML based workflow for design of experiments (DOE) selection and benchmarking data acquisition strategies with simulation models. *Scientific Reports, 14*(1), 32170. Advance online publication. DOI: 10.1038/s41598-024-83581-3

Yang, J., Fan, K., Ma, H., & Zhan, D. (2022). RID-Noise: Towards robust inverse design under noisy environments. *Proceedings of the AAAI Conference on Artificial Intelligence, 36*(4), 4654–4661. DOI: 10.1609/aaai.v36i4.20390

Yang, X., Song, Z., King, I., & Xu, Z. (2023). A Survey on Deep Semi-Supervised Learning. *IEEE Transactions on Knowledge and Data Engineering, 35*(9), 8934–8954. Advance online publication. DOI: 10.1109/TKDE.2022.3220219

Yan, X., Guo, X., He, N., Shi, J., & Zhao, D. (2025). Model reduction of structural mechanical response in the time domain. *Scientific Reports, 15*(1), 10203. Advance online publication. DOI: 10.1038/s41598-025-94681-z

Yarasu, V., & Podgornik, B. (2025). An optimized machine-learning tool to predict heat treatment response of hot-work tool steels. *Results in Engineering, 26*, 105260. DOI: 10.1016/j.rineng.2025.105260

Yeung, K. (2020). Recommendation of the council on artificial intelligence (OECD). *International legal materials, 59*(1), 27-34. DOI: 10.1017/ilm.2020.5

Yeung, R. C., & Fernandes, M. A. (2022). Machine learning to detect invalid text responses: Validation and comparison to existing detection methods. *Behavior Research Methods*, *54*(6), 3055–3070. DOI: 10.3758/s13428-022-01801-y PMID: 35175566

Yin, R. K. (2018). *Case study research and applications* (6th ed.). Sage.

Yoo, Y. (2024). Evolving epistemic infrastructure: The role of scientific journals in the age of generative AI. *Journal of the Association for Information Systems*, *25*(1), 137–144. DOI: 10.17705/1jais.00870

Younisse, R., Ahmad, A., & Abu Al-Haija, Q. (2022). Explaining Intrusion Detection-Based Convolutional Neural Networks Using Shapley Additive Explanations (SHAP). *Big Data and Cognitive Computing*, *6*(4), 126. Advance online publication. DOI: 10.3390/bdcc6040126

Yu, H., & Liu, J. (2024). *Deep insights into automated optimisation with large language models and evolutionary algorithms*. DOI: 10.48550/ARXIV.2410.20848

Yu, X., Xu, S., & Ashton, M. (2023). Antecedents and outcomes of artificial intelligence adoption and application in the workplace: The sociotechnical system theory perspective. *Information Technology & People*, *36*(1), 454–474. DOI: 10.1108/ITP-04-2021-0254

Zăvoianu, A.-C., Lughofer, E., Pollak, R. A., Eitzinger, C., & Radauer, T. (2020). A soft-computing framework for automated optimisation of multiple product quality criteria with application to micro-fluidic chip production. *Applied Soft Computing*, *98*, 106827. DOI: 10.1016/j.asoc.2020.106827

Zeng, X. (2024). A review on design of sustainable advanced materials by using artificial intelligence. *Advanced Materials & Sustainable Manufacturing*, *1*(1), 10006. DOI: 10.35534/amsm.2024.10006

Zhang, B., & Dafoe, A. (2019). Artificial intelligence: American attitudes and trends. *Future of Humanity Institute, University of Oxford*. https://dx.doi.org/DOI: 10.2139/ssrn.3312874

Zhang, J., & Feng, S. (2021). Machine learning modeling: A new way to do quantitative research in social sciences in the era of AI. *Journal of Web Engineering*, *20*(2), 281–302. DOI: 10.13052/jwe1540-9589.2023

Zhang, Q., Zuo, J., & Yang, S. (2025). Research on the impact of generative artificial intelligence (GenAI) on enterprise innovation performance: A knowledge management perspective. *Journal of Knowledge Management*, *29*(7), 2238–2257. DOI: 10.1108/JKM-10-2024-1198

Zhang, X., Wan, F., Liu, C., Ji, X., & Ye, Q. (2022). Learning to Match Anchors for Visual Object Detection. *IEEE Transactions on Pattern Analysis and Machine Intelligence*, *44*(6), 3096–3109. Advance online publication. DOI: 10.1109/TPAMI.2021.3050494 PMID: 33434120

Zhang, Y., He, G., Ma, L., Liu, X., Hjorth, J. J. J., Kozlov, A., He, Y., Zhang, S., Kotaleski, J. H., Tian, Y., Grillner, S., Du, K., & Huang, T. (2023). A GPU-based computational framework that bridges neuron simulation and artificial intelligence. *Nature Communications*, *14*(1), 5798. Advance online publication. DOI: 10.1038/s41467-023-41553-7 PMID: 37723170

Zhang, Y., & Wu, Y. (2021). Introducing machine learning models to response surface methodologies. In *IntechOpen eBooks*. IntechOpen., DOI: 10.5772/intechopen.98191

Zhang, Z., Chen, X., Wang, C., Wang, R., Song, W., & Nie, F. (2025). Structured multi-view k-means clustering. *Pattern Recognition*, *160*, 111113. DOI: 10.1016/j.patcog.2024.111113

Zhou, Y., Kumar, D., Bakshi, S., Mason, J., Miller, A., & Bailey, M. (2018). Erays: Reverse Engineering Ethereum's Opaque Smart Contracts. *USENIX Security Symposium*, 1371–1385. https://www.usenix.org/system/files/conference/usenixsecurity18/sec18-zhou.pdf

Zhou, B., Yang, G., Shi, Z., & Ma, S. (2024). Natural Language Processing for Smart Healthcare. *IEEE Reviews in Biomedical Engineering*, *17*, 4–18. Advance online publication. DOI: 10.1109/RBME.2022.3210270 PMID: 36170385

Zhou, R., Wang, X., Shi, Y., Zhang, R., Zhang, L., & Guo, H. (2019). Measuring e-service quality and its importance to customer satisfaction and loyalty: An empirical study in a telecom setting. *Electronic Commerce Research*, *19*(3), 477–499. DOI: 10.1007/s10660-018-9301-3

Zhuang, F., Qi, Z., Duan, K., Xi, D., Zhu, Y., Zhu, H., Xiong, H., & He, Q. (2021). A Comprehensive Survey on Transfer Learning. In *Proceedings of the IEEE* (Vol. 109, Issue 1). DOI: 10.1109/JPROC.2020.3004555

Zimmer, F., & Debelak, R. (2023). Simulation-based design optimisation for statistical power: Utilizing machine learning. *Psychological Methods*. Advance online publication. DOI: 10.1037/met0000611

Zong, C., Margesin, M., Staudt, J., Deghim, F., & Lang, W. (2022). Decision-making under uncertainty in the early phase of building façade design based on multi-objective stochastic optimisation. *Building and Environment*, *226*, 109729. DOI: 10.1016/j.buildenv.2022.109729

Zouhaier, L., Hlaoui, Y. B. D., & Ayed, L. B. (2021). A reinforcement learning based approach of context-driven adaptive user interfaces. In *2021, IEEE 45th Annual Computers, Software, and Applications Conference (COMPSAC)* (pp. 1463-1468). IEEE. DOI: 10.1109/COMPSAC51774.2021.00217

Zubair, S. W. H., Arafat, S. M., Khan, S. A., Niazi, S. G., Rehan, M., Arshad, M. U., Hayat, N., Aized, T., Uddin, G. M., & Riaz, F. (2024). Coupling Taguchi experimental designs with deep adaptive learning enhanced AI process models for experimental cost savings in manufacturing process development. *Scientific Reports*, *14*(1), 23428. Advance online publication. DOI: 10.1038/s41598-024-73669-1

Zuboff, S. (2019). *The Age of Surveillance Capitalism: The Fight for a Human Future at the New Frontier of Power*. PublicAffairs.

About the Contributors

Mahapara Abbass, a distinguished academician and researcher in the Department of Civil Engineering at Maharishi Markandeshwar (Deemed to be University), Mullana–Ambala, has made significant contributions to the fields of structural engineering and sustainable construction materials. Her academic and research achievements are reflected in a strong portfolio of publications. Dr. Abbass has authored several research papers in prestigious SCI and Scopus-indexed journals, including Multiscale and Multidisciplinary Modeling, Experimentation and Design, Journal of Building Pathology and Rehabilitation, and Industrial Crops and Products. In addition to her journal articles, she has presented her research at leading national and international conferences. Dr. Abbass's expertise also extends to scholarly book publications, where she has authored or co-authored several chapters in critically acclaimed volumes such as Advances in Geotechnics and Structural Engineering and Lecture Notes in Civil Engineering. These contributions have been published by renowned academic publishers, including Elsevier, and appear in SCI and Scopus-indexed volumes. Her research centers on geopolymer concrete, sustainable practices, innovative civil engineering materials, and the application of AI and ML for smart construction solutions.

Munir Ahmad is a seasoned professional in the realm of Spatial Data Infrastructure (SDI), Geo-Information Productions, Information Systems, and Information Governance, boasting over 25 years of dedicated experience in the field. With a PhD in Computer Science, Dr. Ahmad's expertise spans Spatial Data Production, Management, Processing, Analysis, Visualization, and Quality Control. Throughout his career, Dr. Ahmad has been deeply involved in the development and deployment of SDI systems specially in the context of Pakistan, leveraging his proficiency in Spatial Database Design, Web, Mobile & Desktop GIS, and Geo Web Services Architecture. His contributions to Volunteered Geographic Information

(VGI) and Open Source Geoportal & Metadata Portal have significantly enriched the geospatial community. As a trainer and researcher, Dr. Ahmad has authored over 50 publications, advancing the industry's knowledge base and fostering innovation in Geo-Tech, Data Governance, and Information Infrastructure, and Emerging Technologies. His commitment to Research and Development (R&D) is evident in his role as a dedicated educator and mentor in the field.

Shalom Akhai has established himself as a prominent figure in the field of engineering, leveraging over 17 years of academic, research, and technical experience to make significant contributions. With a robust portfolio of over 50 publications in prestigious journals and conferences, Dr. Akhai demonstrates a commitment to advancing knowledge through scholarly work. Additionally, he has authored/co-authored several technical textbooks, further solidifying his expertise and dedication to education in the field. Beyond academia, Dr. Akhai's impact extends to real-world applications through the establishment of IGNIS Technical Solutions, a consultancy firm to translate his knowledge into real-world applications. His skills have resulted in the successful development of over 50 patents+, underscoring his innovative approach and influence in shaping the engineering landscape.

Otmane Azeroual is a Senior Researcher with a Ph.D. in Computer Engineering, specializing in databases, information systems, artificial intelligence, and data science. He teaches at multiple universities and holds executive positions in the private sector. Dr. Azeroual leads international research and consulting projects, bridging the gap between academic research and practical applications. He is also an editor for several prestigious journals and has published extensively in leading academic and applied outlets, contributing to the advancement of global research in computing and AI.

Santosh Durugkar have pursued Ph.D in Software engineering from Amity University Rajasthan Jaipur. He has completed U.G. (B.E.) and P.G. (M.E.) from SPPU, Pune. His area of interests are software engineering, Machine Learning, Artificial Intelligence, Database Management, Data Analysis, Data Mining & Information Retrieval.

Atul Jondhale is pursuing PhD in civil engineering from Sandip University Nashik. His area of interest are application of machine learning techniques in Civil Engineering, data analysis etc.

Tom Kwanya is a Professor of Information and Knowledge Management and Director of the School of Graduate and Advanced Studies at the Technical University of Kenya. His over 18 years of university teaching and 17 years of industry experience, including international consultancies, enables him to blend academic rigour with practical experience to connect theoretical insights with real-world challenges in knowledge management, responsible AI, digital transformation, big data analytics, and digital inclusion. A prolific researcher and keynote speaker, he has authored over 70 journal articles, 31 book chapters, 88 conference papers and multiple scholarly books. Prof. Kwanya's work critically explores who controls, protects, and benefits from knowledge in academic, community and digital realms. His insights inspire scholars and students alike to rethink the future of knowledge management and equity in the emerging global knowledge economy. He sits on the National Technical Committee on Knowledge Management, Kenya, and the Information Science Services Technical Committee of the Kenya Bureau of Standards (KEBS).

Eugene J. M. Lewis completed his Doctor of Philosophy (PhD) in Technology with an emphasis in Marketing in March of 2021 from Capitol Technology University. Dr. Lewis is the former Assistant Professor of Marketing at Oakwood University from 2005 to 2010. Furthermore, he serves as an Adjunct Professor teaching courses in the area Logistics/Supply Chain Management, Marketing, and Information Systems. He has worked in Academia for more than 15 years as a part-time and full-time professor. Currently, he serves as an Adjunct Professor for Edward Waters University, Lindsey Wilson University, and Alabama A&M University. Furthermore, he serves as a United States civil servant as the International Program Manager in the Aviation Industry in the Cargo CH-47 Program Office for the Department of the Army working in the Foreign Military Sales (FMS) arena. He has served the Department of Defense (DOD) for more than 10 years. He has had the opportunity to work with many government officials and dignitaries in several countries all over the world.

Martha Lyaka is a postdoctoral fellow at the University of South Africa (UNISA), where she contributes to advancing scholarship at the nexus of information science and emerging technologies. Her work reflects a commitment to strengthening knowledge ecosystems by integrating traditional epistemologies with contemporary digital innovations. Her research spans knowledge management, artificial intelligence in libraries, research methodology, indigenous knowledge systems, information technology applications, and the enhancement of reference services. Through this interdisciplinary portfolio, she continues to shape conversations on the future of knowledge, its preservation, and its equitable accessibility across diverse communities.

John McCarthy is an academic lecturer and researcher, having worked as a Social Care Professional in the Irish Disability sector since 2007. I am passionate about providing quality, person-centred support services where the voices of people receiving support are central to all activities. John holds a Master's in Management and Leadership in the Community and Public Sector from the University of Galway, as well as a BSc in Management from Trinity College Dublin. He is a registered CORU-registered Social Care Worker and a registered Further Education Teacher with the Teaching Council of Ireland. I am also an active contributor to research in the field of intellectual and developmental disability, with published work on the promotion of self-advocacy and governance in social care.

Valerie McTaggart is an Academic Lecturer & Researcher with Atlantic Technological University (ATU) Sligo Campus. Valerie holds a PhD from DCU and a master's from John Moore's University Liverpool. She is research active in the areas of Leading Digital Transformation, AI and Remote Working. She has published and presented her work both nationally and internationally. Valerie also works with the European Commission as an evaluation expert and is a member of the European Cooperation in Science and Technology (COST) working group REAL - Regaining linkage? Digital technologies are improving civic engagement, political organisations and democracy. Before joining academia full-time in 2018, Valerie worked in the financial services sector for over twenty years as a project manager supporting ICT-enabled transformation.

Bolaji David Oladokun is a lecturer at the Federal University of Technology, Ikot Abasi, Akwa Ibom State, Nigeria. He holds a First-Class degree in Library and Information Science and a Diploma in Mass Communication with distinction. He earned his Master's degree in Library and Information Science from Ignatius Ajuru University of Education, Port Harcourt, Nigeria, graduating with distinction. A Certified Librarian of Nigeria (CLN), he is an active member of the Nigerian Library Association (NLA), the Nigerian Association of Library and Information Science Educators (NALISE), and the Association for Information Science and Technology (ASIS&T). Serving as the Business Editor of the Rivers State NLA Journal, he has authored over 150 articles in accredited journals, conference proceedings, and book chapters.

Nina Olivia Rugambwa is a distinguished educator and information scientist with 16 years of experience in higher education in Uganda. Her academic journey started with a decade at the East African School of Library and Information Science at Makerere University, followed by one and a half years at Kabale University. She currently lectures at Kyambogo University within the Department of Library and

Information Science, part of the School of Computing and Information Science. Holding a PhD in Information Science from Makerere University, which she earned in 2020, Nina's qualifications also include a Master's degree in Information Science and a Bachelor's degree in Education. These credentials provide her with a solid foundation for both teaching and research. Currently, Nina is pursuing postdoctoral studies through Makerere University's Consolidating Early Career Academic Program II, supported by the Carnegie New York Foundation. Nina's research interests focus on information-seeking behavior, Indigenous Knowledge Management, and child-centered research methodologies in school libraries. Her scholarly contributions are noteworthy; she has authored several book chapters and articles in reputable academic journals. One of her significant achievements is her chapter titled "Enhancing African Indigenous Knowledge Collection Management in Ugandan Public University Libraries," published in "Digital Preservation and Documentation of Global Indigenous Knowledge Systems" in August 2023, which has been recognized by SCOPUS for its impact on the field. In addition to her book contributions, Nina has published critical articles such as "Documenting Student Representation of Indigenous HIV/AIDS Information and Integration Into the School Curriculum" in the International Journal of Curriculum Development and Learning Measurement and "Decolonizing African University Pedagogy Through Integrating African Indigenous Knowledge and Information Systems" in the Handbook of Research on Transformative and Innovative Pedagogies in Education. Her other works emphasized the importance of preserving Indigenous knowledge systems, particularly among minority groups of the Batwa in Uganda.

Shaxlo Samanova is a distinguished scholar and educator serving at Mamun University (Khiva, Uzbekistan), where she works as a Professor in the Department of Philosophy. She holds the academic degree of Doctor of Philosophy (PhD) and has built an outstanding career dedicated to advancing philosophical thought, research culture, and academic excellence in higher education. Dr. Samanova's scholarly interests cover a broad spectrum of philosophical disciplines, with a special focus on modern philosophical discourse, ethics, social philosophy, and intellectual history. Her research contributes to the development of contemporary approaches to human values, moral education, and the cultural evolution of society in the digital era. Over the years, she has authored numerous scientific articles, conference papers, and methodological works, published in national and international academic platforms. Dr. Samanova is actively involved in supervising graduate students, participating in academic councils, and engaging in international research collaborations aimed at promoting innovative and interdisciplinary study in the humanities. A respected mentor and leader, Professor Samanova is known for her dedication to developing critical thinking, academic integrity, and deep intellectual culture among her students. Her professional mission is to foster a new generation of scholars who value both traditional philosophical heritage and contemporary global knowledge trends.

Munyika Sibanda is a lecturer at the Great Zimbabwe University; Department of Heritage, Historical and Sustainable Development. He holds a PhD in Information Science from the University of South Africa. He also a]has more than 11 years experience working as an Archivist at the National Archives of Zimbabwe.

Sirojbek Xabibullayevich Sultanov is a distinguished Uzbek academic and researcher specializing in social philosophy, globalization, and social justice. As a PhD holder and Associate Professor in the Department of Social Sciences at Tashkent State Transport University (TSTU), he plays a pivotal role in advancing interdisciplinary education and research within Uzbekistan's transport and humanities sectors. His work bridges philosophical inquiry with contemporary societal challenges, contributing to both academic discourse and practical applications in higher education. Early Education and Academic Journey Born and raised in Uzbekistan, Sultanov began his higher education journey at Urganch State University, where he studied from 2005 to 2009, laying the foundation for his interest in social sciences. He pursued advanced studies at National University of Uzbekistan, completing his master's degree from 2010 to 2012. This period honed his expertise in philosophical methodologies and social theory. Demonstrating a commitment to scholarly excellence, Sultanov advanced to doctoral studies as a PhD candidate at TSTU from 2020 to 2022. In 2022, he successfully defended his PhD dissertation in Philosophy, specializing in Social Philosophy. His doctoral research explored the intersections of globalization and social equity, earning him recognition as a rising voice in Uzbek philosophical scholarship. Professional Career Sultanov has built a robust career at TSTU, one of Uzbekistan's premier institutions for transport engineering and related fields, established in 1931 and restructured in 2020 to emphasize innovation in transportation infrastructure. As Associate Professor in the Department of Social Sciences, he teaches courses in social philosophy, ethics, and interdisciplinary studies, integrating humanistic perspectives into technical curricula. His role involves mentoring students, supervising research, and fostering collaborations that align social sciences with Uzbekistan's transport modernization efforts. Prior to his doctoral completion, Sultanov held teaching and research positions at TSTU, contributing to curriculum development and faculty initiatives. His multilingual proficiency—fluent in Uzbek, Russian, and English—enables him to engage with global academic networks, enhancing cross-cultural dialogues in philosophy.

Adeyinka Tella is a distinguished scholar affiliated with the Department of Library and Information Science at the University of Ilorin, Nigeria, and serves as a research fellow at the Department of Information Science, University of South Africa, Pretoria. He has an impressive academic portfolio, with over 300 articles published in renowned local and international journals in the field of librarianship. In recognition of his outstanding contributions to research, Tella was awarded the prestigious Emerald Literati Award in 2023. He also holds a C2 rating as a researcher, conferred by the National Research Foundation of South Africa. Known for his collaborative spirit, Tella is approachable and open to partnerships across various disciplines (tellayinkaedu@yahoo.com).

Mangubek Tursunbayevich Urazmetov is a faculty member at the Department of Philosophy, Urgench State University, Uzbekistan. He holds a Doctor of Philosophy (PhD) in Philosophy and specializes in modern philosophical thought, ethics, and socio-cultural studies. Dr. Urazmetov has published several scholarly works in national and international academic journals and actively participates in scientific conferences. As a dedicated educator and researcher, he contributes to advancing philosophical education, fostering critical thinking, and guiding young researchers in academic inquiry and intellectual development.

Index

A

B

C

D

E

G

I

T

U

www.ingramcontent.com/pod-product-compliance
Lightning Source LLC
LaVergne TN
LVHW062335300926
849161LV00007B/1695